Bear and Human

The Archaeology of Northern Europe
Volume 3

Advanced Studies on the Archaeology and History of Hunting, edited by the Centre for Baltic and Scandinavian Archaeology (ZBSA), Volume 3

Bear and Human

Facets of a Multi-Layered Relationship from Past to Recent Times, with Emphasis on Northern Europe

Edited by Oliver Grimm,
in cooperation with Daniel Groß, Alexandra Pesch, Olof Sundqvist,
and Andreas Zedrosser

A volume based on papers presented at a conference at Orsa Predator Park, Dalarna, Sweden, Oct. 16th to 18th, 2019

Volume 3.2

BREPOLS

British Library Cataloguing in Publication Data
A catalogue record for this book is available from the British Library.

Copy editor: Gundula Lidke

Layout, typesetting, and image editing: Matthias Bolte and Cornelia Lux-Kannenberg. Cover design by TopicA.

D/2023/0095/164
ISBN: 978-2-503-60611-8 (3 vols)
e-ISBN: 978-2-503-60613-2 (3 vols)
Volume DOI: 10.1484.M.TANE-EB.5.134320
Three-volume set : DOI: 10.1484/M.TANE-EB.5.133678

Printed in the EU on acid-free paper.

Conference sponsors

Vetenskapsrådet (Swedish Research Council)

Kungl. Gustav Adolfs Akademien för svensk folkkultur (The Royal Gustavus Adolphus Academy for Swedish Folk Culture)

Stockholms universitet (Stockholm University)

Zentrum für Baltische und Skandinavische Archäologie (Centre for Baltic and Scandinavian Archaeology)

Book sponsors

Kungl. Gustav Adolfs Akademien för svensk folkkultur (The Royal Gustavus Adolphus Academy for Swedish Folk Culture)

Länsstyrelsen Västerbotten (County Administration of Västerbotten, Sweden)

International Council for Game and Wildlife Conservation (CIC)

Book 3

Chapter 12 – Further reading: Bears in a broader perspective 1063

The occurrence of *Ursus arctos* in relation to other faunal remains in burials during the Late Iron Age (560/70–1050 CE) in Uppland, Sweden

By Hannah Strehlau

Keywords: Late Iron Age, Sweden, bear claws, faunal remains, burial tradition, aristocracy

*Abstract: This chapter deals with the faunal remains of the brown bear (*Ursus arctos*) in burials in relation to other faunal remains. In a study of 83 analysed contexts, bear claws appear in much fewer burials than the remains of other animal species. Specifically, domestic animal remains are deposited at a much higher rate than those of wild animals, including brown bears. In total, four graves contained bear claws, and these burials are presented as comparative case studies. Despite striking differences in the burial type and the equipment of the graves, all case studies are located at three boat-grave cemeteries (Valsgärde, Gamla Uppsala, Tuna in Alsike), which connects them to Vendel Period and Viking Age aristocracies.*

Introduction

During the Vendel Period (560/70–750 CE) and Viking Age (750–1050 CE)[1] in mid-eastern Sweden, graves of women, men, and children often contained an abundance of faunal remains. These bones mostly belonged to the remains of feasting, food gifts, and sacrifices, the latter of which deposited as both complete and incomplete animals. In a few cases, however, the remains do not fit into any of these categories.

Bear claws in prehistoric graves are commonly interpreted as being the only preserved remains of a bear fur (e.g. Petré 1980; Sigvallius 1994, 76), which is supported by recent studies on organically preserved bear hairs from archaeological contexts and skinning practices (Kirkinen 2017).[2] Similar explanations are suggested for the occurrence of phalanges from other fur-bearing animals, such as cat, lynx, squirrel, or marten (Petré 1980; Lindholm/Ljungkvist 2016; Zachrisson/Krzewińska 2019). Even selective depositions of cranial parts, extremities, and vertebrae of horses and cattle

1 This study uses the chronological periods suggested for mid-eastern Sweden by Ljungkvist (2008). He divides the Vendel Period into four periods (Vet 1–4, after Swedish *Vendeltid*) and the Viking Age into six periods (Vit 1–6, after Swedish *Vikingatid*). For further reading on the chronology see also Arrhenius 1983, fig. 6; Nørgård Jørgensen 1999, and Ljungkvist 2015 in particular for the end of the Viking Age in Uppland.

2 Further interpretations of bear claws in graves include: luxury or prestige goods, hunting trophies, magical or cultic objects that are tied to religious or ritual practice (e.g. Iregren 1988, 303–304; Krüger 1988, 361–365; Schönfelder 1994; Wamers 2009).

Bear and Human: Facets of a Multi-Layered Relationship from Past to Recent Times, with Emphasis on Northern Europe, ed. by Oliver Grimm
(Turnhout: Brepols, 2023), pp. 417–428 BREPOLS PUBLISHERS 10.1484.M.TANE-EB.5.134343

were interpreted as the remains of hides (Piggott 1962; Hagberg 1967, 59–60; Russell 2012, 109). Concerning bear claws, some inhumation graves showed, in fact, the undisturbed and original position of the claws, suggesting that the deceased human had been laid on or covered by a fur (e.g. Petré 1980, fig. 2). In addition, Lindholm/Ljungkvist (2016) mention that these bear claws are (almost) never processed, which excludes their use in perforated form as pendants or amulets.[3]

Besides the interpretation of such bear phalanges in prehistoric graves, recent projects focused on the question of how common this practice was, how it was distributed chronologically and geographically, and what impact big game hunting had on the population of the brown bear. In a study on the faunal exploitation of *Ursus arctos* during the Late Iron Age (500–1100 CE) in Sweden, Lindholm/Ljungkvist (2016) presented 323 contexts (including both graves and settlements) in which faunal remains of the Swedish brown bear occur. Earlier, Grimm (2013) had shown that there are approximately 500 known burials during the 1st millennium CE in northern and middle Europe with "bear related furnishings". Does this mean that faunal remains of *Ursus arctos* in graves represent a common phenomenon? Who was equipped with a bear fur in the burial? How can we interpret this type of deposition? This chapter highlights the prevalence of bear claws in relation to other faunal remains in burial contexts from the Late Iron Age.

Case studies

To find out whether the presence of bear remains in the funeral context is a common phenomenon, one has to consider their distribution in relation to the total numbers. Extensive statistical analyses on faunal assemblages in burials that cover a supra-regional area have not been carried out on a large scale for the Iron Age of northern Europe. In this case, small-scale research has been carried out on 83 graves in the province of Uppland in middle-eastern Sweden (Strehlau 2018; cf. Fig. 1).[4] Of the 83 graves, 46 were cremations and 37 were inhumations. These contexts were selected based solely on the quality of osteological reports and dating possibilities, resulting in a cross section from the Vendel Period and Viking Age societies in Uppland. The aim of the project was to find patterns among the deposition of animals in burial contexts and how they shifted in relation to the social status of the dead, the gender of the deceased, and other factors. Which species were present among the faunal remains, which body parts were deposited, in what condition were the bones, and, mainly concerning inhumations, where in the grave were they placed?

Only four graves (4.81 %) held osteological evidence of *Ursus arctos*, and only *phalanges 3*, i.e. the claws of the bear, were preserved as faunal remains (Fig. 2). In contrast, the majority of depositions included domestic animals (dogs, horses, cattle, pigs, sheep/goats, chickens, cats), which appeared in 72 out of 83 graves (86.74 %). Wild animals, on the other hand, showed up in only 21 out of 83 contexts (25.30 %) and only occurred together with domestic animals (Fig. 3). This group includes wild mammals, wild birds, and fish, but excludes geese and rodents.[5] When further excluding fish and all undetermined birds, the number shrinks to 16 graves (16.86 %). Furthermore, excluding 11 graves dating to the Late Viking Age that contain neither faunal remains nor grave goods from the total

3 Exceptions are known, for example, from a Viking Age grave in Rösta (Ås parish, Jämtland), where two single bear claws were found close to the human skeleton. They are interpreted as talismans that might have been placed in a little bag, which would have been attached to the clothes (Jordahl et al., this volume; Petré 1980; cf. Kjellmark 1905, 369).

4 The study and the case studies presented here are part of a Master's thesis in archaeology, defended in 2018 at Uppsala University (cf. Strehlau 2018).

5 Because of the uncertainty whether the geese in these contexts were domestic or wild and the uncertainty whether certain rodents were intentional depositions or ended up in the graves accidentally.

count,[6] the numbers become clearer – in this case domestic animals appeared in 71 out of 72 graves (98.61 %) and wild animals in 21 out of 72 contexts (29.16 %).

Since bears rarely appear in graves compared to domestic animals, we need to consider how remains of wild animals relate to each other. It appears that with an MNI[7] of one each, lynx (*Lynx lynx*), red deer (*Cervus elaphus*), and wild boar (*Sus scrofa*) are the rarest species in the study.[8] All these animals are represented by a few burnt fragments of hooves/lower limbs, and only claws in case of the lynx. Two possible squirrels (*Sciurus vulgaris*) are part of the material, both associated with the biological order Rodentia and mentioned as uncertain in the osteological report.[9] In both graves, these were found as burnt bones in the cremation layer, which suggests their intentional burning as part of the burial ritual. Unlike mice, which were identified among the rodents, squirrels are known as fur-bearing animals. While the depositions of bear claws led to various different interpretations (cf. e.g. Petré 1980; Iregren 1988, 303–304; Krüger 1988, 361–365; Schönfelder 1994; Wamers 2009), the claws of squirrel, marten, lynx, and, in a few cases, cat, need to be considered in the context of the fur trade during that time. However, both potential squirrels were not represented by claws; one is represented by a fragment of *talus 1,* which is a foot bone (Ultuna 4425), and the other by a few fragments of the lower hind extremities (Ultuna 4060).

In contrast to the low number of wild mammals, there is a total of 20 wild birds (MNI) – hunting birds and their prey as well as decoys. Identified species are goshawk (*Accipiter gentilis*), sparrow-hawk (*Accipiter nisus*), eagle-owl (*Bubo bubo*), snowy owl (*Bubo scandiacus*), peregrine falcon (*Falco peregrinus*), merlin (*Falco columbarius*), capercaillie (*Tetrao urogallus*), black grouse (*Lyrurus tetrix*), hazel grouse (*Tetrastes bonasia*), crane (*Grus grus*), and duck (*Anatidae*).[10] Including geese as potential wild birds, the MNI would be 31. Then again, this number is confusing since these animals originate from only six graves (corresponding to 7.22 %), all of which are high-status burials that contain an abundance of both wild and domestic species' remains. Thus, the occurrence of wild birds in graves is, in fact, not much higher than that of bears.[11] These interrelations show that faunal remains from domestic animals are much more common in Upplandic graves than faunal remains from wild animals. Except for wild birds, brown bear remains appear more often than other wild animals in these burial contexts, but are still underrepresented compared to domestic animals.

In the following, we take a closer look at the four bear-related contexts. If the faunal remains of *Ursus arctos* do not occur as frequently as other animal-related depositions, do these graves have anything in common? Were there only certain people buried with bear furs?

The four case studies containing bear claws are two cremation and two inhumation graves. The sites in question are Valsgärde, Gamla Uppsala, and Tuna in Alsike, which are well-known burial sites situated in close vicinity to today's city of Uppsala. These cemeteries, which contain a number of burials with elaborate grave constructions as well as rich grave furnishings, are all part of the Vendel Period and Viking Age upper class boat-grave tradition.

6 Due to the absence of animal depositions and grave goods, such graves from the Late Viking Age can be interpreted as early Christian and thus express a different burial tradition.

7 The Minimum Number of Individuals (MNI) constitutes the least number of individuals of a certain species that occurs in one context. It is a purely anatomical count, in which the appearance of the same bone several times indicates the number of individuals. This number is a minimum, since it cannot be excluded that two different bones that only appear once in the context might actually originate from two different individuals. The absolute number of individuals might only be determined by genetic analyses.

8 Graves 7 and 110 from Odenslunda (cf. Sigvallius 2005), and grave 729 from Inhåleskullen (cf. Ohlsson 2012).

9 Graves 4425 and 4060 from Ultuna (cf. Sjöling/Bäckström 2014).

10 Ducks are here listed among wild animals, but it cannot be ruled out that domesticated ducks are among them.

11 Both wild and domestic birds are probably underrepresented regarding species determination in comparison to bear claws, due to high fragmentation in cremation burials and a lack of sieving, especially on early excavations. A similar source-critical problem exists with the preservation and determination of fish remains.

Grave 65 in Valsgärde is an urn grave with a cremation layer, dating to Vet 2–3 (620/30–700/710 CE) and containing the remains of two human individuals (Ljungkvist 2008). Osteological analyses revealed that the bones belong to a male adult and to a younger, probably juvenile, individual whose sex could not be determined (Bäckström 2001, 9–10). The identification of at least one male individual is supported by the archaeological material: an iron strap mount, a fire steel, a strike-a-light flint, a comb, and melted glass. The osteological report further shows that an MNI of at least five individual animals is present in the grave: one sheep/goat (*Ovis aries/Capra hircus*), one chicken (*Galliformes*), one dog (*Canis familiaris*), one pike (*Esox lucius*), and bear claws (n = 5, *Ursus arctos*; ibid., 10–13). In addition, unburnt long bones (fleshy parts) of an undetermined mammal indicate the presence of a sixth animal species. Except for the bear, pike, and the undetermined mammal, the other animals were deposited as complete carcasses in the grave, and all of them were burnt (ibid.).[12] Due to the relatively sparse grave furnishing, it cannot be classified as a high-status burial. Nevertheless, the damaged status of this context leaves the possibility that the original grave furnishing was somewhat different.

The other cremation from Valsgärde is grave 82, dating to Vet 1 (560/70–620/30 CE), which makes it slightly older than grave 65 (Ljungkvist 2008). This context is a non-concentrated thin find layer, which was found undisturbed. As in the case of grave 65, it contained the remains of an adult. The burial is stratigraphically related to a ship setting, but it is not certain whether burial and ship are related. The bones from this grave have, so far, only been briefly examined by Niels-Gustaf Gejvall,[13] who identified the bones of a human, preliminarily recognised as female,[14] as well as bones from a horse, dogs, sheep, and poultry. However, the grave inventory, consisting of a probable copper-alloy sword sheath mount, a spear rivet, a comb, and a strike-a-light flint, suggests that the deceased was a male. Furthermore, an assembly of iron rivets and nails indicate a metal-bound wooden vessel (Ljungkvist et al. in press). In addition to these finds, the grave also contained bear claws (n = 11, *Ursus arctos*; Bäckström 2001, 18).

The above-mentioned finds in this context make a reconstruction as a weapon grave possible. The deceased was most likely buried with a spatha or seax, indicated by the sheath mount. Other occurring objects are pottery and an iron rod as well as an assembly of small iron rivets and nails. The latter indicate the former existence of a metal-bound wooden vessel. In combination with the weapons and the number of bear claws (11), which strongly suggest the remains of a bear fur, this gives the deceased a fairly high status. The number of animals in the burial has not been determined, but the weight of the bones is *c.* 3,600 g, which is high but not remarkable. It is roughly equivalent of what is expected from a burial with a horse and a few smaller animals from this period (e.g. Prata et al. 2017, tab. 1). Merovingian Period cremation burials with amounts of animals that correspond with the Valsgärde boat burials contain a considerably higher number of bones (cf. Hennius et al. 2016, 85; Hed Jakobsson et al. 2019, 353–363).

Grave 1 from Gamla Uppsala is a Viking Age inhumation burial and boat-grave that has been subject to considerable damage. It held a 35–45-year-old adult whose sex has been osteologically classified as male. The distorted posture of the dead, the accumulation of grave goods, and the skeletons of two dogs placed on top of each other outside the boat give reason to suppose that the grave was reopened and looted not too long after the burial. Thus, the absence of weapons – which should be expected in such a boat-burial – is most likely a consequence of the ancient reopening. After the reopening, the burial still contained an iron knife, 10 arrowheads, a wooden vessel, pottery, horse

12 This is an important observation since unburnt bones would indicate a more complex action in which the unburnt material must have come into the grave after the cremation.

13 Unpublished material by N.-G. Gejvall, Uppsala Universitets Museum för Nordiska Fornsaker (UMF).

14 According to Bäckström's (2001, 18) osteological analysis the sex of the human cannot be determined.

equipment (spikes [to prevent horses from slipping], a bit, trace-hooks [used on horses that pull carts; cf. Nordahl 2001, 26, fig. 23]), a comb, gaming pieces made of bone, a strike-a-light, a piece of flint, a hammer, and a Thor's hammer amulet.

Interestingly, this context includes a relatively high number of animal deposits: the MNI of individual animals is nine, representing two complete dogs (*Canis familiaris*), one complete horse (*Equus caballus*), five claws of a bear (*Ursus arctos*), cranial parts and long bones of sheep (*Ovis aries*) with marrow split, pig (*Sus domesticus*), and cattle (*Bos taurus*), which is an indication of consumption. Furthermore, an undetermined bird (*Aves*) and a fish (*Pisces*) were among the bone material.

In the publication of the boat-graves from Gamla Uppsala, Else Nordahl states that the claws were placed in pairs on the port and starboard sides of the boat (Nordahl 2001, 16). However, the position of the fifth claw is not clear from the publication, and, in the drawing of the excavated grave, the claws (no. 77, no. 305) are depicted on the portside and in the middle of the boat, but not on the starboard side (ibid., pl. 1). According to Nordahl (2001) the deceased person was originally placed on a bear fur, but was then moved to the side in the course of the grave disturbance. However, the degree of disturbance of this grave, in combination with the small number of bear claws that appear to be scattered in the grave rather than in an anatomical position, does not allow for a confident interpretation as a bear fur. The burial of the deceased in a boat in combination with a high number of animal offerings classifies it as a high-status burial.

South of Uppsala lies a cemetery with high-status burials dating to the Viking Age. Grave Alsike II is a much disturbed context with scattered finds and animal bones, but no human remains have been preserved. According to Arne (1934, 25), this context was regarded as originally belonging to grave I in Alsike, an inhumation boat-grave for a deceased male. However, the undisturbed nature of Alsike I, and the fact that the grave goods indicate a chronological deviation of 200 years between the two burials, gives reason to regard Alsike II as a separate construction, dating to the early 9th century (personal communication, John Ljungkvist; cf. Arne 1934, 70–71). Due to the unburnt state of the faunal remains, this context is here regarded as the very disturbed remnants of a former inhumation grave in which the human body was either not preserved or was removed during reopening.[15]

The archaeological remains found at Alsike II are a horse bridle, a double-edged iron sword broken into several parts, an iron whip handle fitting, a gaming piece made of glass, an iron chain from a kettle, a green clay bead, spikes, and other iron fragments. The faunal remains are represented by seven bear claws, the back half of a horse as well as the tibia (shin) and metatarsus (foot bone) of another horse, and the back half of a dog. Five of the bear phalanges are still anatomically connected (Fig. 4), despite the high degree of disturbance of this grave. This indicates that the disruption happened when the organic decomposition of the bear remains was not yet advanced. And, more importantly, it supports the idea of a bear paw and possibly an entire fur in the grave.

Discussion

What these burials have in common is their placement in three boat-grave cemeteries that are tied to economic wealth and political power. So far, women are underrepresented among these graves with ursine remains, which correlates with a study of Migration Period burials from mid-eastern Sweden (Bennett 1987, tab. 81). Other studies, however, mention equal numbers of female graves with bear claws (Petré 1980; Klos 2007; Grimm 2013, 291; Gustavsson/Ljungkvist, this volume;

15 The absence of human remains in Vendel Period and Viking Age inhumation graves, as well as the comparatively bad preservation of faunal remains in close proximity to the estimated original placement of the deceased human, are also known from the cemeteries at Vendel and Valsgärde in Uppland (cf. e.g. Stolpe/Arne 1912; Arwidsson 1977).

Jordahl et al., this volume). Their affiliation to important cemeteries makes them seem to be part of the same burial tradition, yet the graves themselves show striking differences: there are two inhumation boat-graves and two cremation graves, and the number of animal depositions varies from an MNI of three (Alsike II) to an MNI of nine (Gamla Uppsala 1). Even though the two cremation burials containing bear claws (graves 65 and 82 from Valsgärde) cannot be regarded as high-status graves on the basis of the remaining grave goods, the fact that they are situated right next to some of the most elaborate boat-graves in Sweden gives reason for reconsidering this classification. One could argue that there might have been an affiliation to the aristocracy, otherwise they would not have been buried in such close proximity. The number of deposited animals and the species representation are common for graves of both higher and lower social status in Vendel Period Uppland (cf. Strehlau 2018).

Considering the effort and danger of hunting and killing a life-threatening predator like the brown bear, it appears logical that the possession of a bear fur created a certain reputation in a hunting society (cf. Oehrl 2013). At the same time, it most likely had a high value on an economic scale and was thus regarded as a status symbol in the densely populated Uppsala region, where large wild mammals, including bears, were probably already very rare or even extinct (cf. Lindholm/Ljungkvist 2016). However, horses are commonly regarded as status symbols as well, but they occur in 50 out of 83 graves (60.24 %) in the study, which is a much higher number compared to the occurrence of bear claws. There is an even higher number of dogs, which occur in 63 out of 83 graves (75.90 %), but they were probably not seen as status symbols.

Looking beyond this study, it appears that six out of 95 graves in Valsgärde contain bear claws (Tab. 1), of which five are cremations and one is a Late Roman Iron Age chamber grave with an inhumation (cf. Ljungkvist 2008). This corresponds to 6.31 %. In Valsgärde, there are four inhumation graves from the Pre-Roman Iron Age (*c.* 500 BCE–0), 29 inhumation graves dating from the Roman Iron Age (*c.* 0–400 CE) to the Viking Age, and 62 cremation graves from the Late Iron Age. None of the boat-graves in Valsgärde held any faunal remains of brown bear, and it seems to be the same situation in Vendel. Admittedly, the cemetery in Vendel was excavated at the end of the 19th century, and osteological analyses might not have been up to today's standards. However, if it was possible to determine the claws of different types of birds of prey, it seems unlikely that bear phalanges would have been missed (cf. Stolpe/Arne 1912). Cremation graves from the Migration Period to the Viking Age from Vendel contained bear claws in 15 out of 191 cases, which corresponds to 7.85 % (cf. Seiler 2001).

Seven out of 135 osteologically analysed graves in Gamla Uppsala contain bear claws (Tab. 1). This burial area includes the three so-called king's mounds, four boat-graves and seven cremation graves in Prästgården, a grave field located northwest of the royal mounds. Bear claws appeared in the east and west mounds (the middle mound remains unexcavated to date; cf. Sten/Vretemark 1999), in Grave 1, which is mentioned above, and in cremation grave 24 (cf. Nordahl 2001). The remaining three bear-related burials derive from a Late Iron Age cemetery, situated in the village area with 122 excavated cremation graves (cf. Prata et al. 2017). Seven out of 135 graves correspond to a percentage of 5.18 %.

In the Viking Age cemetery of Tuna in Alsike, one grave (discussed above) contained bear claws (7.69 %). With the exception of two burials (XIII, XIV), the excavated graves from this site date to the Viking Age, with some of these dating to the 11th century (Arne 1934). Burials from this time often follow early Christian practices, which would explain the absence of both faunal remains and grave goods. Moreover, Lindholm/Ljungkvist (2016) have shown that the number of bear-related graves started to decrease in the late 7th century. Similar observations exist concerning the Iron Age cemetery of Spånga in Stockholm County. Of the 488 cremation burials, 29 contained *phalanges 3* of *Ursus arctos* (5.9 %), with a peak during the Migration Period and a decline in the Late Iron Age

(Sigvallius 1994, 74–75). The lowest number of graves with bear claws was documented on the island of Adelsö in Birka; only eight out of 1,000 graves with remains of *Ursus arctos* were found here (Swedish History Museum, Digital Archive, accessed online, August 05, 2020). In total, there are 1,922 burials in the six cemeteries mentioned above, of which 66 contain bear claws (3.43 %; Tab. 1). This matches largely with the results of the study on the 83 graves in Uppland (cf. Strehlau 2018). The slightly higher percentage among the material of that study (4.81 %) can be explained by the more substantial proportion of high-status graves.

The comparably rare frequency of bear-related graves and their connection to an upper social class can be compared to the occurrence of animal depositions that are related to falconry. Swedish archaeologist Ann-Sophie Gräslund regards the combination of large dogs (greyhounds), horses, and birds of prey in burials as an indication for falconry (Gräslund 2014, 37). Based on their osteological studies, Sten/Vretemark (1988) note that birds of prey often occur in graves together with other wild birds that are interpreted as decoys and prey. These species typically occur in high-status graves, for instance, in the boat-graves in Vendel and Valsgärde, in the chamber graves from Tuna in Alsike, and also in the richly furnished cremation graves at Rickeby and Gnista in Uppland. In this study, bear furs do not appear in graves that show indications for falconry or contain remains of other wild mammals, but the simultaneous occurrence of the two is known from other very richly equipped cremation graves (cf. ibid.). Moreover, there is a dominance of male individuals both in burials with bear claws and in those with indications for falconry. This largely matches with other contexts outside this study (ibid.).

A difference between falconry and the bear claw burials, however, is the deposition type. While *Ursus arctos* is only represented by *phalanges 3*, animals that are connected to falconry, including dogs and horses, are typically deposited as complete carcasses. Complete animals deposited in burial contexts are normally interpreted as sacrifices (e.g. Kaliff 2004, 28; Mansrud 2004, 94–95; Vretemark 2013, 381). Incomplete depositions, to which bear claws belong, are open to much more varied interpretations. They can represent the remains of a funeral feast, food gifts for the deceased, objects such as furs, amulets, or other symbolic depositions (cf. Russell 2012). The deposition of bear, and also lynx claws is neither the sacrifice of a complete animal nor can it be interpreted as the remains of feasting or food offerings. Assuming that the claws are the remains of furs, these finds, in fact, need to be regarded as a product, a crafted item, possibly tied to social status, rather than an animal deposition, which indicates certain actions in the burial ritual, like feasting or the killing of an animal as a potential sacrifice and as a symbol in itself.

The different frequencies of bear remains compared to those of dogs and horses might be explained by their meaning in the burial ritual. The high number of complete dogs and horses deposited in Upplandic graves suggests that they played an important role in the burial ritual, regardless of the social status of the deceased (Strehlau 2018). Bear furs, on the other hand, seem to have been in the possession of a much smaller number of people in the Late Iron Age, i.e. the Vendel Period and Viking Age. At the least, the rather rare occurrence of bear claws in graves suggests that bear furs should be regarded as an unusual object and grave good rather than as an animal that played an important part in the burial ritual. The differences between the graves containing bear claws indicate that, possibly, social status during the Vendel Period and Viking Age cannot be explained solely on the basis of the burial type or the grave goods that remain visible in the archaeological record today. Likewise, it is possible that those buried with a bear fur were part of a sort of allegiance, who, although they acquired a certain social status, never gained the level of wealth and power that could have been reflected in a lavish burial with a monumental mound or a boat-grave.

Conclusion

The study presented here has shown that the faunal remains of *Ursus arctos* appear with relatively low frequency in the analysed graves of the Vendel Period and Viking Age. The vast majority of faunal remains are those of domestic animals. However, in comparison with other wild mammals, including fur-bearing animals, brown bear remains occur more often in burial contexts. Wild birds appear with a similar frequency in graves, but are often deposited as complete animals in graves with a higher MNI. In addition, bear claws representing bear furs should be considered as grave goods rather than animal depositions. But, even if the total number of graves with ursine evidence is low in several cemeteries, it is striking that, in Vendel and Viking Period central Uppland, they are clearly present in burial places that are strongly connected to the Upplandic upper class. The absence of bear claw finds in boat-graves from Vendel and Valsgärde might be explained by poor preservation conditions and plundering. At the same time, the highest number of graves with bear claws, compared to the total number of graves in a cemetery, is reached in cremation burials in the Vendel cemetery. Its location, 35 km north of Uppsala, and the character of some burials were most recently discussed as being influenced by the northern Swedish regions, which were the brown bear's home and also the area where it was hunted (Ljungkvist/Hennius 2020). It seems that the affiliation to a certain place or group of people was more important than the burial type. Likewise, it is possible that it was easier to get possession of a bear fur in an environment of intensive trade and production and of economic wealth. As is evident from the case studies, social status and reputation might not solely be explained by the sheer quantity of animal depositions, grave goods, and an elaborate burial.

Acknowledgements

This paper was written within "The Viking Phenomenon" project, funded by the Swedish Research Council (grant 2015-00466), and "The Bear in the Grave" project, both Uppsala University.

Bibliography

Arne 1934 T. J. Arne, Das Bootsgräberfeld von Tuna in Alsike, Uppland. KVHAA Arch. Monogr. 20 (Stockholm 1934).

Arrhenius 1983: B. Arrhenius, The chronology of the Vendel graves. In: J. P. Lamm/H.-Å. Nordström (eds.), Vendel Period Studies. Transactions of the Boat-grave symposium in Stockholm, February 2–3, 1981 (Stockholm 1983) 39–70.

Arwidsson 1977: G. Arwidsson, Valsgärde 7. AMASRUU 5 = Gräberfunde Valsgärde 3 (Uppsala, Stockholm 1977).

Bäckström 2001: Y. Bäckström, Osteologisk analys. Brandgravar. SAU Rapp. 2001:8 (unpublished) (Uppsala 2001).

Bennett 1987: A. Bennett, Graven. Religios och social symbol. Strukturer i folkevandringstidens gravskick i Mälaromradet. Theses and Papers in North-European Archaeology 18 (Tierp 1987).

Gräslund 2014: A. Gräslund, En hund begraven. Hunden från stenålder till medeltid. In: A.-S. Gräslund/I. Svanberg (eds.), Från renhållningshjon till modeaccessoar. 10000 år av relationer människa-hund i Sverige. Acta Academia Regiae Gustavi Adolphi 133 (Uppsala 2014) 31–48.

Grimm 2013: O. Grimm, Bear-skins in northern European burials and some remarks on other bear-related furnishings in the north and middle of Europe in the 1st millennium AD. In: O. Grimm/U. Schmölcke (eds.), Hunting in northern Europe until 1500 AD. Old traditions and regional developments, continental sources and continental influences. Papers presented at a workshop organized by the Centre for Baltic and Scandinavian Archaeology, ZBSA. Schriften des Archäologischen Landesmuseums, Ergänzungsreihe 7 (Neumünster 2013) 277–296.

Hagberg 1967: U. E. Hagberg, The Archaeology of Skedemosse II. The Votive Deposits in the Skedemosse Fen and their Relation to the Iron-Age Settlement on Öland, Sweden (Stockholm 1967).

Hed Jakobsson et al. 2019: A. Hed Jakobsson/C. Lindblom/L. Lindwall, Husfruar, Bönder och Odenkrigare. Kumla i Östra Fyrislund från Romersk Järnålder till Vikingatid. Rapp. från Arkeologikonsult 2019:2901,3042 (Stockholm 2019).

HENNIUS et al. 2016: A. HENNIUS/E. SJÖLING/S. PRATA, Människor kring Gnistahögen och tidig medeltid. Begravningar från vendeltid, vikingatid och tidig medeltid (Uppsala 2016).

IREGREN 1988: E. IREGREN, Finds of Brown Bear (Ursus arctos) in Southern Sacandinavia. Indications of Local Hunting or Trade? In: B. Hårdh/L. Larsson/D. Olausson/ R. Petré (eds.), Trade and Exchange in Prehistory. Studies in Honour of Berta Stjernquist. Acta Arch. Lundensia, Series in 8° no. 16 (Lund 1988) 295–308.

KALIFF 2004: A. KALIFF, Offerritual och gravplatser. Några kommentarer kring sambandet mellan arkeologisk terminologi och tolkning. In: L. Melheim/L. Hedeager/ K. Oma (eds.), Mellom himmel og jord: Foredrag fra et seminar om religionsarkeologi, Isegran, 31. Januar – 2. Februar 2002. Oslo Archaeological Ser. 2 (Oslo 2004) 18–31.

KIRKINEN 2017: T. KIRKINEN, "Burning pelts" – Brown bear skins in the Iron Age and Early Medieval (0–1300 AD) burials in south-eastern Fennoscandia. Estonian Journal of Archaeology 21(1), 2017, 3–29.

KJELLMARK 1905: K. KJELLMARK, Ett gravfält från den yngre järnåldern i Ås i Jämtland. Ymer 25 (Stockholm 1905) 351–371.

KLOS 2007: L. KLOS, Lady of the rings. Järnålderns kvinnor mellan makt och kult. Iron Age women between power and cult. In: I. Nordgren (ed.), Kult, Guld och Makt. Ett tvarvetenskapligt symposium i Gotene. Historie Västra Götaland, Ser. B: Vetenskapliga rapporter och smaskrifter 4 (Göteborg 2007) 70–86.

KRÜGER 1988: S. H. KRÜGER, Bjørneklør fra vestlandske graver. In: S. Indrelid/S. Kaland/B. Solberg (eds.), Festskr. A. Hagen. Arkeologiske Skrifter 4 (Bergen 1988) 357–366.

LINDHOLM/LJUNGKVIST 2016: K.-J. LINDHOLM/J. LJUNGKVIST, The Bear in the Grave. Exploitation of Top Predator and Herbivore Resources in First Millennium Sweden – First Trends from a Long-Term Research Project. European Journal of Archaeology 19(1), 2016, 3–27.

LJUNGKVIST 2008: J. LJUNGKVIST, Valsgärde. Development and change of a burial ground over 1300 years. In: S. Norr (ed.), Valsgärde Studies. The place and its people, past and present (Uppsala 2008) 13–56.

LJUNGKVIST 2015: J. LJUNGKVIST, Gravar i en övergångsperiod. De yngsta kammargravarna och brandgravarna i Uppland under 1000- och 1100-talet. META Historiskarkeologist Tidskrift 2014–2015, 2015, 21–45.

LJUNGKVIST/HENNIUS 2020: J. LJUNGKVIST/A. HENNIUS, The dating of Ottarshögen and the emergence of monumental burial mounds in Middle Sweden. In: C. Hillerdal/ K. Ilves (eds.), Re-imagining Periphery. Archaeology and text in northern Europe from Iron Age to Viking and Early Medieval Periods (Oxford 2020) 91–101.

LJUNGKVIST et al. in press: J. LJUNGKVIST/A. GRÄSLUND/ K. OJALA (eds.), Valsgärde. The cremation and chamber burials (in press).

MANSRUD 2004: A. MANSRUD, Dyrebein i graver. En kilde til jernalderens kult og forestillingsverden. In: L. Melheim/ L. Hedeager/K. Oma (eds.), Mellom himmel og jord. Foredrag fra et seminar om religionsarkeologi, Isegran, 31. Januar – 2. Februar 2002. Oslo Archaeological Ser. 2 (Oslo 2004) 82–112.

NORDAHL 2001: E. NORDAHL, Båtgravar i Gamla Uppsala. Spår av en vikingatida högreståndsmiljö. Med Bidrag af A. Malmius, P. Molnar, A. Kjellström och B. Schönbäck (Uppsala 2001).

NØRGÅRD JØRGENSEN 1999: A. NØRGÅRD JØRGENSEN, Waffen und Gräber. Typologische und chronologische Studien zu skandinavischen Waffengräbern 520/30 bis 900 n. Chr. (Kopenhagen 1999).

OEHRL 2013: S. OEHRL, Bear hunting and its ideological context (as a background for the interpretation of bear claws and other remains of bears in Germanic graves of the 1st millennium AD). In: O. Grimm/U. Schmölcke (eds.), Hunting in northern Europe until 1500 AD. Old traditions and regional developments, continental sources and continental influences. Papers presented at a workshop organized by the Centre for Baltic and Scandinavian Archaeology, ZBSA. Schriften des Archäologischen Landesmuseums, Ergänzungsreihe 7 (Neumünster 2013) 297–332.

OHLSSON 2012: A. OHLSSON, Bilaga 5. Osteologisk Analys. In: A. Seiler/K. Appelgren (eds.), Inhåleskullen. Ett mångtydigt gravfält från yngre bronsålder – äldre vikingatid. UV rapport 2012:158 (Stockholm 2012) 112–125.

PETRÉ 1980: B. PETRÉ, Björnfällen i begravningsritualen. Statusobjekt speglande regional skinnhandel? Fornvännen 75, 1980, 5–14.

PIGGOTT 1962: S. PIGGOTT, Heads and hoofs. Antiquity 36, 1962, 110–118.

PRATA et al. 2017: S. PRATA/E. SJÖLING/R. GUSTAVSSON, Brandgravar vid Storby backe – Osteologisk analys. Utbyggnad av Ostkustbanan genom Gamla Uppsala. Rapport 2017:1_13: Arkeologisk undersökning (Stockholm 2017).

RUSSELL 2012: N. RUSSELL, Social Zooarchaeology. Humans and animals in prehistory (New York 2012).

SCHÖNFELDER 1994: M. SCHÖNFELDER, Bear-Claws in Germanic Graves. Oxford Journal of Archaeology 13(2), 1994, 217–227.

SEILER 2001: A. SEILER, I skuggan av båtgravarna. Landskap och samhälle i Vendel socken under yngre järnålder. Theses and Papers in Archaeology B/7 (Stockholm 2001).

SIGVALLIUS 1994: B. SIGVALLIUS, Funeral Pyres. Iron Age Cremations in North Spånga. Theses and Papers in Osteology (Stockholm 1994).

SIGVALLIUS 2005: B. SIGVALLIUS, Bilaga 5. Osteologisk analys. In: M. Olausson (ed.), Bytomten vid Odenslund. Bebyggelselämningar från folkvandringstid. Nyare tid samt delar av ett gårdsgravfält från vendeltid – vikingatid (Hägersten 2005) 55–74.

SJÖLING/BÄCKSTRÖM 2014: E. SJÖLING/Y. BÄCKSTRÖM, Osteologisk analys. In: H. Hulth, Den skyddande logen. Brandgravar från yngre järnålder samt en och annan medeltida och efterreformatorisk grop på Ultuna. Arkeologisk för- och slutundersökning. Med bidrag av Y. Bäckström, E. Sjöling, J. Ljungkvist and E. Pettersson. SAU rapport 2014:1 (Uppsala 2014) 48–57.

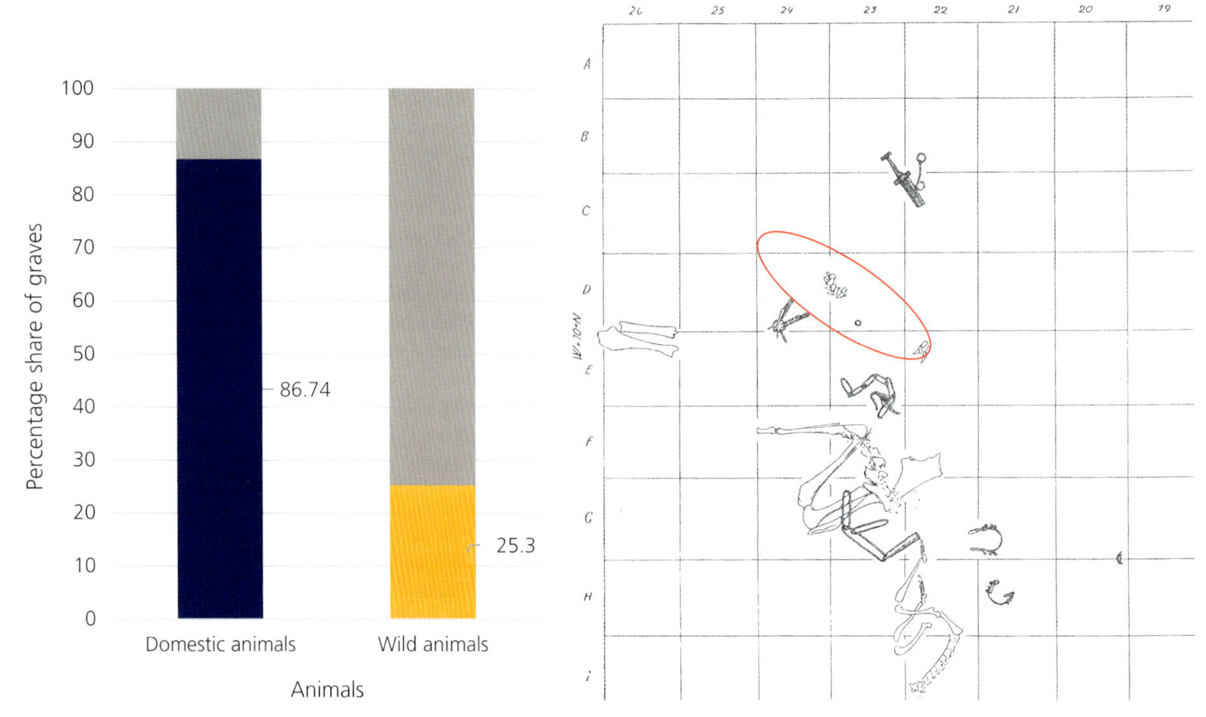

Fig. 3. Percentages of domestic (dark blue) and wild (yellow) animals found among the 83 graves from Uppland in middle-eastern Sweden. Wild animals only occur in combination with domestic animals.

Fig. 4. Grave II from Tuna in Alsike in middle-eastern Sweden, 9th century, with position of bear claws marked by red circle (modified after Arne *1934, pl. 23).*

Table 1. The frequencies of graves containing bear claws compared to the total number of analysed graves from a certain cemetery or area during the Late Iron Age in Uppland, middle-eastern Sweden.

Cemetery/area of study	Graves containing bear claws	Total number of graves	Percentages of bear claws in graves (in %)
Vendel	15	191	7.85
Tuna in Alsike	1	13	7.69
Valsgärde	6	95	6.31
Spånga	29	488	5.9
Gamla Uppsala	7	135	5.18
Birka	8	1,000	0.8
Total	**66**	**1,922**	**3.43**

Bear bones from the Viking Age cult place at Frösö church – the unifying factor in bear-human relationships in Viking Age Jämtland, northern Sweden

By Ola Magnell

Keywords: Brown bear, Viking Age, hunting, body-part frequency, butchering

Abstract: The role and significance of the brown bear in hunting, ritual practice and inter-cultural relationships during the Viking Age in Middle Sweden have been studied through the osteological remains from the cult place at Frösö church in Jämtland, northern Sweden. Bear hunting has been examined through age distribution, with age determination based on epiphyseal closure or tooth wear. The results indicate a harvest focused on animals of about 2–4 years of age or adults over four years old. The body-part distribution and butchering marks reflect the transport, utilisation and treatment of bear bones at the site, with furs and possibly amulets made from canines exported from Frösö to other regions. The bear's role in Viking culture and its association with Sami ritual practices is discussed. The bear-human relationship seems to have had a unifying role in the social interactions and networks between the local Viking aristocracy on Frösö and the Sami hunters in the surrounding hunting grounds.

Introduction

In September 1984, excavations at Frösö church in the county of Jämtland in northern Sweden revealed a pre-Christian cult site and, simultaneously, one of the largest brown bear bone assemblages recovered from an archaeological site in Sweden. Below the choir of the church, the remains of a partly decomposed birch tree stump, surrounded by a dark layer with bones and fire-cracked stones, were found (Fig. 1). The composition of the bone assemblage and the archaeological context with the remains of the birch tree indicate that it most likely represents a ritual site dating to the Viking Age. The find is the most striking evidence of place continuity between pre-Christian and Christian cult in Scandinavia, and of a church built on top of a cult place.

Since the first paper was published in *Populär Arkeologi* in 1985, the finds from Frösö church have been presented and dealt with in several earlier papers and publications, mainly focusing on different aspects of the site as a pre-Christian cult place from the Viking Age (Hildebrandt 1985; 1989; Brink 1990; Näsström 1996; Andrén 2002; Welinder 2003; 2008). The osteological analysis of the faunal remains was first presented by Elisabeth Iregren and has been followed by other papers dealing with aspects of ritual practices concerning the animal bones (Iregren 1989; Magnell/Iregren 2010; Magnell 2013). For this study, the osteological brown bear remains from Frösö church have been reconsidered in order to highlight the significance of the brown bear for the societies and their

Bear and Human: Facets of a Multi-Layered Relationship from Past to Recent Times, with Emphasis on Northern Europe, ed. by Oliver Grimm (Turnhout: Brepols, 2023), pp. 429–452 BREPOLS PUBLISHERS 10.1484.M.TANE-EB.5.134344

cult practices in the northern parts of Sweden during the Viking Age. Using earlier unpublished age distribution data, an effort has been made to study aspects of bear hunting during the Viking Age. Re-analysis of body-part frequency and bone modifications, along with the spatial distribution of bear bones, has been carried out in order to allow additional perspectives on bear hunting and on the role of the brown bear at the cult site of Frösö church.

The site and its chronology

Frösö is an island covering 41 km^2 in Lake Storsjön, which is situated in the county of Jämtland in the northern part of Sweden (Fig. 2). The area around the lake is characterised by fertile calcareous soils suitable for agriculture and several groups of Iron Age burial mounds. Frösö is situated centrally in this landscape and was, during the Iron Age and in the Middle Ages, the economic and administrative centre of Jämtland (Hemmendorff 2010). The area around Frösö and Lake Storsjön has been described as a sacred landscape, and many of its place names are associated with the Old Norse religion. The place names Frösö, Odensala, Norderön and Ullvi can be associated with the gods Freyr, Odin, Njord and Ull, while the place names Vi and Hov refer to cult sites (Brink 1990; Vikstrand 1993). Frösö means "the island of Freyr". Freyr was a fertility deity and one of the most important gods in the Old Norse mythology.

The place where Frösö church stands is called Hov. The exact meaning of the Old Norse word *hof* is not precisely clear, but it usually refers to a building with a sacred function, which was probably also the seat of the elite and local rulers (Vikstrand 2001, 253; Sundqvist 2007, 159). No archaeological finds of cult houses or large halls have been made near Frösö church. This is as expected since larger archaeological excavations, which are usually required to find these kinds of archaeological features, have not taken place in the area. However, burial mounds in the churchyard and the name "Hov" of the village near the church indicate that the area by Frösö church was important in the Late Iron Age society in the Lake Storsjön area and was probably the seat of a regional aristocracy (Hemmendorff 2010).

Before the renovation of Frösö church in 1984, an excavation was carried out by Jämtlands läns museum. Underneath the floor in the choir, a layer of pulverised wood representing the remains of an earlier floor in the church was found. The stratigraphy below the layer of the wooden floor included a black cultural layer with bones, fire-cracked stones and charcoal covering an area of 3 x 3 m that pre-dates the church. The remains of a stump and the roots of a birch tree were found in the middle of the choir (Fig. 1). Bones were found on top of the roots and not beneath or on the tree stump. Besides an iron pin from a buckle and an iron crook, no other artefacts were recovered. The cultural layer was disturbed in the western part due to the construction of graves and a sepulchral chamber during the 18th century. The eastern and southern walls of the choir also cut into the cultural layer. It is not known whether the layer with bones continues outside the church or not, and thus the original extent of this layer is most uncertain. The excavation resulted in no indications of an earlier wooden church or other structures of buildings beneath today's church (Hildebrandt 1989, 162–163).

Radiocarbon dating results of the tree remains and bones show that the depositions of bones took place in the later Viking Age. Two datings of charcoal to the 7th–9th centuries indicate earlier activities at the place or, alternatively, are a result of the old wood effect. All datings of the bones and the birch tree could be assigned to the 10th and 11th centuries AD, indicating that the tree was standing when the bones were deposited at the site. Three dates of bones from brown bear and one each from cattle, pig and deer all overlap. The longest period of use, based on the radiocarbon dating of six animal bones, is indicated from AD 780–1190 (95.4 % probability), but a combination of all six radiocarbon datings in OxCal suggests that the bones could have been deposited within a short period of 45 years between AD 980–1025. In addition, one human bone has been dated to the 10th century and three others to the 11th–12th centuries (Fig. 3).

The end of the depositions of animal bones at Frösö church correlates well with the Christianisation process in Jämtland. The area around Lake Storsjön was christianised in the first half of the 11th century. The youngest pre-Christian graves of the area are from AD 1020–1030, and a rune stone on Frösö dating to AD 1050–1080 states that Östman (*Austmaðr*), the son of Gudfast (*Guðfastar*), christianised the county of Jämtland (Gräslund 1996, 22). It has been assumed that the men mentioned on the rune stone belonged to a lineage of chieftains and the local aristocracy in Jämtland (Welinder 2003, 522–523). An alternative interpretation of the rune inscription is that *Guðfastar*, the name of the father of Östman, is a compilation of the words *goði* and the name Faste. *Goði* was the Old Norse title for chieftains and religious leaders of cult places as well as men of great political and social importance (Sundqvist 2016, 182).

The Frösö church standing today was rebuilt in the 18th century; it was first built in the late 12th to early 13th century with the choir, the most sacred part of a church, placed on top of the pre-Christian cult place. It is uncertain whether the church had any wooden predecessor or not (Hildebrandt 1989, 153–154).

Methods

The frequency of taxa in bone assemblages has been assessed by a calculation of NISP (Number of Identified Specimens) and MNI (Minimum Number of Individuals). There are limitations and problems associated with both methods, causing a bias in quantification; as many fragments may originate from one single individual or bone element, differential fragmentation can occur between taxa and sample size (Grayson 1984; Marshall/Pilgram 1993). By using both NISP and MNI, it is possible to better evaluate the relative importance of different taxa rather than relying on a single quantification method. For quantification of taxa and a comparison between Frösö church and other sites, NISP has been used; it is the most commonly used method for inter-site comparisons in zooarchaeology.

Body-part frequency was quantified by NISP, but also by deriving MNE (Minimum Number of Elements) and MAU (Minimum Animal Units). Quantification of body-part frequency based on NISP can be problematic, since the fragmentation of bones generally results in several identifiable specimens that vary between different bone elements. For instance, a skull is usually fragmented into several identifiable fragments, while carpals and tarsals are less often fragmented. By deriving MNE (Minimum Number of [bone] Elements) the problem with differential fragmentation and the risk of counting the same bone several times is avoided (Binford 1978, 50).

Because the number of bone elements varies between different anatomical regions – for example, a bear skeleton only has one skull and two scapulas, but 20 proximal phalanges (phalanx 1) and 26 ribs – MAU has been calculated. The quantification of MAU takes into consideration the anatomical differences in the number of bones in the skeleton in different taxa. By dividing the MNE value with the number of bone elements that is found in the skeleton of a single individual, MAU is calculated. Commonly the minimum animal unit is presented as %MAU and is derived by dividing the MAU values with the greatest MAU value in the bone assemblage (Binford 1978, 51).

The age estimation has been based on epiphyseal closure for grizzly bear (*Ursus arctos horribilis*; Weinstock 2009) and tooth eruption for Eurasian brown bear (*Ursus arctos arctos*; Guskov 2015). Due to large individual variations in tooth wear, the age assessment has not been categorised into absolute ages, but into five groups following a study of Japanese black bear (Shimoinaba/Oi 2015): I – unworn, II – slightly worn (subadult/young adults), III – those with distinct wear facets on the molariform of adult teeth, IV – teeth with the enlarged wear facets of mature bears, and V – the heavily worn teeth of old animals.

Age determination based on thin-section and dental-cementum has not been used since it is a destructive method and fragile archaeological material often also requires embedding. Further, the application of this age determination method to archaeological material has relatively often shown to be problematic, due to diagenesis of the dental cementum. Diagenetic changes in archaeological teeth, such as the leaching of collagen and the growth of apatite crystallites, can result in the degradation of microscopic structures and may cause bands that mimic increments in the dental cementum (Stutz 2002).

The bone assemblage

The osteological material from the cult place at Frösö church consists of about 5 kg and 588 bone fragments (NISP) that have been taxonomically identified (Table 1). A few bones from bat, small rodents, frogs and western jackdaw are from animals that may reside in churches and have thus been interpreted to represent later, secondary intrusions into the faunal assemblage.

The composition of the faunal remains, which include a large proportion of wild animals, shows that it is a special site and that it differs strikingly from other settlements from the Late Iron Age and Early Middle Ages in the region. Wild animals, mainly bear and elk, make up 58 % of bone fragments from Frösö church, while the corresponding frequencies are 0–3 % at three sites on Frösö island and 25 % at the site of Kyrklägdan on the mainland around Lake Storsjön (Fig. 4). The rather high frequency of wild animals from Kyrklägdan indicates that hunting played a significant role in the subsistence of the settlements around Lake Storsjön, in contrast to Viking Age and early medieval settlements in southern Scandinavia, where the bones of wild game make up only a small percentage of the identified bones.

The most striking feature of the faunal remains from Frösö church is the large amount of 256 brown bear bones, which comprise 45 % of the NISP. This is a species usually represented by just a few single (if any) bones at Viking Age settlements. The other wild game is mainly elk, but also red deer, red squirrel and western capercaillie. Based on MNI, the proportion of wild game is lower, but still high at 40 %. The bear bones originate from a minimum of seven individuals. Brown bear is never the most frequently occurring animal at settlements from any period in Sweden. Two specimens of brown bear were found at the settlement at Kyrklägdan. Otherwise, there are no finds of brown bear at other Late Iron Age and medieval sites in the area around Lake Storsjön.

Further, the composition of the faunal remains of livestock from Frösö church is unusual, with a large proportion of pig bones, which are not common in the northern parts of Sweden where cattle and sheep are usually more common. From Frösö church, bones of pigs make up 61 % of the remains of domestic animals, while, on settlements from this period by Lake Storsjön, cattle, sheep and goats dominate and the frequency of pigs ranges from 13–26 %.

Among the bones are a total of 29 human ones from a minimum of four individuals; two adults, one child aged about 3–5 years, and one infant aged 0–6 months. When the human bones were first identified it was uncertain if they were contemporary with the animal bones or if they originated from a later burial in the church. As mentioned earlier, radiocarbon dating results of four human bones show that at least one is contemporary with the animal bones while the dates of the other three partially overlap, but this could also be interpreted to represent later burials from the end of the 11th or the 12th century. Thus, it is possible that the human bones represent human sacrifices or other ritual treatments of human remains at the site.

Age determination based on epiphyseal closure indicates only a few bones of bears less than 2 years old and a larger proportion of bones that originate from animals killed between 2–4 years of age and about one third having been 4–8 years old, respectively. One fifth are bears older than 8 years (Fig. 5). The kill-pattern based on epiphyses shows large similarities with recent hunting in Sweden, besides a few indications of bears less than two years old from Frösö church. The age distribution according to the epiphyses indicates a killing and hunting pattern directed towards young bears aged 2–4 years of age. However, a large part of the data set for epiphyses may originate from one single subadult individual aged 2–3 years of age that may have skewed the age distribution towards young adults in the 2–4 years age group.

Age determination based on tooth eruption and wear has resulted in an age distribution of one mandible from a subadult animal aged 18–24 months, one young adult, five mandibles of adults and one premaxilla of an old animal, which differs slightly from the age estimation that was based on epiphyses (Fig. 6; Table 2). The age distribution by tooth wear shows that bears of different age groups were hunted, but it differs from age data based on epiphyses, which indicate the hunting of only a few subadults or young adults less than four years old; the most hunted animals are adult ones, which rather indicates that hunting leaned towards full-grown bears.

It is uncertain which age estimation method provides the most accurate picture of the hunting and age profile. Age determination based on epiphyses is based on a large data set but, as mentioned, it may be biased because most of the bones may originate from one individual. The data set based on tooth wear is small, but on the other hand it is based for certain on several different individuals. Both age estimation methods are in accordance with the finding that bears younger than two years old occur in low frequencies at Frösö church, which possibly reflects hunting practices that avoided killing females with cubs or family groups.

Further, it is important to consider that the age distribution of brown bears at Frösö church reflects the animals brought to the cult place in Frösö and it may not reflect the hunted population as a whole, but rather a selection of specific animals.

Based on tooth wear and tooth development of the canine with an open root apex, one mandible has been age determined to 18–24 months. Based on the knowledge that brown bears are usually born in January–February, this mandible is from a bear that was most likely killed in the period July–December. This shows that bear hunting during the Viking Age was not limited to hunting in dens in spring as is often assumed.

Sex estimation has not been possible to assess on the basis of the osteological material. The presence of a baculum (*Os penis*) could have been used to confirm males, but this particular bone element has not been identified. However, to consider this as indicating an absence of males is highly uncertain. If the faunal remains represent leftovers from consumption, it is possible that the genitals of bears were not considered food. Other explanations could be that the bacula have been used for ritual or magical purposes and because of this have not been included with the other bones. Among the Sami, the bacula of bears were used and hung on drums used by the *noadi* (ritual specialist/shaman), and among the Yenisey Ostjaks in Siberia, the reproductive organ of the bear was used as a cult object (Paproth 1962, 70–73; Zachrisson/Iregren 1974, 81).

Body-part representation

The body-part frequency indicates the presence of bones from most regions of the body (Table 3; Fig. 7). The anatomical distribution of the bear is different from that of the other animals, such as domestic ones and elk in the location from which almost all bones originate; from the head and mainly the mandibles (Fig. 6). This shows that bears were treated differently from other species and indicates that the bear had a specific significance to the humans and in the cult practices at the site. A large proportion of the bear bones also originate from the head, mainly mandibles, but there is also a high frequency of bones from the lower limbs and the paws of both anterior and posterior limbs (Table 3; Fig. 7).

Due to the anatomy of bears, the paws with five digits result in several smaller bones in comparison with the large limb bones. Further, fragmentation may result in a large proportion of bone specimens from a single bone, for example a skull. A quantification (Minimum Animal Unit [MAU]) that considers anatomical differences and fragmentation gives a more correct picture of the body part representation. The MAU shows that bones of different body regions were indeed brought to Frösö; in most cases it was heads and especially mandibles, along with lower limbs below the elbow and knee with paws (Fig. 7).

The meaty upper parts of the limbs show a low representation and, strikingly, no scapula or pelvis has been found, which are large and relatively robust bones that would be expected to have been preserved at the site. Possibly, these body parts were consumed at ritual meals in other areas, such as cult buildings (*hof*) or nearby halls and, because of this, more rarely deposited in the cult place.

In many cultures around the world, there are rules concerning the sharing of body parts of hunted animals and which part belongs to the hunter who kills the animals and which part to the owner of the hunting grounds, or how they are shared between men and women (Altman/Peterson 1988; Rosman/Rubel 1989). Medieval laws from Norway, such as *Gulatingslovi*, state that the shoulder (*skotbogen*), i.e. the scapula, was the part that was allotted to the hunter who killed an animal (Indrelid/Hufthammer 2011). So, were the shoulder and pelvis body parts reserved to the hunters who killed the bears?

The osteological remains from the head have a particular distribution, with several loose teeth from both upper and lower dentition as well as mandibles, but the nine bone fragments from the skull are all from the anterior parts (premaxillare, maxillare, palatinum) by the nose. No cranial parts of the maxilla with molars and posterior part of the viscerocranium and neurocranium have been found. This indicates that the skulls without mandibles and anterior parts could have been placed elsewhere, possibly on poles or in the birch tree.

At ritual sites from the Iron Age in Scandinavia, such as depositions in wetlands and wells, parts from the head and lower limbs are often found very frequently, in comparison with bones from the trunk or upper limbs (Wikborg/Magnell 2017; Magnell 2019). The less meaty parts, the head and lower limbs, could be viewed as the gods' share, while the mortals took care of the meaty parts. Further, written sources from the Viking Age imply rituals with the heads. Ibn Fadlan's narrative of his meeting with the Rus and Al-Tartuschi's description of his travels to Haithabu mention that the heads of the sacrificed animals were placed on poles, and the description of the cult at Uppsala by Adam of Bremen can be interpreted as that the heads of the killed animals were given to the gods by hanging them in the trees of the sacrificial grove (Birkeland 1954, 103–104; Ibn Fadlan 1978, 65; Hultgård 1997, 32).

Ritual treatment of head parts is also a characteristic of bear ceremonialism among several circumpolar societies (Hallowell 1926; Rockwell 1991). The placing of bear skulls in trees by the Finns is an example of a ritual practice that could explain the missing cranium parts from Frösö church (cf. Keinänen, this volume; Piludu, this volume).

No claws (distal phalanges) but only the proximal and middle phalanges of brown bear have been found at the site (Table 3; Fig. 7). This indicates that either the furs with distal phalanges were not brought to the site, or rather that the furs were not deposited at the ritual site and instead were utilised and brought to the houses, or that they were taken from Frösö to other regions.

Among the loose bear teeth there are also remarkably few canines (especially considering that the large canines would not be expected to be lost during excavation, as could be the case with smaller premolars or incisors): There is only one canine, while 21 incisors, eight premolars and 13 molars have been identified. This indicates that canines might have been taken away from the cult place.

The brown bears found at the site were most likely not killed on Frösö. The area of the island is too small to hold a population of bears, and the animals were probably killed on hunting grounds on the mainland around Lake Storsjön. The body-part representation indicates that, while occasionally whole carcasses were brought to Frösö, in other cases it was only specific body parts, such as paws and heads or mandibles.

Butchering and utilisation

Butchering marks have been identified on 61 bear bones, which gives a detailed picture of the butchering process and utilisation of the bear carcasses. The high frequency (29 %) of the bones with cut and chop marks shows that the carcasses were thoroughly utilised.

Cut marks on a mandible, metacarpal and several proximal and middle phalanges from skinning indicate that the furs were an important resource provided by the bears (Fig. 8). Bear claws have been found in burials from the Late Iron Age in the southern parts of Sweden and even in regions that did not hold any bear populations, like Gotland, which indicates that bear skins were a coveted commodity during this period (Petré 1980; Iregren 1988; cf. Grimm, this volume; Lindholm/Ljungkvist, this volume).

Chop marks by the alveoli of the canines of both skulls and mandibles also show that the canines were extracted and removed, probably to be used as amulets and ritual objects (Fig. 9). Bear canines with drilled holes in root apexes used as pendants have been found in burials as well as at settlements and towns like Sigtuna (Wikell 2015; cf. Magnell, this volume, on humans and bears in Sweden). According to folklore from as late as the 19–20th centuries, bear teeth were used for various magical purposes, such as for protection or to bring luck in hunting and logging or as a cure for toothache (Björklöf 2010, 276–277).

Chop and cut marks show that both axes and knives were used in the dismembering of the carcasses. The dismembering marks are found on most major joints of the bear skeleton from the head down to the proximal and middle phalanges (Fig. 8). This further confirms a thorough utilisation of the bear carcasses and that the meat was cut up into smaller pieces. Filleting marks from the cutting of meat from the bones are also found on different parts of the bears, but mainly on the meatier parts of the trunk (Fig. 8). Chop marks indicate that the mandibles were broken for the extraction of bone marrow (Fig. 9). The traces of butchery and cooking reveal an intense utilisation of the carcasses and indicate that feasting on meat from the animals was important.

The finds of fire cracked stones and charcoal in the cultural layer among the animal bones are most likely the remains of food preparation near the tree. All the bear bones are unburned – as are most of the animal bones from Frösö church. However, there is evidence of burning and exposure of fire on the teeth of the piglet and sheep mandibles, most likely from cooking. The absence of traces of burning on the teeth and mandibles of bears could reflect differences in the cooking practices used for various animals, with indications of the roasting of domestic animal mandibles over the fire and possibly the cooking of bear meat in vessels.

The deposition

Gnaw marks made by dogs or other carnivores are only found on five bear bones (2 %), which is a low frequency in comparison with settlements from the Viking Age, where often up to 25 % of the bones are gnawed (Magnell 2017). This indicates that dogs had only restricted access to bones at the site. Either the bones were placed in the tree out of reach of scavenging dogs and other carnivores, or possibly some kind of enclosure protected the cult place. Weathering on the bones, showing as cracks and eroded surfaces, indicates that the bones were exposed for a while, possibly lying on the ground around the tree before the bones became covered by soil formed from decaying leaves and refuse.

The spatial distribution shows that the bear bones were scattered around the birch stump beside bones of elk, sheep, pig and cattle with a concentration to the south of the tree (Fig. 10). Tarsals, metatarsals and phalanges from the posterior paw and carpals, metacarpals and phalanges from the anterior paws, which could be refitted into anatomical units, were found in concentrations indicating that whole paws were deposited at the site. In the northern parts of the excavated area, bones with unfused epiphyses, a mandible from a subadult bear and possibly bones of one whole individual have been found, while most of the bones from adult bears and other animals are concentrated in the opposite area (Fig. 10). The spatial distribution indicates that the main area for ritual depositions was on the southern side of the birch tree, while the bones from the subadult bear on the other side of the tree probably reflect a single event.

Discussion

The large amount of osteological remains of brown bear, along with bones of elk and pigs, from the cult place at Frösö church is an apparent reflection of the importance of the bear to humans during the Viking Age in Jämtland. The bear bones from the site have been used to study different aspects of the bear-human relationship from hunting, butchering and utilisation to ritual practices and regional social interactions between different cultural groups.

Bear hunting

The study of hunting in the past, from periods with or without limited written sources, is problematic, but is made possible by the zooarchaeological analysis of taxa as well as age and sex distribution. This usually requires larger faunal assemblages, and, since bear bones are usually found in low numbers at archaeological sites, bear hunting in the past is often difficult to study. In the light of this, the osteological remains from Frösö church provide us with an unusual opportunity to study bear hunting during the Viking Age, even though we must take into account that the age distribution may not be fully representative, due to their origins in a ritual site.

Based on epiphyses, hunting was directed towards young bears aged 2–4 years and older animals, with only a few younger than two years, which is a harvest similar to present-day hunting in Sweden. At about two years of age the brown bear leaves its mother, so possibly the age distribution based on epiphyses could reflect hunting focused on young, inexperienced bears. Hunting biased towards young bears is usually explained by the fact that this age group has greater mobility, which increases the probability of encountering a hunter (Litvaitis/Kane 1994; Kohlmann et al. 1999; Bischof et al. 2008).

The age estimation based on tooth wear provides a somewhat different picture of the age distribution, with the hunting of mainly adults, about four years or older. This is in accordance with the age of the animals from the Sami bear burials, where most of the individuals were also adults (Zachrisson/Iregren 1974, 66–68). In relation to hunting in Sweden during 1981–2004, the age estimation based on teeth differs to a certain extent and shows more similarities with bear hunting in North America, with a larger proportion of older animals (Litvaitis/Kane 1994; Kohlmann et al. 1999; Frank et al. 2017).

Even though both age determination methods have provided somewhat different results, both are in accordance with the fact that bears younger than two years old occur in low frequencies at Frösö church. This possibly reflects hunting practices that avoided killing females with cubs or family groups, which would be similar to present-day Swedish bear hunting regulations (Bischof et al. 2008).

Frösö is a too small an island to hold a bear population, and the animals were probably mainly hunted in the forests outside the agricultural core area around Lake Storsjön. The anatomical distribution with higher frequencies of body parts from the head and lower limbs could be interpreted as that it was mainly these parts that were brought to the cult site at Frösö church and are therefore seen as an indication of hunting at a larger distance from the site. It has not been possible to identify who the bear hunters were, they could either have been farmers from Frösö that went on hunting trips, or groups of hunters living in the forested areas around the Lake Storsjön area that belonged to the Sami culture.

Bears and Vikings

The cult place at Frösö church at a place named Hov and the presence of a Viking Age settlement that has been partly excavated, along with Iron Age mounds on the church yard, clearly associated the find with the Viking society on Frösö. Osteological remains of different kinds of livestock, mainly piglets, and especially mandibles, are also a typical feature of ritual sites from the Viking Age (Jennbert 2002; Magnell 2019).

Certain aspects of the faunal remains from Frösö church have been interpreted to reflect rituals that could be linked to the Old Norse religion. The finds of deer teeth and squirrel metapodials from Frösö church can be associated with the Old Norse mythology and the descriptions in *Edda* of the world-tree Yggdrasil, where the squirrel *ratatǫskr* runs up and down the tree and the four deer *Dáinn, Dvalinn, Dúneyrr* and *Duraþrór* graze on its twigs. It has been suggested that the remains of squirrel and deer were used in rituals that were staging the mythology in a symbolic transformation of the birch tree into the world-tree (Iregren 1988). Pigs could have been specifically selected as sacrificial animals on the basis of the association between pigs, fertility symbolism and *Freyr*, based on Old Norse sources (Näsström 2001, 161).

The seasonality of the sacrifices at the site also seems to correlate with the Old Norse calendar ritual feasts known as *blót* from Old Norse sources. An analysis of the seasonality, based on the age of death of juvenile animals studied by age estimation, which was based on tooth development and tooth wear in elk, sheep, pigs, cattle and goat, indicate three annual events; one between October–November, in the winter, one in March near the vernal equinox, and one around the summer solstice at end of June (Magnell/Iregren 2010).

The human remains found among the animal bones could also be interpreted as representing human sacrifices. Written sources and archaeological finds indicate that human sacrifices were a living tradition in Old Norse religion and during the Viking Age, even though this has been a subject of debate (Wikström af Edholm 2020). The human bones could thus be associated with the Old Norse religious practice.

Even though the Old Norse mythology is full of different kinds of animals and several were associated with different gods, the bear seems to have played a minor role (cf. Sundqvist, this volume). Thus, the bear does not seem to be a sacred animal in Viking society and, in the Sagas, the bear is actually described as a commodity (DuBois 2012). However, this does not mean that the bear was an animal of insignificant symbolic meaning to the Vikings. The bear occurs in Scaldic poetry and the Sagas, and it was common in personal names (cf. Lombardi, this volume; Ney, this volume; Nedoma,

Altman/Peterson 1988: J. Altman/N. Peterson, Rights to game and rights to cash among contemporary Australian hunters-gatherers. In: T. Ingold/D. Riches/J. Woodburn (eds.), Hunters and Gatherers 2. Property, Power and Ideology (Oxford 1988) 75–99.

Andrén 2002: A. Andrén, Norrön ritual och kultplatskontinuitet. In: K. Jennbert/A. Andrén/C. Raudvere (eds.), Plats och praxis. Studier av nordisk förkristen ritual (Lund 2002) 299–342.

Binford 1978: L. R. Binford, Nunamiut Ethnoarchaeology (New York 1978).

Birkeland 1954: H. Birkeland, Nordens historie i middelalderen etter arabiske kilder (Oslo 1954).

Bischof et al. 2008: R. Bischof/R. Fujita/A. Zedrosser/A. Söderberg/J. E. Swenson, Hunting patterns, ban on baiting, and harvest demographics of brown bears in Sweden. Journal of Wildlife Management 72, 2008, 79–88.

Björklöf 2010: S. Björklöf, Björnen i markerna och kulturen (Möklinta 2010).

Brink 1990: S. Brink, Cult sites in Northern Sweden. In: T. Ahlbäck (ed.), Old Norse and Finnish Religions and Cultic Place-Names. The Donner Institute for Research in Religious and Cultural History (Åbo 1990) 458–489.

DuBois 2012: T. A. DuBois, Diet and deities. Contrastive livelihoods and animal symbolism in Nordic Pre-Christian religions. In: C. Raudvere/J. P. Schjødt (eds.), More than mythology. Narratives, ritual practices and regional distribution in Pre-christian Scandinavian religions (Lund 2012).

Edsman 1994: C.-M. Edsman, Jägaren och makterna. Samiska och finska björnceremonier. Skrifter utgivna genom Dialekt- och folkminnesarkivet i Uppsala Serie C, Lapskt språk och lapsk kultur 6 (Uppsala 1994).

Frank et al. 2017: S. C. Frank/A. Ordiz/J. Gosselin/A. Hertel/J. Kindberg/ M. Leclerc/F. Pelletier/S. M. J. G. Steyaert/O.-G. Støen/J. Van de Walle/A. Zedrosser/J. E. Swenson, Indirect effects of bear hunting: a review from Scandinavia. Ursus 28(2), 2017, 150–164.

Gräslund 1972/1973: A.-S. Gräslund, Barn i Birka. Tor 15, 1972/1973, 161–179.

Gräslund 1996: A.-S. Gräslund, Kristna inslag i Jämtlands vikingatid. In: S. Brink (ed.), Jämtlands kristnande. The Christianization of the province of Jämtland, Projektet Sveriges kristnande, Publikationer 4 (Uppsala 1996) 21–44.

Grayson 1984: D. K. Grayson, Quantitative zooarchaeology (London 1984).

Guskov 2015: V. Y. Guskov, Skull-Based Method of Age Determination for the Brown Bear, *Ursus arctos* Linnaeus, 1758. Achievements in the Life Sciences 8(2), 2015, 137–141.

Hallowell 1926: A. I. Hallowell, Bear Ceremonialism in the Northern Hemisphere. American Anthropologist (new series) 28, 1926, 1–175.

Hemmendorff 2010: O. Hemmendorff, Långsiktigt hållbara lokaliseringar. Platser för kult och makt på Frösön. Arkeologi i Norr 12, 2010, 37–81.

Hildebrandt 1985: M. Hildebrandt, En kyrka byggd på hednisk grund? Populär arkeologi 3(4), 1985, 9–13.

Hildebrandt 1989: M. Hildebrandt, Frösö kyrka på hednisk grund. In: O. Hemmendorff (ed.), Arkeologi i fjäll, skog och bygd 2. Järnålder – medeltid (Östersund 1989) 153–166.

Hildebrandt 1991: M. Hildebrandt, Rapport över arkeologiska undersökningar i och vid Frösös kyrka, Prästbordet 1:1, Frösö sn, Östesunds kommun, Jämtlands län 1984. Jämtlands läns museum, Kulturhistorisk utredning 41 (Östersund 1991).

Holmgren 1985: Å. Holmgren, Analys av benmaterialet från fornlämning 45 Ås sn, Jämtland. Appendix C. In: M. Olausson, Kyrklägdan i Ås. Arkeologisk undersökning av en boplats från folkvandringstid till medeltid (Östersund 1985) 50–52.

Hultgård 1997: A. Hultgård, Från ögonvittnesskildring till retorik. Adam av Bremens notiser om Uppsalakulten i religionshistorisk belysning. In: A. Hultgård (ed.), Uppsala och Adam av Bremen (Nora 1997) 9–50.

Hultkrantz 1996: Å. Hultkrantz, A new look at the world pillar in Arctic and sub-Arctic religion. In: J. Y. Pentikäinen (ed.), Shamanism and Northern Ecology (Berlin 1996) 31–49.

Ibn Fadlan 1978: Ibn Fadlan, Araber Vikingar Väringar. Translation S. Wikander (Lund 1978).

Indrelid/Hufthammer 2011: S. Indrelid/A. K. Hufthammer, Medieval mass trapping of reindeer at the Hardangervidda mountain plateau, South Norway. Quaternary International 238, 2011, 44–54.

Iregren 1988: E. Iregren, Finds of brown bear (*Ursus arctos*) in Southern Scandinavia – Indications of local hunting or trade? In: B. Hårdh/L. Larsson (eds.), Trade and exchange in Prehistory. Studies in honour of Berta Stjernquist. Acta Archaeologica Lundensia Series 8° (Lund 1988) 295–308.

Iregren 1989: E. Iregren, Under Frösö kyrka – ben från en vikingatida offerlund. In: L. Larsson/B. Wyszomirska (eds.), Arkeologi och religion (Lund 1989) 119–133.

Jennbert 2002: K. Jennbert, Djuren i nordisk förkristen ritual och myt. In: K. Jennbert/A. Andrén/C. Raudvere (eds.), Plats och praxis – studier av nordisk förkristen ritual. Vägar till Midgård 2 (Lund 2002) 105–133.

Kohlmann et al. 1999: S. G. Kohlmann/R. L. Green/C. E. Trainer, Effects of collection method on sex and age composition of black bear (*Ursus americanus*) harvest in Oregon. Northwest Science 73, 1999, 34–38.

Lindholm/Ljungkvist 2016: K.-J. Lindholm/J. Ljungkvist, The bear in the grave. Exploitation of top predator and herbivore resources in 1st millennium Sweden, first trends from a long-term research project. European Journal of Archaeology 19(1), 2016, 3–27.

Litvaitis/Kane 1994: J. A. Litvaitis/D. M. Kane, Relationship of hunting technique and hunter selectivity to composition of black bear harvest. Wildlife Society Bulletin 22, 1994, 604–606.

Magnell 2004: O. Magnell, Djurben från vikingatida lager nordväst om Frösö kyrka. Unpublished osteologisk rapport, Jämtlands län museum (Östersund 2004).

Magnell 2013: O. Magnell, The taphonomy of ritual bone depositions. An approach to the study of animal bones and ritual practice with an example from Viking Age Frösö, Sweden. In: G. Ekroth/J. Wallensten (eds.), Bone, behaviour and belief. The zooarchaeological evidence as a source for ritual practice in ancient Greece and beyond (Athens 2013) 115–127.

Magnell 2017: O. Magnell, Gårdarnas djur – osteologisk analys. Utbyggnad av Ostkustbanan genom Gamla Uppsala. Arkeologerna Rapport 2017:1(12) (Stockholm 2017).

Magnell 2019: O. Magnell, Animals of sacrifice. Animals and the *blót* in the Old Norse sources and ritual depositions of bones. In: O. Sundqvist (ed.), Myth, Materiality and Lived Religion in Merovingian and Viking Scandinavia (Stockholm 2019) 303–337.

Magnell/Iregren 2010: O. Magnell/E. Iregren, Veitstu hvé blóta skal. The Old Norse *blót* in the light of osteological remains from Frösö church, Jämtland, Sweden. Current Swedish Archaeology 18, 2010, 223–250.

Marshall/Pilgram 1993: F. Marshall/T. Pilgram, NISP vs. MNI in quantification of body-part representation. American Antiquity 58(2), 1993, 261–289.

Mulk 2009: I.-M. Mulk, From metal to meat: Continuity and change in ritual practices at a Saami sacrificial site, Viddjávarri, Lappland, Northern Sweden. In: T. Äikäs (ed.), Máttut-máddagat: The roots of Saami ethnicities, societies and spaces/places (Oulu 2009) 116–133.

Näsström 1996: B.-M. Näsström, Offerlunden under Frösö kyrka. In: S. Brink (ed.), Jämtlands kristnande. Projekten Sveriges kristnande (Uppsala 1996) 65–85.

Näsström 2001: B.-M. Näsström, Blot. Tro och offer i det förkristna Norden (Stockholm 2001).

Oehrl 2013: S. Oehrl, Bear hunting and its ideological context (as a background for the interpretation of bear claws and other remains of bears in Germanic graves of the 1st millennium AD). In: O. Grimm/U. Schmölcke (eds.), Hunting in northern Europe until 1500 AD. Old traditions and regional developments, continental sources and continental influences. Papers presented at a workshop organized by the Centre for Baltic and Scandinavian Archaeology, ZBSA. Schriften des Archäologischen Landesmuseums, Ergänzungsreihe Band 7 (Neumünster 2013) 297–332.

Pales/Garcia 1981: L. Pales/M. Garcia, Atlas ostéologique pour servir à l'identification des Mammifères du Quaternaire 2 – Carnivores/Homme (Paris 1981).

Paproth 1962: H.-J. R. Paproth, Das Bärenfest der Ketó in Nordsibirien in Zusammenhang gebracht mit den Bärenceremonien und Bärenfesten anderer Völker der nördlichen Hemisphäre. Anthropos 57, 1962, 55–88.

Petré 1980: B. Petré, Björnfällen i begravningsritualen, statusobjekt speglande regional skinnhandel. Fornvännen 75, 1980, 5–14.

Rockwell 1991: D. Rockwell, Giving Voice to Bear: North American Indian Myths, Rituals and Images of the Bear (Niwot 1991).

Rosman/Rubel 1989: A. Rosman/P. G. Rubel, Stalking the wild pig: hunting and horticulture in Papua New Guinea. In: S. Kent (ed.), Farmers as Hunters. The Implications of Sedentism (Cambridge 1989) 27–36.

Salmi et al. 2011: A.-K. Salmi/T. Äikäs/S. Lipkin, Animating rituals at Sámi ritual sites in northern Finland. Journal of Social Archaeology 11(2), 2011, 212–235.

Shimoinaba/Oi 2015: S. Shimoinaba/T. Oi, Relationship between tooth wear and age in the Japanese Black Bear in Hiroshima prefecture, Japan. Mammal Study 40(1), 2015, 53–60.

Stutz 2002: A. J. Stutz, Polarizing microscopy identification of chemical diagenesis in archaeological cementum. Journal of Archaeological Science 29, 2002, 1327–1347.

Sundqvist 2007: O. Sundqvist, Kultledare i fornskandinavisk religion (Uppsala 2007).

Sundqvist 2016: O. Sundqvist, An arena for higher powers. Ceremonial buildings and religious strategies for rulership in Late Iron Age Scandinavia (Leiden, Boston 2016).

Sundström 1982: J. Sundström, Ett "storkök" från vikingatiden. Jämten 168, 1982, 168–169.

Thilderqvist 2005: J. Thilderqvist, Analys av det osteologiska materialet från den arkeologiska undersökningen vid Västerhus på Frösön 2005. In: B. Oskarsson, Arkeologisk undersökning RAÄ 33 mfl, Västerhus kapell, Frösö sn, Östersunds kn, Jämtlands län, Jämtlands läns museum (Östersund 2005) 1–23.

Vikstrand 1993: P. Vikstrand, Förkristna sakrala ortnamn i Jämtland. Namn och bygd 81, 1993, 49–84.

Vikstrand 2001: P. Vikstrand, Gudarnas platser. Förkristna sakrala ortnamn i Mälarlandskapen (Uppsala 2001).

Wallin/Martinsson-Wallin 1990: P. Wallin/H. Martinsson-Wallin, Osteologisk analys av brända och obrända ben från Frösö kyrkogård, Frösö sn. Östersunds kn. i Jämtland. In: M. Pagoldh, Rapport över arkeologisk undersökning inom Prästbordet 1:1, Frösö socken, Östersunds kommun, Jämtlands län. Arkeologiska undersökningar utförda av Jämtlands läns museum 436, Jämtlands läns museum (Östersund 1990).

Weinstock 2009: J. Weinstock, Epiphyseal Fusion in Brown Bears: A Population Study of Grizzlies (*Ursus arctos horribilis*) from Montana and Wyoming. International Journal of Osteoarchaeology 19(3), 2009, 416–423.

Welinder 2003: S. Welinder, Christianity, politics and ethnicity in Early Medieval Jämtland, Mid Sweden. In: M. Carver (ed.), The cross goes North. Process of conversion in Northern Europe AD 300–1300 (York 2003) 509–530.

Welinder 2008: S. Welinder, Welinder, Jämtarna och samerna kom först (Östersund 2008).

Wikborg/Magnell 2017: J. Wikborg/O. Magnell, Händelser kring stolpar. En analys av stolpfundamentens fyndinnehåll. In: L. Beronius Jörpeland/H. Göthberg/A. Seiler/J. Wikborg, (eds.), *at Upsalum* – människor och landskapande. Utbyggnad av Ostkustbanan genom Gamla Uppsala. Arkeologerna, Statens historiska museer, Rapport 2017:1-1 (Stockholm 2017) 293–312.

Wikell 2015: R. Wikell, En björnkvinna eller bärsärk i Sigtuna? In: A. Söderberg/L. Tamm/R. Edberg/N. Källström/J. Runer (eds.), Situne Dei. Årsskrift för Sigtunaforskning och historisk arkeologi 2015 (Sigtuna 2015) 6–13.

Wikström af Edholm 2020: K. Wikström af Edholm, Människooffer i myt och minne. En studie av offerpraktiker i fornnordisk religion utifrån källtexter och arkeologiskt material (Åbo 2020).

Zachrisson/Iregren 1974: I. Zachrisson/E. Iregren, Lappish bear graves in Northern Sweden. Early Norrland 5 (Stockholm 1974).

Dr. Ola Magnell
Arkeologerna, National Historical Museums
Lund
Sweden
ola.magnell@arkeologerna.com

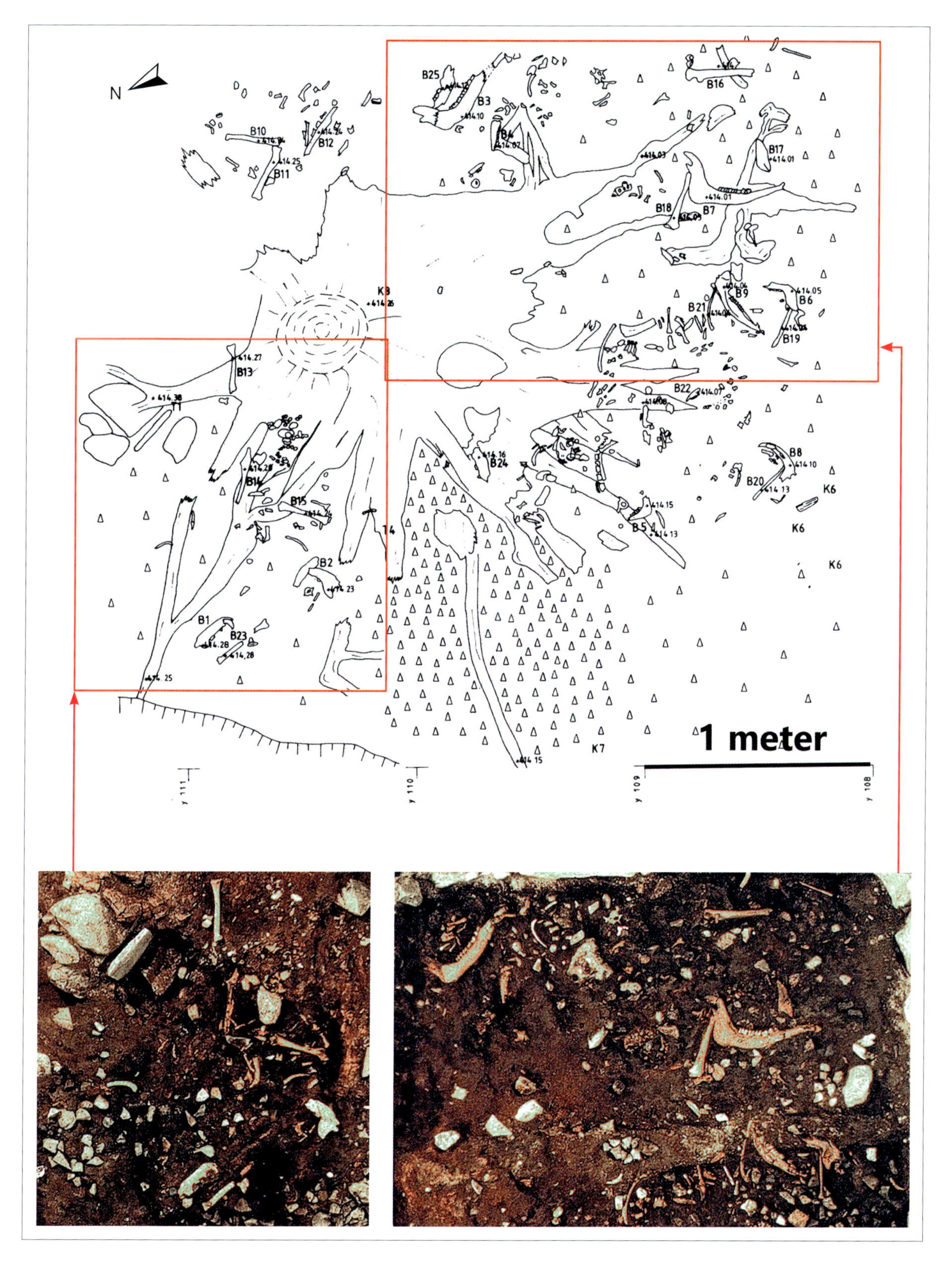

Fig. 1. Stump of birch tree surrounded by layer containing bones and the inner choir walls of Frösö church (photos and plan M. Hildebrandt, Jämtlands läns museum).

Fig. 2. Frösö church and other sites with zooarchaeological analyses from the Lake Storsjön area in Jämtland, Sweden. 1: Frösö church and Prästbordet, Frösö; 2: Västerhus, Frösö; 3: Trusta, Norderön; 4: Kyrklägdan, Ås (map GIS department, ZBSA).

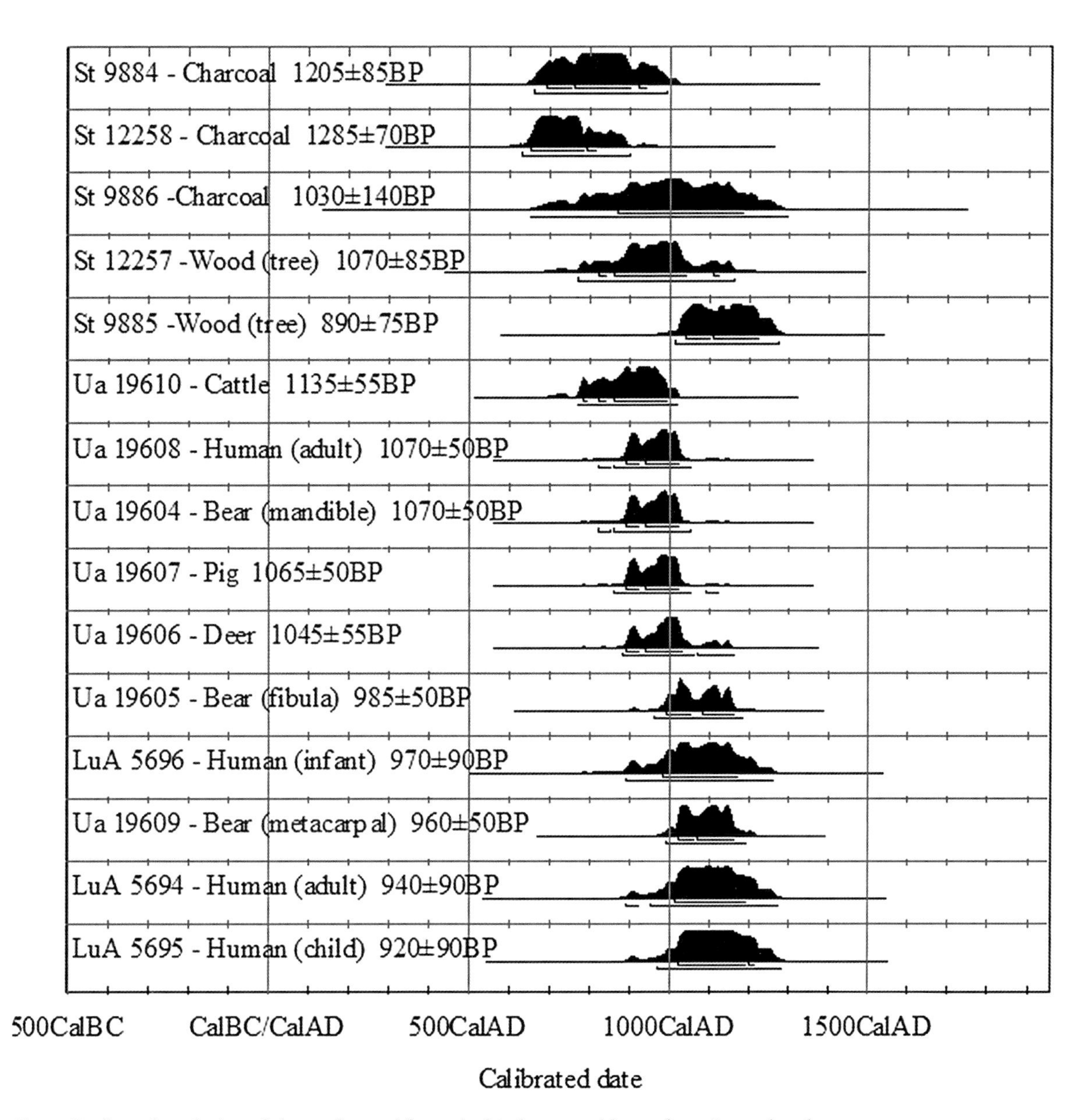

Fig. 3. Radiocarbon dating of charcoal: wood from the birch tree and bones from Frösö church.

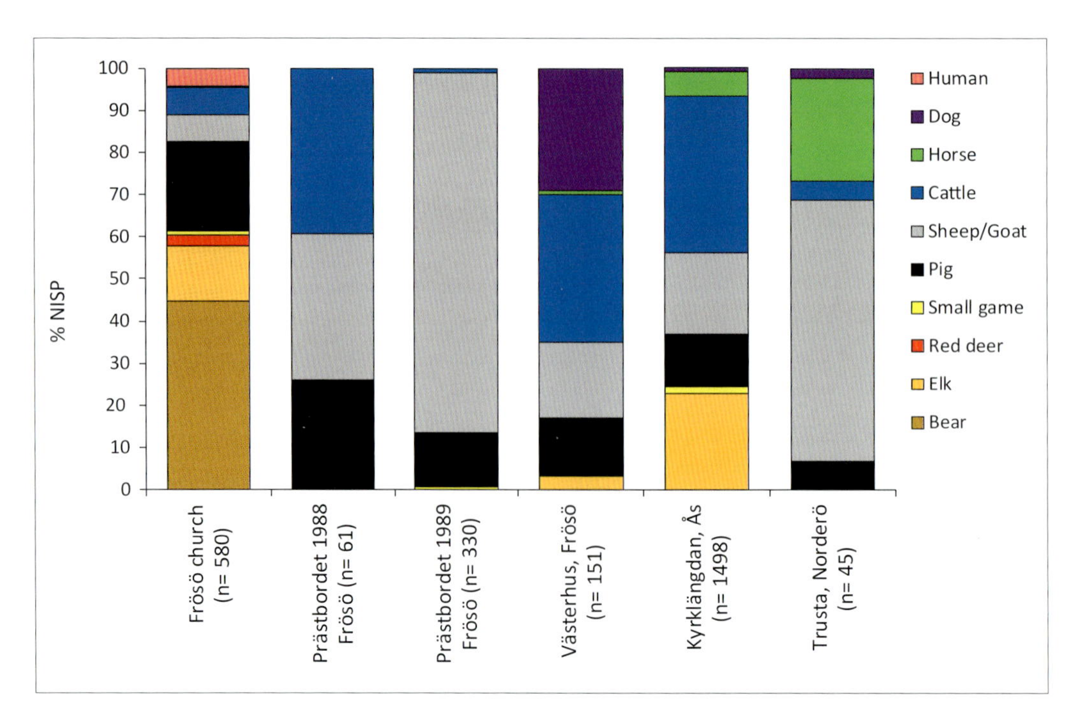

Fig. 4. Frequency of bones (NISP) from Frösö church in comparison with other sites in the Lake Storsjön area: Prästbordet, excavation from 1988, Frösö, 9–11th centuries (after Magnell *2004); Prästbordet excavation from 1990, Frösö, 7–13th centuries (after* Wallin/Martinsson-Wallin *1990); Västerhus, Frösö, 12–14th centuries (after* Thilderqvist *2005); Kyrklägdan, Ås, 11–16th centuries (after* Holmgren *1985); Trusta, Norderön, 8–10th centuries (after* Sundström *1982).*

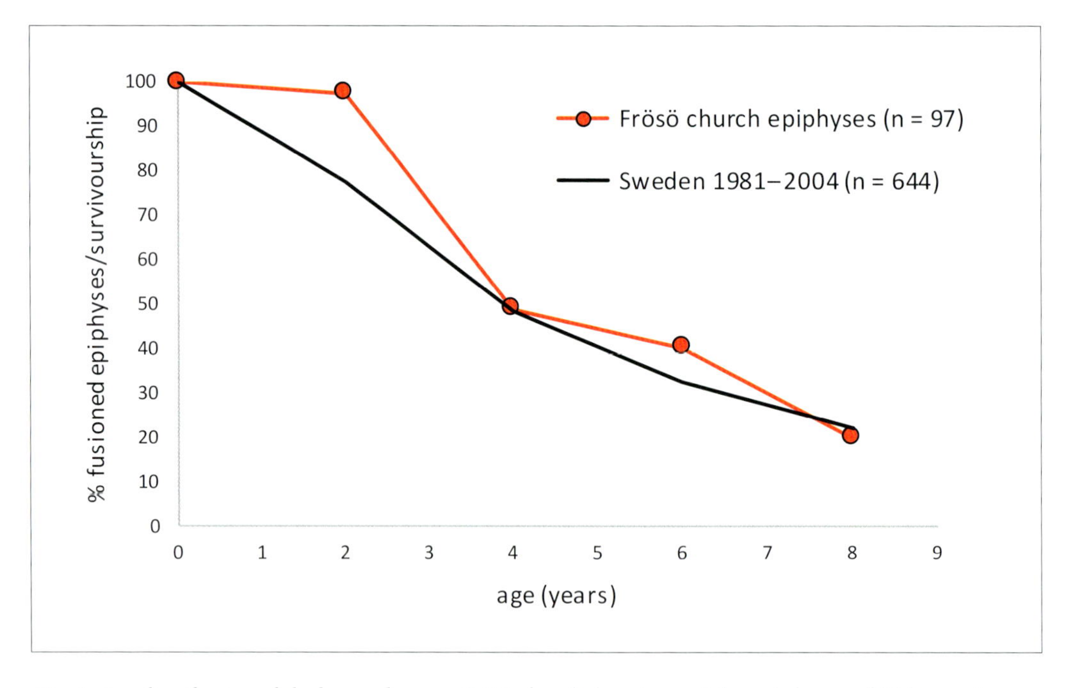

Fig. 5. Age distribution of the brown bears at Frösö church, based on epiphyseal closure of 97 bones. In comparison, that of the total of brown bears hunted and killed in Sweden 1981–2004 (after Bischof *et al. 2008).*

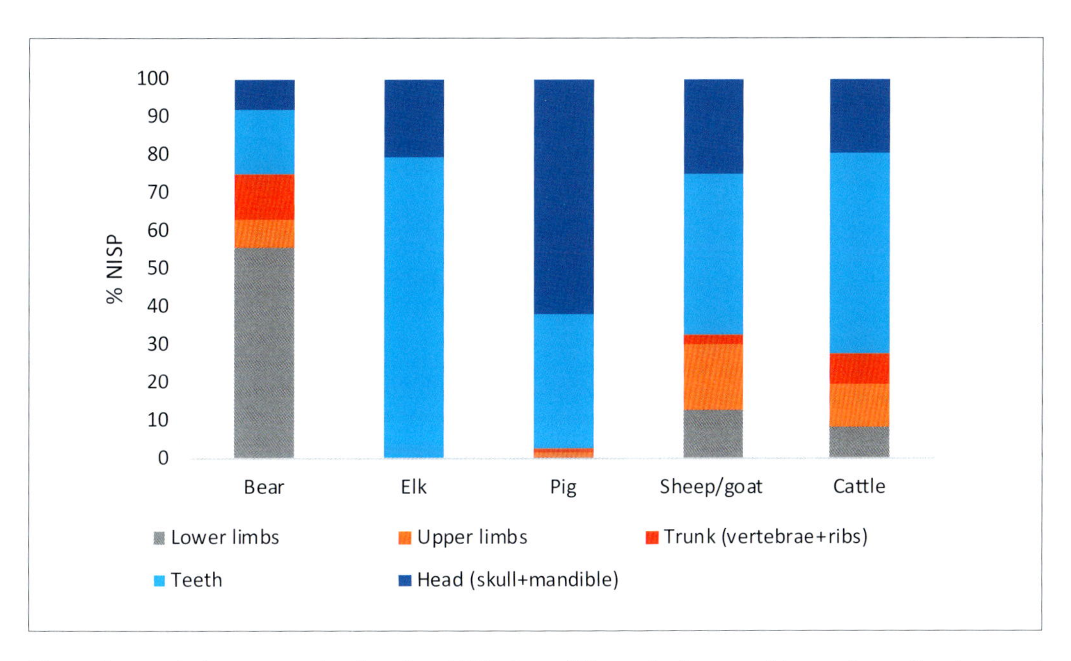

Fig. 6. Anatomical representation based on NISP from different body parts of brown bear, elk, pig, sheep/goat, and cattle from Frösö church.

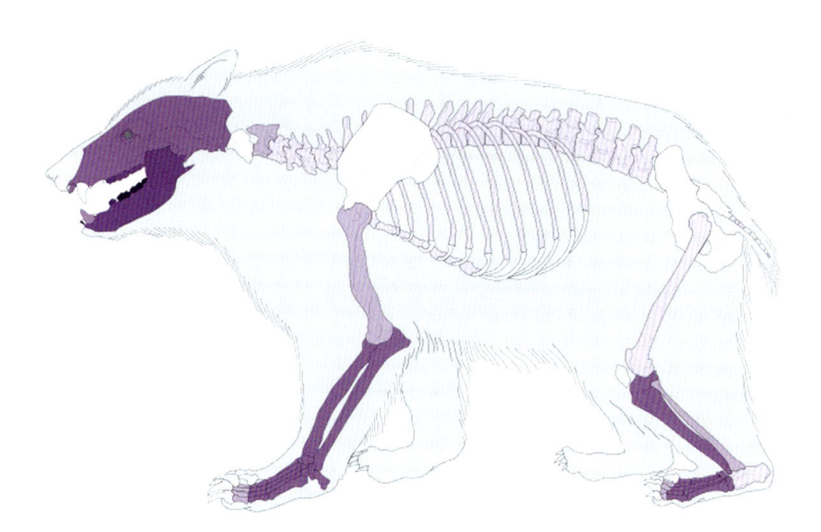

Fig. 7. Anatomical representation of brown bears from Frösö church based on MAU (Minimum Animal Unit) (© 2003 ArcheoZoo.org; after Pales/Garcia *1981, pl. 13)*

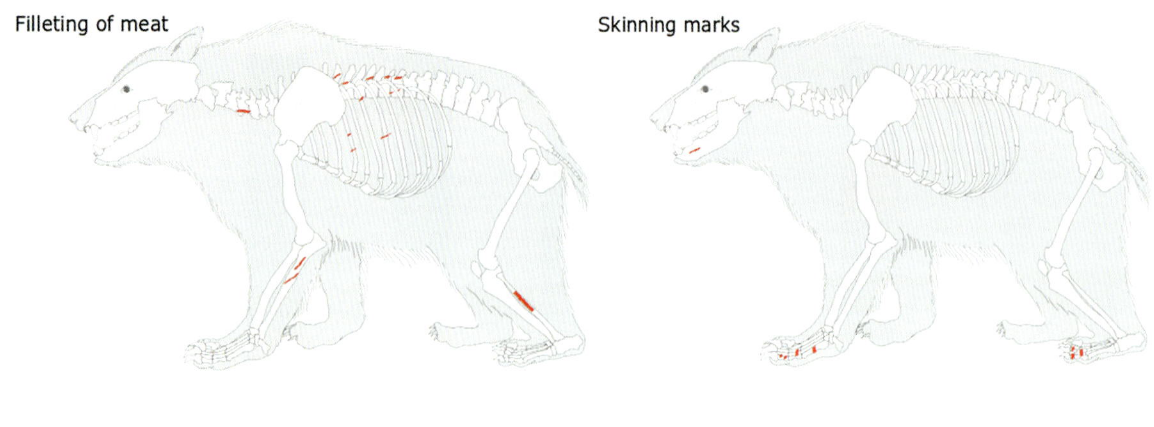

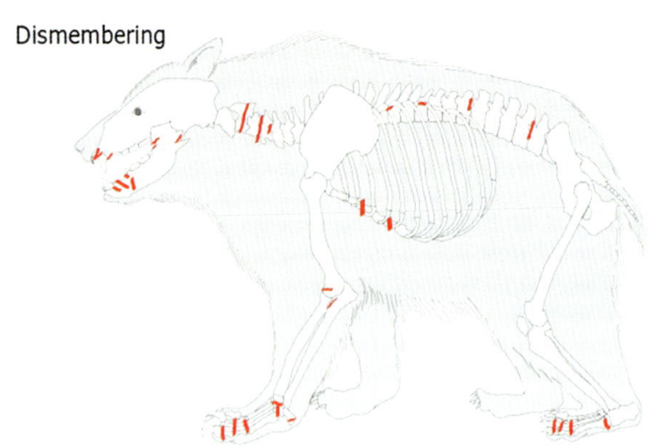

Fig. 8. Anatomical distribution of butchering marks with different functions (skinning, dismembering and filleting of meat) on brown bear bones from Frösö church (© 2003 ArcheoZoo.org; after Pales/Garcia *1981, pl. 13).*

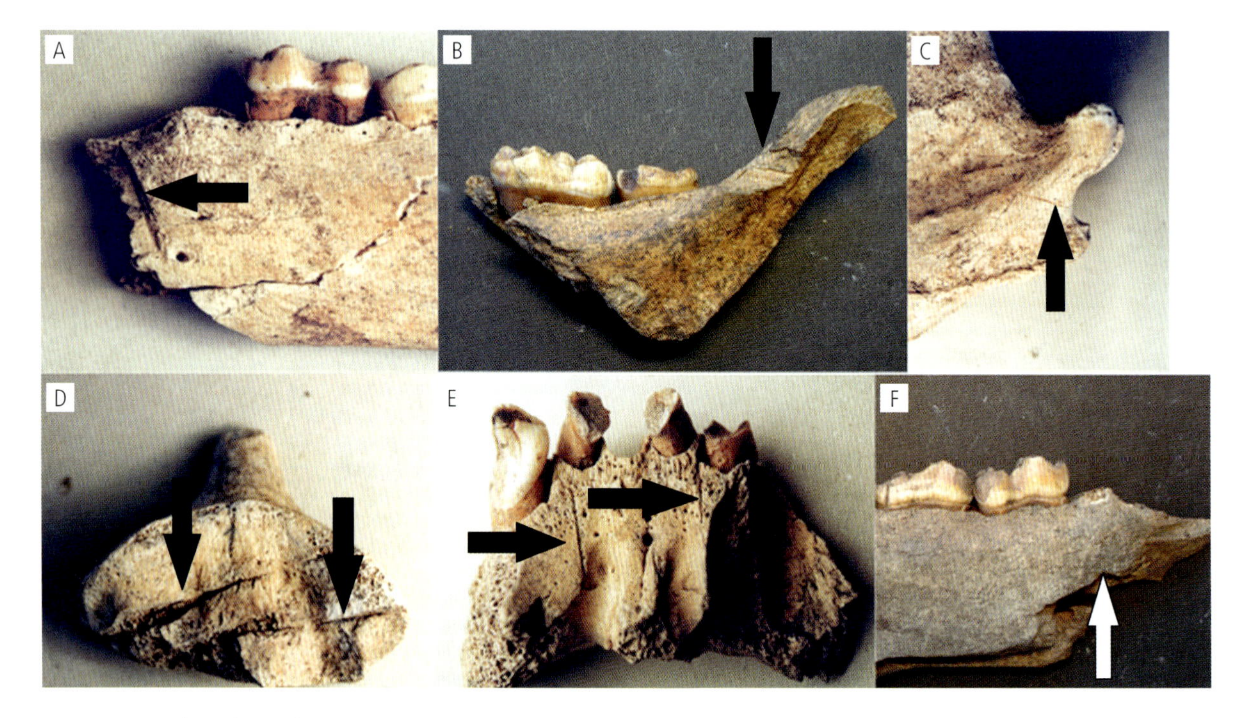

Fig. 9. Brown bear bones from Frösö church with bone modifications from butchering. a; f: Chop mark and impact mark on corpus of mandible caused by the extraction of a canine and bone marrow; b; c: Chop mark and cut mark on ramus of mandible from the dismembering and separation of mandibles from the skull; d: Chop marks on axis (2nd cervical vertebra) caused by the parting of the head from the trunk; e: Chop marks on premaxillare from the extraction of upper canines from the skull (photos O. Magnell).

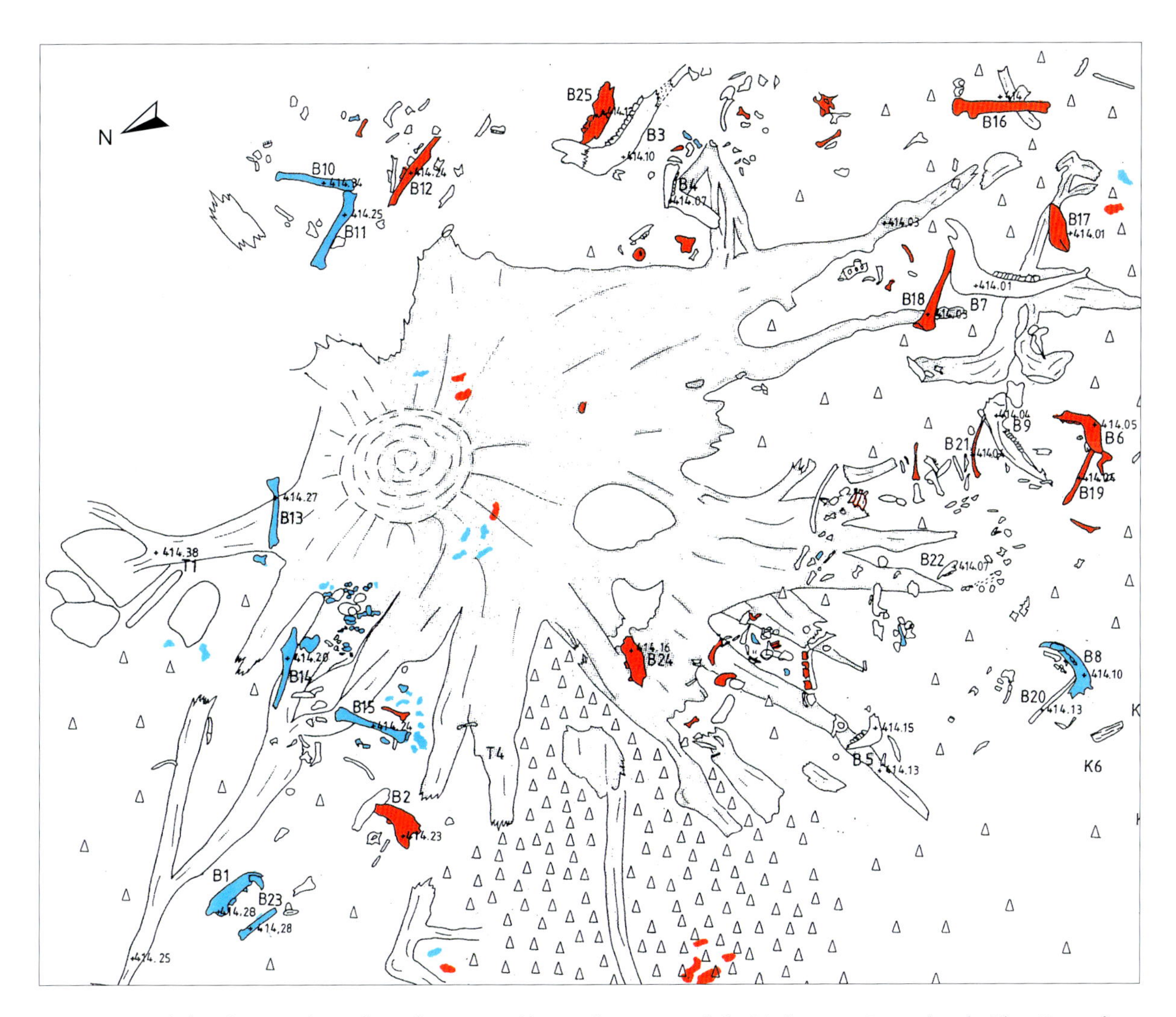

Fig. 10. Spatial distribution of osteological remains of brown bears around the birch tree at Frösö church. Blue: Bones from subadult individual (c. 2–4 *years); red: Bones from adult animals (graphics O. Magnell, based on* Hildebrandt *1991).*

Bear claws in Iron Age burials on Gotland, Sweden – a first survey

By Jane Jordahl, John Ljungkvist and Sabine Sten

Keywords: Bear, claws, phalanges 3, grave, Gotland, Sweden, Iron Age, database

Abstract: This article examines the remains of bear claws discovered in Iron Age burials (c. 500 BC–AD 1100) on Gotland, Sweden. Bears do not occur naturally on this island, meaning that every single find is a result of contacts and trade with surrounding forested regions around the Baltic Sea. This study builds on previous works, and has extended them by surveying the digital records from the Swedish History Museum in Stockholm. It has resulted in the discovery of new finds, so that these are more numerous than previous compilations have revealed. Today, the total number of graves containing bear remains is 148. Almost all finds are third phalanges related primarily to furs, but also to a few other interesting contexts. The presented results are still preliminary, as they will be further examined in J. Jordahl Master's thesis. As the material is of a considerable size, it forms a solid basis for more in-depth studies, involving both a look at patterns and contexts in the practice of depositing bear remains in Gotlandic graves, and also at osteological studies of the actual claws in the burials.

Introduction and general remarks

This article deals with the occurrence of bear claws (phalanges 3) in Iron Age graves in Gotland, Sweden (Fig. 1; Tables 1–2). The island of Gotland is one of the large islands in the Baltic Sea. As Öland, Saaremaa, Bornholm, and the Åland archipelago, it is characterised by its wealth during the 1st millennium AD. This wealth can either be restricted to a certain period of time, as on Åland in the Viking Age (cf. Gustavsson/Ljungkvist, this volume), or may have been prevalent throughout the entire millennium, as in the case of Gotland. Burials from the latter island contain, nearly continuously, a wealth of imports that separates them from burials in other regions around the Baltic. Alongside continental imports such as glass and bronze vessels and beads, there is also evidence for permanent contact and trade with adjacent regions in the Baltic.

The largest group of regional import material in the museum collections which originates from the forest regions is probably represented by combs, or rather the antlers they were made of from elk, reindeer, or red deer (Ashby et al. 2015; Luik et al. 2020). Antler combs were extremely common imports, but forest products of a more exclusive character, particularly the claws of bear and lynx, are also found in burials across the island (Zachrisson/Krzewińska 2019, 104–105).

There is no current evidence that species that, for example, might provide antlers for combs, ever migrated to Gotland after the last glacial period. The only recorded deer on the island were imported for the royal hunting grounds in the 17th century (Nyrén 2012, 173). In the same way, large predators never occupied this island. The reason for this lack is Gotland's isolation, 80 kilometres from the Swedish mainland and about 140 kilometres from the nearest Baltic territory (Latvia). Furthermore,

Bear and Human: Facets of a Multi-Layered Relationship from Past to Recent Times, with Emphasis on Northern Europe, ed. by Oliver Grimm (Turnhout: Brepols, 2023), pp. 453–468 BREPOLS PUBLISHERS 10.1484.M.TANE-EB.5.134345

Gotland is surrounded by the deepest waters of the Baltic, which hardly ever freeze, even in the coldest of winters.

The considerable quantity of burials furnished with bear claws on Gotland is highly interesting, not least in relation to the numerous graves with claws as burial goods from the mainland of present-day Sweden, Norway, Denmark, Finland (Åland), the Baltic countries, and central Europe (Petré 1980, 5; Grimm 2013, 290–291; Beermann 2016; Lindholm/Ljungkvist 2016, 10–12; Kirkinen 2017, 4). A primary purpose of this article is to highlight further research potential for bear claws in the Gotland burials by compiling current research on the subject and summarising the contexts of hitherto registered bear claws on the island.

This research is partly based on the data and previous work of Lindholm/Ljungkvist (2016). In this analysis, it was possible to see a specific chronology of wild animal skeletal remains as grave furnishings and, for example, a correlation with the use of pit trap systems. By examining graves and pit traps collectively, it is possible to compare the chronologies in two different but concurrent contexts in order to understand the use of animal resources. The present study presents the first results of J. Jordahl's Master's thesis. It has resulted in the discovery of dozens of additional burials with bear claws from Gotland, which makes it possible to approach more solid and independent studies of the material from the island.

Most finds belong to the collections of the Swedish History Museum (SHM). This examination is primarily based upon information that is available digitally and openly on the SHM's website: http://mis.historiska.se/mis/sok/sok.asp. Further information about the contexts can be gained by studies of drawings and excavation reports from the The Antiquarian Topographic Archive (ATA), the Swedish National Heritage Board, and of course in-depth studies of bones from single burials. A more thorough analysis and a further detailed review will be carried out within the framework of J. Jordahl's Master's thesis. So far, no bear claws have been recorded among the collection from Gotland Museum, but additional finds can quite probably be found there.

The survey for this paper, as well as the previous analysis, all refer to the pioneer study by Bo Petré (Petré 1980). During the early 1970s, bone materials with Iron Age date (in Sweden, *c.* 500 BC–AD 1100) were transferred from the SHM's archive to the Osteological Research Laboratory at Stockholm University. Therefore, the archaeologists Bo Petré and Margareta Wigardt made an inventory of the bone material that the SHM administered, and they found a total of 103 burials containing bear claws, distributed over 14 (out of 25) Swedish landscapes. The inspection of these archival materials led Petré (1984a; b) to continue the analysis of burials with bear claws.

Petré's initial and overall comparison indicated that the bear claws occurred in periods during the Iron Age in which regions had an economic upswing. He stated himself that this research was merely an overview, but his early conclusions are still central in demonstrating how contacts and trade can be traced through the analysis of animal bones. Later studies, which included far more contexts, have altered the distribution pattern substantially, which opens up possibilities to have a look at the characteristics and changes in specific regions, not least Gotland (Petré 1980, 10–12; Grimm 2013, 283–285; Lindholm/Ljungkvist 2016, 4–5).

The island is important for this discussion as the many finds of bear claws reveal that obviously a substantial degree of bear hunting was conducted in order to satisfy market demands outside the hunting grounds. Both Petré (1980) and Lindholm/Ljungkvist (2016) agree that the highest concentrations of bear claws in archaeological contexts are found in the central parts of inland Sweden, in the eastern coastal areas around Lake Mälaren, and on Gotland. Various historical and archaeological pieces of evidence show a continuous connection between Lake Mälaren and Gotland over time, which makes these concentration of bear claws seem reasonable. The material culture of Gotland also yields strong evidence for frequent contacts with multiple areas in the Baltic. Therefore, it is highly likely that bear furs could have originated in either present-day Finland, the Baltic countries,

Russia, Belarus, or Poland. Currently, studies on the DNA of ancient bears in Sweden, including finds from Gotland, are performed by PhD student Erik Ersmark, Swedish Museum of Natural History. Broader aDNA studies have also been carried out on European bear finds (Sommer/Benecke 2005, 161).

The distribution of Gotland bear claws in time and space

One of the purposes behind the work with the database is an attempt to make a spatial analysis of the distribution of graves with bear claws in the Gotlandic landscape, and also to try to understand how the bear claws occur over time. At present, not all sites have coordinates; therefore, the spatial analysis is limited to parish level (Fig. 2). Accordingly, due to the limitations of coordinates, this spatial analysis ought to be considered as preliminary.

Since Petré's study, the numbers of registered bear claws in the SHM collections have increased substantially due to new excavations and new finds in the museum's collections. These changes in the volume of data apply to Sweden in general and Gotland in particular. In general, the new materials validate and strengthen the result of Petré's study, but there are also notable differences. For example, according to Petré, there was a significantly lower amount of Migration Period graves (375/400–560/570) with bear claws compared to the previous period, the Roman Iron Age (1–375/400; cf. Petré 1980, 9–12). However, the number of burials for Gotland and Sweden on the whole is actually higher in the Migration Period than in the Roman Iron Age (Lindholm/Ljungkvist 2016, table 2; cf. Lindholm/Ljungkvist, this volume), especially if we consider the length of each period (*c.* 150 years, versus *c.* 400 years). Under these circumstances, it is relevant to establish a more precise dating for the Roman Period graves in the future, and to clarify whether the number of bear claws in graves increases during the latter part of that period. The assemblage of registered bear claws demonstrates that on Gotland the number of these objects gradually increased from the Roman Iron Age to their most frequent occurrence during the Vendel Period. During the following Viking Era, the claws became significantly rarer (Table 1).

A large proportion of the graves might potentially be dated more precisely, which may enable us to discuss trade fluctuations or burial practices in more detail. There was probably a decline in the depositing of bear claws, and other imports, in the mid-7th century, which could mean that the Late Vendel Period (560/570–750/800) has more in common with the Viking Age (750/800–1050/1100), regarding the frequency of claws in burials. Furthermore, some graves are related only to broad time spans: Iron Age (IA) 500 BC– AD 1100, Early Iron Age (EIA) 500 BC– AD 560/70, or Late Iron Age (LIA) 560/70–1050/AD 1100. Potentially, a closer examination of the finds and the funeral practices, plus the conducting of ^{14}C-datings, may lead to a more precise chronological allocation. Most likely, there is some degree of under-representation of Early Iron Age graves (from the Pre-Roman to the Migration Period). In comparison with Vendel and Viking Age burials, these contain fewer datable finds and thus have a higher probability of falling into the broad Iron Age (IA) or Early Iron Age (EIA) categories. Graves from the later periods do, on average, contain more datable finds, such as glass beads and metal objects, and also combs. Even those that are damaged or partially plundered often yield such objects (Table 1).

When it comes to burial customs, 94 graves on Gotland are registered as cremations and 40 as inhumation burials. Another ten graves contain both cremated and uncremated bones. For the latter, a closer analysis may find an explanation for the composition of the finds, but so far these graves are classified only as burials with a mixed content of bones (Table 2).

The spatial distribution illustrates that the graves are scattered around the central parts of the island, with slightly sparser numbers in the northwestern parts. The northernmost graves have been recorded for Lärbro and the southernmost for Grötlingbo parish. No bear claws have so far come to light in the northernmost or the southernmost parishes of Gotland (Fig. 2). The concentrations reflect many excavated graves, as e.g. the two burial sites Barshalder and Rojrhage (SHM 3262; 334401; cf. Rundkvist 2003), located in the Grötlingbo parish. Furthermore, there is also a high proportion of osteologically analysed burials, which probably led to a substantial number, or rather an overrepresentation, of identified bear claws (Rundkvist 2003, 58–59). Other concentrations are known from the Broe (SHM 35335) burial ground in Halla parish and finally the Trullhalsar (SHM 8555) burial site in Anga parish (Fig. 3).

The recording and study of the graves with bear claws is challenging in different respects. A source-critical aspect concerning the registration from the SHM is whether the species assessment regarding the phalanges is correct. According to previous studies, there is a certain misjudgement regarding the claws that have been found. Claws have, for instance, been mistaken as belonging to seals or dogs (Petré 1980, 8; Lindström 1995, 5; Lennblad 2008, 25). One such example is a burial in Asarve (375/400–560/570), in Hemse parish (SHM 8859:9) that does not contain a seal claw, as stated in the original catalogue, but one of a bear, as a re-analysis has shown.

Another lesson can be learnt from the analysis of animal bones during the so-called "Storgravsprojektet" ("Chieftain's graves project") from 1988 (Sten/Vretemark 1988, 145–156; Sten 2013, 223–231). After their study had been published, they learned that they had missed some bear claws in their analysis. Since these had originally been interpreted as evidence for furs, the claws had been categorised as archaeological objects and placed in the archaeological collections. This illustrates the problems in finding objects in museum archives, due to different mind sets about the organisation of the collection (pers. comm., S. Sten, August 3, 2020).

Digital records from the SHM contain descriptions that vary considerably from the original excavations reports, which affects the quality of the data. In the framework of the present paper, there has not been enough time to gain supplementary excavation data from the ATA archive (Lindström 1995, 5; Grimm 2013, 279; Lindholm/Ljungkvist 2016, 10). Distinguishing individuals in double graves can be troublesome, too, if the individuals have been placed close to each other (Johansson 2007, 35; Lennblad 2008, 24–25). A similar case to that effect is the presumed double inhumation with the remains of an actual bear skin, found in a large cist from Nedre Aure, Voss, in inner Hordaland, Norway (Grimm 2013, 282).

So far, third phalanges are the only identified bear bones in Gotlandic graves (Sigvallius 1994, 76; Jennbert 2003, 142–144; Lindholm/Ljungkvist 2016, 13). This largely follows a pattern that can be observed for other parts of Scandinavia. As bears are large animals, it is highly unlikely that additional bones have been missed, perhaps with the exception of single first or second phalanges. All excavations following the SHM number 30,000 (except those that are connected to Rundkvist's study of Barshalder [Rundkvist 2003]) cannot be studied via the SHM, but require archive work at the ATA.

Drilled bear claws are absent in the material from Gotland, and neither are there remains of mounting devices to which bear claws could have been attached, as are known from Lithuania and Poland (Petré 1980, 8).

At least in the case of inhumations burials, it should be possible to prove the presence of actual furs. On Gotland, the preservation conditions for organic material, particularly bones, are often good, primarily due to the limestone bedrock. Further, several inhumation burials contain many copper alloy objects which, in general, means favourable preservation conditions for textiles and potential parts of furs that have been in contact with that metal. The conditions are, in other words, partly similar to those that have been successfully examined in Finland by Kirkinen (2015).

A source-critical problem for estimating the amount of Gotlandic bear claw finds lies in the fact that most inhumation burials were excavated during the late 19th and early 20th century. The excavation procedure was, back then, often considerably cruder, at least from a modern, purely technical point of view, which most likely has affected the number of claws discovered in many burials. One cannot expect the archaeological records to be of present-day quality standard. In the ATA's archives, however, there are several illustrations of inhumation burials with detailed information about the positions of the bodies and finds, including the bear claws in cists, chambers, etc. An additional concern is that many inhumation burials have been plundered or disturbed. This raises uncertainties regarding the original location and number of the bear phalanges. Related source-critical problems apply to the cremation burials; the burnt claws are not found *in situ*, and, for various reasons, many (or most) bones are often not deposited in the burial. In addition, we know that some early excavators did not prioritise the recovery of all burnt bones from the cremation burials.

Furs or not?

One of the most conventional arguments for furs is that the phalanges are remains of paws that can be identified when a number of phalanges are found closely grouped *in situ* (Jennbert 2003; Grimm 2013, 284; Kirkinen 2015, 109; 2017, 3–7). These arguments for furs are of course more difficult for the Gotlandic cremation burials, even if the number of claws is substantial in several cases. In some inhumation burials, the evidence for furs is much stronger. One case to that effect is burial 1/53 from Hallvede (550–750/800) in Eke parish (SHM 25133); the bear claws (paws) there are clearly represented by four groups around the buried individual (Fig. 4). In a second case, a weapon burial from Broa (550–750/800) in Halla parish (SHM 20517:13), the grave contained eight bear claws. These were located in two groups, one by the feet and one above the head of the deceased (Fig. 5). This grave seems to be undisturbed, thus one can take as given a fur with only two paws attached. The third case is an earlier inhumation from Smiss (*c.* 160/180–325) in Eke parish (SHM 16113; Lund Hansen 1987, 446). Likewise, the grave displays an example with claws distinctly positioned in the four corners of the cist: three times five and one time three (Almgren/Nerman 1923, fig. 156a; Fig. 6). Moreover, the grave's overall preservation degree is high.

In two undisturbed burials of women, each with a complete set of jewellery, we see another way of depositing bear claws. In the first case, grave 280D from Ihre Hallvede (550–750/800) in Hellvi parish (SHM 20826), only two bear claws were found in a surprising location, namely in a bronze cauldron (Fig. 7). The location of this vessel is not shown precisely on the excavation plan of the grave's bottom level, but it may be vaguely marked as a circle at the top to the right of the cranium. Furthermore, the preservation conditions for the skeleton and probably for other animal bones in this grave were limited. However, from the same grave, there is a well-preserved Cypraea (cowrie shell), which suggests that claws might also have been preserved (Nerman 1955, 211; Ljungkvist 2010, 427–428). In the other case, a burial from Allekvia (550–750/800) in Endre parish (SHM 24277:3), the preservation conditions are even better. According to the original collection description, there were two bear claws in the grave, one of which is clearly depicted on the left side of the skull according to the field drawing. However, the other registered bear claw is difficult to review in the drawn plan (Fig. 8).

These two female burials reveal that there are perhaps some alternative explanations for claws in burials instead of furs. So far, the claws have only been interpreted on the basis of the grave plans, and it is still difficult to tell if they show any marks or have been modified in some way. However, they have no drilling holes, which would indicate a use as pendants, and the cauldron in grave 280D was too small to house a complete bear fur. In this respect, grave IV from Rösta, Ås parish in Jämtland, middle Sweden (10th century AD), is worth mentioning, too (Fig. 9). Here, however, two claws were found in relation to a bag, placed by the waist of the buried person (Kjellmark 1905, 369).

So, if the claws in these cases do not represent furs, what alternative explanations are there? It is still too early to tell if these cases represent a broader pattern, but they raise a number of questions. For example, were parts of bear paws with attached claws deposited in graves, and might the bronze vessel from Ihre 280D or the bag from Rösta IV contain preserved hair fragments from a paw? These graves might explain why a number of burials with bear claws contain only 1–5 claws. A small number of claws is thus perhaps not always related to post-cremation activities or the plundering of inhumation burials. Another tempting alternative is that the claws represent an amulet with similarities to much later specimens from the Siberian Khanty culture in western Siberia, or even 19th century objects from Swedish Lapland (Kirkinen 2017; cf. Fig. 10).

Gender and social aspects of the burials with bear claws on Gotland and in Norway

According to the registered burials that have been gender-assessed, it seems that bear claws are nearly evenly distributed between male and female graves on Gotland. The same pattern seems to appear among the Norwegian graves (Grimm 2013, 282–283). Burials with bear and lynx claws on Gotland do not belong exclusively to a distinct higher social stratum in the population. Previous studies show that bear claws are frequently found in rich burials, but, depending on the period, also in burials from the middle levels of society (Lindholm/Ljungkvist 2016, 12). In the near future, J. Jordahl will perform an in-depth study on whether bear claws can be found in specific gender-related funerals, and to what degree bear claws are related to individuals of specific social groups.

With regards to Norway, both male and female graves have yielded bear claws, and the burial furnishings range from very wealthy to rather poor (Grimm 2013; cf. different contributions for Norway in this volume). These burials, mostly cremations, are most common in the Migration Period and are thus older than the majority of the graves on Gotland. The source situation is special for Norway, since roughly a dozen burials, mostly inhumations, have yielded actual bear skins with preserved hair found jointly with claws.

Summary and conclusions

This study strengthens and underlines patterns observed in previous research; bear remains were deposited in Gotlandic graves throughout the entire Iron Age (*c.* 500 BC– AD 1100). The remains are so far entirely represented by phalanges 3, and the majority are probably related to furs. In some cases, however, pairs of claws in well-documented inhumation burials do not appear to be related to complete furs. They seem rather to have served another purpose, perhaps as amulets.

The burials in this data set can be subject to more detailed analyses regarding the deposition patterns in the landscape, but they can also be used for studies of their social context, their gender relation, their spatial location at specific cemeteries, and their dating. Further, more detailed analyses of the actual claws can potentially reveal details about the size of the animals, the handling of furs, and perhaps even changes in import patterns.

Some developments have previously been noticed in the frequency of bear claw burials in the Swedish material as a whole (Lindholm/Ljungkvist 2016). The burials on Gotland, like those on the Swedish mainland, were most common during the Migration and Early Vendel Periods (*c.* 400–650), while a significant decrease began in the Late Vendel Age and continued into the Viking Age (*c.* 650–1,000). In mainland Sweden, it appears that bear parts became more exclusive and increasingly related to particularly high-status burials. This might indicate unsustainable hunting in mainland Sweden and perhaps even other regions around the Baltic, which made bears rarer and pelts more exclusive as well as valuable due to an increased scarcity. Comparable in relation to prices and increased pressure, this is even today a far too common phenomenon, from the tuna and cod fishery to the hunt for tigers, elephants, and rhinoceros.

Gotland is indeed a place where the sheer volume of various imports is massive. Petré (1980, 12–13) paid considerable attention to Gotland and its relation to an early fur trade. The high frequency of bear claws shows that this animal, together with lynx (claws; cf. Zachrisson/Krzewińska 2019, 104–105) and wild herbivores (antlers; cf. Ashby et al. 2015; Luik et al. 2020) – i.e. products of the forested regions of Scandinavia and regions on the other side of the Baltic – played a vital role in the broad and complex trading webs during the 1st millennium AD. It also indicates that the remains of large predators were displayed by a broad range of the population of Gotland, who had never seen a large mammal in the wild of their island. As mentioned earlier in this article, there is no evidence of large wild animals on Gotland from prehistoric periods. Together with other large mammals, such as the deer depicted on the Gotlandic picture stones (most recently Oehrl 2019, 161–168, 179, tables 159–167, 183c), bears probably played a role in myths, religion, and storytelling.

In mythology, sagas, and perhaps even for present-day archaeologists, bears are primarily associated with the male sphere. This is hardly surprising due to their size, strength, and potential ferocity in a conflict situation. However, according to the burial records, there is no striking difference between the occurrences of claws in male or female burials on Gotland. This makes the bear in burial connections even more complex and interesting. Nevertheless, we should not rule out that more variations can be revealed from the burial records by analyses that involve the age and status of the buried individuals. Other important variables are related to the potential size of the animals, the number of deposited claws in burials, and the question of how and where they were placed in inhumations. In short, there is far more to explore about the remains of this animal found on an island in the middle of the Baltic.

Acknowledgements

John Ljungkvist's work on this paper was undertaken within "The Viking Phenomenon" project at Uppsala University, funded by the Swedish Research Council (grant 2015-00466). We would like to thank Senior Curator Johnny Karlsson at the Department of Collections and Research, the National Historical Museum, Stockholm, the ATA (The Antiquarian Topographic Archive), and the Swedish National Heritage Board for providing grave plans. Also, thanks to Oliver Grimm and Sigmund Oehrl for their essential input and feedback. Language reviewing was carried out by Dostan Rashid.

Bibliography

Almgren/Nerman 1923: O. Almgren/B. Nerman, Die ältere Eisenzeit Gotlands 2 (Stockholm 1923).

Ashby et al. 2015: S. Ashby/A. N. Coutu/S. N. Sindbæk, Urban Networks and Arctic Outlands: Craft Specialists and Reindeer Antler in Viking Towns. European Journal of Archaeology 18, 2015, 679–704.

Beermann 2016: S. Beermann, Bärenkrallen und Bärenfelle in Brand- und Körpergräbern der vorrömischen Eisenzeit bis Völkerwanderungszeit in Mittel- und Nordeuropa. Universitätsforschungen zur Prähistorischen Archäologie 279 (Bonn 2016).

Grimm 2013: O. Grimm, Bearskins in northern European burials and some remarks on other bear-related furnishings in the north and middle of Europe in the 1st millennium AD. In: O. Grimm/U. Schmölcke, Hunting in northern Europe until 1500 AD. Old traditions and regional developments, continental sources and continental influences. Papers presented at a workshop organized by the Centre for Baltic and Scandinavian Archaeology (ZBSA) Schleswig, June 16th and 17th, 2011. Schriften des Archäologischen Landesmuseums, Ergänzungsreihe 7 (Neumünster 2013) 277– 296.

Jennbert 2003: K. Jennbert, Animal Graves – Dog, Horse and Bear. Current Swedish Archaeology 11, 2003, 139–152.
Johansson 2007: I. Johansson, Brandgravar – Yngre järnålder i Broe, Halla socken, på Gotland. En studie i olika metoder, med fokus på kremeringen. Unpubl. Master's thesis in osteology, Gotland University (Visby 2007).

Kirkinen 2015: T. Kirkinen, The role of the wild animals in death rituals: Furs and animals' skins in the late Iron Age inhumation burials in South-eastern Fennoscandia. Fennoscandia archaeologica XXXII, 2015, 101–120.
Kirkinen 2017: T. Kirkinen, "Burning pelts" – Brown bear skins in the Iron Age and Early Medieval (1–1300 AD) burials in south-eastern Fennoscandia. Estonia Journal of Archaeology 21(1), 2017, 3–29.
Kjellmark 1905: K. Kjellmark, Ett gravfält från yngre järnåldern i Ås Jämtland. Ymer 25, 1905, 351–371.

Lennblad 2008: A. Lennblad, Brandgravsmaterial från Linde socken på Gotland – En osteologisk studie av ett brandgravsmaterial från yngre järnåldern. Unpubl. Bachelor's thesis in osteology, Gotland University (Visby 2008).
Lindholm/ljungkvist 2016: K. Lindholm/J. Ljungkvist, The Bear in the Grave: Exploitation of Top Predator and Herbivore Resources in First Millennium Sweden – First Trends from a Long-Term Research Project. European Journal of Archaeology 19(1), 2016, 3–27.
Lindström 1995: T. Lindström, Björnklor i gotländska järnåldersgravar – En studie av gravförhållanden. Unpubl. Bachelor's thesis in archaeology, Stockholm University & Gotland University (Stockholm, Visby 1995).
Ljunkvist 2010: J. Ljungkvist, Influences from the Empire: Byzantine-related objects in Sweden and Scandinavia – 560/70–750/800 AD. In: F. Daim/J. Drauschke (eds.), Byzanz – Das Römerreich im Mittelalter 3: Peripherie und Nachbarschaft. Monographien des Römisch-Germanischen Zentralmuseums 84,3 (Mainz 2010) 419–441.
Luik et al. 2020: H. Luik/J. Peets/J. Ljungkvist/R. Maldre/L. Maldre/R. Allmäe/M. Muñoz-Rodríguez/K. McGrath/C. Speller/S. Ashby, Antler combs from the Salme ship burials – find context, origin, dating and manufacture. Estonian Journal of Archaeology 24(1), 2020, 3–44: doi.org/10.3176/arch.2020.1.01.
Lund Hansen 1987: U. Lund Hansen, Römischer Import im Norden. Warenaustausch zwischen dem Römischen Reich und dem freien Germanien während der Kaiserzeit unter besonderer Berücksichtigung Nordeuropas. Nordiske Fortidsminder B 10 (Copenhagen 1987).

Nerman 1955: B. Nerman, Elfenben och snäckor i Gotländska Vendeltidsgravar. Fornvännen 50, 1955, 209–221.
Nyrén 2012: U. Nyrén, Rätt till jakt: En studie av den svenska jakträtten ca 1600–1789 Unpubl. PhD thesis in Philosophy, Gothenburg University (Gothenburg 2012).

Oehrl 2019: S. Oehrl, Die Bildsteine Gotlands – Probleme und neue Wege ihrer Dokumentation, Lesung und Deutung. Studia archaeologiae medii aevi 3 (Friedberg 2019).

Petré 1980: B. Petré, Björnfällen i begravningsritualen – Statusobjekt speglande regional skinnhandel? Fornvännen 75, 1980, 5–14.

Rundkvist 2003: M. Rundkvist, Barshalder 2. Studies of Late Iron Age Gotland. (Stockholm 2003).

Sigvallius 1994: B. Sigvallius, Funeral Pyres, Iron Age cremations in north Spånga. Thesis and Papers in Osteology 1 (Stockholm 1994).
Sommer/Benecke 2005: R. S. Sommer/N. Benecke, The recolonization of Europe by brown bears *Ursus arctos* Linnaeus, 1758 after the Last Glacial Maximum. Mammal Review 35(2), 2005, 156–164: doi.org/10.1111/j.1365-2907.2005.00063.x
Sten 2013: S. Sten, Sacrificed animals in Swedish Late Iron Age monumental mound burials In: G. Ekroth/J. Wallensten (eds.), Bones, Behaviour and Belief. The Zooarchaeological Evidence as a Source for Ritual Practice in Ancient Greece and Beyond (Athens 2013) 223–231.
Sten/Vretemark 1988: S. Sten/V. Vretemark, Storgravsprojektet – osteologiska analyser av yngre järnålderns benrika brandgravar. Fornvännen 83(3), 1988, 145–156.

Zachrisson/Krzewińska 2019: T. Zachrisson/M. Krzewińska, The »Lynx Ladies«: Burials Furnished with Lynx Skins from the Migration and Merovingian Periods found in Present-day Sweden. In: M. Augstein/M. Hardt (eds.), Sächsische Leute und Länder: Benennung und Lokalisierung von Gruppenidentitäten im ersten Jahrtausend. Neue Studien zur Sachsenforschung 10 (Braunschweig 2019) 103–119.

PhD student Jane Jordahl
Department of Archaeology and Ancient History
Uppsala University
Sweden
janejordahl@hotmail.com

Associate Professor John Ljungkvist
Department of Archaeology and Ancient History
Uppsala University
Sweden
john.ljungkvist@arkeologi.uu.se

Prof. Sabine Sten
Department of Archaeology and Ancient History,
Uppsala University, Campus Gotland
Sweden
sabine.sten@arkeologi.uu.se

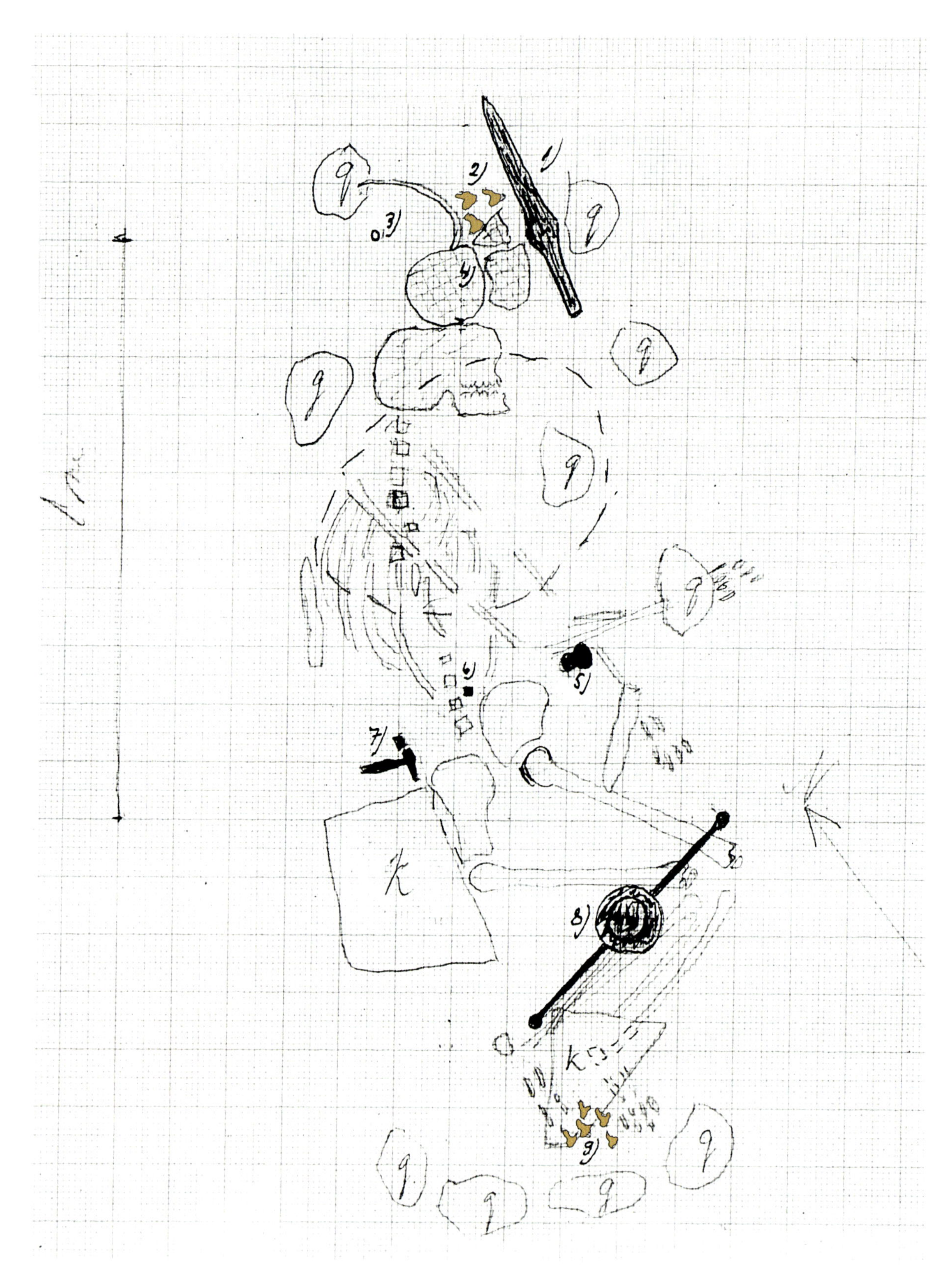

Fig. 5. A field drawing of grave 13 from Broa (550–750/800) in Halla parish (SHM 20517, ATA: Go 2363 F; unknown artist; image modified, with the two concentrations of claws by the feet and above the head [left of the spearhead] of the deceased highlighted in brown, by J. Jordahl).

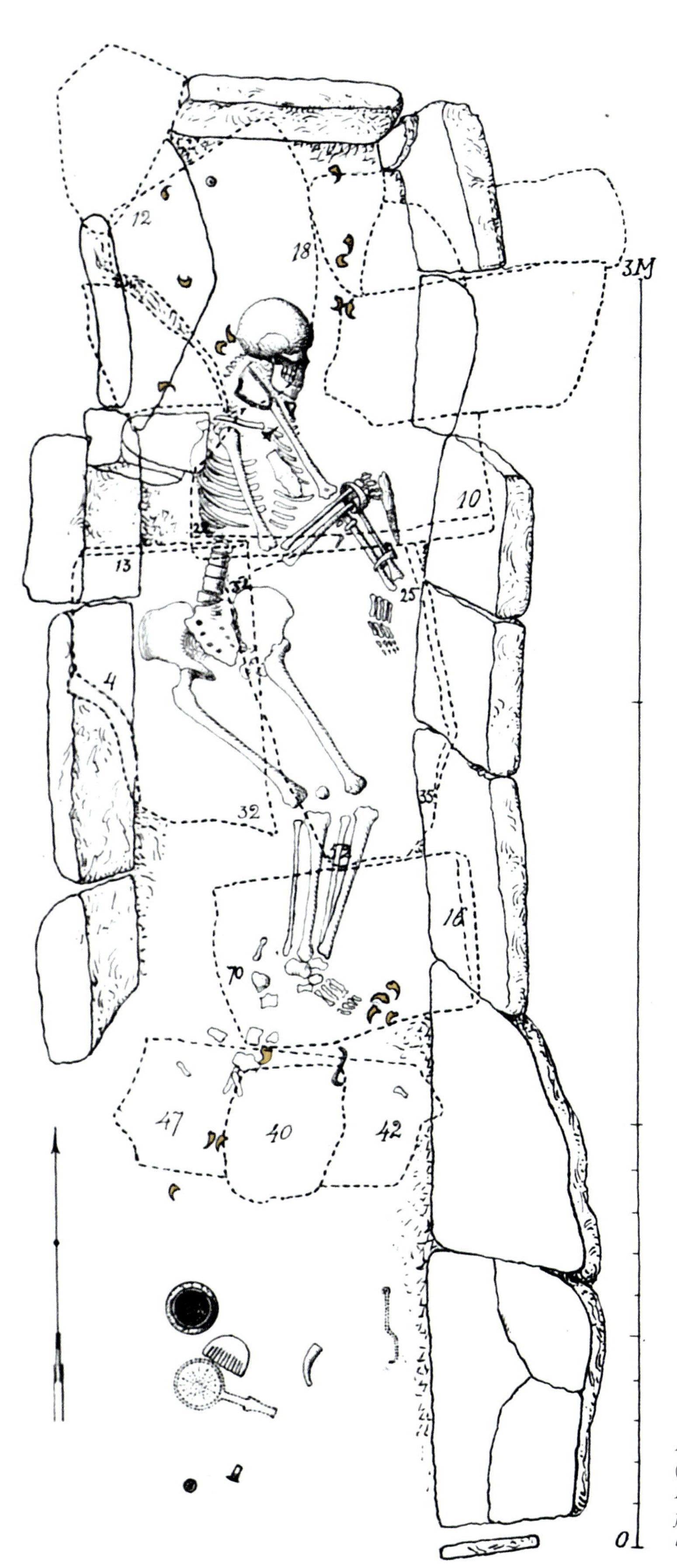

Fig. 6. Depiction of the grave from Smiss (160/180–325) in Eke parish (SHM 16113, ATA: 1875F; unknown artist; image modified, with bear claws highlighted in brown, by J. Jordahl).

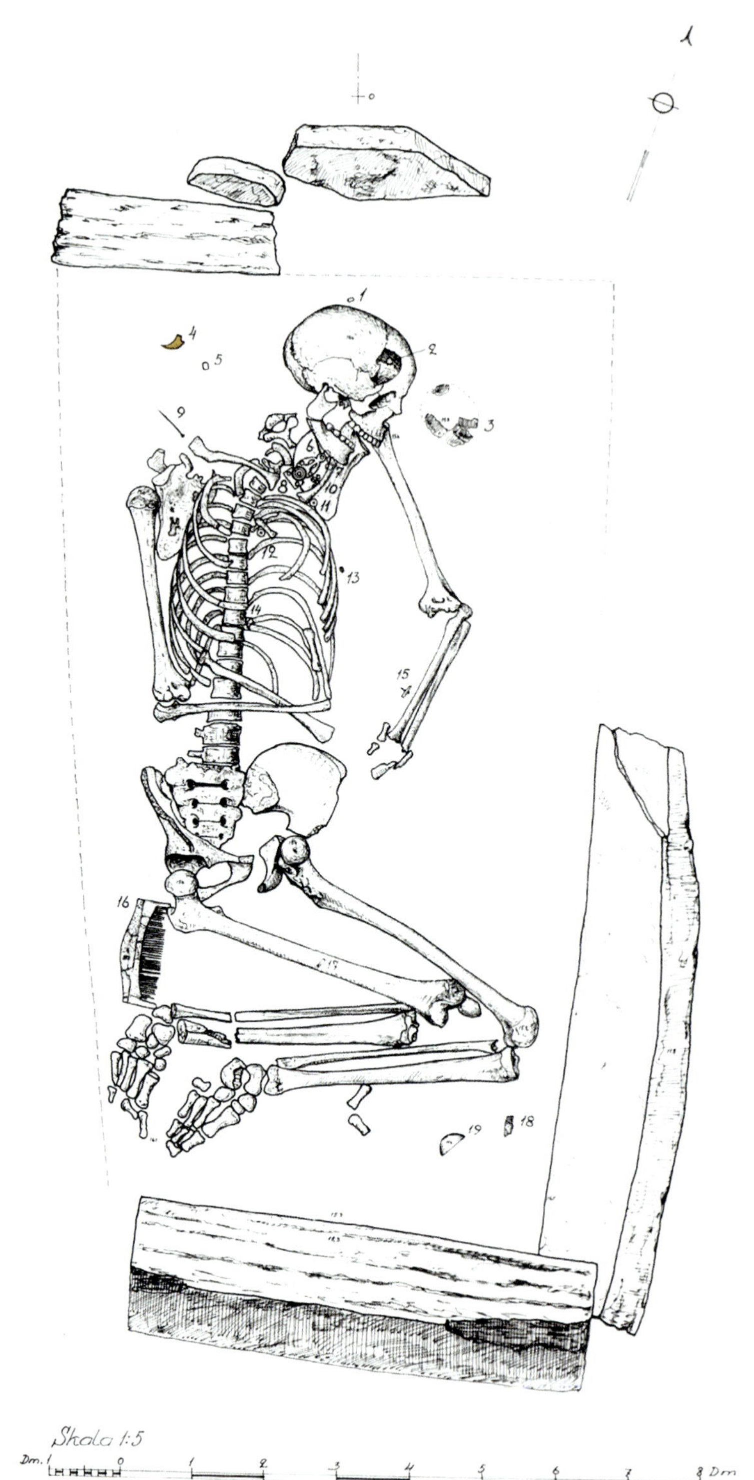

Fig. 7. A bear claw from the bronze cauldron in grave 280D (550–750/800), Ihre Hellvi parish (SHM 20826, picture ID 224283; photo J. Ljungkvist).

Fig. 8. Grave 3 in Allekvia (550–750/800) in Endre parish (SHM 24277:3, ATA: Go 1927F; unknown artist; image modified, with the bear claw highlighted in light brown [no. 4], by the left side of the cranium, by J. Jordahl).

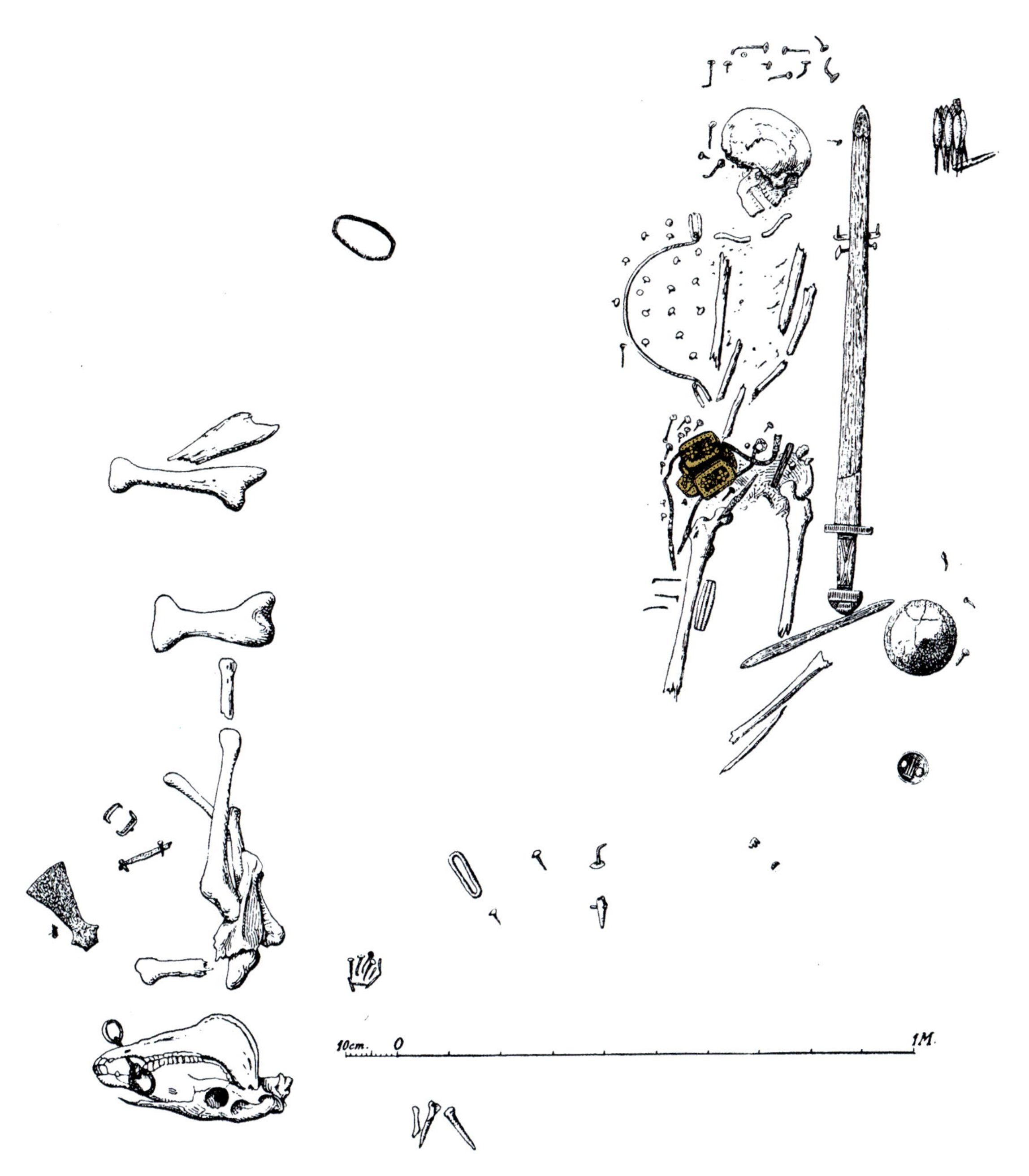

Fig. 9. A field drawing of grave IV in Rösta, Ås parish, in Jämtland, 10th century (unknown artist, image modified, with the bag at the waist highlighted, by J. Jordahl).

Fig. 10. Sami amulet in the shape of a partial bear paw with three claws, attached to belt straps (Nordiska Museet, ID: NM.0181624A-B, photo B. Wreting).

Table 1. Total number of graves on Gotland with bear claws associated with a specific period.

Period	Total number of graves
Pre-Roman Iron Age (PRIA)	2
500 BC – AD 1	
Roman Iron Age (RIA)	16
AD 1–400	
Migration Period (MP)	13
AD 400–560/570	
Migration Period – Vendel Period (MP – VET)	6
AD *c.* 400–750/800	
Vendel Period (VET)	36
AD 560/570–750/800	
Vendel Period – Viking Age (VET – VIT)	2
AD 560/570–1050/1100	
Viking Age (VIT)	7
AD 750/800–1050/1100	

Table 2. Number of graves containing bear claws according to funeral rites.

Funeral rites	Total number of graves
Cremation burial	94
Inhumation burial	40
Both cremation and inhumation burials	10

Claws in Late Iron Age graves (*c.* 550–1100 CE) and bones in a castle (post 1500) – *Ursus arctos* in the Åland archipelago

By Rudolf Gustavsson and John Ljungkvist

Keywords: Åland Islands, Late Iron Age, burials, bears, zooarchaeology

Abstract: Bears have never been a part of the fauna in the Åland Islands. Yet they are present in the archaeological record in a number of ways. Most finds of claws have come from 30 Late Iron Age (550–1100 CE) cremation burials. At least one so-called bear tooth pendant also shows a symbolic presence of bears in its shape. Another type of object that is characteristic of Åland burials and is perhaps also related to bears, is represented by the numerous clay paws. The pre-modern finds of bears on the archipelago are not limited to prehistoric grave finds. A few bear bones have also been found at the medieval castle of Kastelholm. This find reveals a way of using bears that is rather different from that seen in earlier burial contexts. Even if bears were not living on the archipelago, they are clearly present in the archaeological records through several centuries. The volume of the data is much smaller than in other areas around the Baltic, but it is enough to see that the Ålandic people chose a partially independent way of using bear products and bear symbolism.

Landscape, faunal history and local economy[1]

The landscape of the Åland Islands and southwestern Finnish archipelago differs considerably from the other large island landscapes of the Baltic (Fig. 1). Today, the islands and skerries number around 27,000 and their total landmass is around 1,553 km^2. The main island, or mainland Åland, covers an area of *c.* 1,000 km^2, the eastern archipelago an area of about 500 km^2 (ÅSUB 2019).

This is a landscape characterised by rocky, broken terrain with larger differences in elevation in the northern parts of the islands and flatter, sandy moraine landscapes in the southeastern parts. Between the rocky areas in the north are valleys and plains with sand and clay-based soils suitable for the growing of crops. Due to the ongoing land uplift, this is a constantly changing landscape. When the first settlers arrived in *c.* 5500 BCE, only a few rocky islands with a steep topography protruded from the sea (Núñez 1986; Jaatinen et al. 1989, 25–34). The calculated landmass at this point covered about 24 km^2, and in *c.* 1500 BCE the landmass had increased to some 500 km^2. By the end of the Iron Age, about 1100 CE, the landmass occupied some 970 km^2 of broken terrain. It had, in other words, grown considerably, but it was still too small to house a population of bears, especially alongside a growing human population.

1 In addition to the works mentioned in the bibliography, further sources from the archives of Ålands Museum have been used: Kastelholm castle excavations' catalogue of bones (Bnr catalog) and archaeological field documentation; Ålands Museum's catalog of archaeological finds.

Bear and Human: Facets of a Multi-Layered Relationship from Past to Recent Times, with Emphasis on Northern Europe, ed. by Oliver Grimm (Turnhout: Brepols, 2023), pp. 469–488 BREPOLS PUBLISHERS 10.1484.M.TANE-EB.5.134346

The early faunal history of the islands is still more or less unexplored. In archaeological assemblages from the Stone Age, *c.* 5500–1800 BCE, only a few fragments of terrestrial animals can be found among the numerous bones of marine mammals, fish, and seabirds. It has been estimated that the islands, at least theoretically, could have supported a local population of elk from *c.* 1800 BCE, but mountain hare is the only wild terrestrial mammal that is repeatedly identified in the assemblages from the Neolithic up to the Late Iron Age (Núñez 1986; Storå 2000). Analysed bone assemblages from the Bronze Age, 1800–500 BCE, and Early Iron Age, 500 BCE–550 CE, are limited, and only from the 5th century CE onwards do the finds get larger and the number of species increases, including occasional finds of roe deer (Larsson 1999; Andersson manuscript). The assemblages from the Late Iron Age consist almost exclusively of sheep, goat, cattle, and seals. Pigs are present in the Neolithic and in very limited quantities also during the Late Iron Age (Storå 2000; Gustavsson et al. 2014). They only start to appear frequently in medieval find layers. In short, this archipelago was not particularly suitable for large wild mammals.

Bears are noted as a missing species in the faunal historical literature from the 19th–20th centuries (Radloff 1795; Nordberg 1939a; b). According to Nordberg (1939a), both Tuneld (1741) and Bergstrand (1852) mention that bears were occasionally found in the Åland islands, but Radloff had already disputed this. In particular, the data on wild animal species published by Bergstrand have been considered dubious. Bergstrand has, for example, claimed that the arctic fox was present in the Åland Islands, something that has to be viewed as extremely unlikely. Eventually, populations of elk and roe deer established themselves on the archipelago, but they were hunted to extinction in the early modern period, the former in 1778 and the latter already in the early 16th century. Wild boars have never been part of the local fauna, but the Åland Islands apparently had a prosperous population of lynx, together with pine marten, fox, and wolf until the mid-19th century (Nordberg 1939a). One reason for this is that the animal populations on the Åland archipelago, in contrast to other Baltic islands like Gotland and Bornholm (see Jordahl et al., this volume), could have been supplemented by new animals from the Finnish mainland during wintertimes, over the frozen sea.

The economy of the islands was adapted to the local environment, and sediments suitable for crop farming are mostly found below 5 m above the present shoreline. Thus, this type of land was not available until historic times (post *c.* 1100 CE), and even in the 18th century the cultivated lands were only a third of today's (Jaatinen et al. 1989, 27). Subsistence was mainly based on livestock herding, fishing, fowling, and sealing for grey seal, ringed seal, harp seal and sometimes also harbour seal (Storå et al. 2012; Kivikero et al. 2020). The archipelago, with its unique landscape in constant change presented different conditions, especially for farming and crop growing, in comparison with other Baltic islands like Gotland, Öland, Dagö, or Ösel/Saaremaa. As an effect of these conditions, the marine economy, including seafaring and trade, became essential for the population of the islands.

The main basis for studies of contacts and trade in prehistoric Åland comes from the Late Iron Age burial mounds. It is a norm that these contain rivets that originated from boats. Other artefacts indicate close connections with the outside world, where especially Finnish, Swedish, Gotlandic, and Baltic objects are present in the burial assemblages (Kivikoski 1959; Gustavsson et al. 2014; Tomtlund 2014). Bear products, primarily furs, are another foreign product, indicating contacts, travel, and trade with the surrounding regions. For the Åland community, the role of bears clearly seems to have reached beyond the role of signalling status, power and wealth. The material also clearly indicates that bears had a broad presence in symbology and burial practices and therefore probably had a lively role in myth and folklore.

Burial contexts with bear finds

The known graves with bear claws originate from some of the *c.* 400 Late Iron Age cemeteries in the Åland Islands, which have a total of *c.* 10,400 visible mounds (Tomtlund 2014). They are far more common than Early Iron Age graves, which number some 2,200 cairns. However, some of the larger Bronze Age cairns are known to include several secondary burials dating up to the Migration Period. Only few excavations have been undertaken on this type of monument, and the find material is scarce, which might be one reason for the lack of bear bones from this period (Gustavsson 2008). Some 740 Late Iron Age burial mounds have been excavated since the creation of the Ålands Museum in the 1930s, and bones from 286 of these have been analysed by osteologists/zooarchaeologists. The burial composition and the construction of the individual monuments in the cemeteries highly resemble the middle Swedish cremation burial practice. But the composition of artefacts in the burials constitutes a mix of Finnish, Swedish, Gotlandic, and Baltic objects, with some local varieties (Kivikoski 1959; 1963; 1980; Rundkvist 2010; Gustavsson et al. 2014; Tomtlund 2014).

Five cemeteries have been completely excavated, and three of these have also been fully published: Kvarnbacken in Saltvik (Kivikoski 1963), Långängsbacken in Sund (Kivikoski 1980), and Bol in Finström (Landin/Rosborg 1984). All remaining excavations are only available as grey literature in the shape of reports, catalogues and analyses in the Ålands Museum archives.

The publications by Kivikoski (1959; 1963; 1980) form the basis for almost all studies of Late Iron Age burials on the archipelago. The sites are located relatively close to each other in the northeastern part of mainland Åland and can be characterised as relatively rich in contrast to the material from e.g. Bol in the middle of mainland Åland, which is much sparser. This inequality is interesting as it indicates certain differences between farms and communities on Åland, and perhaps even regional differences (cf. Tarsala 1998; Tomtlund 1999; Gustavsson et al. 2014). It is still uncertain if the examples given by Kivikoski represent the burial practice and social structure of Åland as a whole.

An osteological analysis of burials is only included in the publication of Bol in Finström (Landin/Rosborg 1984), whereas the remaining osteological studies are available in the form of unpublished theses and reports in the archives of the Ålands Museum (Vormisto 1981a; b; 1982; Landin 1982; [without year] a; b; Martinsson 1984b; c; 1986; Martinsson-Wallin 1986; Martinsson-Wallin/Wallin 1986; Wallin 1986; Aunér 2012; Gustavsson et al. 2014; Gustavsson 2015; Prata 2018; Ahlström-Arcini/Magnell [without year]).

Frequency and chronology

From the 286 osteologically analysed graves, bear claws or, more accurately, third phalanges, have been identified in 18, or 6.2 % (Fig. 2; Table 1). The number of individual claws in a single mound varies from one to 17 and adds up to a total of 119 individual claws. In addition to previously analysed materials, some cremated remains from mounds have been checked in the collections of the Ålands Museum. This rudimentary survey can in no way replace an osteological analysis, but it has resulted in another 12 burials with bear claws. One of the "new" sites is the cemetery of Knösbuskarna in Svartsmara, Finström. Here, a total of 20 mounds were excavated in the 1930s and 1940s, and bear claws were identified from three mounds (1, 26, 32), i. e. 15 %. The other nine graves with bear claws are scattered over mainland Åland and originate from archaeological investigations in which single mounds from cemeteries have been excavated to give way for other land use. The number of burials with bear claws thus adds up to at least 30, and they were found in at least 16 different cemeteries across mainland Åland. Systematic osteological analyses of bones in the museum collections, carried out in the future, will undoubtedly result in more finds.

The burials with bear claws have been dated using artefact typology, coins, horizontal stratigraphy on cemeteries, and, in only two cases, radiocarbon dating. For this compilation, a revised typological dating of the artefacts has also been made. However, in many cases it is only possible to relate the burials to the Late Iron Age (550–1100 CE).

At least 23 of the 30 burials can be dated to a specific chronological period or to an even more narrow timespan (Table 2). In comparison with studies of Swedish and Norwegian findings, it is surprising that not a single bear claw burial has been dated to the Early Iron Age (Petré 1980; Grimm 2013; Lindholm/Ljungkvist 2016). Further, only a probably female burial, in mound 6, from Mellangårds in Rangsby (Sa 23.7), and a male burial in Sålis (Sa 29.1) date to the Early/Mid-Merovingian Period (*c.* 550–700 CE). However, as many as 21 burials date to the Viking Age (800–1100 CE). Of the remaining undated burials with bear claws, two from Kvarnbacken in Saltvik (Sa 2.4) are located in a part of the cemetery with only Viking Age burials. They are therefore assumed to be of the same date, bringing the Viking Age bear claw burials to 23 (Wallin 1986; Gustavsson et al. 2014). Some of the mounds with bear claws contain very few objects, and it might be argued that these burials relate to the final part of the Viking Age, when Christian funerary customs started to influence the burial rites.

There also appears to be a substantial difference in how the Viking Age bear claw burials are dated within this period. Only mound 19 from Hästhägnan in Syllöda (Sa 28.6), also a probably female burial, dates to the 9th century, and, among the remainder, as many as 15 date to the 10th century. This pattern is highly interesting as it deviates considerably from the general pattern in mainland Sweden, Gotland, and Norway (see below). Burying bear claws in Ålandic graves was obviously a tradition that covered several centuries, but the intensity in one specific late phase is rather special from a Scandinavian point of view. The peak frequency of claws in burials appears 300–500 years later than on Gotland as well as in middle and northern Sweden and in Norway (see various articles, this volume).

Sex assessment

Conducting sex assessments of the cremated bone assemblages has only been possible in a few cases due to their highly fragmented state. In nine out of ten burials, the deceased has been interpreted as male, and one has been classified as a "probable female". This indicates a strong connection between bear claws and male burials. An assessment based on the artefacts gives a slightly different picture with seven female burials and an additional six males. In total, there are eight female and 15 male burials, in other words about one third women and two thirds men. Interestingly, two of the three oldest burials (see above) are those of women, which may point to a stronger dominance of male graves later in the Viking Age.

Bear claws in a material context

As mentioned previously, the chronology of the Ålandic burials that include bear claws differs from the general pattern in Scandinavia. As the material is based upon only 30 burials, it is difficult to make any reliable interpretations about burial practice and bear claws. The fact that all burials are cremations means that many of the artefacts are fragmented or in other ways damaged, not only from the cremation, but also from post-cremation burial practices as well as from later land use.

Bear claws have been found with some rare objects primarily among the dominating 10th-century burials (see above). The finds include coins, weights, dress ornaments, weaponry, and beads. Among a total of 113 burial mounds from Långängsbacken, four weights were found in three different burials (39, 43, 78), all of which have also yielded claws. Mound 78 also contained one out of two Arabic silver coins recorded from the burial ground. At Knösbuskarna, mound 32 contained two Arabic coins, minted around 900 and 920 CE, respectively, out of a total of four from the cemetery (Dreijer 1945). Out of 140 burials at Kvarnbacken, a single possible weight was found in mound 80, together with bear claws and the tip of a sword. Another four burials in this study contain objects relating to a warlike sphere. Shield rim mounts were found in Kvarnbacken 94, a complete shield boss in Godby 1, a spearhead in mound 42 at Långängsbacken, and a scabbard chape in mound 78.

Small whetstones with a hole for suspension were found in four mounds. Three stones, from Strömshagen, Finström (Fi 20.2), and Ödkarby, Saltvik (Sa 35.6), are made of banded slate with cross sections of 12 x 15 and 12 x 12 mm, respectively. Some whetstones of this type have been interpreted as functioning as touchstones for testing the quality of precious metals, as streaks of metals have been found on the surface (Ježek 2013). It should also be emphasised that the deposition of this type of stone is a phenomenon that is typical for the 10th century, which is when we also have the most graves with bear claws on Åland.

A characteristic type of object with eastern provenance, or rather a sign of eastern influence in the male dress fashion, is represented by buttons used for either caftans or purses. These have been found in five mounds with bear claws, and they represent all known bronze buttons of the type found on the Åland islands. This garment has links to the fashionable dress style of the Byzantine empire during the 10th century (Kivikoski 1980, 32; Hedenstierna-Jonson/Holmquist-Olausson 2006; Androshchuk 2013, 220).

That several bear claw burials contain grave goods in the shape of silver coins, weights, caftan buttons, and faceted beads made from carnelian rock not only indicates trade, but also suggests contacts with regions in the east, together with an Ålandic participation in the same networks as neighbouring regions (see below). In addition, there are so far no existing holistic studies on social structures on Viking Age Åland, but the find combinations mentioned above also indicate some relation between bear claws and, from a material sense, leading persons in households or communities, irrespective of gender.

Many of the Ålandic burials have, as mentioned above, a fragmentary content that is largely a consequence of the cremation burial rite. Despite this, it is possible to see strong parallels to late burials at Birka, such as graves 581, 644, or 944, and other rich burials containing weaponry and foreign imports/influences in the Mälaren region, such as Valsgärde 12 and 15, and Vendel IX (Stolpe/Arne 1912; Arbman 1940; Schönbäck/Thunmark-Nylén 2002). These 10th-century burials contain a typical assemblage of objects consisting of, for example, full sets of weapons, including the characteristic 10th-century battle knives and axes, the slate whetstones with holes for suspension, and the dress accessories that relate to areas further to the east in the shape of the caftan buttons, purses, and silk objects. In addition to the eastern connections, there are also other Ålandic objects with close parallels to the Birka material. One peculiar find is a pair of very unusual bronze hooks (for garters) found in grave Birka 905, with almost exact parallels in mound 78 at Långängsbacken (Kivikoski 1980, pl. 11).

The female burials on Åland are not characterised by the same varied composition of finds with eastern connections as their middle Swedish counterparts. But at least two of them contain the characteristic mid-10th-century sets of carnelian and rock crystal beads, frequently found in contemporary middle Swedish burials. There is strong evidence that the Ålanders participated in a broad material symbolic language, but the bear claws from the 10th century indicate an Ålandic narrative in which bears appear more frequently in comparison with Birka. From the more than 1,000 excavated burials on Björkö Island, bear claws have only been recorded in six graves according to the Swedish History Museum's digital bone catalogue. This means less than 1 % of the excavated graves from the island. All these burials are cremations, of which three can be securely dated to the 10th century (graves 93, 148, 676). For source-critical reasons, e.g. the few osteologically analysed cremation assemblages and the often poor preservation of inhumations, more bear claws may be hidden among the cremated bones in the collection. But the number of burials with bear claws from Birka is not assumed to be substantially larger. In comparison, it seems that the bear, represented by the claws, was present in a far stronger way on Åland during the 10th century.

Most of the bear claws in this study are assumed to represent bear hides cremated with the deceased on the pyre. In addition to the claws, a bear tooth pendant in bronze (Åm 10:87) of Finnish type has been found in mound 7 at the Sjöhagen cemetery in Hammarland (Ha 22.11), a place where mound 3 also includes bear claws. In mound 7, a number of beads of different types, including faceted carnelian ones, as well as a bronze chain classified as a Swedish type were also found. The beads date the burial to the Late Viking Age. The bear tooth pendant can be assumed to have a different symbolic meaning than the furs on the pyre; as a pendant, it was intended to be worn and to communicate something about the wearer within its society (Asplund 2005; Kivisalo 2008). In recent times, the link between bronze pendants and bear claws has been strongly disputed (see Mannermaa et al., this volume). Whatever link these pendants had to bears, we can conclude that the evidential value is stronger for the actual bear claws.

Bear tooth pendants from the Finnish mainland are more often found to be made of bronze instead of actual bear teeth (cf. Mannermaa et al., this volume). Pendants are exclusively known from female burials dating from the Viking Age and up to the late 12th century (Asplund 2005; Kivisalo 2008, 264). A future osteological analysis of the Hammarland burial may shed light on whether this is true for the Åland Islands as well. The number and types of beads from mound 7 suggest the burial is that of a female, and the lack of beads in mound 3 might indicate a male burial, even if this method does not replace an osteological analysis.

Differences between Åland and the eastern and western mainlands

For comparison with the situation in the west, a study of cremation burials in the Stockholm area can be used (Sigvallius 1994). In this study, based on 488 cremation burials, bears are represented in 29 graves, or 5.9 % of the total (Sigvallius 1994, 74), a frequency close to the situation on the Åland Islands. The chronology, however, differs in comparison to eastern Middle Sweden, where bear claws turn up primarily in the Migration and Early Merovingian Periods, becoming less common during the 8th century and onwards (Sigvallius 1994, 74, 147; Lindholm/Ljungkvist 2016). As shown above, the claws turn up primarily in Late Viking Age burials on Åland. The sex distribution in the Ålandic burials that include bear claws is strongly linked to men, which also differs from Sigvallius' Middle Swedish material, where 14 sex-assessed burials are distributed between six men and eight women (Sigvallius 1994, 75).

In Finland, chronology, frequency and sex distribution are more difficult to determine, as most of the bear claws originate from collective so-called "cremation cemeteries under level ground", which were in use mainly from *c.* 600–1000 CE (Mannermaa et al., this volume). In these cemeteries, the individual burials are very difficult, if not impossible, to identify. Two ^{14}C-datings of bear claws from such cemeteries have given results pointing to the Migration Period and the Viking Age, respectively (Kirkinen 2017; Mäntylä-Asplund/Storå 2010), which only shows that bear hides were included in the funerals throughout the time period.

As mentioned at the beginning of this paper, the evidence of bears is not exclusively related to Late Iron Age burials but also to Kastelholm, the only medieval castle on the Åland archipelago (Table 3). The construction date of the castle has been debated through the years and has been set between the 12th to 15th centuries. It most likely existed in the later part of the 14th century, as it is mentioned in the will of Bo Johansson Grip in 1388. The castle was in use until the end of the 18th century, when fires left it in ruins and most of the administration was moved to Turku (cf. CARLSSON 1993; KIVIKERO 2020).

Excavations have been carried out since the late 19th century and, all in all, over five tons of animal bones have been collected, though only a small portion has so far been analysed (KIVIKERO 2020, 25–26). During the 1980s and 1990s, archaeologist and osteologist Ronnie Carlsson supervised the packing and cataloguing of the bones from the excavations. Lists of observed species in each context were noted, and general comments were attached to these lists. The lists are to be considered as indicative only and do not replace a full osteological analysis, nor can it be assumed that each individual bag has been carefully checked. However, six bear bones are noted in the lists.

The identified bear bones include four claws, a single tooth, and a mandible. The single tooth was found together with one of the claws. Three of the four claws probably represent furs, while the fourth has a drilled hole and was probably worn as a pendant. The latter find is a relatively large claw from the front limb, probably a personal object that could, for example, represent the memory of a bear hunt or be an object of folk religion.

The mandible is perhaps the most interesting specimen of them all (Fig. 3). It is a right-hand mandible with the canine and second molar still in place and some severe pathological features. At some point it was broken between molars 1 and 2; it has healed in a slightly bent shape. On the buccal side of the mandible, an 8 x 5 mm cloaca drained the infected bone; it is connected to a 5 x 2 mm opening just behind the alveola of the posterior root of the first molar. The alveola has some build-up of secondary bony tissue, but the tooth seems to have been lost post mortem, as is the case for all the other missing teeth.

The second molar is only worn through the enamel to the dentine at a 1 x 1 mm patch on the anterior lingual cusp, and the apical aperture of the canine is wide open. This individual was young, less than six years of age (GUSKOV 2014), but can still be considered as fully grown. The canine shows an enamel hypoplasia 18 mm below the tip that can perhaps be related to the stress caused by the breaking of the mandible. The canine grows from *c.* 8 to 20 mm in the second year of life (ZAVATSKY 1976), and the trauma may have occurred in the second autumn or winter. Some thin cut-marks below the alveolae of the premolars indicate that the individual was skinned.

The DNA of the bear mandible has been analysed by Erik Ersmark from the Museum of Natural History in Stockholm, and this individual is from the same population as a Finnish bear from the 19th century in the collections of the same museum. The genome differs from the northern Swedish bear population from that time (pers. comm. E. Ersmark, May 2020). The genetics might probably also indicate a Russian origin.

All Kastelholm bear bones date to the timespan 1550–1700 CE. By this time wild animals, mainly deer, were kept in the vicinity of the castle in a kind of a simple menagerie, and the whole of Åland was declared royal hunting grounds. A number of reindeer from Norrland, Sweden, was also kept at the menagerie but did not reproduce at the site (STORÅ 2009, 117). In this period, court life at the castle reached its most extravagant phase, and a dancing bear or just a living bear used for showing off would fit this setting. The bear with the broken jaw may have been captured during its second winter, injuring the jaw in the process, and kept in captivity until reaching sexual maturity at five to six years of age. The infection of the mandible would have caused a bad temper, and with growing strength the individual would have been harder to contain and control, something that might have led to its ultimate death.

The other bear bones from the castle probably represent a continuation of the Late Iron Age fur trade, which was in some degree coordinated from the castle. The written records tell us about lynx and seal hunters, employed to conduct local hunting on Åland (Nordberg 1939a; Kivikero 2020, 72), but the bear hides come from areas beyond the islands. The court bear suggests that the catchment area in this case was Finland.

Clay paws

Beyond the actual remains of bears, there is a characteristic and peculiar type of artefact from the Åland Islands in the shape of the so-called clay paws (Fig. 4). These items, made of untempered clay, are found only in cremation burials, where they were probably fired on the cremation pyre. A few paws have been estimated to have been fired twice (Kivikoski 1980, 34). These objects show great individual differences, yet are an unmistakable type of artefact, ranging in size from a few centimeters up to *c.* 13 cm in length.

The clay paws are found in cremation burials that range in date from the 7th century up until the transition to Christianity, which ended the practice of cremation. According to Ilves (2019, tab. 2), clay paws have been found in 113 of 952 excavated mounds on the islands. The quantification of excavated mounds prior to the creation of the Ålands Museum in the 1930s is, however, complicated for a number of reasons, such as lack of preserved documentation from the 19th century. The Ålandic clay paws have an almost identical equivalent in the Jaroslavl' area in present-day Russia, dating from the 9th to 11th centuries. In this area, the practice was also abandoned during the transition to Christianity. Such objects are missing from mainland Finland and Sweden; only one specimen has been found in Södermanland (Callmer 1994; Ilves 2019).

Over the years, the clay paws have primarily been related to either bear or beaver. Alternative species, such as humans and dogs, have been suggested but not been given wider acceptance. Both beaver and bear are missing from the faunal history of the Åland Islands, but bear is much more strongly represented in Germanic, Finno-Karelian, and Sami mythology than the beaver, which generally is of marginal importance (Frog 2014, 385–395). Seals – as a source of hides, food, and seal oil – have been of great importance to the Ålandic economy from the Mesolithic to the 20th century, and they have been proposed as an alternative to bear or beaver (Gottberg 2018; Ilves 2019). For more on clay paws, see Callmer (1994), Tarsala (1998), Frog (2014), and, most recently, Ilves (2019).

A highly relevant question is, of course, whether there is a connection between clay paws and bear claws in burials. Of the 30 graves with bear claws in this study, clay paws are included in five, and a possible fragment has been found in yet another. There is, in other words, no obvious connection. Based on the burials in Långängsbacken and Kvarnbacken, it has previously been stated that clay paws and bear claws are not present in the same burials (Gustavsson et al. 2014, 181). But this study, which covers a bigger sample from a larger geographical setting, shows that this is not the case. Callmer (1994) has noted that the cemetery of Knösbuskarna in Svartsmara, Finström, has a high proportion of clay paws. The same cemetery, as noted above, also has a high proportion of bear claws. Two of the three burials with bear claws also contain clay paws.

The clay paws, and other objects in several of the burials with bear claws, have parallels in the Magyar/Meryan area in central Russia, where sites with clay paws, such as Petroskoe in the Jaroslavl' area, reveal strong Scandinavian connections in the Viking Age (Callmer 1994, 32–39). This is an area with Fenno-Ugric speaking populations, whose contacts with regions in Finland can be traced back hundreds of years before the Viking Age, for example in the shape of the Permian/Nevolino belts

(Goldina/Goldina 2018). These people had direct access to vast hunting grounds for various fur bearing animals. In Fennoscandia, bears have a greater magic aura than most animals. The ritualised performance of *karhunpeijaiset* (bear wakes) in order to reincarnate the killed bear was performed until at least the 17th century among the Finns (Sarmela/Poom 1982). Part of that magic and the idea of reincarnation was perhaps transferred into a representation in clay. Wolves, ravens, and eagles played a major role in old Norse mythology and poetry, in contrast to bears (cf. Sundqvist, this volume), although the latter and lynx are represented in the form of claws or furs. Bears may not be frequently represented in battle descriptions among the scavengers of corpses (Jesch 2002) or in the iconography of Scandinavia as a whole, but, on the other hand, it is interesting to explore their role later on, especially on Åland and in Middle Sweden which, culturally and geographically, were positioned closer to both the Sami and Karelo-Finnish spheres. It is also primarily in the middle Swedish and Finnish regions that we find depictions of bears on spear sockets, sword pommels, and belt mounts (see Oehrl, this volume; Lamm/Rundkvist 2005). In other words, regional religious traits of this area did not perhaps leave such a large imprint on the old Norse literature as did those of the southern and western Scandinavian regions. Åland is right in the middle of the area, and is a meeting point between the Finnish and Scandinavian traditions. It is impossible to decide, based on morphology, which particular species the clay paws represent, but the bear is a good candidate.

Conclusion

The remains of bears on the Åland Islands can be interpreted in different ways. Were the Ålanders themselves engaged in bear hunting expeditions in areas away from the islands to acquire furs, or were they involved in the trade as middlemen? From medieval sources we know that the Ålanders bought goods elsewhere to sell them in other markets, and also transported and sold goods for third parties (Friberg 1983, 70–71; Sandström 1990). The Åland Islands' subsistence is, to some extent, dependent on the trade in foodstuffs to diversify the mainly marine resources. This form of trade is in itself often not lucrative, but more of a zero-sum economy. Several of the burials that include bear claws also contain artefacts which point to contacts with areas further to the east, and some also contain weaponry. Weapons and bear hides are often associated with the berserks of the Norse saga literature, although both may on their own represent wealth and not necessarily indicate a warrior elite (see various contributions, this volume). The clay paws and other foreign objects connected Åland with different regions that were hunting grounds for a variety of fur-bearing animals, meaning that the Ålanders had contacts with several regions that held bear populations, or had close contacts with such areas. These included present-day Sweden, Finland, Russia, and the Baltic countries. The chronology of burials with bear claws on the Åland Islands differs from the general pattern on the Swedish mainland and Gotland, and they may represent a partially different network. The bear hides from the islands' Late Viking Age are part of a package of eastern objects and clothing that appear to represent a special group with regular interaction with people living along the eastern European rivers.

There are indications that bear symbolism is more strongly represented in certain cemeteries or local communities on the archipelago, such as at Knösbuskarna in Svartsmara, Finström, or at Sjöhagen in Hammarland. In the latter case, a bear hide and a bear tooth pendant were found in two adjacent burials, those of a man and a woman, respectively.

The medieval bear bones at Kastelholm also indicate that the region continued to be involved in fur-trading and -handling after the Iron Age. The findings of bear claws indicate the importation of furs, and the court, which was one of the northernmost in Europe, may also have been entertained by one of the dancing bears known from other late medieval and Renaissance courts on the European continent. The bear from Kastelholm has its origin on the Finnish mainland, an area where hunting for furs has been of great importance to the local economy.

Although the archipelago has never been inhabited by bears, this creature still played a significant role in the material culture. As Frog (2014) and Kirkinen (2017) have discussed, the bear cult is of a different nature in the Germanic versus the Sami and Fenno-Karelian areas, and Åland is an interesting meeting point between these cultural spheres (cf. also various contributions, this volume). The material culture on the Åland Islands is characterised by a strong mix of artefact types and shapes, into which the islanders selectively incorporated parts from mostly mainland Finland and Sweden, but also areas further away. Few communities in the Baltic area have chosen to mix their material culture from so many diverse areas as the Ålandic, and, if the material culture is a reflection of oral stories, religion and myths, these islands had a very interesting culture indeed, which probably appeared both familiar and confusing to visitors from the eastern and western mainlands. For an area without bears, the animal is exceptionally well represented in the material culture. Perhaps the lack of living bears has only added to their mythical value.

Acknowledgements

Rudolf Gustavsson's work was funded by Societas Archaeologica Upsaliensis' forskningsfond, and John Ljungkvist's work on this paper was undertaken within "The Viking Phenomenon" project at Uppsala University, funded by the Swedish Research Council (grant 2015-00466).

Bibliography

Ahlström-Arcini/Magnell (without year): C. Ahlström-Arcini/O. Magnell, Osteologisk analys av brända ben från Rangsby, Saltvik, Åland. Unpubl. report, Arkeologerna, SHMM (Mariehamn, without year).

Andersson (manuscript): J. Andersson, Rapport, arkeologisk undersökning av skärvstenshög i Finby, Su 6.18. Unpubl. report, manuscript, Ålands landskapsstyrelse, museibyrån (Mariehamn, without year).

Androshchuk 2013: F. Androshchuk, Vikings in the East. Essays on contacts along the road to Byzantium. Acta Universitatis Upsaliensis. Studia Byzantina Upsaliensa 14 (Uppsala 2013).

Arbman 1940: H. Arbman, Birka 1. Die Gräber 1–2. 1: Text. 2: Tafeln (Stockholm 1940).

Asplund 2005: H. Asplund, The bear and the female. Bear-tooth pendants in Late Iron Age Finland. In: S. Mäntylä (ed.), Rituals and relations. Studies on society and material culture of the Baltic Finns. Suomalaisen Tiedeakatemian toimituksia. Humaniora 336 (Helsinki 2005) 13–30.

Aunér 2012: M. Aunér, Gravmönster under yngre järnålder – en jämförelse mellan åländska och svenska gravfält. Unpubl. Master's thesis, Högskolan på Gotland (Visby 2012).

Bergstrand 1852: C. E. Bergstrand, Ålands däggdjur, foglar, amphibier och fiskar (Västerås 1852).

Callmer 1994: J. Callmer, The Clay Paw Burial Rite of the Åland Islands and Central Russia: A Symbol in Action. Current Swedish Archaeology 2, 1994, 13–46.

Carlsson 1988: R. Carlsson, KS 5, Arkeologisk undersökning, öster om Norra längan och Östra längans norra del. Kastelholms slott, arkeologiska undersökningar 1982 och 1983; KS 1 – KS 14. Rapport museibyrån 1988(1), 1988, 141–267.

Carlsson 1991: R. Carlsson, KS 33/34, Arkeologisk undersökning, söder om slottet. Kastelholms slott, arkeologiska undersökningar 1985 och 1989; KS 30 – KS 52. Rapport museibyrån, Kastelholm 1991(1), 1991, 61–192.

Carlsson 1993: R. Carlsson, Kastelholms slott: dateringen av Kastelholms slott. Museibyrån 1993,1 (Mariehamn 1993).

Dreijer 1935: M. Dreijer, Berättelse över arkeologiska undersökningar i Finström, Svartsmara, i oktober 1935 (Fi 18.1 / ÅM50). Unpubl. report, Ålands Museum (Mariehamn 1935).

Dreijer 1936: M. Dreijer, Berättelse över grävningar som landskapsarkeolog M. Dreijer utfört I Svartsmara by av Finströms socken år 1935 (Fi 18.1 / ÅM45). Unpubl. report, Ålands Museum (Mariehamn 1936).

Dreijer 1937: M. Dreijer, Berättelse över arkeologiska grävningar som av landskapsarkeologen utförts å Knösbuskarna benämnda gravfält i Svartsmara by av Finströms socken år 1937 (Fi 18.1 / ÅM70). Unpubl. report, Ålands Museum (Mariehamn 1937).

Dreijer 1945: M. Dreijer, En åländsk fågelnål. Finska fornminnesföreningens tidskrift XLV (Helsingfors 1945) 204–210.

Dreijer 1948: M. Dreijer, Berättelse över arkeologiska grävningar å gravfältet i Finströms gårds prästbol år 1948 (Fi 13.3 / ÅM169). Unpubl. report, Ålands Museum (Mariehamn 1948).

Dreijer 1951: M. Dreijer, Berättelse över utgrävningar på gravfältet på Mellangårds tomt i Rangsby by av Saltviks socken år 1950 (Sa 23.7 / ÅM187). Unpubl. report, Ålands Museum (Mariehamn 1951).

Drejer 1953: M. Dreijer, Arkeologiskt nytt från Åland 1952. Åländsk Odling 1952 (Mariehamn 1953) 146–148.

Dreijer 1957: M. Dreijer, Berättelse över en arkeologisk undersökning av gravfältet Strömhagen, Torrbolsta, år 1957 (Fi 20.2 / ÅM306). Unpubl. report, Ålands Museum (Mariehamn 1957).

Dreijer 1966: M. Dreijer, Rapport över utgrävningar på gravfältet mellan Ollas och Mickels ekonomibyggnader i Ödkarby av Saltviks socken 1966 (Sa 35.6 / ÅM396). Unpubl. report, Ålands Museum (Mariehamn 1966).

Friberg 1983: N. Friberg, Stockholm i bottniska farvatten: Stockholms bottniska handelsfält under senmedeltiden och Gustav Vasa: en historisk-geografisk studie (Stockholm 1983).

Frog 2014: Frog, From mythology to identity and imaginal experience: an exploratory approach to the symbolic matrix in Viking Age Åland. In: J. Ahola/Frog/J. Lucenius (eds.), The Viking Age in Åland. Insights into Identity and Remnants of Culture. Annales Academiae Scientiarum Fennicae. Humaniora 372 (Helsinki 2014) 349–414.

Goldina/Goldina 2018: E. Goldina/R. Goldina, On North-Western contacts of Perm Finns in VII-VIII. centuries. Estonian Journal of Archaeology 22, 2018, 163–180.

Gottberg 2018: V. Gottberg, De åländska lertassarna och sällabbarna. Unpubl. Students' thesis, Uppsala universitet (Uppsala 2018).

Grimm 2013: O. Grimm, Bear-skins in northern European burials and some remarks on other bear-related furnishings in the north and middle of Europe in the 1st millennium AD. In: O. Grimm/U. Schmölke (eds.), Hunting in northern Europe until 1500 AD – Old traditions and regional developments, continental sources and continental influences. Papers presented at a workshop organised by the Centre for Baltic and Scandinavian Archaeology, Schleswig, June 16th and 17th, 2011. Schriften des Archäologischen Landesmuseums, Ergänzungsreihe 7 (Schleswig 2013) 277–296.

Guskov 2014: V. Yu. Guskov, Skull-Based Method of Age Determination for the Brown Bear Ursus arctos Linnaeus, 1758. Achievements in the Life Sciences 8(2), 2014, 137–141: doi.org/10.1016/j.als.2015.04.002.

Gustavsson 2008: R. Gustavsson, Personal notes on artefacts and animals species found in bronze age and early Iron Age cairns on Åland. Unpubl. report, Ålands landskapsstyrelse, museibyrån (Mariehamn 2008).

Gustavsson 2015: R. Gustavsson, Osteologisk analys. Osteologisk kontroll av brända ben från en brandgrav i Jomala Överby, Åland. Åm 769, Jo 37.7, år 2014. Unpubl. report, SAU rapport 2015:30 (Mariehamn 2015).

Gustavsson et al. 2014: R. Gustavsson/J.-E. Tomtlund/J. Kennebjörk/J. Storå, Identities in transition in Viking Age Åland. In: J. Ahola/Frog/J. Lucenius (eds.), The Viking Age in Åland. Insights into Identity and Remnants of Culture. Annales Academiae Scientiarum Fennicae. Humaniora 372 (Helsinki 2014) 159–186.

Hedenstierna-Jonson/Holmquist Olausson 2006: C. Hedenstierna-Jonson/L. Holmquist Olausson, The Oriental Mounts from Birka's Garrison. Antikvariskt arkiv 81 (Stockholm 2006).

af Hellström 1945: O. af Hellström, Berättelse över arkeologiska undersökningar i Hammarland, Postad by, Österängsbacken (Ha 17.3 / ÅM125). Unpubl. report, Ålands museum (Mariehamn1945).

Ilves 2019: K. Ilves, Seals as a ritual signifier: re-evaluating the Ålandic clay paw burial rite. Fennoscandia Archaeologica XXXVI, 2019, 32–52.

Jaatinen et al. 1989: S. Jaatinen/A. Peltonen/J. Westerholm, Ålands kulturlandskap – 1700-talet (Helsingfors 1989).

Jesch 2002: J. Jesch, Eagles, Ravens and Wolves: Beasts of Battle, Symbols of Victory and Death. In J. Jesch (ed.), The Scandinavians from the Vendel Period to the Tenth Century: An Ethnographic Perspective. Studies in Historical Archaeoethnology 5 (Woodbridge 2002) 251–271.

Ježek 2013: M. Ježek, Touchstones of archaeology. Journal of Anthropological Archaeology 32, 2013, 713–731: doi.org/10.1016/j.jaa.2013.04.004.

Karivieri 1991: A. Karivieri, KS 47, Arkeologisk undersökning, nordöst om slottet. Kastelholms slott, arkeologiska undersökningar 1985 och 1989; KS 30 – KS 52. Rapport museibyrån, Kastelholm 1991(1), 1991, 423–472.

Kirkinen 2017: T. Kirkinen, "Burning pelts" – Brown bear skins in the Iron age and early medieval (1–1300 AD) burials in south-eastern Fennoscandia. Estonian Journal of Archaeology 21(3), 2017: 10.3176/arch.2017.1.01.

Kivikero 2020: H. Kivikero, The Economy of Food: Tracing food production and consumption in the Castles of Kastelholm and Raseborg from the 14th to the 16th centuries. University of Helsinki (Helsinki 2020).

Kivikero et al. 2020: H. Kivikero/R. Gustavsson/J. Storå, Sealing Economy: Exploring seals as resources in the Åland Islands ca. 1100–1700 CE through zooarchaeology and account books. Journal of Archaeological Science: Reports 29, 2020: doi.org/10.1016/j. jasrep.2019.102011.

Kivikoski 1933: E. Kivikoski, Undersökning av gravar i Hammarland sn, Torp (Östanträsk) by, Norrgårds hn, Sjöhagen sommaren 1933 (Ha 22.11 / ÅM10). Unpubl. report, Ålands Museum (Mariehamn 1933).

Kivikoski 1934: E. Kivikoski, Undersökning av två gravhögar i Hammarlands socken, Mörby by, på Julius Mattsons (Engströms) mark. Högarna tillhöra det stora gravfältet på Södergårds (Ha 15.1 / ÅM40). Unpubl. report, Ålands Museum (Mariehamn 1934).

Kivikoski 1959: E. Kivikoski, En kedjeuppsättning från Ålands Vendeltid. Åländsk Odling 1959, 3–14.

Kivikoski 1962: E. Kivikoski, Berättelse över Ella Kivikoskis grävning på Stehagen, Kastelholm, Sund d. 26.6.–6.7.

1961, vid vilken högarna nris 3, 5, 6, 8, 10 och 15 undersöktes (Su 12.14 / ÅM 356). Unpubl. report, Ålands Museum (Mariehamn 1962).
Kivikoski 1963: E. Kivikoski, Kvarnbacken: ein Gräberfeld der jüngeren Eisenzeit auf Åland (Helsinki 1963).
Kivikoski 1980: E. Kivikoski, Långängsbacken: ett gravfält från yngre järnåldern på Åland (Helsingfors 1980).
Kivisalo 2008: N. Kivisalo, The Late Iron Age Bear-Tooth Pendants in Finland: Symbolic Mediators between Women, Bears, and Wilderness? Temenos 44, 2008, 263–291: doi.org/10.33356/temenos.4591

Lamm/Rundkvist 2005: J. P. Lamm/M. Rundkvist, Björnen i ägget – en vapengrav i Eds socken, Uppland och vendeltidens vapen med djurfiguriner. Fornvännen 100, 2005, 101–113.
Landin 1982: M. Landin, Osteologisk analys från Långängsbacken (Su 12.7). Unpubl. report, Ålands landskapsstyrelse, museibyrån (Mariehamn 1982).
Landin (without year) a: M. Landin, Osteologiskt analys PM (Sa 14.1 A123). Unpubl. report, Ålands landskapsstyrelse, museibyrån (Mariehamn, without year).
Landin (without year) b: M. Landin, Osteologiskt analys PM (Sa 14.1 A125). Unpubl. report, Ålands landskapsstyrelse, museibyrån (Mariehamn, without year).
Landin/Rosborg 1984: M. Landin/B. Rosborg, Under Godbyvägen: rapport från en arkeologisk undersökning av ett höggravfält/boplatsområde (Mariehamn 1984).
Larsson 1999: Å. Larsson, Osteologisk analys av material från Finby 6.18. Unpubl. report, Ålands landskapsstyrelse, museibyrån (Mariehamn 1999).
Lindholm/Ljungkvist 2016: K.-J. Lindholm/J. Ljungkvist, The Bear in the Grave: Exploitation of Top Predator and Herbivore Resources in First Millennium Sweden – First Trends from a Long-Term Research Project. European Journal of Archaeology 19(1), 2016, 3–27: doi.org/10.1179/1461957115Y.0000000010.

Mäntylä-Asplund/Storå 2010: S. Mäntylä-Asplund/J. Storå, On the archaeology and osteology of the Rikala cremation cemetery in Salo, SW Finland. Fennica Archaeologica XXVII, 2010, 53–68.
Martinsson 1984a: H. Martinsson, Rapport. Arkeologisk undersökning av yngre järnåldersgravfält, Saltvik; Sålis 29.1.1984. Unpubl. report, Ålands landskapsstyrelse, museibyrån (Mariehamn 1984).
Martinsson 1984b: H. Martinsson, Osteologisk analys av bränt benmaterial från Saltvik sn. Sålis 29.1. Unpubl. report, Ålands landskapsstyrelse, museibyrån (Mariehamn 1984).
Martinsson 1984c: H. Martinsson, Osteologisk rapport. Sa Borgboda 1984 (Sa 3.3, 3.4, 3.5). Unpubl. report, Ålands landskapsstyrelse, museibyrån (Mariehamn 1984).
Martinsson 1986: H. Martinsson, Osteologisk analys av bränt benmaterial från Saltvik sn. Sålis 29.1.1985. Unpubl. report, Ålands landskapsstyrelse, museibyrån (Mariehamn 1986).
Martinsson-Wallin 1986: H. Martinsson-Wallin, Rapport. Arkeologisk undersökning av en yngre järnålders grav Godby 8.4, Finström sn. (Fi 8.4 / ÅM586). Unpubl. report, Ålands landskapsstyrelse, museibyrån (Mariehamn 1986).
Martinsson-Wallin/Wallin 1986: H. Martinsson-Wallin/P. Wallin, Osteologisk analys. Rapport över brända ben från A1 Sa Lagmansby 18.3, Högtomt. Unpubl. report, Ålands landskapsstyrelse, museibyrån (Mariehamn 1986).

Nordberg 1939a: S. Nordberg, Den Åländska däggdjursfaunans utveckling. Acta Societas pro Flora et Fauna Fennica 62(3), 1939, 1-39.
Nordberg 1939b: S. Nordberg, Några drag ur den Åländska faunans historia. Åländsk Odling 1939(2), 1939, 19–29.
Núñez 1986: M. Núñez, Om bosättningen på Ålandsöarna under stenåldern. Åländsk Odling 1986, 13–28.

Petré 1980: B. Petré, Björnfällen i begravningsritualen – statusobjekt speglande regional skinnhandel? Fornvännen 75, 1980, 5–14.
Prata 2018: S. Prata, med bidrag av R. Gustavsson och E. Sjöling, Osteologisk analys. Brandgravar från yngre järnålder i Geta, Vestergeta by. Unpubl. report, SAU rapport 2018:28O (Mariehamn 2018).

Radloff 1795: F. W. Radloff, Beskrifning öfver Åland (Åbo 1795).
Ramsdahl 1937: C. Ramsdahl, Arkeologiska grävningar verkställda på Åland av t.f. landskapsarkeologen mag. C. Ramsdahl (Fi 13.1 / ÅM 58). Unpubl. report, Ålands Museum (Mariehamn 1937).
Rosborg 1987: B. Rosborg, Rapport. Arkeologisk undersökning på ett yngre järnåldersgravfält. Saltvik 28.6, Syllöda. Unpubl. report, Ålands landskapsstyrelse, museibyrån (Mariehamn 1987).
Rundqvist 2010: M. Rundqvist, Ålands tidiga spännbucklor: Modesmycken från sen vendeltid och tidigvikingatid. Finskt museum 116, 2010, 55–62.

Sandström 1990: Å. Sandström, Mellan Torneå och Amsterdam. En undersökning av Stockholms roll som förmedlare av varor i regional- och utrikeshandel 1600–1650. Stockholms universitet (Stockholm 1990).
Sarmela/Poom 1982: M. Sarmela/R. Poom, Death of the Bear: An Old Finnish Hunting Drama. The Drama Review – TDR 26(3), 1982, 57–66: doi.org/10.2307/1145415.
Schönbäck/Thunmark-Nylén 2002: B. Schönbäck/L. Thunmark-Nylén, De vikingatida båtgravarna vid Valsgärde – relativ kronologi. Fornvännen 97, 2002, 1–8.
Schröder 2016: T. G. S. Schröder, Geta, Västergeta, fornlämning Ge 15.4. Arkeologisk undersökning av fyra gravar från yngre järnålder. Unpubl. report, Ålands landskapsregering, kulturbyrån (Mariehamn 2016).
Sigvallius 1994: B. Sigvallius, Funeral Pyres: Iron Age Cremations in North Spånga. Theses and Papers in Osteology 1 (Stockholm 1994).
Stolpe/Arne 1912: H. Stolpe/T. J. Arne, Graffältet vid Vendel (Stockholm 1912).
Storå 2000: J. Storå, Sealing and animal husbandry in the Ålandic Middle and Late Neolithic. Fennoscandia archaeologica XVI(I), 2000, 57–81.
Storå 2009: N. Storå, Åland som kunglig jaktpark och allmogejaktens villkor. Åländsk Odling 2008/2009, 2009, 104–136.

Storå et al. 2012: J. Storå/J. Hillborg/J. Kennebjörk/R. Lindblad/J. Lindeberg/J. Mosseby/R. Gustavsson, Late Iron Age Subsistence in the Åland Islands in Light of Osteoarchaeology. In: S. Niinimäki/A.-K. Salmi/J.-M. Kuusela/J. Okkonen (eds.), Stones, Bones & Toughts: Festschrift in Honour of Milton Núñez (Oulu 2012) 175–189.

Tarsala 1998: I. Tarsala, Öar i strömmen: Den yngre järnåldern på Åland. Aktuell arkeologi VI. Stockholm archaeological reports 35, 1998, 107–123.

Tomtlund 1999: J.-E. Tomtlund, Åland under vikingatiden: Den första glimten av vikingarnas värld. SFV-Kalendern årgång 113, 1999, 18–31.

Tomtlund 2014: J.-E. Tomtlund, The Viking Age in Åland: An Introduction. In: J. Ahola/Frog/J. Lucenius (eds.), The Viking Age in Åland. Insights into Identity and Remnants of Culture. Annales Academiae Scientiarum Fennicae. Humaniora 372 (Helsinki 2014).

Tuneld 1741: E. Tuneld, Inledning till Geographien öfwer Swerige upstäld af Eric Tuneld (Stockholm 1741).

Vormisto 1981a: T. Vormisto, Osteologisk analys av brända ben från Åland. Unpubl. Master's thesis, Stockholms universitet (Stockholm 1981).

Vormisto 1981b: T. Vormisto, Osteologisk analys. Saltvik, Bertby, Larsas Kvarnbacke 2.4. Gravhög 66 och 140. Unpubl. report, Ålands landskapsstyrelse, museibyrån (Mariehamn 1981).

Vormisto 1982: T. Vormisto, Osteologisk analys. Godby 8.11. 525 och 533. Unpubl. report, Ålands landskapsstyrelse, museibyrån (Mariehamn 1981).

Wallin 1986: P. Wallin, Kvarnbacken. Osteologisk analys. Rapport över brända ben från Larsas Kvarnbacke Bertby 2.4, Saltvik sn, Åland. Unpubl. report, Ålands landskapsstyrelse, museibyrån (Mariehamn 1986).

Zavatsky 1976: B. P. Zavatsky, The Use of the Skull in Age Determination of the Brown Bear. In: Bears, their biology and management, vol. 3. A selection of papers from the third International Conference on Bear Research and Management held at the 54th Annual Meeting of the American Society of Mammalogists, Binghamton, New York, USA and first International Theriological Congress, Moscow, USSR, June 1974. IUCN Publications, New Ser. 40 (Morges 1976) 275–279.

Åqvist 1991: C. Åqvist, KS 42, Arkeologisk undersökning, öster om slottet. Kastelholms slott, arkeologiska undersökningar 1985 och 1989; KS 30 – KS 52. Rapport museibyrån, Kastelholm 1991(1), 1991, 361–394.

ÅSUB 2019: Statistical Yearbook of Åland 2019. Ålands statistik- och utredningsbyrå (Mariehamn 2019).

PhD student Rudolf Gustavsson
Mariehamn
Åland
Finland
rudolf.gustavsson@gmail.com

Associate Prof. John Ljungkvist
Department of Archaeology and Ancient History
Uppsala University
Uppsala
Sweden
john.ljungkvist@arkeologi.uu.se

Fig. 1. The Åland Islands in the Baltic region (map GIS department, ZBSA).

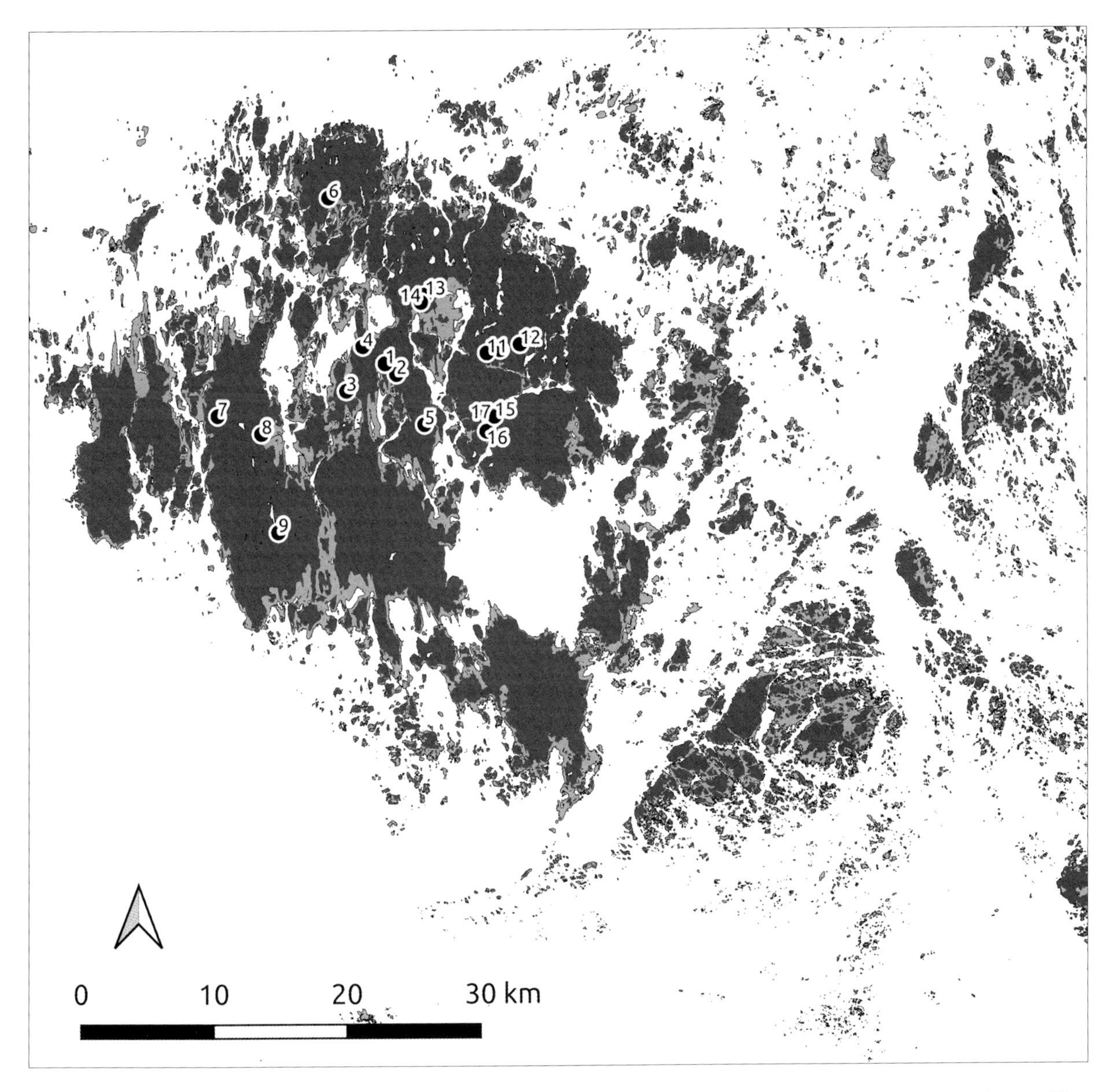

Fig. 2. Sites with burials including bear bones. Land shaded in light grey emerged only after c. *1000. Numbers refer to Table 1 (map R. Gustavsson, using elevation data from the National Land Survey of Finland Topographic Database 9/2016).*

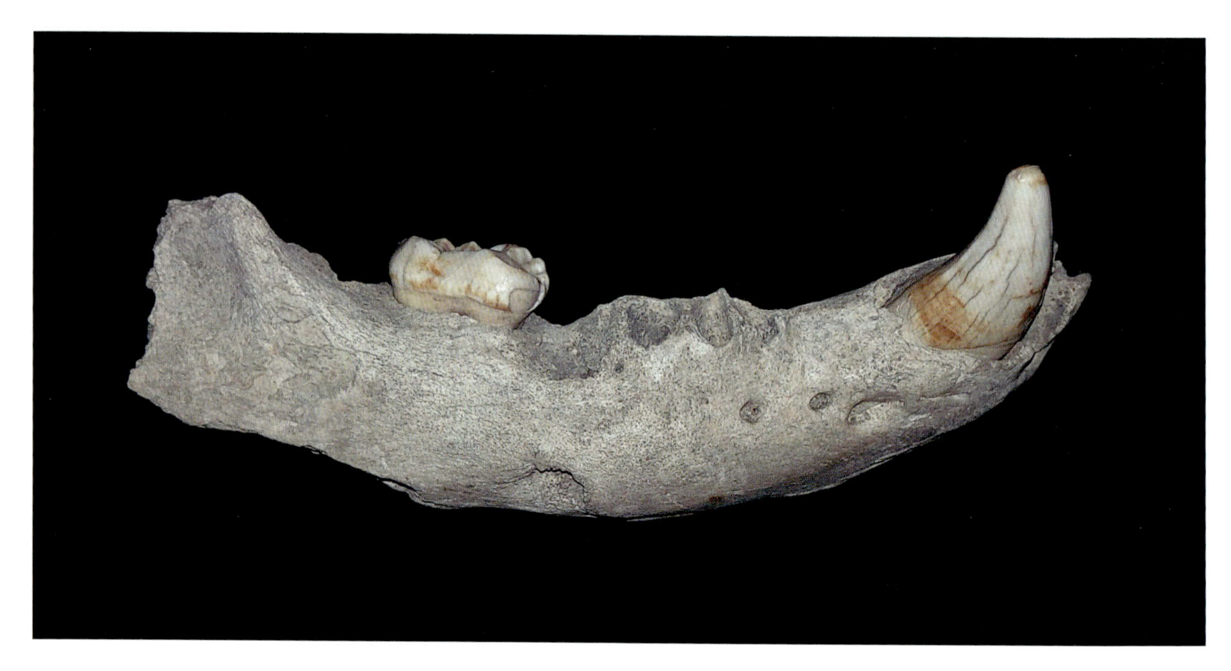

Fig. 3. Bear mandible from the late 16th century. Kastelholm on Åland. Find no. KS33: Bnr 398 in Ålands Museum (photo R. Gustavsson).

Fig. 4. Example of an intact clay paw from the Åland Islands. This paw, find no. ÅM233:2, was discovered in the archaeological excavation of burial mound no. 5 in Tjudnäs, Kastelholm, in 1952; it is dated to the 5th century (DREIJER 1953; photo A. Mendes, Ålands Museum).

Table 1. Burials with bear claws in this study. Map numbers refer to Fig. 2. Sex assessment in parenthesis is based on artefacts. Dates are based on a revised typological dating for this study except for: [1] – based on horizontal stratigraphy (Dreijer 1945), [2] – coins, terminus post quem 920 CE (Dreijer 1945), [3] – ^{14}C-dated, laboratory ID Ua-62161 (Schröder 2016), [4] – ^{14}C-dated, laboratory ID Ua-60672 (Ilves 2019).

Map	Site	Name	Mound	Date	Sex	Clay paw	Find catalog	Report	Osteology
1	Fi 13.1	Östra riåkersbacken	23	900–1000			ÅM 58	Ramsdahl 1937	–
2	Fi 13.3	Kyrkbol	36	–	(F)		ÅM 169	Dreijer 1948	–
3	Fi 18.1	Knösbuskarna	1	900–1000	(F)	x	ÅM 71	Dreijer 1937	–
		Knösbuskarna	26	c. 1000[1]	(?)		ÅM 50	Dreijer 1935	–
		Knösbuskarna	32	920–1000[2]	(M?)	x	ÅM 45	Dreijer 1936	–
4	Fi 20.2	Strömhagen	2	550–1100	(M)	x	ÅM 306	Dreijer 1957	–
5	Fi 8.4	Godby	1	900–950	M	?	ÅM 586	Martinsson-Wallin 1986	Martinsson-Wallin 1986
6	Ge 15.4	Geta	2	990–1025[3]	?		ÅM 775	Schröder 2016	Prata 2018
7	Ha 15.1	Hemängen	4	500–1100	(?)		ÅM 40	Kivikoski 1934	–
8	Ha 17.3	Söderåkersbacken	34	900–1100	(?)		ÅM 125	af Hellström 1945	–
9	Ha 22.11	Sjöhagen	3	800–1100	(M?)		ÅM 10	Kivikoski 1933	–
		Sjöhagen	7	900–950?	(F)		ÅM 10		
10	Sa 2.4	Kvarnbacken	80	800–1100	(M)		ÅM 337	Kivikoski 1963	Wallin 1986
		Kvarnbacken	94	800–1100	(M)		ÅM 345		
		Kvarnbacken	97	900–950	(F?)		ÅM 345		
		Kvarnbacken	136	500–1100	(?)		ÅM 188		
		Kvarnbacken	139	500–1100	(?)		ÅM 345		
11	Sa 23.7	Mellangårds	6	550–700	(F)		ÅM 187	Dreijer 1951	–
12	Sa 28.6	Hästhägnan	19	800–850	(F)	x	ÅM 604	Rosborg 1987	–
13	Sa 29.1	Sålis	5	640–720[4]	M	x	ÅM 560	Martinsson 1984a	Martinsson 1984b
14	Sa 35.6	Ödkarby	16	900–1000	(M?)		ÅM 396	Dreijer 1966	–
15	Su 12.14	Stenhagen	5	–	M		ÅM 356	Kivikoski 1962	Vormisto 1981a

Cont. Tab. 1. ▶

Cont. Tab. 1.

Map	Site	Name	Mound	Date	Sex	Clay paw	Find catalog	Report	Osteology
16	Su 12.7	Långängsbacken	11	550–1100	M		ÅM 362	Kivikoski 1980	Landin 1982
		Långängsbacken	19	900–950	(F)		ÅM 367		
		Långängsbacken	38	900–1000	F?		ÅM 406		
		Långängsbacken	39	800–1100	M		ÅM 404		
		Långängsbacken	42	900–1000	M		ÅM 376		
		Långängsbacken	43	900–1000	M?		ÅM 404		
		Långängsbacken	44	950–1000	M		ÅM 376		
		Långängsbacken	52	800–1100	(F)		ÅM 376		
		Långängsbacken	78	950–1000	M		ÅM 386		

Table 2. Chronology of burials with bear claws.

Phase	Number of burials
Late Iron Age (550–1100)	7
550–700	2
800–1100	5
800–850	1
900–1000	15
Total	30

Table 3. Bear bones from Kastelholm on Åland.

Excavation	Find no.	Grid	Element	Date	Report
KS 5	Bnr 423	AB4, +5,72	Phalanx 3	KS 5, phase VII: 1550–1600	Carlsson 1988
KS 5	Bnr 127	C5, +5,61	Phalanx 3 and tooth	KS 5, phase VIII: 17th century	Carlsson 1988
KS 33	Bnr 398	D2, Pl. 5	Mandibula, dexter	KS 33, phase VIb: 1550–1600	Carlsson 1991
KS 42	Bnr 39	K16, Pl. 2, layer VIII	Phalanx 3 with drilled hole	KS 42, phase 2, demolition layer: 17th century	Åqvist 1991
KS 47	Bnr 66	M10, Pl. 3, layer S6/S8	Phalanx 3	KS 47, phase XVII: late 17th century	Karivieri 1991

The power of the paw. Multi-species perspectives on the bear claw burial tradition in a long-time perspective in South Norway

By Anja Mansrud

Keywords: Bear lore, cremation, bear claws, transhumance, protection, amulet, Early Iron Age Norway

Abstract: This chapter explores the issue of object and animal agency through a contextual analysis of bear claws in Iron Age cremations in South Norway (Sør-Norge). Bear claws were identified in 130 cremations, mainly dated within the Roman and Migration periods (1–550 CE). The presence of bear claws is independent of economic status, age, or gender. They occur in male, female, and mixed cremations, occasionally also with children. Most burials contain only one claw. Rather than representing chiefs, shamans, or warriors as previously assumed, the archaeological evidence suggests that individuals cremated with bear claws were also farmers, herders, and hunters. Drawing on Norwegian folklore and a multi-species perspective, I employ a retrospective approach and investigate relations between humans, livestock and bears in the context of hunting and transhumance, arguing that bear claws were powerful agents, utilised for ritual and safeguarding purposes.

Introduction

People are not to do things with stones or fill them with magic power with the idea of tying them on people or on people's cattle. If a man puts trust in stones for his own health or that of cattle, the penalty is lesser outlawry (Grágás, 39[1]*).*

Do objects have agency? The present chapter aims to explore the issue of object and animal agency through a contextual analysis of bear claws deposited in Iron Age cremations in South Norway.[2] Burials with bear remains have puzzled archaeologists for more than a hundred years (cf. Shetelig 1912, 112, 144–147; Møhl 1977; Petré 1980; Ström 1980; Krüger 1988; Schönfelder 1994; 2011, 95; Mansrud 2004a; b; Frederiksen 2006; Wamers 2009; Grimm 2013; Oehrl 2013; Klokkervoll 2015; Skogstrand 2016; Kirkinen 2017). Early finds of preserved bear skin and pelt in richly furnished Migration-period inhumations in Norway indicated that bodies had been laid to rest on top of bear skins (Shetelig 1912, 112, 144–147; Petersen 1940). Since bear claws are rarely perfo-

1 Translation after Dennis et al. 2006, 39.

2 South Norway refers to the whole of Norway south of Nordland. Southeastern Norway encompasses the counties of Viken, Oslo, Telemark & Vestfold, and Innlandet. Southern Norway (Sørlandet) encompasses the county of Agder, and Western Norway includes the county of Vestland.

Bear and Human: Facets of a Multi-Layered Relationship from Past to Recent Times, with Emphasis on Northern Europe, ed. by Oliver Grimm (Turnhout: Brepols, 2023), pp. 489–532 BREPOLS PUBLISHERS 10.1484.M.TANE-EB.5.134347

Mortuary treatment and regional variation

The mortuary assemblage from South Norway (Appendix 1) represents a variety of landscapes, ranging from the coastal zone to alpine areas, and subsequently also different groups of people with diverse subsistence economies and cultural traditions. Iron Age mortuary customs differed chronologically and between regions (Solberg 2000). Overall, mounds, cairns, and stone settings with successive burials are common in all areas of South Norway during the Roman and Migration periods. Cremation and inhumation often occur in the same mound. In several cases unburnt bear skins were also utilised as a cover for the urn, as a cover for the cremated bones deposited within the urn, and as part of the furnishing of the grave (Appendix 1; cf. also Grimm 2013, 283–284). Many of the graves with bear claws were documented during early excavations. The documentation of these contexts is variable and sometimes lacking (Appendix 1). Some of the graves were not excavated by archaeologists but discovered during farm work and handed in by local farmers, and many burials have further been robbed or disturbed in other ways. These source-critical problems complicate the contextual analysis.

The burials from western and southern Norway are constructed according to strict conventions (Shetelig 1912, 110; see Appendix 1). Stone chambers/cists are common. The cremated remains are usually deposited in a cauldron, a ceramic vessel, or an organic container, more rarely as an ash layer scattered with charcoal and burnt bones. Cauldrons and so-called bucket-shaped pots are frequently used as urns. The burnt assemblages commonly consist of human and animal bones mixed with burnt artefacts which have been collected from a pyre site. Bones and teeth from horses, sheep/cattle and birds are noted in some cases.

The mortuary customs in southeastern Norway appear more diverse (Appendix 1). This part of Norway can be further divided into geographic areas with regional variation in mortuary practices, but generally the cremations in mounds occur in ash layers, pit burials, or deposited in urns. Ceramic vessels with handle and organic containers made of birch bark are frequent, bucket-shaped pots and cauldrons less common. There are also several examples of cremated remains wrapped in an unburned piece of hide (variously determined as cow, horse, or bear) before being deposited into the urn (C2707, C8999, C23256, C35805). Animal bones mixed with human remains are common (Holck 1986). Cattle, horse, and dog are the most frequently identified species, but only a limited number of the faunal remains have been analysed so far (Mansrud 2004a; b; 2006, 2008).

Gender and age distribution

Figure 6 summarises the most frequent artefacts appearing together with bear claws in the investigation area. Objects commonly considered indicative of wealth or gender, like weapons, silver/gold objects and jewellery have been merged into categories (Fig. 6, for further description of the grave goods, see Appendix 1). Following Sellevold et al. (1984), Petré (1993), and Kristoffersen (2004), the following criteria were used for archaeological gender assignment for female burials: textile tools (spindle whorls, sword [weaving] batons), bone hairpins, keys, jewellery, more than five beads; for male burials: weapons (swords, axes, spearheads, lances, arrowheads, shields), stirrups, simple ring pins, weights, singular beads, blacksmith and carpenter tools (hammer, anvil, pliers, drill, saw). Several types of grave gods like knives, combs, or gaming pieces occur with both genders (Sellevold et al. 1984, 232; Kristensen 2007). While male burials only comprise one or few beads, necklaces occur exclusively with females (Johansen 2004, 104). Based on these criteria, 35 burials were assigned as female, three as possibly female, nine as male, and four have objects associated with both genders. The lack of proper documentation and excavation techniques, together with disturbances of graves, often makes it difficult to decide whether the latter category represents double graves or mixed graves. In most cases, 79, it is not possible to determine whether the interred individuals were male or female (Fig. 7).

The cremated human remains from southernmost and southeastern Norway have additionally been morphologically sexed by osteologists (HOLCK 1986). Twenty-eight individuals were identified as male and nine as female (Fig. 8). Several cremations suggest that more than one individual is present, and three burials comprise a male as well as a female. There is also a discrepancy between the osteological sex-determination and the grave goods in several cases.

Determining the sex of burnt human remains is by no means a straightforward task (MCKINLEY 2000, 412). The methodological issues inherent in osteological sex-determination as well as the normative categorisations imposed by archaeologists classifying grave goods as male or female (HJØRUNGDAL 1991; PETRÉ 1993; SOLLI 2002; LUND/MOEN 2019; PRICE et al. 2019) make it difficult to quantify the relationship between gender and bear claws, but arguably, bear claws occur with both genders. With regards to age, the number of children or subadult persons burials is low (as among Iron Age burials in general), amounting to only three cases in western Norway (KRÜGER 1988, 361) and three in southeastern Norway (C32757, C52083, C560077). The cremated remains were identified as children aged 1–8 years old, and in all cases, they were buried together with adult individuals (Appendix 1).

GRAVE GOODS, WEALTH, AND IDENTITY

Inhumations with bear skins are commonly linked with economic wealth and high-status individuals (PETRÉ 1980; KRISTOFFERSEN 2000; GRIMM 2013; LINDHOLM/LJUNGKVIST 2016, 5, 10–12; ZACHRISSON/KRZEWIŃSKA 2019). An exception is the burial from Døsen with rather meagre furnishings (GRIMM 2013, 282). How are wealth and identity expressed in the cremations with bear claws? Jewellery is present in 24 % of the burials, gold/silver objects are present in 16 %. These two categories are overlapping, since identifying burnt objects is often difficult. Certain jewellery types indicating the highest status are present in two graves. C28986 contained a gold berlock and B5931 a gold finger-ring with a serpent head, an artefact considered as insignia of the uppermost elites (REIERSEN 2018). Brooches are present in eleven burials. Of these, five are cruciform, one is equal-armed, and one is beak-shaped. The remaining ones are undefined. Fragments of various fibulas of silver or bronze are identified in 19 graves. In cases with disturbed graves, gold/silver objects and weapons would have likely been looted and could thus be underrepresented. In addition to items which were part of the dress, like fibulas, belt buckles, and various fittings, bear claws most often appear in combination with combs[6] (51 %), gaming pieces/dice (26 %) and bone pins, commonly interpreted as hair pins (11 %). Such objects most likely represent personal belongings. Several graves also contained objects interpreted as amulets or as having protective functions: beads, a copy of a Roman emperor gold medallion with animal figures (B11546), a solidus coin (B8983), and a bone knife with a runic inscription (B6700; SAMDAL 2000, 62; RØSTAD 2008). In 29 cases (27 %) the cremated remains have been deposited in a cauldron (Fig. 6). Cauldrons as burial containers are often regarded as a high-status phenomenon, and these burials are often richly equipped (HAUKEN 2005), but not all the cauldron-burials with bear claws belong to this category.

Weapons occur in 8 % of the burials. Weapons like arrows and spears may have been used for warfare as well as for hunting. Arrowheads occur in seven burials. Two of the arrowheads are made of iron, the remaining five are bone arrowheads. Seven graves contain spearheads, four of them are dated

6 These are commonly referred to as bone combs in the database, but the osseous raw material used for comb-making was antler, not bone.

to the Roman/Migration period. C27077, C17305, C22231 and C28980 can be classified as "warrior graves". They contain sets of weapons including spears, swords and shield bosses and are also similarly constructed with burnt bones deposited in a cauldron (C26791 lacks contextual information). In C28980 stirrups for riding were present – this is the only Early Iron Age grave in the total assemblage containing such equipment. Two more finds of stirrups are from the interior regions of southeastern Norway and dated to the late Iron Age. C20314 is presumably a double grave, containing a full set of weapons, riding equipment (bits) as well as a weaving baton, keys and two spindle whorls, whereas C32693 only contains weapons associated with hunting. This context will be returned to below.

Swords are present in three graves (C27077, C52092, C22231). C52092 is classified as a weapon burial from the pre-Roman Iron Age (Martens 2008, 309). Arrowheads were identified in seven graves, of which five are dated to the early Iron Age (C37631, B6475, B6691, B8872, B11546). In B6475 an iron arrowhead was present, in the remaining graves the arrowheads are made of bone. Thus, none of these assemblages can be classified as "warrior burials". Moreover, these contexts are interpreted as male because arrowheads are present, but they also include objects which are not gender-specific, or which point to the presence of a female. C37631 contained a fibula, bone pin and spindle whorls, and B6691 contained a bone pin.

Several burials are also "simple" in content as well as in construction. For example, C55949, a cremation pit ^{14}C-dated to 240–420 CE, contained skull and joint fragments of a slender, adult individual, 153 g of burnt animal bone, a bone needle and a bear claw (Fyllingen 2008). Another example is C60497, i.e. a pit filled with bones (originally deposited in an organic container), probably those of a female, 29–40 years old, buried with an iron ring, six bear claws and 34 fragments of gaming pieces, dated to 400–540 CE (Ødegård 2017). Cremated bone deposits are not always associated with artefacts, and some bone deposits do not contain human bones at all (Holck 1986). Seven human individuals had cut-marks according to Holck (1986, 178–185), who interprets this as intentional destruction of the bodies as a part of the mortuary rituals, suggestibly indicating anthropophagy or some form of ritual de-fleshing. The results are thus in accordance with Krüger's (1988, 359–361) observation that cremations with bear claws range from rich to simple and do not differ in this respect compared to cremations without bear claws. So, if bear claws are not associated with warriors, shamans, or elite individuals in general, what does their presence in a grave indicate? It is for example interesting to note that in five graves, fragments of wooden buckets or trays are present (C8999, B4591, B6756, B8853, B12046). Cattle remains, buckets, dung forks, a trough, barrels, and shovels were also part of the furnishing of the Oseberg Viking Age ship burial (Hedeager/Østmo 2005, 283). The farming equipment points to the importance of maintenance activities and farm work. Yet, archaeological interpretation tends to focus on the potential status of the buried woman as a queen (Christensen et al. 1992; Pedersen 2017). Grave goods and mortuary rites potentially have many meanings. Social identity encompasses more than gender – most individuals likely fulfilled multiple social roles and functions in life, and the various items in a grave may express these composite identities (Pedersen 2017; Lund/Moen 2019, 142). In the following, these different meanings and identities will be explored.

Transformation or protection?

As initially stated, bear claws in burials are commonly linked to the idea that the dead were cremated with bear skins. Animal antlers, skins and paws were used for rituals and ceremonies and as media for undertaking human-animal transformations, in particular as part of hunting-rituals or shamanistic seances (Ingold 2000, chapter 6; Willerslev 2007), and shamanism has been an important frame of reference for interpreting these finds.

Human-animal metamorphosis and hybrid beings characterise the symbolic iconography of Roman and Migration period animal art (Kristoffersen 2010), and the perception of the human-animal boundary as fluid and permeable, rather than made up by fixed and static identities, has been suggested as an essential feature of the pre-Christion ontology by archaeologists (Mansrud 2006; 2008; Danielsson 2007; Hedeager 2011; Armstrong Oma 2015) as well as by historians of religion (Guðmundsdóttir 2007; Samson 2011; Dale 2014; Wallenstein 2015; Nordberg/Wallenstein 2016). Price (2002) makes a connection between *seiðr* and shamanistic practices in the circumpolar north. In the Norse tradition, human-animal transformation is referred to as *hamskipta*, which means "to change one's shape", or *hamrammr*, "able to change one's shape". These compounds include the noun *hamr*, meaning "form, shape", "skin", or "shape assumed by a disembodied spirit", as their first element. The latter term has the adjective *rammr* "strong, powerful mighty" as a second element (Fritzner 1954; Heggstad et al. 2012). A person who had the ability to change shape was also called *hamhleypa* "ham-leaper", i.e. had the ability to act out-of-body (Raudvere 2001). In archaeology, such practices have been related to the idea of perspectivism (Viveiros de Castro 1998), involving an ontological understanding that an animal's perspective and properties can be accessed through the use of animal materials (Conneller 2004; Lahelma 2019).

However, the concept of human-animal transformation is neither confined to the pre-Christian ritual practice, nor to circumpolar traditions. Similar understandings are also documented among hunters, herders and fishermen in Norway in the 19th and early 20th century (Hermundstad 1967; Hodne 1997, 101), a period otherwise characterised by an ontology focused upon ideas of a hierarchical ordering of Gods, humans and animals (analogism; see Descola 2010, 338). Ontologies may vary within cultures – folklorists have for example identified shamanistic influences in Norwegian folktales (Grambo 1975) – and mastering the art of shape-shifting was an important part of the Norse *trolldóm* (Raudvere 2001, 86). The vernacular Iron Age ontology may have been hybrid, encompassing animistic/shamanistic and analogistic elements. Within both ontological systems, interaction with parallel dimensions, inhabited by the dead, animals, and supranormal beings, was not limited to shamans or other religious experts, but a natural and integrated part of everyday life (Price 2002; Østigård/Kaliff 2020). Importance attributed to bear skins is also known from medieval and early modern churches. Six radiocarbon-dated skins from churches in South Norway show a range in dates from the high Middle Ages around 1300 to the 18th century (Jahnsen 2012, 88–89 ; see also Jahnsen, this volume). The temporal continuity of the bear skin tradition is also remarkable.

Price (2002) interprets *seiðr* as an aggressive battle-magic, associated with Oðinn and crucial to the martial, predatory way of thinking in a warrior society, involving sexual magic, shape-shifting, and animal transformation of warriors into *berserkir* and *ulfhednar*. But human-bear transformation is, except for the tale of Bǫđvar Bjarki in the saga of Hrolf Krake, who was able to transform into a bear during battle, unusual in the Germanic texts (Pentikäinen 2007, 26). Overall, origin-stories in which non-human beings are entangled in social and sexual relations with humans are much more widespread (Gilhus 2006). Such stories appear in Germanic as well as classic myths, in documents like Olaus Magnus' history of the Nordic people (*Historia de gentibus septentrionalibus*, 1555), in circumpolar ethnography, and Norwegian and Finnish folklore. Legends of encounters between bear-humans and other supranatural beings are also noted in later lore. In the Norwegian folk tradition specifically, a shape shifter (*hamløper*) always refers to a man who turns into a bear[7] (Norw. *mannbjørn*). Stories of man-bears are long-lived and were recorded in various regions of Norway well into the 19th century. The transformation was undertaken using a piece of a bear skin, a foetal membrane,

7 In classical and historical legends, this phenomenon is termed lycantrophy, derived from Greek *lykos* "wolf" and *anthropos* "man", and most often concerns a human being turning into a wolf (Lid 1950, 86; Edsman 1994; Kuusela 2012; Pluskowski 2015).

or a belt (Reichborn-Kjennerud 1933; 1947, 113–115, 123, 134; Lid 1950, 82, 86–87; Nordland 1958; Hermundstad 1967, 69–108; 1985, 96–106; Hodne 1997, 121–136). In the Norwegian bear tradition, transforming into a bear or encountering such a creature is not described as something desirable – rather it was the result of magic and something to be avoided. The core motif in Norwegian bear legends is an encounter between a woman and a bear taking place in the outfields (Reichborn-Kjennerud 1947, 128; Hermundstad 1985, 99). These legends are linked with the idea that bears were dangerous for pregnant women, because the enchanted man-bear was intent to rip out the unborn child and raise it as his own, to break the spell and become human again (Reichborn-Kjennerud 1933, 58–59; 1947, 129–131; Nordland 1958, 143).

Holck (1986, 173–174) has argued against the idea that bear skins were cremated with the dead; maintaining that the bear skin with thick fur would have protected the body and hampered the cremation. Another possibility to consider is whether the claws could be related to clothing. Bear skins are heavy, rather unsuited for making clothes, but have been used for making capes, mittens, and furnishing for sleds and beds (Grimm 2013, 278; Kirkinen 2017, 15). In one case (C21945) a bear phalanx had cut marks, indicating that it had been cut loose from the paw. This agrees with the observation that during skinning, the skin is detached by cutting the second phalanges, while the third remains with the skin (Kirkinen 2017, 7). Bear skin collars are mentioned as part of the outfit of Germanic tribes (Hatt 1969, 26), and the presence of capes in burials has recently been suggested for Iron Age burials in Sweden (Zachrisson/Krzewińska 2019[8]). One of the burials (C29912) contained 12 complete and 13 fragments of phalanges and metapodials from pine marten (Mansrud 2004b, 48, 130, see also Appendix 1). This find might be from a marten cape or collar, with the feet still attached. Ethnographic collections encompass Sami bear collars where the front paws have been sewed onto the front, perhaps as decoration, or for protection (Fig. 9).

Literature on fur clothing among Eurasian groups rarely mentions claws, teeth or paws as part of everyday dress (Hatt 1969). Such items are more commonly used as pendants or attachments. From most cremations it is not possible to infer how the Iron Age bear claws were used, kept, or worn, but an unusually well-preserved female inhumation (B4234) indicates that they may have been attached to clothing. Preserved fragments of a woven belt with bronze fittings were found in this grave, with a piece of animal skin and a bear claw still attached to the belt-fittings. This suggest that the woman wore a claw in the belt, perhaps alongside other personal items, as noted in contemporary Iron Age contexts in cultures in Finland and Latvia, which also have a rich bear tradition. Here, pendants interpreted as bear teeth replicated in bronze, occasionally combined with real bear teeth, ornaments and various everyday utensils, were attached to the female dress, suggestively functioning as protective items[9] (Kivisalo 2008, 265; Herva/Lahelma 2019, 84; for further discussions of bear pendants and trophies cf. also Henriksen 2009; Girininkas/Daugnora 2013). Thus, the use of bear claws as protective items will be contextualised within the framework of vernacular belief, and the analytical concept of apotropaic magic.

Amulets, magic and agency: Objects for protection, manipulation, and luck

Magic is a complex notion and there is no unanimously agreed definition (Otto/Stausberg 2013, 1). In this chapter I focus on magic as a vernacular, supranormal practice. The most often used term in

8 Their re-evaluation of Swedish bear claw burials further identified several cases where claws assumed to be bear turned out to be birds of prey, wolf, or lynx (Zachrisson/Krzewińska 2019). This might also be the case in C29610 and B6691.

9 The interpretation has been challenged by Jonuks (2017) arguing that the bronze pendants do not resemble bear canines. Alternative interpretations and species involve dogs, wolves, or pigs, and he also speculates that the bronze pendants can represent fangs of fantastic creatures like dragons or serpents.

Norwegian is *trolldom* (ALVER 2008, 13–33, 38).[10] FRAZER's (1994) seminal work, originally published in 1890, outlining magic as a primitive form of science rooted in a human desire to master nature, has remained influential. His view of magic as form of primitive superstition, a precursor to religion, and a stage in an evolutionary development from irrational beliefs towards rational science, made the notion problematic. It became a negative marker of otherness, relegating non-Western subjects, past peoples, and rural populations to primitives (OTTO/STAUSBERG 2013). As noted by GILCHRIST (2019, 27): "Historians grapple with ambiguities in the definition of magic, but their starting point can be found in normative categories of magic as defined by the authors and critics of magic texts. The starting point for archaeologists is in the material record, which has no direct voice; the subtleties of meaning, intention and agency can only be unlocked by developing theoretical frameworks for interpreting archaeological evidence."

The ontological turn, with the concepts of animal and object agency, allows for approaching the concept of magic in new ways (GELL 1998; HENARE et al. 2007). In vernacular understanding, magic was perceived as something very concrete, a force or power affecting animals and humans, which could be harnessed and transmitted from a person to another, from a person to an animal or a thing by actions, words, thoughts, or objects (TILLHAGEN 1958, 116–117; STRÖM 1967, 221; ALVER 1971, 175). Magical remedies and spells were used for coping with everyday challenges, illness, and bad luck.[11] The conception of *trolldom* remained strongly integrated in the Norwegian folk tradition, despite protestant Christianity and the efforts of the church to fight superstition and discipline the congregation into accepting the correct faith (ALVER 2008, 288).

The question of object agency has recently been brought to the fore of Iron Age archaeology (LUND 2017; VEDELER 2018; ERIKSEN 2020). Previous archaeological writings on the notion of magical items in burials depart from the understanding that a relationship between the object or material in question should be documented in written sources (MEANEY 1981; GILCHRIST 2008; 2019). In the Old Norse religion various types of magical practices are described including techniques for communicating with spirits, such as *seiðr* and *galdr* (a form of *trolldom)*, incantations, curses (verbal magic), rune magic (word magic; cf. STRÖMBÄCK 1935; STRÖM 1967, STEINSLAND/SØRENSEN 1990, DILLMANN 2006; TOLLEY 2009; PRICE 2019; RØTHE 2020). In the Old Norse literature artefacts with magical properties like "life stones" and "victory stones" (CHRISTIE 1837; RINGSTAD 1988, 328) are generally referred to as *hlutr/lutr* (objects). In archaeology, tooth and claw pendants, lead crosses, bracteates with runic inscriptions, beads and egg-shaped stones found in burials have been interpreted as amulets (LINDERHOLM 1918, 49; OLSEN 1907; GRIEG 1954, 163; BØE 1956; BLINDHEIM 1959; MAGNUS 1986, 349; FUGLESANG 1989; RØSTAD 2008; GILCHRIST 2019). A study of grave goods from western and southwestern Norway resulted in 185 burials with such artefacts (SAMDAL 2000).

An amulet can be defined as an object kept or worn on the body for preserving against affliction; its protective power can be apotropaic, therapeutic, or exorcistic (GILCHRIST 2008, 124). Originating in the Latin *amulêtum*, or the Old Latin *amoletum*, meaning "protection", the word is used in Pliny's *Naturalis Historia* from the 1st century CE to describe objects that offer protection, are used in medical or prophylactic treatment, or as a substance in medicine. Pliny offers several examples of protective amulets, for example to hang a piece of amber or horn on an infant to protect it against sorcery and evil forces (BUDGE 1961, 12–14). The use of items like "thunderstones", i.e. stone axes believed to be conjured by lightning, is described in classical as well as Old Norse literature, and widely documented in European folklore, underscoring the continuity of the custom.

10 "Magic" is derived from the Greek word *magos*, referring to the intellectual priesthood in ancient Persia, researching subjects like medicine, astrology, dream interpretation, oracle interpretation, etc. (STORM 1975, 28).

11 It is commonly distinguished between scholarly magic (chemistry, astronomy, etc.) and folk magic (practical knowledge of pregnancy, medicine; cf. ALVER 2008).

2020). The practice was further intensified through the Late Iron Age and Middle Ages. After the demographic decline following the Black Death (*c.* 1348 or 1349) archaeological evidence shows a hiatus in the use of the mountain areas, lasting until *c.* 1600 (Pilø et al. 2020, 448). From 1600 until the industrialisation *c.* 1850, farms were unable to sustain their livestock without outfield grazing (Reinton 1961; Visted/Stigum 1971, 146–148).

Outdoor grazing made livestock particularly vulnerable to predating bears, and the fear of losing cattle to predators, disease, supranormal beings or harmful magic is a recurring theme in the Norwegian folk tradition (Solheim 1952, 8–9; Nordland 1958, 134; Visted/Stigum 1971, 214; Bø 1980). Livestock would be taken out around the same time as the bears leave their dens in spring. The brown bear hibernates between October and April, and during the summer it needs large amounts of proteins and carbohydrates to build muscles and a generous layer of fat. Bears normally shy away from people and feed on plants and carrion. But occasionally one is attracted to human habitation and turns into a hunter (Norw. *slagbjørn*), attacking sheep and cattle (Sørensen 1990). People adapted to predators by herding domestic animals, using guard dogs and building "bear-proof" stalls. When bears were spotted in an area, large fires were built to keep the predators away (Bø 1980, 92–93). A direct reference to magic protection against predators is stated in a grimoire from Nord-Hordaland: "To prevent wolves and bears taking your livestock; take the bones from wolves and bears, crush them and smear them onto your bovids, then no predator can harm them [...]" (Solheim 1952, 10, author's translation). Another grimoire recipe recommends drying the predator's meat, grinding it into powder and ingesting it, as a measure to defuse the power of the marauder (ibid.). In Sweden seal blubber was smeared on cattle to protect them from bears (Edsman 1994, 174–174), and Pentikäinen (2007, 120–123) gives many examples of the use of bear remains as protective amulets in Finnish folklore. The body parts contained the *väki*, or bear's strength. Additionally, many practices related to the Norwegian shieling tradition concerned healing or protection of cows, and protection of the equipment used for storing milk-products. Quartz crystals were seen as helpful to protect cows and humans against *trolldom* and supranormal beings (Solheim 1952, 99–107, 199–228; Alver 1971, 163, 194).

The physical outline of the farm differed between the early Iron Age and later periods. Yet, traditional farming in prehistorical and historical periods presumably shared great similarities, since keeping livestock involves a certain human-animal rhythm which is controlled by the needs of the animals and therefore remains largely constant in time and space. Attending to the livestock, securing enough fodder for them to survive the winter, and protecting them from predators and disease was equally important for survival in earlier as well as in later historic periods. In the early Iron Age, the permanent farm was organised around the three-aisled longhouse, which humans shared with their herds, suggesting a close and intimate bond between humans and livestock during this period (Armstrong Oma 2017). Excavations at Nyset-Steggje in Sogn further revealed a direct connection between the permanent farms in the lowland and the summer shielings. Burnt bones from the hearths, identified as sheep/goat, cattle and pig, show that pasturing in the mountains was practiced, and marine species like seal, cod, coalfish, eel and herring were brought up from the coast. In a retrospective, multi-species perspective, practices related to farming and transhumance are relevant for interpreting human-bear relations in the Iron Age. Finds of projectile points, reindeer and ptarmigan bones as well as large pitfall hunting systems nearby demonstrate the importance of hunting (Bjørgo 2005, 223–227). Importantly, bear bones were also found, indicating that bears were hunted in the shieling area.

In Scandinavia, tending to livestock and milking was traditionally the work of women (Svensson 2015). Finds of beads, spinning whorls and loom weights used for textile production indicate the presence of women at Iron Age shielings (Bjørgo 2005, 224; Skrede 2005, 38). In the dwelling structures excavated at Nyset-Steggje, flakes of flint/quartzite and bifacial projectile points, dated to the Neolithic and Bronze Age, were found on top of the floor layers and in the walls, and beads were found in many of the dwellings (Bjørgo 2005, 215–216, 227). As noted by Røstad (2008), beads were presumably also perceived as magic items. Rather than ending up in the houses accidentally as part of building materials these finds might, together with amber objects, represent protective amulets intentionally deposited there, like it was done in later periods. A small bear-shaped figurine of soapstone, interpreted as an amulet, was found in an Early Iron Age pit dwelling at Modvo in Sogn (Fig. 13).

I have suggested that bear claws were attached to the female dress. SAMDAL's (2000) study of amulets in western Norway also reveals a connection between women and protective charms – objects interpreted as amulets occur about twice as often in female than in male burials. In the folk tradition and the sagas there is a relationship between women, bears, magic, medicine and healing (REICHBORN-KJENNERUD 1927; 1933; MEANEY 1981; ALVER 2008, 291; HEDEAGER 2011, 123–126; RØTHE 2020). Avoiding accidents and surviving pregnancy and birth were fundamental issues, and the use of amulets was related to vernacular understandings of illness – how disease was inflicted and how it could be avoided. Folk traditions distinguish between natural and unnatural death. Disease and unnatural illness came about because of *trolldom* (ALVER 2008, 291). Body parts like hair and claws were perceived as a source of life force, susceptible to magic, and in folk tradition, combs were typically agentive objects with ability to harm as well as protect humans and cattle (HOFTUN 1993, 57; GILCHRIST 2019). Figure 14 shows a skinned bear paw with the claws removed. Such objects were used for generations to assist at difficult births and cure mastitis. During a difficult labour the paw would be used as a remedy by stroking the abdomen and the genitals, and a woman suffering from mastitis would sleep with the paw on her breast. Bear paws were also used for curing and protecting cows in a similar manner (REICHBORN-KJENNERUD 1933, 66; NORDLAND 1958; HOLCK 1986). The use of bear claws for apotropaic purposes is not stated in the Norse sagas. The only mentioning of bear claws is in the poetic Edda verse *Sigrdrífumál*, stanza 16, where the valkyrie Sigrdrífa reveals the magic properties of runes (SCHULZ 2019, 47):

> On the paws of the bear,
> and on Bragi's tongue,
> On the wolf's claws bared,
> and the eagle's beak,
> On bloody wings,
> and bridge's end,
> On freeing hands
> and helping foot-prints.[13]

Archaeological evidence such as the bear claws discussed here reveals material practices which are not documented in texts, including the placement of bear remains with the dead and objects like beads concealed in house remains interpreted as shielings. 40 % of the early Iron Age cremations with bear claws are found in the county of Vestland. Furthermore, seven bear claws have been found under the floor of Urnes church in Sogn. In 12 Norwegian churches bear paws or claws, interpreted as amulets for healing, have been discovered beneath church floors. According to JAHNSEN (2012, 69; cf. JAHNSEN, this volume) these items were used for magical purposes, secretly placed under the floor of the holy place in the hope to "charge them" with power. More than any other large predator, the bear resembles human beings in its behaviour and anatomical features (SØRENSEN 1990, 64–65). Like humans, the bear has five fingers and toes and walks on a flat foot, and a skinned paw is remarkably similar to a human hand (Fig. 14). Thus, it seems that the tradition of using of bear remains for protection fits into a long-term vernacular tradition which was specifically focused on the agency of the bear paw.

13 *Sigrdrífumál* (The Ballad of the Victory-Bringer), transl. H. A. Bellows, The Poetic Edda (Princeton 1936).

In this reassessment of early Iron Age burials in South Norway I have shown that the presence of bear claws is independent of economic status, age, or gender. Bear claws occur in male, female, and mixed cremations, occasionally also with children. In rare cases, a large number of claws are present, but most burials contain only one claw. There is also great variety in the mortuary contexts where bear claws appear. Occasionally, they are found together with weapons and hunting equipment, but most often the claws are deposited in combination with dress equipment, combs, gaming pieces and bone pins, spindle whorls, knives, beads and other personal belongings. Bear claws may have been part of burnt clothing, like collars or capes, used as pendants or worn as amulets or charms, attached to the belt. I have argued that the archaeological evidence suggests that individuals cremated with bear claws were ordinary farmers, herders, and hunters rather than chiefs, shamans, or warriors.

The reverence for the brown bear as a charismatic and powerful agentive being with supernatural powers seems to represent a cross-cultural and cross-temporal tradition throughout the geographic regions inhabited by bears (Nordland 1958, 142; Edsman 1995; Pentikäinen 2007). Folklorists, archaeologists and historians of religion alike have maintained that the circumpolar bear tradition is an ancient practice of long continuity, possibly originating in the Palaeolithic (Ström 1980; Pentikäinen 2007, 9–15; Herva/Lahelma 2019, 80). This evokes the question of how to deal with the relationship between general phenomena and historical situations in archaeological analysis. I have addressed this issue by employing a retrospective approach, arguing for a continuity in landscape utilisation, ritual practice, and bear lore over an extended time span from *c.* 400 to 1850. Similarities in practice do not necessarily constitute evidence for a direct continuity of beliefs across time, but they perhaps indicate a longstanding, common repertoire of ritual actions (Gilchrist 2019). Arguably, the perception of bears among Germanic, circumpolar, and early modern peoples were grounded in different epistemologies, relationships, and interactions with bears. Although the concept of transformation in general, and human-animal boundary crossing in particular, was integrated in the early Iron Age ontology, I maintain that the multi-species engagements involving bears and livestock confronting the Iron Age farmers resemble the experiences of early modern farmers. As noted by Pentikäinen (2007, 24) the Germanic texts do not refer to the bear as a sacred animal in the way it is perceived in the circumpolar bear traditions. Based on evidence from Finland, it has been argued that attitudes towards bears transformed as the subsistence base changed from hunting to an agro-pastoral economy. For agricultural societies the bear became a harmful predator, killing cattle and eating grain. A similar change in attitude is documented in poetry and bear rituals, which acquired new functions in an agricultural context – to protect the cattle and the whole human sphere (Kivisalo 2008, 276–277). Against this background bear claws can productively be conceptualised as protective amulets, rather than devices for transformation or symbols of shamanic practices. By drawing on the notion of apotropaic magic as an analytic framework, bear claws in the local context of early Iron Age South Norway can be perceived as powerful objects, imbued with a potent essence of "bearness" which was utilised for therapeutic, safeguarding and protective purposes.

Why were the bear claws burnt and finally deposited in early Iron Age burials? Overall, great care seems to have been undertaken in gathering the claws and placing them in the grave along with the cremated human bones. This indicates that the deposition of claws with these dead individuals was an important part of the mortuary rite (Ström 1980; Krüger 1988). Perhaps the bear claws were intended to keep on protecting their owners in the world beyond, or they were considered an effective mediator, facilitating the transition to the realm of the dead. The bear hibernates through the winter and awakes from its den in spring. Its seasonal behaviour has been associated with liminal states. Its material remains have been interpreted as objects that stand in contact with other dimensions and therefore have transcendental properties which facilitate transformation (Hedeager 2011, 84, 92).

Artefacts with strong agentive powers and a long history of use and interaction with their owners could also be perceived as dangerous, and the performance of magic often involves the deliberate destruction or mutilation of such objects (Gilchrist 2019, 17). Cremation has been suggested as a way of releasing the spirit of the dead (Williams 2015). As described above, the remains of seven individuals interred with bear claws had intentional cut marks indicating the deliberate destruction of their bodies (Holck 1986, 178–185). Iron Age cremations in South Norway further contain a significant number of deliberately destroyed items like swords and gold rings (Shetelig 1912, 179–201; Aannestad 2018; Reiersen 2018), and similar ideas have been discussed in relation to human remains (Oestigaard 1999). When a person dies, his or her belongings must either be destroyed, burnt or purified ritually, or deposited in the grave (Viveiros de Castro 1998; Kouljok 1999, 105). In this perspective, the destruction of bear remains through fire could also be an act of transformation – a way of harnessing the powerful capacities inherent in these objects (Fig. 15).

Acknowledgements

I wish to thank Ingunn Marit Røstad for commenting on an earlier version of the paper and helping me with the English artefact terminology, Kyrre Kverndokk for numerous discussions and invaluable help with literature about the Norwegian shieling and bear traditions, and Grethe Moell-Pedersen for making the map.

Bibliography

Aannestad 2018: H. L. Aannestad, Charisma, violence and weapons. In: M. Vedeler/I. M. Røstad/S. Kristoffersen/A. Z. T. Glørstad (eds.), Charismatic Objects: From Roman Times to the Middle Ages (Oslo 2018) 147–166.

Alver 1971: B. G. Alver, Heksetro og trolddom (Oslo 1971).

Alver 2008: B. G. Alver, Mellem mennesker og magter. Magi i hekseforfølgelsenes tid (Oslo 2008).

Andrews 1994: C. Andrews, Amulets of ancient Egypt (London 1994).

Armstrong Oma 2015: K. Armstrong Oma, [...] det treet, som ingen veit kvar det av rotom renn. Viking 2015, 189–206.

Armstrong Oma 2017: K. Armstrong Oma, Long time, long house. In: F. Iversen/H. Petterson (eds.), The Agrarian Life of the North 2000 BC to AD 1000: Studies in Rural Settlement and Farming in Norway (Oslo 2017) 11–26.

Armstrong Oma 2018: K. Armstrong Oma, The sheep people: the ontology of making lives, building homes and forging herds in Early Bronze Age Norway (Sheffield 2018).

Armstrong Oma/Goldhahn 2020: K. Armstrong Oma/J. Goldhahn, Introduction: Human-animal relationships from a long-term perspective. Current Swedish Archaeology 28, 2020, 11–22.

Birch 2018: S. E. P. Birch (ed.), Multispecies archaeology (London, New York 2018).

Bjørgo 2005: T. Bjørgo, Iron Age house remains from mountains areas in inner Sogn, Western Norway. UBAS Nordisk 1. Fra funn til samfunn. Jernalderstudier tilegnet Bergljot (Bergen 2005).

Blindheim 1959: C. Blindheim, En amulett av rav. Universitetets Oldsakssamlings Årbok 1958/1959, 1959, 78–98.

Bø 1980: O. Bø, Bjørnen i folkedikning og folketru. Norveg 23, 1980, 89–100.

Bøe 1956: T. Bøe, Amulettar. In: J. Danstrup (ed.), Kulturhistorisk leksikon for nordisk middelalder. Fra vikingtid til reformationstid (København 1956) 129–134.

Budge 1961: E. A. Budge, Amulets and Talismans (New York 1961).

Byggstøyl 2012: I. Byggstøyl, Fangstmarksgraver i Hedmark: En analyse av kulturelle markører i gravmateriale og gravform. Master's thesis, University of Oslo (Oslo 2012).

Christensen 1986: A. E. Christensen, Reinjeger og kammaker, en forhistorisk yrkeskombinasjon. Viking 49, 1986, 113–133.

Christensen et al. 1992: A. M. Christensen/A. S. Ingstad/B. Myhre, Osebergdronningens grav: vår arkeologiske nasjonalskatt i nytt lys (Oslo 1992).

Christie 1837: W. C. Christie, Om noen merkelige stenringe. Urda 1–2 (Bergen 1837).

Conneller 2004: C. Conneller, Becoming deer. Corporeal transformations at Star Carr. Archaeological Dialogues 11(1), 2004, 37.

Conneller 2011: C. Conneller, An Archaeology of Materials: Substantial Transformations in Early Prehistoric Europe (London, New York 2011).

Dale 2014: R. T. D. Dale, Berserkir: A Re-Examination of the Phenomenon in Literature and Life (Nottingham 2014).

Danielsson 2007: I. M. B. Danielsson, Masking moments: the transitions of bodies and beings in Late Iron Age

Scandinavia. Unpubl. PhD thesis, Institutionen för arkeologi och antikens kultur (Lund 2007).
DENNIS et al. 2006: A. DENNIS/P. FOOTE/R. PERKINS (transl.), Laws of Early Iceland. Grágás I. The codex regius of Grágás with material from other manuscripts (Winnipeg ²2006).
DESCOLA 2010: P. DESCOLA, Cognition, perception and worlding. Interdisciplinary Science Reviews 35(3/4), 2010, 334–340.
DESCOLA 2013: P. DESCOLA, Beyond Nature and Culture (Chicago, London 2013).
DILLMANN 2006: F. X. DILLMANN, Les magiciens dans l'Islande ancienne: Études sur la représentation de la magie islandaise et de ses agents dans les sources littéraires norroises. Acta Academiae regiae Gustavi Adolphi XCII (Uppsala 2006).
DOMMASNES 2016: L. H. DOMMASNES, Introduction. The farm as a social arena. In: L. H. Dommasnes/D. Gutsmiedl Schümann/A. T. Hommedal (eds.), The Farm as a Social Arena (Münster, New York 2016) 9–22.

EDSMAN 1956: C. M. EDSMAN, Björnfest. In: Kulturhistorisk leksikon for nordisk middelalder. Fra Vikingtid til reformationstid (København 1956) 129–134.
EDSMAN 1994: C. M. EDSMAN, Jägaren och makterna. Samiska och finska björnceremonier. Skrifter utgivna genom dialekt- och folkminnesarkivet i Uppsala (Uppsala 1994).
ERIKSEN 2020: M. H. ERIKSEN, 'Body-objects' and personhood in the Iron and Viking Ages: processing, curating, and depositing skulls in domestic space. World Archaeology 52(1), 2020, 103–119.
ESTES 1989: J. W. ESTES, The Medical Skills of Ancient Egypt (Massachusetts 1989).

FLØYSTAD et al. 2020: I. FLØYSTAD/H. BRØSETH/B. BANKEN/H. G. BAKKE/S. B. HAGEN, Populasjonsovervåking av brunbjørn. DNA-analyse av prøver innsamlet i Norge i 2019. NINA Rapport 1808. Norsk institutt for naturforskning (Trondheim 2020).
FRAZER 1994: J. G. FRAZER, The Golden Bough. A Study in Magic and Religion (Oxford 1994; originally published 1890).
FREDERIKSEN 2006: P. D. FREDERIKSEN, Moving closer to the fire: heat transformations and bucket-shaped pots in burials. Norwegian Archaeological Review 39(2), 2006, 126–137.
FRITZNER 1954: J. FRITZNER, Ordbog over det gamle norske sprog 1–3 (Oslo ²1954; originally published 1883–1896).
FUGLESANG 1989: S. FUGLESANG, Viking and medieval amulets in Scandinavia. Fornvännen 84, 1989, 16–27.
FYLLINGEN 2008: H. FYLLINGEN, Bosetningsspor fra jernalderen. E18-Spydeberg, Molle Østre 2/1, Molle Vestre 2/2 og Skøyen 5/1, Spydeberg kommune, Østfold. Rapport Arkeologisk utgravning, Kulturhistorisk museum (2008).

GELL 1998: A. GELL, Art and Agency: An Anthropological Theory (Oxford 1998).
GILHUS 2006: I. S. GILHUS, Animals, Gods and Humans: Changing Attitudes to Animals in Greek, Roman and Early Christian Thought (New York 2006).
GILCHRIST 2008: R. GILCHRIST, Magic for the dead? The archaeology of magic in later medieval burials. Medieval Archaeology 52(1), 2008, 119–159.
GILCHRIST 2019: R. GILCHRIST, Magic and archaeology: Ritual residues and "odd" deposits. In: S. Page/C. Rider (eds.), The Routledge History of Medieval Magic (Abingdon 2019) 383–401.
GIRININKAS/DAUGNORA 2013: A. GIRININKAS/L. DAUGNORA, Hunting in the territory of Lithuania from the Late Palaeolithic to the Middle Ages. In: O. Grimm/U. Schmölcke eds.), Hunting in Northern Europe until 1500 AD – Old traditions and regional developments, continental sources and continental influences. Papers presented at a workshop organised by the Centre for Baltic and Scandinavian Archaeology, Schleswig, June 16th and 17th, 2011. Schriften des Archaologischen Landesmuseums Ergänzungsreihe 7 (Neumünster 2013) 567–596.
GOLDHAHN 2018: J. GOLDHAHN, Älvornas Arkeologi. Fornvännen 113(4), 2018, 210–232.
GRAF 1997: F. GRAF, Magic in the Ancient World (London 1997).
GRAMBO 1975: R. GRAMBO, Traces of shamanism in Norwegian folktales and popular legends. Fabula 16(1), 1975, 20.
GRIEG 1954: S. GRIEG, Amuletter og gudebilder. Viking XVIII, 1954, 157–209.
GRIMM 2013: O. GRIMM, Bear-skins in northern European burials and some remarks on other bear-related furnishings in the north and middle of Europe in the 1st millennium AD. In: O. Grimm/U. Schmölcke eds.), Hunting in Northern Europe until 1500 AD – Old traditions and regional developments, continental sources and continental influences. Papers presented at a workshop organised by the Centre for Baltic and Scandinavian Archaeology, Schleswig, June 16th and 17th, 2011. Schriften des Archaologischen Landesmuseums Ergänzungsreihe 7 (Neumünster 2013) 277–296.
GUÐMUNDSDÓTTIR 2007: A. GUÐMUNDSDÓTTIR, The Werewolf in Medieval Icelandic Literature. Journal of English and Germanic Philology 2007, 277–303.

HATT 1969: G. HATT, Arctic Skin Clothing in Eurasia and America an Ethnographic Study. Arctic Anthropology 5(2), 1969, 3–132.
HAUKEN 2005: Å. D. HAUKEN, The Westland cauldrons in Norway. AmS-Skrifter 19 (Stavanger 2005).
HEDEAGER 2011: L. HEDEAGER, Iron Age Myth and Materiality. An Archaeology of Scandinavia AD 400–1000 (London 2011).
HEDEAGER/ØSTMO 2005: L. HEDEAGER/E. ØSTMO (eds.), Norsk arkeologisk leksikon (Oslo 2005).
HEGGSTAD et al. 2012: L. HEGGSTAD/F. HØDNEBØ/E. SIMENSEN, Norrøn ordbok. Det norske samlaget (Oslo 2012).
HENARE et al. 2007: A. HENARE/M. HOLBRAAD/S. WASTELL (eds.), Thinking Through Things: Theorising Artefacts Ethnographically (London, New York 2007).
HENNIUS 2020: A. HENNIUS, Towards a refined chronology of prehistoric pitfall hunting in Sweden. European Journal of Archaeology 23(4), 2020, 530–546.
HENRIKSEN 2009: M. B. HENRIKSEN, Brudager Mark – en romertidsgravplads nær Gudme på Sydøstfyn. Fynske jernaldergrave 6,1–2 (Odense 2009).
HERMUNDSTAD 1967: K. HERMUNDSTAD, Truer om villdyr, fangst og fiske 8 (Oslo 1967).
HERMUNDSTAD 1985: K. HERMUNDSTAD, Truer om dyr. Norsk Folkeminnelags Skrifter (Oslo 1985).

HERVA/LAHELMA 2019: V. P. HERVA/A. LAHELMA, Northern archaeology and cosmology: a relational view (London, New York 2019).

HILL 2011: E. HILL, Animals as agents: hunting ritual and relational ontologies in prehistoric Alaska and Chukotka. Cambridge Archaeological Journal 21(3), 2011, 407–426.

HILL 2013: E. HILL, Archaeology and animal persons: toward a prehistory of human-animal relations. Environment and Society 4(1), 2013, 117–136.

HJØRUNGDAL 1991: T. HJØRUNGDAL, Det skjulte kjønn. Acta archaeologica Lundensia 9 (Lund 1991).

HODNE 1997: Ø. HODNE, Fiske og jakt. Norske Folketradisjoner (Oslo 1997).

HOFTUN 1993: O. HOFTUN, Kammene fra elder jernalder i Vest-Norge og trosforestillinger tilknyttet kammer. Hovedoppgave i arkeologi, Universitet i Bergen (Bergen 1993).

HOLCK 1986: P. HOLCK, Cremated Bones. A Medical-Anthroplogical Study of an Archaeological Material on Cremation Burials. Antropologiske Skrifter 1 (Oslo 1986).

INGOLD 2000: T. INGOLD, The Perception of the Environment: Essays on Livelihood, Dwelling and Skill (London, New York 2000).

JAHNSEN 2012: J. B. JAHNSEN, Bjørnen i kirken. Om bjørneskinn, bjørnelabber, bjørneklør og bjørnefigurer i norske kirker (Vallset 2012).

JOHANSEN 2004: L. M. B. JOHANSEN, Perler i jernalder – kilde til mote i kvinnegraver og magi i mannsgraver. In: L. Melheim/K. Oma/L. Hedeager (eds.), Mellom himmel og jord. Foredrag fra et seminar om religionsarkeologi. Oslo Archaeological Series 2 (Oslo 2004) 468–490.

JONUKS 2017: T. JONUKS, Bronze tooth pendants from the Late Iron Age: Between real and fictional zooarchaeology. Norwegian Archaeological Review 50(2) 2017, 135–148.

KIRKINEN 2017: T. KIRKINEN, Burning pelts – brown bear skins in the Iron Age and Early medieval (1–1300 AD) burials in south-eastern Fennoscandia. Eesti Arheoloogia Ajakiri 21(1), 2017, 3–29.

KIVISALO 2008: N. KIVISALO, The Late Iron Age Bear-Tooth Pendants in Finland: Symbolic Mediators between Women, Bears, and Wilderness? Temenos-Nordic Journal of Comparative Religion 44(2), 2008, 263–292.

KLOKKERVOLL 2015: A. KLOKKERVOLL, Dyrenes rolle i germansk gravskikk. En analyse av dyrenedleggelsene i nordnorske jernaldergraver. Master's thesis, UiT Norges arktiske universitet (Tromsø 2015).

KRISTENSEN 2007: S. KRISTENSEN, Brettspillet – Jernalderkrigerens virtuelle arena: Strategispill i sørøst-Norge i jernalder. Master's thesis, University of Oslo (Oslo 2007).

KRISTOFFERSEN 2000: S. KRISTOFFERSEN, Expressive objects. In: D. S. Olausson/H. Vankilde (eds), Form, Function and Context. Material Culture Studies in Scandinavian archaeology. Acta Archaeologica Lundensia. Series in 8°, No 31,275–297.

KRISTOFFERSEN 2004: S. KRISTOFFERSEN, Symbolism in rites of transition in Iron Age Norway. In: M. Wedde (ed.), Celebrations, Sanctuaries and the Vestiges of Cult Practices (Bergen 2004) 287–303

KRISTOFFERSEN 2010: S. KRISTOFFERSEN, Half beast – half man: hybrid figures in animal art. World Archaeology 42(2), 2010, 261–272.

KRÜGER 1988: S. H. KRÜGER, Bjørneklør fra vestlandske graver. In: S. Indrelid/S. Kaland/B. Solberg (eds.), Festskrift til Anders Hagen. Arkeologiske Skrifter Universitetet i Bergen 4 (Bergen 1988) 357–366.

KOULJOK 1999: K. E. KOULJOK, Moder jord och andra mödrar: Föreställningar om verkligheten bland folken i norr och vår syn på den (Stockholm 1999).

KUUSELA 2012: T. KUUSELA, Varulven i fornnordisk tradition. In: E. Odstedt (ed.), Varulven i svensk folktradition (Stockholm 2012) 320–352.

LAHELMA 2019: A. LAHELMA, Sexy beasts: animistic ontology, sexuality and hunter-gatherer rock art in Northern Fennoscandia. Time and Mind 12(3), 2019, 221–238.

LEV 2003: E. LEV, Traditional Healing with Animals (Zootherapy): Medieval to Present-Day Levantine Practice. Journal of Ethnopharmacology 85, 2003, 107–118.

LID 1935: N. LID, Folketru, Nordisk kultur XlX (Oslo 1935).

LID 1950: N. LID, Til varulvens historie. Trolldom. Festskrift til Nils Lid til sekstiårsdagen (Oslo 1950).

LINDERHOLM 1918: E. LINDERHOLM, Nordisk magi. Svenska landsmål ock svenskt folkeliv, B.20 (Stockholm 1918).

LINDHOLM/LJUNGKVIST 2016: K. J. LINDHOLM/J. LJUNGKVIST, The bear in the grave: exploitation of top predator and herbivore resources in first millennium Sweden – first trends from a long-term research project. European Journal of Archaeology 19(1), 2016, 3–27.

LUND 2017: J. LUND, Connectedness with things. Animated objects of Viking Age Scandinavia and early medieval Europe. Archaeological Dialogues 24(1), 2017, 89–108.

LUND/MOEN 2019: J. LUND/M. MOEN, Hunting identities: Intersectional perspectives on Viking Age mortuary expressions. Fennoscandia Archaeologica 36, 2019, 142–155.

MAGNUS 1986: B. MAGNUS, Iron Age Exploitation of High Mountain Resources in Sogn. Norwegian Archaeological Review 19(1), 1986, 44–50.

MANSRUD 2004a: A. MANSRUD, Dyrebein i graver – en kilde til jernalderens kult og forestillingsverden. In: L. Melheim/K. Oma/L. Hedeager (eds.), Mellom himmel og jord. Foredrag fra et seminar om religionsarkeologi. Oslo Archaeological Series 2 (Oslo 2004) 82–111.

MANSRUD 2004b: A. MANSRUD, Dyret i jernalderens forestillingsverden. En studie av forholdet mellom mennesker og dyr inordisk jernalder, med utgangspunkt i dyrebein fra graver. M.A.-thesis, University of Oslo (Oslo 2004).

MANSRUD 2006: A. MANSRUD, Flytende identiteter? – Dyrebein i graver og førkristne personoppfatninger. In: T. Østigård (ed.), Lik og ulik. Tilnærminger til variasjon i gravskikk. Universitetet i Bergen Ark. Skr. 2 (Bergen 2006) 133–158.

MANSRUD 2008: A. MANSRUD, "Stykkevis og delt": noen refleksjoner omkring forholdet mellom kropp, identitet og personoppfatninger i det førkristne samfunnet. In: K. Chilidis/J. Lund/C. Prescott (eds.), Facets of Archaeology: Essays in Honour of Lotte Hedeager on her 60th Birthday. Oslo Archaeological Series 18 (Oslo 2008) 385–395.

MARTENS 2008: J. MARTENS, A princely burial at Bøli vestre, Østfold. The (re-)introduction of weapon burial rites in Iron Age Norway. In: K. Chilidis/J. Lund/C. Prescott (eds.), Facets of Archaeology: Essays in Honour of Lotte Hedeager on her 60th Birthday. Oslo Archaeological Series 18 (Oslo 2008) 309–322.

Mauss 2001: M. Mauss, A General Theory of Magic (London, New York 2001; originally published 1902).

McKinley 2000: J. McKinley, The analysis of cremated bone. In: M. Cox/S. Mays (eds.), Human Osteology in Archaeology and Forensic Science (Cambridge 2000) 403–421.

Meaney 1981: A. L Meaney, Anglo-Saxon Amulets and Curing Stones. BAR, Brit. Ser. 96 (Oxford 1981).

Mjærum 2020: A. Mjærum, Jordbruk i grenseland. Utviklingen av jordbruket på Hedmarken og i Østerdalen frem til ca. 570 e.Kr. In: C. L. Rødsrud/A. Mjærum (eds.), Et fortidens grenseland-landskap, ferdsel og kulturminner i Løten og Elverum gjennom 10 000 år (Oslo 2020) 137-155.

Møhl 1977: U. Møhl, Bjørneklør og brandgrave. Dyreknogler fra germansk jernalder i Stilling. Kuml 1977, 119–129.

Nóbrega et al. 2013: A. Nóbrega/R. Rômulo Romeu/L. Ierecê (eds.), Animals in Traditional Folk Medicine: Implications for Conservation (London, New York 2013).

Nockert 1991: M. Nockert, The Högom find and other Migration period textiles and costumes in Scandinavia. Archaeology and Environment 9. Högom II (Uddevalla 1991).

Nordberg/Wallenstein 2016: A. Nordberg/F. Wallenstein, "Laughing I Shall Die!": The Transformations of Berserkers and Úlfheðnar in Old Norse Society. In: P. Haldén/P. Jackson (eds.), Transforming Warriors: The Ritual Organization of Military Force (New York 2016) 49–65.

Nordeide 2015: S. W. Nordeide, Scener fra vikingtid som kilde til kosmologi. Viking LXXVIII, 2015, 207–219.

Nordland 1958: O. Nordland, Mannbjørn. Ein studie i heimfestingsproblemet ved vandresegner. By og bygd 11, 1958, 133–156.

Oehrl 2013: S. Oehrl, Bear hunting and its ideological context (as a background for the interpretation of bear claws and other remains of bears in Germanic graves of the 1st millennium AD). In: O. Grimm/U. Schmölcke eds.), Hunting in Northern Europe until 1500 AD – Old traditions and regional developments, continental sources and continental influences. Papers presented at a workshop organised by the Centre for Baltic and Scandinavian Archaeology, Schleswig, June 16th and 17th, 2011. Schriften des Archäologischen Landesmuseums, Ergänzungsreihe 7 (Neumünster 2013) 297–332.

Oestigaard 1999: T. Oestigaard, Cremations as transformations: When the dual cultural hypothesis was cremated and carried away in urns. European Journal of Archaeology 2(3), 1999, 345–364.

Olsen 1907: M. Olsen, Valby-amulettens runeindskrift. Christiania Videnskabs-Selskabs Forhandlinger 1907, Nr. 6 (Christiania 1907).

Orton 2010: D. Orton, Both subject and object: herding, inalienability and sentient property in prehistory. World Archaeology 42(2), 2010, 188–200.

Otto/Stausberg 2013: B. C. Otto/M. Stausberg, Introduction. In: B. Otto/M. Stausberg (eds.), Defining Magic: A reader (London, New York) 1–13.

Overton/Hamilakis 2013: N. J. Overton/Y. Hamilakis, A manifesto for a social zooarchaeology. Swans and other beings in the Mesolithic. Archaeological dialogues 20(2), 2013, 111–136.

Pedersen 2017: U. Pedersen, Vikingtidskvinner i maktens innerste sirkel. In: K. Kjeserud/N. Løkka (eds.), Dronningen i vikingtid og middelalder (Oslo 2017) 98–125.

Pentikäinen 2007: J. Pentikäinen, Golden King of the Forest. The Lore of the Northern Bear (Helsinki 2007).

Petersen 1940: T. H. Petersen, En bjørnegrav. Trekk av lappisk folketro. Viking 1940, 253–167.

Petré 1980: B. Petré, Björnfällen i begravningsritualen – statusobjekt speglande regional skinnhandel? Fornvännen 75, 1980, 5–14.

Petré 1993: B. Petré, Male and female finds and symbols in Germanic Iron Age graves. Current Swedish Archaeology 1(1), 1993, 149–154.

Pilø et al. 2020: L. Pilø/E. Finstad/J. H. Barrett, Crossing the ice: An Iron Age to medieval mountain pass at Lendbreen, Norway. Antiquity 94, 2020, 437–454.

Price 2002: N. Price, The Viking Way. Magic and Mind in Late Iron Age Scandinavia. (Oxford, Philadelphia ²2002).

Price et al. 2019: N. Price/C. Hedenstierna-Jonson/T. Zachrisson/A. Kjellström/J. Storå/M. Krzewińska/A. Götherström, A Viking warrior women? Reassessing Birka chamber grave Bj. 581. Antiquity 93, 2019, 181–198.

Pluskowski 2015: A. Pluskowski, Before the Werewolf Trials: Contextualising Shape-Changers and Animal Identities in Medieval North-Western Europe. In: W. Blécourt (ed.), Werewolf Histories (London 2015) 82–120.

Prøsch-Danielsen et al. 2020: L. Prøsch-Danielsen/C. Prescott/E. D. Fredh, Land cover and exploitation of upland resources on the Høg-Jæren Plateau, southwestern Norway, over the last 6500 years. Journal of Archaeological Science: Reports 32, 2020. https://doi.org/10.1016/j.jasrep.2020.102443.

Raudvere 2001: C. Raudvere, Trolldóm in Early Medieval Scandinavia. In: K. Jolly/C. Raudvere/E. Peters (eds.), Witchcraft and Magic in Europe. The Middle Ages (Philadelphia 2001) 73–171.

Reichborn-Kjennerud 1927: I. Reichborn-Kjennerud, Vår gamle trolldomsmedisin. Skrifter utgitt av Det Norske Videnskaps-Akademi i Oslo, II Hist. Filos. klasse 1927, no. 6 (Oslo 1927).

Reichborn-Kjennerud 1933: I. Reichborn-Kjennerud, Vår gamle trolldomsmedisin. Skrifter utgitt av Det Norske Videnskaps-Akademi i Oslo, II Hist. Filos. klasse 1933, no. 2 (Oslo 1933).

Reichborn-Kjennerud 1947: I. Reichborn-Kjennerud, Vår gamle trolldomsmedisin. Skrifter utgitt av Det Norske Videnskaps-Akademi i Oslo, II Hist. Filos. klasse 1947, no. III (Oslo 1942).

Reiersen 2018: H. Reiersen, The Death of Serpent-Head Rings. Ritual destruction of elite insignia from the Roman Period. In: M. Vedeler/I. M. Røstad/E. S. Kristoffersen/Z. T. Glørstad (eds.), Charismatic Objects: From Roman Times to the Middle Ages (Oslo 2018) 31–55.

Reinton 1961: L. Reinton, Sæterbruket i Norge III. Institutt for sammenlignende Kulturforskning, Serie B XLVIII (Oslo 1961).

Resi 1986: H. G. Resi, Gravfeltet på Hunn i Østfold. Norske Oldfunn XII (Oslo 1986).

Resi 2005: H. G. Resi, Archaeological finds of jet from Norway. Collegium Medievale 18, 2005, 86–105

Ringstad 1988: B. Ringstad, Steiner brukt som amuletter i forhistorisk tid – et eksempel fra Kvåle i Sogndal. In: S. Indrelid/S. Kaland/B. Solberg (eds.), Festskrift til Anders Hagen. Arkeologiske Skrifter Universitetet i Bergen 4 (Bergen 1988) 325–342.

Rygh 1885: O. Rygh, Norske oldsager: ordnede og forklarede 1 (Christiania 1885).

Røstad 2008: I. M. Røstad, En liten perle. Om perler og magi i folkevandringstid. In: K. Chilidis/J. Lund/C. Prescott (eds.), Facets of Archaeology: Essays in Honour of Lotte Hedeager on her 60th Birthday. Oslo Archaeological Series 18 (Oslo 2008) 439–450.

Røthe 2020: G. Røthe, Magi og mirakler fra vikingtid til middelalder (Trondheim 2020).

Samdal 2000: M. Samdal, Amuletter. Gjenstander med amulettkarakter i vestnorske graver i tidsrommet 350–1000 e.Kr. Unpubl. M.A.-thesis, University of Bergen (Bergen 2000).

Samson 2011: V. Samson, Les Berserkir. Les guerriers-fauves dans la Scandinavie ancienne, de l'âge de Vendel aux Vikings (VIe-XIe siècle) (Villeneuve d'Ascq 2011).

Schulz 2019: K. Schulz, Inscriptions in Old Norse Literature. In: R. Wagner/Ch. Neufeld/L. Lieb (eds.), Writing Beyond Pen and Parchment (Berlin, Boston 2019) 41–62. https://library.oapen.org/handle/20.500.12657/23521.

Schönfelder 1994: M. Schönfelder, Bear-claws in Germanic graves. Oxford Journal of Archaeology 13, 1994, 217–227.

Sellevold et al. 1984: B. J. Sellevold/U. L. Hansen/J. B. Jørgensen, Iron Age Man in Denmark. Prehistoric Man in Denmark III. Nordiske fortidsminner B,8 m (København 1984).

Shetelig 1912: H. Shetelig, Vestlandske graver fra jernalderen. Bergens museums skrifter, ny række II,1 (Bergen 1912).

Sjöling 2017: E. Sjöling, Osteologisk analys av en brandgrav från Fallaveien, Falla 33/2, 57, Fet kommune, Akershus SAU (Societas Archaeologica Upsaliensis). SAU rapport 2017:4 O (Uppsala 2017).

Skjølsvold 1969: A. Skjølsvold, En fangstmanns grav i Trysil-fjellene. Viking 33, 1969, 139–199.

Skogstrand 2016: L. Skogstrand, Warriors and other men: notions of masculinity from the late Bronze Age to the early Iron Age in Scandinavia (Oxford 2016).

Skrede 2005: M. A. Skrede, Shielings and landscape in western Norway In: I. Holm/S. Innselset/I. Øye (eds.), 'Utmark'. The outfield as industry and ideology in the iron age and the middle ages. UBAS International 1 (Bergen 2005) 31–41.

Solberg 1999: B. Solberg, «Holy white stones». Remains of fertility cult in Norway. In: Völker an Nord- und Ostsee und die Franken. Akten des 48. Sachsensymposiums in Mannheim vom 7. bis 11. September 1997. Mannheimer Geschichtsblätter, Neue Folge, Beiheft 2 (Bonn 1999) 99–106.

Solberg 2000: B. Solberg, Jernalderen i Norge: ca. 500 f. Kr. – 1030 e.Kr. (Oslo 2000).

Solheim 1952: S. Solheim, Norsk sætertradisjon. Instituttet for sammenlignende kulturforskning, Ser. B: Skrifter XLVII (Oslo 1952).

Solli 2002: B. Solli, Seid. Myter, sjamanisme og kjønn i vikingenes tid (Oslo 2002).

Steinsland/Sørensen 1990: G. Steinsland/P. Sørensen, Før kristendommen. Diktning og livssyn i vikingtiden (København 1990).

Storm 1975: G. Storm, Anglo-Saxon Magic (Nijmegen 1975).

Ström 1967: F. Ström, Nordisk hedendom. Tro og sed i førkristen tid (Lund 1967).

Ström 1980: Å. V. Ström, Björnfällar och Oden-religion. Fornvännen 75, 1980, 206–210.

Strömbäck 1935: D. Strömbäck, Sejd (Stockholm 1935).

Svensson 2015: E. Svensson, Upland living: The Scandinavian shielings and their European sisters. In: I. Baug/J. Larsen/S. Samset Mygland (eds.), Nordic Middle Ages – Artefacts, Landscapes and Society. Essays in Honour of Ingvild Øye on her 70th Birthday. University of Bergen Archaeological Series 8 (Bergen 2015) 289–300.

Sørensen 1990: O. J. Sørensen, Bjørnen. In: A. Semb-Johansen (ed.), Norges pattedyr (Oslo 1990).

Tillhagen 1958: C. H. Tillhagen, Folklig läkekonst (Stockholm 1958).

Tolley 2009: C. Tolley, Shamanism in Norse Myth and Magic. Folklore Fellows Communications 296 (Helsinki 2009).

Vedeler 2018: M. Vedeler, The charismatic power of objects. In: M. Vedeler/I. M. Røstad/S. Kristoffersen/A. Z. T. Glørstad (eds.), Charismatic Objects: From Roman Times to the Middle Ages (Oslo 2018) 9–29.

Visted/Stigum 1971: K. Visted/H. Stigum, Vår gamle bondekultur (Oslo 1971).

Viveiros de Castro 1998: E. V. Viveiros de Castro, Cosmological deixis and Amerindian perspectivism. Journal of the Royal Anthropological Institute 1998, 469–488.

Walker 2013: H. Walker. Under a Watchful Eye: Self, Power, and Intimacy in Amazonia (Berkeley 2013).

Wallenstein 2015: F. Wallenstein, Varggudens raseri: extas och djurblivande hos bärsärkar och úlfheðnar. Aiolos – Tidskrift för litteratur teori och estetik 49, 2015, 25–44.

Wamers 2009: E. Wamers, Von Bären und Männern. Berserker, Bärenkämpfer und Bärenführer im frühen Mittelalter. Zeitschrift für Archäologie des Mittelalters 37, 2009, 1–46.

Wiseman/Ellis 1996: N. Wiseman/A. Ellis, Fundamentals of Chinese Medicine (Brookline 1996).

Willerslev 2007: R. Willerslev, Soul Hunters: Hunting, Animism, and Personhood among the Siberian Yukaghirs (California 2007).

Williams 2015: H. Williams, Towards an archaeology of cremation. In: C. W. Schmidt/S. A. Symes (eds.), The Analysis of Burned Human Remains (Amsterdam ²2015) 259–293. https://doi.org/10.1016/C2013-0-18935-0.

Zachrisson/Krzewińska 2019: T. Zachrisson/M. Krzewińska, The «lynx ladies»: Burials furnished with lynx skins from the Migration and Merovingian Periods found in present-day Sweden. In: M. Augstein/M. Hardt (eds.), Sächsische Leute und Länder: Benennung und Lokalisierung von Gruppenidentitäten im ersten Jahrtausend. 66. Internationales Sachsensymposium = Neue Studien zur Sachsenforschung 10 (Braunschweig 2019) 103–119.

ØDEGÅRD 2017: M. ØDEGÅRD, Del av gravfelt; branngrav, mulig fotgrøft og grav. Falla 33/2, 57, Fet kommune, Akershus. Rapport Arkeologisk utgravning, Kulturhistorisk Museum, 2017).

ØDEGÅRD et al. 2017: M. ØDEGAARD/T. WINTHER/L. E. GJERPE/J. K. HELLAN, Gårder fra bronse- og jernalder. Foreløpige resultater fra dobbeltspor Dilling. Primitive Tider 19, 2017, 31–42.

ØSTIGÅRD/KALIFF 2020: T. ØSTIGÅRD/A. KALIFF, Likbrud og dødsbryllup: Sjelen, sykdommer og oldnordiske gravskikker (Uppsala 2020).

Online source

https://digitaltmuseum.no/011022725479/bjorneklo

Associate Prof. Anja Mansrud
Arkeologisk museum
Universitetet i Stavanger
Stavanger
Norway
anja.mansrud@uis.no

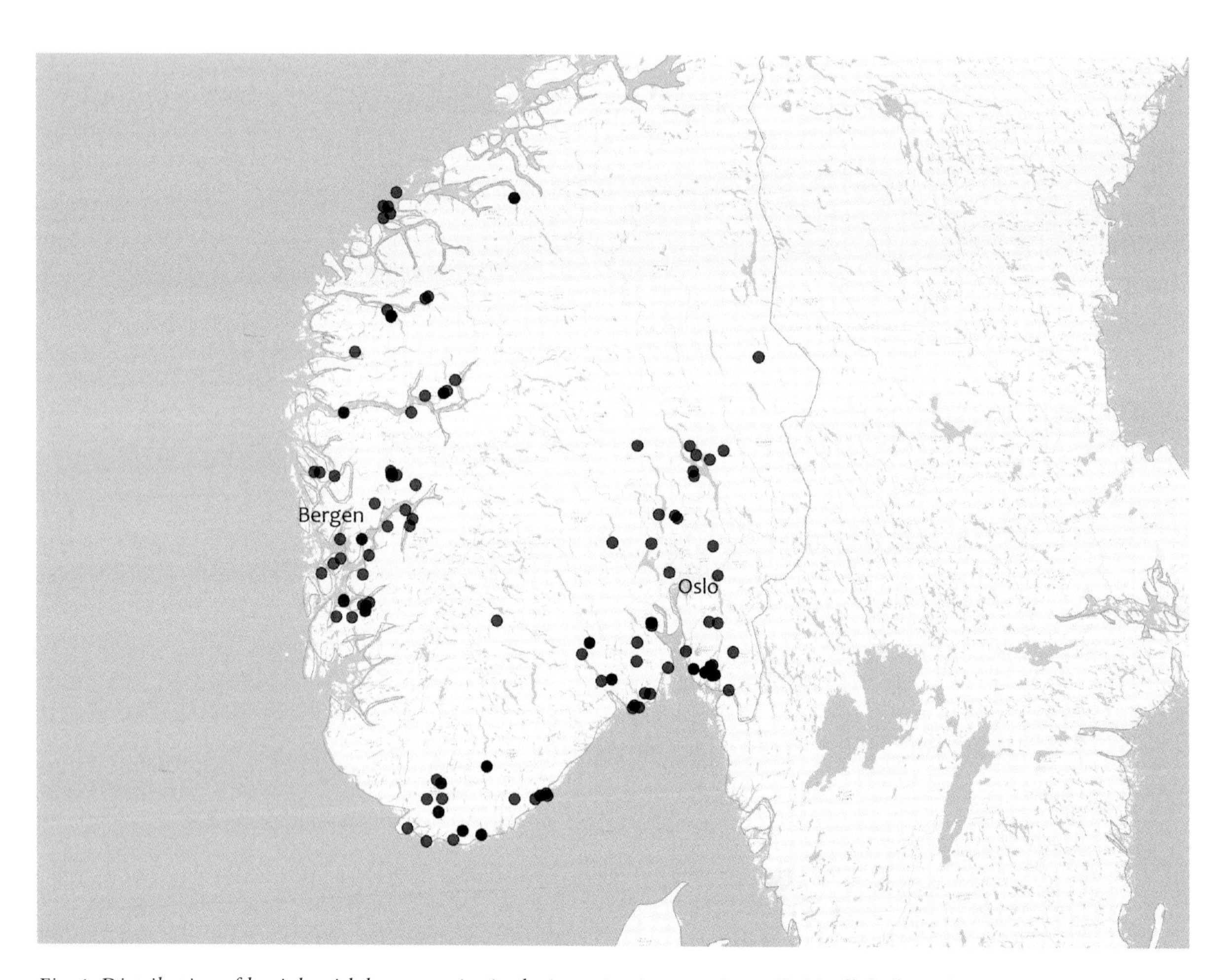

Fig. 1. Distribution of burials with bear remains in the investigation area (map G. Moell-Pedersen).

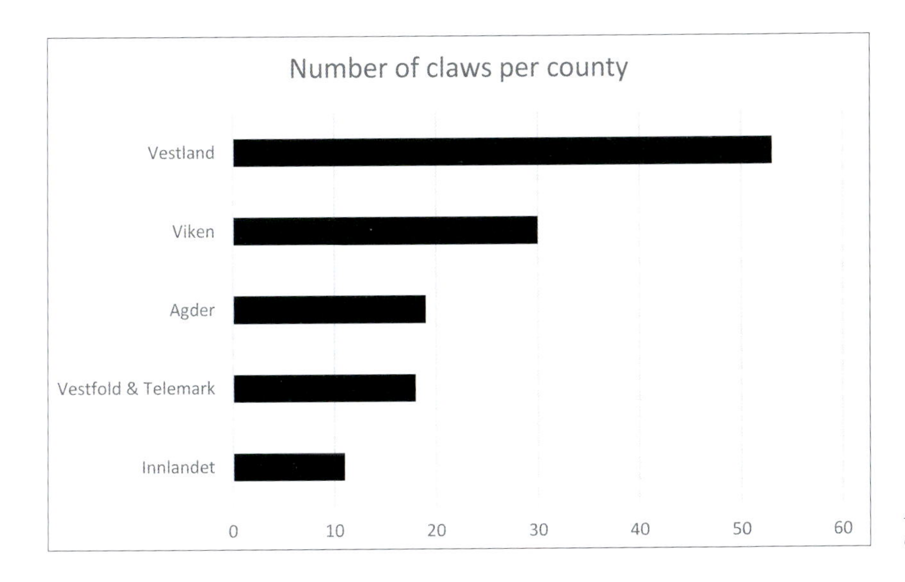

Fig. 2. Number of burials with bear claws per county in South Norway.

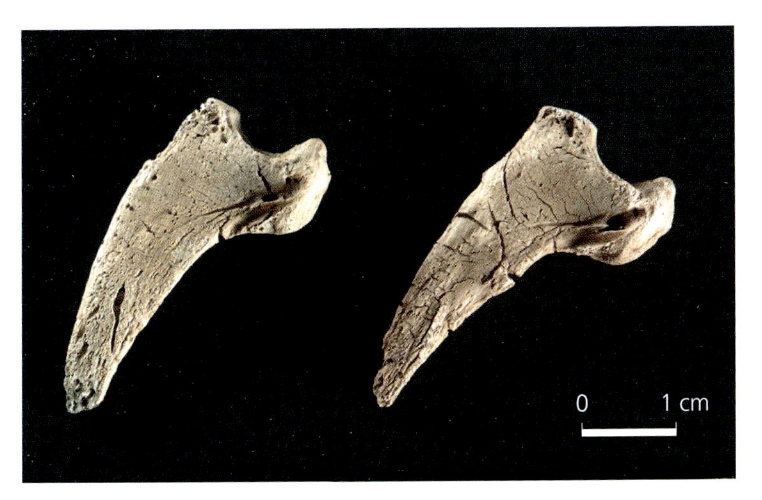

Fig. 3. Two unperforated bear claws from burial C22231 (photo K. Helgeland, Museum of Cultural History, University of Oslo; CC BY-SA 4.0).

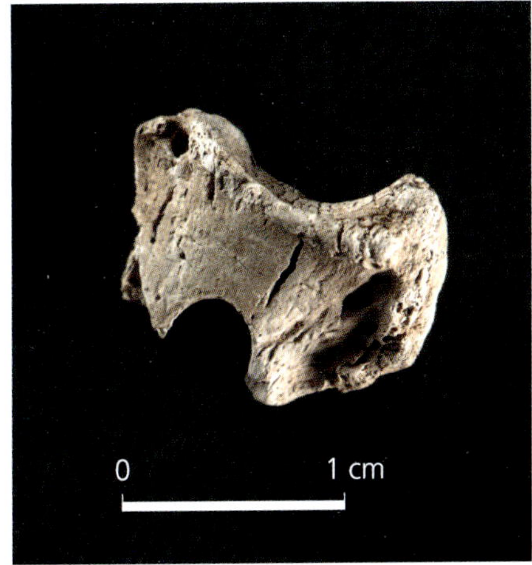

Fig. 4. Perforated bear claw, used as a pendant or amulet, from C35844 (photo K. Helgeland, Museum of Cultural History, University of Oslo; CC BY-SA 4.0).

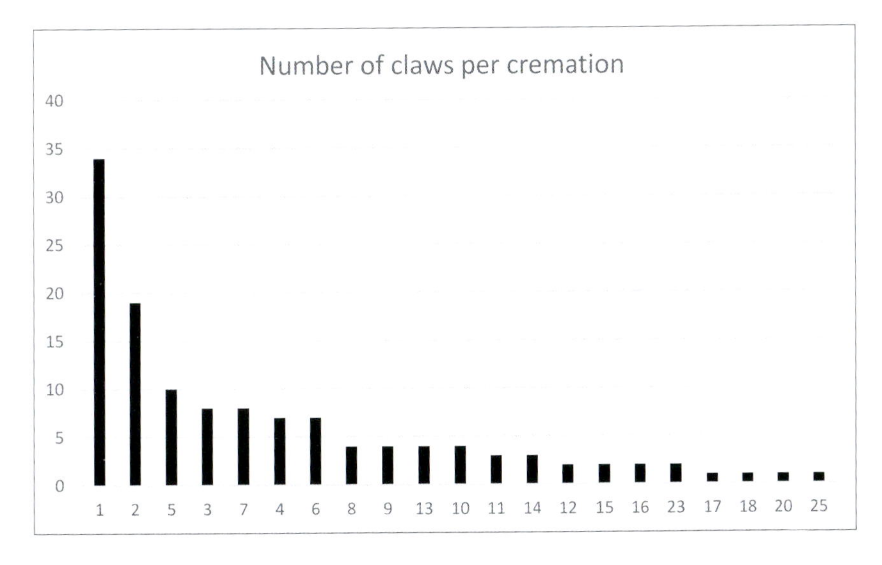

Fig. 5. Number of bear claws identified in the 130 cremations of South Norway. 80 % contain less than 10 claws, 40 % contain one or two claws.

Fig. 12. Bear claw once belonging to Daniel Andersen Tyskeberget (1778–1856) who allegedly shot around 100 bears (photo B. Løken, Anno Norsk Skogmuseum).

Fig. 13. Small bear figurine found in an Iron Age pit dwelling at Modvo in Sogn, western Norway (B11436; photo O. Espevoll, Bergen University Museum; CC-BY-NC-ND-3.0).

Fig. 14. Skinned bear paw with the claws removed (c. 1890–1894) from Veitestrand in Hafslo, used for assisting at difficult births and for healing mastitis in women and in cows (photo L. Asle Vold, De Heibergske Samlinger).

Fig. 15. The power of the paw? Five bear claws anatomically arranged as a paw by Ellen C. Holte (photo E. C. Holte, Museum of Cultural History, University of Oslo).

Appendix 1. Overview of Iron Age cremations with bear claws. Chronological periods – EIA: Early Iron Age; LIA: Late Iron Age. Chronological phases – PRIA: Pre-Roman Iron Age (500 BCE – 1 CE); RP: Roman period (1–400 CE); ERP: Early Roman period (1–200 CE); LRP: Late Roman period (200 CE–400); MP: Migration period (400–550 CE); MVP: Merovingian period (550–800 CE); VP: Viking period (800–1030 CE). Information about finds and context gathered from Unimus-portalen (Universitetetsmuseenes kulturhistoriske samlinger). B – Catalogue-numbers referring to Bergen University Museum, Bergen; C – Catalogue-numbers referring to the Museum of Cultural history, Oslo. Object type R. refers to Rygh *1885. Information about age and sex of the human skeletal remains mainly after* Holck *1986. Additional sources:* Shetelig *1912;* Resi *1986;* Samdal *2000;* Mansrud *2004a; b;* Fyllingen *2008;* Martens *2008;* Ødegård *2017;* Ødegård *et al. 2017.*

Museum no.	Site/region	Claws	Period	Phase	Description of burial and finds	Sex/age (mainly after Holck 1986)	Gender (acc. to grave goods)
C959	Fossum, Viken	5	EIA		Human and animal bone, no information about context.	Adult male	Unknown
C2707	Halsa, Agder	1	EIA		Submitted find, not excavated. Bronze cauldron filled with human and animal bones, wrapped in animal hide. Iron axe, glass mosaic bead, gaming pieces, fragments of iron sword, knife, spear.	Adult male	Male
C4941	Solum, Vestfold og Telemark	1	EIA		Mound containing a bronze cauldron filled with burnt human and animal bones, wrapped in a calf hide.	Adult male and female	Unknown
C5578	Kraby, Innlandet	2	EIA		Mound with ash layer, containing a pot with handle and spout, burnt human bones and fragments of a bone comb.	Adult and subadult (teeth), unknown gender	Unknown
C7340	Fjære, Agder	23	EIA		Mound containing a double grave, burnt human and animal bones; no information about grave goods.	Adult male and female	Unknown
C7353	Fjære, Agder	1	EIA		Mound; no information about grave goods.	Adult male	Unknown
C8999	Lyngdal, Agder	1	EIA	MP	Mound with stone chamber containing three bucket-shaped urns and a wooden bucket. Bronze cauldron filled with burnt human bones, wrapped in horse skin. Quartz whetstone, skin pouch containing an unidentifiable fragment of bronze, an iron celt with wooden shaft, an iron knife, a golden ring, five hazelnuts.	Adult male	Unknown
C11092	Evje verk, Evje, Agder	16	EIA		Mound with stone chamber, containing a bronze cauldron, covered in birch bark. Burnt human and animal bones, fragments of a bone comb.	Adult male	Unknown
C17305	Hvannes, Sauherad, Vestfold og Telemark	1	EIA		Mound with stone chamber, containing a bronze cauldron with burnt human bones, two iron spears, a shield boss with fittings, iron fittings and an unidentified bone artefact.	Adult male	Male

Cont. Appendix 1 ▶

Museum no.	Site/region	Claws	Period	Phase	Description of burial and finds	Sex/age (mainly after Holck 1986)	Gender (acc. to grave goods)
C17342	Sauherad, Vestfold og Telemark	3	EIA		Presumably from the same grave as C17305. Fragments of pottery and burnt human and animal bones.	Adult, unknown	Unknown
C19172	Opstad, Viken	7	EIA		Damaged mound with irregular stone chamber, scattered bones and artefacts. Pottery vessel, type R. 361 or R. 364. No human bones identified.	Not analysed	Not analysed
C20314	Arstad, Stange, Innlandet	1	LIA	VP	Mound with ash layer, burnt human and animal bones. Single-edged sword, spear, axe, three shield bosses, two bits, three knives, three rattles, iron cauldron, weaving baton, sickle, three scissors, two spindle whorls, iron chisel, casket handle, casket fittings, iron key, skewers, iron nails, four combs, a perforated bear claw.	Adult male	Male, female?
C20754	Hunn, Fredrikstad, Viken	1	EIA		Cremation containing burnt human bones; no information about context.	Two adult females	Unknown
C20755	Hunn, Fredrikstad, Viken	14	Unknown		Cremation containing burnt human bones; no information about context.	Adult male	Unknown
C21508	Store Dal, Viken	6	EIA	MP	Mound with ash layer, burnt human bones. Bucket-shaped pot, pottery vessel type R. 361, small clay cup, silver fittings, bronze fragments, comb.	Adult male	Unknown
C21556	Store Dal, Viken	2	EIA		Mound with ash layer, burnt human bones. Comb, fragments of bronze, fragments of pottery vessels, nails.	Unknown	Unknown
C21706	Tanum, Bærum, Viken	1	EIA	LRP	Mound with scattered burnt human and animal bones. Pottery vessels, types R. 368, 18R360, R. 364, pottery vessel with spout, fragments of a green glass cup.	Adult, unknown	Unknown
C21850	Vinje, Vestfold og Telemark	1	Unknown	Unknown	Large collection of burnt human bones, no info about context.	Adult male	Unknown
C21927	Bjerknes, Sigdal, Viken	2	EIA	MP	Mound with ash layer, burnt human and animal bones. Iron knife, iron fittings, three whetstones, comb, flint piece.	Adult male	Unknown

Cont. Appendix 1 ▶

Cont. Appendix 1

Museum no.	Site/region	Claws	Period	Phase	Description of burial and finds	Sex/age (mainly after Holck 1986)	Gender (acc. to grave goods)
C21945	Ø. Egeland, Kvinesdal, Agder	7	EIA	MP	Mound with burnt human and animal bones deposited in a cauldron type R. 353. Pottery vessel type R. 369, comb, bone knife. Bear claws with cutmarks.	Unknown	Unknown
C21957	Ski, Rakke-stad, Viken	1	EIA	RP	Collection of burnt human and animal bones. No info about context.	Adult male	Unknown
C22123	Stor-Stav, Ringsaker, Innlandet	2	EIA		Mound with ash layer, two assemblages of burnt human bones. Rounded bone pin, iron lock, fragment of iron bands and iron pins, slate whetstone.	Adult, unknown	Female
C22136	Løiten, Innlandet	5	Unknown		Mound with ash layer, burnt human and animal bones. Slate whetstone, knapping stone, piece of flint.	Adult, unknown	Unknown
C22231	Vennolum, Gran, Innlandet	2	EIA	RP	Damaged grave. Bronze cauldron (type R. 352), filled with burnt human bones and a bronze shield boss, bronze fittings for a shield handle (type R. 222), various fittings, bronze belt-end fittings, fragments of bronze/tin belt buckle, two spears, iron band and handle belonging to the cauldron. Beside the cauldron, a double-edged sword was deposited.	Not analysed	Male
C22466	Sande, Sem, Vestfold og Telemark	1	EIA		Mound with ash layer, burnt human bones. Beak-shaped brooch, bronze belt-end fittings, fragments of comb, bronze fragments, equal-armed brooch.	Adult male	Female
C22767	Gile, Innlandet	1	EIA		Mound with ash layer, burnt human and animal bones. Bronze celt, comb, stone spindle whorls.	Adult male	Male/female?
C23256	Voldhaugen, Søgne, Agder	9	EIA	MP	Damaged mound with stone chamber, containing a bronze cauldron type R. 353, filled with burnt human and animal bones, comb, two gaming pieces, iron handle, fittings and nails. The bones were wrapped in felt and bear pelt.	Two adults (one younger, one older), unknown gender	Unknown
C23295	Gjevre, Nordre Land, Innlandet	1	EIA	LRP	Mound with scattered, cleaned cremated human and animal bones. Equal-armed brooch, bronze fibula, two bronze needles, two gold spiral finger rings, two iron knives, fragment of curved knife, iron fittings for a wooden container, bone spindle whorls, bone pin, iron nails.	Adult female, adult of unknown gender	Female

Cont. Appendix 1 ▶

Museum no.	Site/region	Claws	Period	Phase	Description of burial and finds	Sex/age (mainly after Holck 1986)	Gender (acc. to grave goods)
C23928	Søgne, Agder	13	EIA	RP	Mound with a small cairn built up around a cauldron, which was deposited on a flat stone slab. The cauldron was filled with burnt human bones, a bone pin, and comb. Cutmarks on the adult skeletal remains.	Adult male, infans II	Unknown
C25078	Gvarv, Sauherad, Vestfold og Telemark	1	EIA	Unknown	Assemblage of burnt bones found under a stone slab; submitted find, not excavated.	Adult male	Unknown
C26307	Y. Vatne, Hægebostad, Agder	3	EIA	MP	Mound with ash layer, covered by a layer of birch bark. Small amounts of burnt human bone. Sherds of bucket-shaped pot and type R. 361, fragments of a comb and 12 gaming pieces, silver ring, bronze belt buckle, fragment of a brooch, iron fragments, animal tooth.	Adult, unknown.	Female
C26398	Birkeland, Agder	2	EIA	RP	Mound with cauldron (type R. 352), covered with birch bark and a stone slab. Cauldron filled with cleaned, burnt human bones with cutmarks, a silver fibula (type R. 231), a bronze fibula, a bone comb.	Adult male	Female
C26791	Veinstein, Askim, Viken	5	Unknown		Submitted find; no information about context. Iron spear, pottery sherds, burnt human bones.	Adult, unknown	Unknown
C27077	Raknehaugen, Ullensaker, Viken	1	EIA		Burnt bones, unidentifiable.	Unknown	Unknown
C28980	Hunn, Borge, Viken	1	EIA	RP	Mound with multiple burials. C28980 was deposited at the bottom layer. Bronze cauldron type R. 352, filled with burnt human bones, two fragments of bronze stirrups, iron shield boss, fragments, or iron fittings, presumably for a shield, iron spear (type R. 207), iron spear (type R. 206).	Adult female, infans II	Unknown
C28985	Hunn, Borge, Viken	1	EIA		Mound with ash layer, six small silver balls, a small bronze ball, silver object, bronze needle, pottery sherds, flint flakes.	Adult, unknown	Female?
C28986f	Hunn, Borge, Viken	4	EIA	RP	Mound with several burials. The grave contained a gold berlock (type R. 2839), a comb (type R. 158), fragments of a spindle whorl, four small silver balls, a fragment of a silver pin.	Adult male	Unknown

Cont. Appendix 1 ▶

Cont. Appendix 1

Museum no.	Site/region	Claws	Period	Phase	Description of burial and finds	Sex/age (mainly after Holck 1986)	Gender (acc. to grave goods)
C32757	Store Tune, Tune, Viken	13	EIA		Primary burial in a mound containing a pottery vessel with a long neck, belt fittings, fragments of green, smelted glass, 10 gaming pieces, a comb, two silver fragments.	Infans II	Unknown
C34549	Lardal, Vestfold og Telemark	10	EIA		Disturbed mound with two burials. The primary burial contained three pottery vessels placed in a thick charcoal layer. The burnt bones (from several individuals) were partly deposited in one of the vessels but also scattered in the charcoal layer.	Adult female; young individual of unknown gender	Unknown
C34692	Kathrineborg, Brunlanes, Vestfold og Telemark	13	EIA		Mound with burial containing several round bone pins, five blue beads, sherds from several pots, resin for several birch bark containers, burnt human bones.	Adult male	Female
C34814	Tingelstad, Gran, Innlandet	5	LIA	MVP	Mound containing an iron cauldron, burnt human bones, fragments of several combs, fragments of a bronze fibula or fittings, decorated with an eagle, a pair of tweezers. Later investigations have revealed a one-edged knife, presumably a weapon-knife, which might indicate that this is a double burial (I. M. Røstad, pers. comm.).	Not analysed	Female?
C34884a	Fuskeland, Mandal, Agder	11	EIA		Mound with three burials. Burial I contained burnt human bones and bear claws.	Adult male	Unknown
C34884f	Fuskeland, Mandal, Agder	4	EIA		Mound with three burials. The bear claws belonged to Burial II, an organic container filled with burnt bones, 14 fragments of gaming pieces and 15 comb fragments.	Adult male	Unknown
C35042	Manvik, Brunlanes, Vestfold og Telemark	5	EIA	MP	Disturbed mound containing a cauldron filled with large amounts of burnt human and animal bones, two fragments of a bone arrowhead, fragments of bone artefacts, probably gaming pieces and a comb, sherds of a bucket-shaped pot.	Adult male	Male
C35109	Skjervum øvre, Gran, Innlandet	2	EIA	RP	Mound with ash layer containing a spiral ring of bronze, a bronze ring, a needle container or shaft of bronze, fragments of bronze fittings and various fragments of bronze. Comb and round bone pin, pottery sherds, burnt flint piece, an unburnt tooth (horse?).	Unknown	Female
C35237	Berg, Brunlanes, Vestfold og Telemark	6	Bronze Age		Pit with no marking above ground, containing burnt human bones, resin from an organic container, and bear claws. Uncertain dating; assumed: Bronze Age.	Unknown	Unknown

Cont. Appendix 1 ▶

Cont. Appendix 1

Museum no.	Site/region	Claws	Period	Phase	Description of burial and finds	Sex/age (mainly after Holck 1986)	Gender (acc. to grave goods)
C35805	Gaalaas, Ringsaker, Innlandet	1	EIA		Disturbed mound, the primary burial was a richly furnished female inhumation. Structures with burnt bones and artefacts were scattered in and under the mound. The bear claws belonged to scatter llll.	Unknown	Unknown
C35844	Fjære, Fevik, Agder	1	EIA		Burial in a pit under a stone circle with burnt human bones. Green glass bead, shaped like a ring, two blue glass beads, fragments of green glass, fragments of a smelted silver object, pottery sherds, flat and round fragments of bronze, iron piece with a bronze nail, fragments of a bone pin, fragments of ornated bone, possibly a comb. Fragment of bear claw with a drilled hole.	Adult male (identified as such by P. Holck, 22.01.1983).	Unknown
C35859	Landvik, Grimstad, Agder	6	EIA	MP	Disturbed burial, with min. two burials. The cremation contained a cruciform fitting of gilded bronze, a fibula (type Shetelig 1912, fig. 37), white glass bead shaped like a ring, 15 narrow iron tubes, fittings for textiles, fragments of comb.	Unknown	Female
C36400	Evje, Agder	8	EIA		New objects from mound previously catalogued under C34327 and C34328. Contained burnt human bones, glass cup, pottery sherds, fragments of iron and bronze, nine gaming pieces, two dice, a comb, resin from an organic container.	Unknown	Unknown
C36679	Revøy Midtre, Lyngdal, Agder	1	EIA		Mound with ash layer, containing burnt human bones, a simple bronze fibula, bronze belt buckle, bronze fittings, a pottery vessel with handle, pottery sherds, nine gaming pieces, two dice, comb.	Adult female	Female
C37631k	Bratsberg, Skien, Vestfold og Telemark	2	EIA	ERP	Mound with several burials. Grave I: a scatter containing a burnt human bone, a bronze fibula, comb, a bone pin, spindle whorls, bone arrowheads, pottery sherds, resin from an organic container.	Unknown	Male/female?
C37631q	Bratsberg, Skien, Vestfold og Telemark	15	EIA	ERP	Mound with several burials, see above. Grave II: an ash layer at the bottom, containing a comb, a bronze container, a knife.	Unknown	Unknown
C38506	Berg Skole, Halden, Viken	4	EIA		Mound containing a glass cup, belt buckle and belt fittings of bronze, iron needle, several pottery vessels, comb, flint strike-a-light, whetstone.	Unknown	Female
C52083	Aas Østre, Sande, Vestfold og Telemark	1	EIA	RP	Mound with two burials. The bear claws belonged to a secondary cremation, a birch bark container with burnt human bones, several pottery vessels, fragments of gold jewellery, fragments of a fibula, several glass beads.	Adult female (Bones analysed by P. Holck, 24.11.1998)	Female

Cont. Appendix 1 ▶

Cont. Appendix 1

Museum no.	Site/region	Claws	Period	Phase	Description of burial and finds	Sex/age (mainly after Holck 1986)	Gender (acc. to grave goods)
C52092	Bøli, Råde, Viken	10	EJA	PRIA	An imported bronze pan used as urn. It contained burnt human bones, a folded single-edged sword, scabbard fittings, a lancehead, a spiked shield boss, a plain golden finger ring, claws from at least two bear paws, and a pair of slender knife blades (cf. Martens 2008, 309).	Not analysed	Male
C55949	Molle østre, Spydeberg, Viken	1	EIA	RP	Burial pit marked by a stone. Burnt human bones from a young individual and an ornamented bone pin. Dated to 240–420 CE (Beta Analythic Inc., Lab. no. 227486).	Adult, unknown gender (Bones analysed by H. Fyllingen, 2006)	Female
C56077	Rødbøl, Vestfold og Telemark	3	EIA		Burial pit containing human bones from two individuals, a pottery vessel with spout, resin from an organic container.	Adult; Infans II (bones analysed by P. Holck)	Unknown
C58714	Veien kultur-minnepark, Viken	1	EIA		Mound with several burials. The bear claws were found with scattered burnt human bones and pottery under a flat stone.	Adult, unknown gender	Unknown
C60497	Fet, Viken	6	EIA	MP	Burial pit, presumably covered by a mound removed by ploughing. Human and animal bones, several gaming pieces (cf. Ødegård 2017).	Adult female	Unknown
C61166	Dilling, Viken	10	EIA	PRIA	Burial ground located close to a cluster of houses dated to the Pre-Roman Iron Age. The pit contained burnt human bones and bear claws (cf. Ødegård et al. 2017).	Adult, unknown gender	Unknown
B3358	Borgund Ø, Vestland	16	EIA	RP	Mound with several burials; the primary one represented by a cauldron covered by a slab and a small cairn, containing burnt human bones, bird bones, comb, ornamented bone plate, a Roman gold coin used as a pendant. Inside the cauldron: traces of resin and birch bark. Another structure contained pottery, a smelted silver brooch with spiral and dragon ornaments, a bronze ring, bronze fittings, iron nails and unidentified iron fragments, unburnt horse bones/teeth.	Not analysed	Female
B3855	Øvrabø Søre, Kvinnherad, Vestland	9	EIA		No information.	Not analysed	Unknown
B4003	Sogn, Vestland	3	EIA		Mound with chamber containing an iron cauldron placed on a slab. Cauldron filled with burnt bones, fragments of a comb, a green glass bead. Inside the bottom of the cauldron decoration in form of a hexa-gonal star surrounded by two incised double circles.	Not analysed	Unknown

Cont. Appendix 1 ▶

Cont. Appendix 1

Museum no.	Site/region	Claws	Period	Phase	Description of burial and finds	Sex/age (mainly after Holck 1986)	Gender (acc. to grave goods)
B4095	Eide, Fjælberg, Vestland	2	EIA		Mound with a stone chamber, containing a bucket-shaped pot, filled with large amounts of burnt bones, sand and charcoal, bronze rivets, comb fragments.	Not analysed	Unknown
B4207	Jondal, Hardanger, Vestland	6	EIA		Mound with a small stone chamber containing a cauldron (type R. 352), wrapped in birch bark, filled with burnt bones, min. 15 gaming pieces, two bone pins, two combs, a small bone plate, a piece of glass, fragments of teeth.	Not analysed	Female
B4259	Naustdal i Søndfjord, Vestland	1	EIA		Mound with a kernel of stone surrounding an assumed wooden chamber containing a cauldron (type R. 352) filled with burnt bones, gaming pieces, pottery, fragments of birch bark, a piece of pyrite used for making fire.	Not analysed	Unknown
B4367	Alne, Vikevik, Vestland	5			Mound with a scatter of burnt bones and fragments of a comb.	Not analysed	Unknown
B4591	Gloppen, Vestland	4	EIA		Mound with pyramid-shaped chamber, burnt human bones. Bronze fibula type R. 247, fragments of a comb and another bone artefact, fragment of resin, fragment of wooden bucket (perhaps not part of the burial).	Not analysed	Female
B4593	Evebø, Gloppen, Vestland	2	EIA		Mound with stone chamber, burnt human and animal bones. Pottery (types R. 367 and 369), iron fragment, fragments of a comb, clay spindle whorls.	Not analysed	Female
B4866	Etne, Vestland	1	EIA		Mound with urn covered by stones and a slab on top. Pottery vessel type R. 363 with burnt bones, no artefacts.	Not analysed	Unknown
B4877	Kvam, Vestland	1	EIA		Mound (cairn) with a cauldron (type R. 352) placed in an opening between stones, on top of a birch bark layer covered by a slab. The cauldron was filled with burnt bones and pieces of birch bark.	Not analysed	Unknown
B4978	Vinnes, Fuse, Vestland	1	Unknown		Cairn with a small stone chamber, containing burnt bones, and an iron fragment.	Not analysed	Unknown
B5060	Spangereid, Vestland	1	EIA		Submitted find, uncertain context. Silver buckle with animal heads, two gilded knobs of fibulas, fragments from two bucket-shaped pots and several pots with handles, some unburnt human and animal bones (horse?), flint blade.	Not analysed	Female

Cont. Appendix 1 ▶

Cont. Appendix 1

Museum no.	Site/region	Claws	Period	Phase	Description of burial and finds	Sex/age (mainly after Holck 1986)	Gender (acc. to grave goods)
B5693	Voss, Vestland	4	EIA		Disturbed mound with small square stone chamber, covered by a slab. Contained sherds of a bucket-shaped pot and burnt bones. Other artefacts may have been removed from the grave.	Not analysed	Unknown
B5838	Etne, Vestland	2	EIA		Disturbed mound with small stone chamber. In the chamber, a bucket-shaped pot filled with burnt bones, placed on top of a charcoal layer. Fragments of bone objects, probably pins, a fragment of a round bone pin.	Not analysed	Female
B5931	Kvinnherad, Vestland	4	EIA	LRP	Mound with ash layer at the bottom and cremation deposit, both belonging to the same burial. Contained large assemblage of burnt bones, a gold finger ring with a serpent head terminal, fragments of a bronze chain, a bronze needle, at least six smelted beads, a round bone pin (type R. 277), fragments of flat bone pins, fragments of a comb, bone plate with ornaments, curved knife, an iron pole, some iron nails and fittings. Fragments of min. three pottery vessels (one bucket-shaped and one with handle), slate whetstone.	Not analysed	Female
B5963	Rimbereid, Fitje, Vestland	5	EIA	RP	Partly disturbed mound with a low chamber consisting of a long slab resting on stones at the ends. Under the slab an ash layer with burnt bones, sherds from at least two pottery vessels, comb, three dice, nine gaming pieces.	Not analysed	Unknown
B6086	Manger, Vestland	2	EIA	MP	Mound with a small stone chamber, a cauldron filled with burnt bones, and fragments of a bone comb.	Not analysed	Unknown
B6103	Kvinherad, Vestland	?	EIA	MP	Mound with complex stratigraphy. In a small room covered by a slab was an assemblage of cleaned burnt bones, wrapped in birch bark. In the layer below a small drop of smelted gold, fragments of a bronze plate, a bronze needle, a piece of smelted bronze, fragments of a comb and smelted glass, burnt bones from a small mammal. Number of claws not noted.	Not analysed	Female
B6111	Luster, Sogn, Vestland	11	EIA	MP	Mound with finds of an iron plate, iron fitting with a bullet-shaped head, iron fragments, large pottery vessel with two handles, burnt bones, fragments of a comb, a gaming piece (type R. 177), fragments of a thin bone plate, ornamented at both sides.	Not analysed	Unknown
B6197	Stryn, Vestland	2	EIA	RP	Disturbed mound. Submitted find. Cauldron (type R. 352) placed on a stone slab beside a large boulder, surrounded by a layer of birch bark. Burnt bones, no artefacts.	Not analysed	Unknown

Cont. Appendix 1 ▶

Cont. Appendix 1

Museum no.	Site/region	Claws	Period	Phase	Description of burial and finds	Sex/age (mainly after Holck 1986)	Gender (acc. to grave goods)
B6200	Etne, Vestland	?	EIA	RP	Mound (cairn) with several burials. Grave II: ash layer, covered by slabs, containing burnt human bones and bear claws, a strike-a-light, an oval pebble of reddish quartzite with use wear, two flat bone pins, fragments of a round bone pin, a comb and an unidentified bone tool, possibly a meat knife. Sherds of pottery, type R. 360. Number of claws not noted.	Not analysed	Female
B6233I	Etne, Vestland	7	EIA		Mound with several cremations. The primary one was a cauldron wrapped in birch bark and filled to the brim with burnt human bones, with fragments of a comb.	Not analysed	Unknown
B6233IIa	Etne, Vestland	6	EIA		Mound with several ash layers with burnt bones and artefacts. Grave IIa was located north of the cauldron, and contained sherds of a bucket-shaped pot, resin from a wooden container, fragments of a comb.	Not analysed	Unknown
B6475	Vangen, Voss, Vestland	?	EIA	MP	Disturbed mound, built around a boulder, with an ash layer containing a small round red bead, fragments of a comb and a gaming piece (type R. 177), two iron arrowheads, fragments of iron, pottery sherds from a bucket-shaped pot and a pot with handle. Number of claws not noted.	Not analysed	Male
B6539	Stryn, Vestland	6	EIA		Mound with a large stone chamber, filled with a gravel layer and containing unburnt bones, wood, and teeth (probably horse), but no artefacts. At the outskirt of the mound a cauldron lined with felt, filled with a large assemblage of burnt bones and artefacts: fragments of two glass cups, types R. 337 and R. 338c, fragments of several combs and two gaming pieces, covered by a slab. The bones were wrapped in birch bark, the package was tied together with a ribbon of bast.	Not analysed	Unknown
B6597	Ullensvang, Vestland	7	EIA	MP	Small mound with a cauldron (type R. 353) covered by a slab. The cauldron was surrounded by a layer of charcoal and burnt bones; it was filled with burnt human and animal bones (some identified as sheep/goat), a knob from a bronze fibula, a bronze fragment from another brooch, a belt ring of bronze, fragments of two or three combs, a bone knife (type R. 450), two quartz crystals.	Not analysed	Female
B6691	Vik, Vestland	25	EIA	MP	Mound with large stone chamber, three cremations. The bear claws belonged to a cremation placed on top of the primary inhumation. Large assemblage of burnt bones, two combs, four bone pins, a bone plate (possibly a knife), bone arrowheads, fragments of burnt bronze.	Not analysed	Male and female?

Cont. Appendix 1 ▶

Cont. Appendix 1

Museum no.	Site/region	Claws	Period	Phase	Description of burial and finds	Sex/age (mainly after Holck 1986)	Gender (acc. to grave goods)
B6700	Tysnes, Vestland	17	EIA	MP	Mound with a simple urn filled with large amounts of burnt human and animal bones (some identified as bird), fragments of smelted silver, fragments of a comb, fragments of a bone spoon, and fragments of a bone knife with runic inscriptions.	Not analysed	Unknown
B6756k	Mundheim, Strandebarm, Vestland	10	EIA	MP	Large mound with one inhumation and three cremations. The bear claws belonged to the primary burial, Grave 3, which was a combination of ash layer and stone chamber. Among the burnt bones were fragments of a comb, three bone dice (type R. 176), one cubic bone die, eight gaming pieces, small fragments of smelted bronze and two sherds of a bucket-shaped pot.	Not analysed	Unknown
B6756h	Mundheim, Strandebarm, Vestland	1	EIA	MP	Grave 4 in the mound, a secondary cremation placed over Grave 3 at the outskirts of the mound. The burnt bones were probably deposited in a wooden bucket with traces of resin and iron fittings, several bone pins, a comb, bone fragments (possibly a knife), and sherds of a bucket-shaped pot.	Not analysed	Female
B6763	Ulvik, Hardanger, Vestland	5	EIA	MP	Mound with stone chamber, containing large amounts of burnt bones, fragments of a comb, two flat bone pins, a drilled plate of bone, fragments of a bone knife, some drops of smelted bronze, some pottery sherds, tar for a wooden container.	Not analysed	Female
B6807	Norheim, Kvam, Vestland	9	EIA	MP	Mound built around a low stone hill containing an ash layer covered by slabs. Together with the burnt bones there were small lumps of smelted bronze, smelted blue glass, an iron needle, hook and awl, iron nails. Fragments of a comb, four bone pins, fragments of an ornamented bone knife, iron celt, pottery sherds.	Not analysed	Female
B7608	Vangen, Voss, Vestland	3	EIA	MP	Disturbed mound with stone chamber containing cleaned burnt bones, an oval bronze belt buckle, an iron fragment (knife?), two fragments of a comb.	Not analysed	Unknown
B7908	Støle, Etne, Vestland	7	EIA	MP	Mound with a stone chamber covered by charcoal, presumably from a fire on top of the grave. Floor layer was also burnt. Burnt human and animal bones (one pig tooth identified) scattered in the chamber and the charcoal layer. One bucket-shaped pot, one vessel with handle, one rounded vessel. 21 gaming pieces (type R. 177), fragments of a comb.	Not analysed	Unknown

Cont. Appendix 1 ▶

Cont. Appendix 1

Museum no.	Site/region	Claws	Period	Phase	Description of burial and finds	Sex/age (mainly after Holck 1986)	Gender (acc. to grave goods)
B7956	Støe, Etne, Vestland	3	EIA	RP	Mound with long stone chamber, cauldron type R. 353 placed on top of a slab deposited in a layer of birch bark. An ash layer was stretching outside the chamber. Six pottery sherds from two vessels, some slag and burnt hazelnut shells.	Not analysed	Unknown
B8200	Kinsarvik, Ullensvang, Vestland	7	EIA	MP	Mound with long stone chamber, containing burnt bones, a bronze brooch, a knob from a cruciform brooch, a drop of smelted silver, fragments of a comb and gaming pieces (type R. 177), sherds of bucket-shaped pots, a belt stone of grey/white quartzite (type R. 155).	Not analysed	Female
B8697	Eknes, Hosanger, Vestland	9	EIA		Mound with square stone chamber. Gaming pieces and burnt bones were placed on a slab and covered by soil and another slab. East of the chamber an ash layer containing more burnt bones, gaming pieces and claws.	Not analysed	Unknown
B8853	Oppedal, Lavik, Vestland	12	EIA	MP	Large mound with a stone chamber containing burnt bones, fragments of bone pin, faceted glass cup, fragments of a bucket-shaped pot, fittings from a wooden bucket, birch bark used as lid, possibly also as wrap for the wooden container.	Not analysed	Unknown
B8871	Haram, Sunnmøre, Vestland	14	EIA	MP	Disturbed mound with stone chamber, containing fragments of a comb and burnt bones.	Not analysed	Unknown
B8872	Haram, Sunnmøre, Vestland	8	EIA	MP	Ash layer under stone circle, containing burnt bones, pottery sherds, fragments of a comb, a large amount of bone arrowheads, two gaming pieces, fragments of burnt glass and bronze, resin and iron handles from a wooden container, strike-a-light stone of quartzite.	Not analysed	Male
B8983	Hamre, Leikanger, Vestland	12	EIA	MP	From mound 6333. Cauldron placed in a stone chamber, contained burnt and unburnt human and animal bones, a glass cup, a solidus (Theodosius I), two combs (type R. 159), bronze fragments, pottery sherds, fragments of birch bark.	Not analysed	Unknown
B10042	Uggdal, Tysnes, Vestland	7	EIA		Disturbed mound built around a boulder. Burnt bones and pottery sherds in red-burnt layer of clay.	Not analysed	Unknown
B10097	Indre Vereide, Gloppen, Vestland	3	EIA		Pit found in the lower layers of a big mound, containing cleaned burnt bones, fragments of a comb.	Not analysed	Unknown

Cont. Appendix 1 ▶

Cont. Appendix 1

Museum no.	Site/region	Claws	Period	Phase	Description of burial and finds	Sex/age (mainly after Holck 1986)	Gender (acc. to grave goods)
B10790	Giske, Borgund, Vestland	18	EJA		Mound with a cremation and an inhumation (previously B 894–B 900, re-catalogued). The bear claws belonged to the cremation, containing two bronze cauldrons (type R. 353), a ring of gold, smelted drops of gold, silver and bronze, various fittings, tacks and tubes of bronze, two dice, fragments of gaming pieces (type R. 177), and a comb, pottery, snail shell (*Patella vulgata*), burnt human and animal bones, unburnt bird bones, a fragment of birch bark with seam, probably a cover for the cauldron.	Not analysed	Unknown
B11546	Giske, Summøre, Vestland	8	EIA	LRP	Mound with stone chamber, cauldron filled with burnt human and animal bones (some identified as goat/sheep), a gold medallion (copy of a Roman emperor medallion with animal figures). Payment-ring and tack of silver, two silver cups, bronze fragments, fragments of a comb, five bone arrowheads, pottery, iron rivets.	Not analysed	Male
B12046	Gjerstad, Voss, Vestland	3	EIA		Mound with circular ash layer at the bottom, surrounded by stones. Burnt human and animal bones (some identified as bird) and artefacts scattered in the layer. Fragment of a bronze fibula, bronze ribbon, several bronze fittings, some of them presumably for a wooden bucket. Fragments of 46 gaming pieces, fragments of a comb and smelted green glass. Nails and ribbon of silver. Iron buckle with needle. Bear and bird claws.	Not analysed	Female
B12048	Haram, Sunnmøre, Vestland	15	EIA	LRP	Disturbed mound, artefacts not in situ except a fingerring of gold placed on a flat stone. A bronze plate, with burnt bones and artefacts, had been wrapped in skin and placed on an ash layer at the bottom of the grave. An armring and three fingerrings of gold, a double bracteate, with Roman emperor portrait on both sides. Fragments of glass, comb, gaming pieces, a piece of skin with a visible seam and brown animal hair (a pouch or bag?), birch bark, flint strike-a-light.	Not analysed	Unknown
B13367	Ullensvang, Vestland	1	EIA		Disturbed mound with a stone chamber, small amount of burnt bones.	Not analysed	Unknown
B13955	Kvåle, Sogndal, Vestland	4	EIA	MP	No information about context. Animal head from the foot of a cruciform bronze brooch, bronze buckle, fragment of silver fibula, bronze fragments, smelted silver, belt stone of quartz/quartzite, iron rivets and nails, bucket-shaped pottery and pottery with a handle, ornamented bone plate, a large amount of gaming pieces, ornamented bone piece, flint strike-a-light.	Not analysed	Female
B13956	Kvåle, Sogndal, Vestland	11	EIA		No information about context. Fragments of a bone comb and burnt human bones.	Not analysed	Unknown

Appendix 2. Sites with catalogue numbers and coordinates.

Coordinate system	East	North	Catalogue no.	Site
EU89-UTM, zone 32	556973	6592360	C959	Fossum
EU89-UTM, zone 32	404371	6434491	C2707	Halsa
EU89-UTM, zone 32	527510	6561698	C4941	Solum
EU89-UTM, zone 32	602810	6725525	C5578	Kraby
EU89-UTM, zone 32	476288	6469547	C7340	Fjære
EU89-UTM, zone 32	476288	6469547	C7353	Fjære
EU89-UTM, zone 32	391908	6456399	C8999	Lyngdal
EU89-UTM, zone 32	432042	6492732	C11092	Evje verk, Evja
EU89-UTM, zone 32	517514	6591699	C17305	Sauherad
EU89-UTM, zone 32	517514	6591699	C17342	Sauherad
EU89-UTM, zone 32	616441	6573868	C19172	Opstad
EU89-UTM, zone 32	602651	6571407	C20755	Hunn
EU89-UTM, zone 32	615916	6738448	C20314	Arstad, Stange
EU89-UTM, zone 32	602651	6571407	C20754	Hunn
EU89-UTM, zone 32	620367	6567069	C21508	Store Dal
EU89-UTM, zone 32	620367	6567069	C21556	Store Dal
EU89-UTM, zone 32	582757	6640758	C21706	Tanum, Bærum
EU89-UTM, zone 32	440424	6609213	C21850	Vinje
EU89-UTM, zone 32	536606	6671933	C21927	Bjerknes, Sigdal
EU89-UTM, zone 32	382643	6466676	C21945	Ø. Egeland, Kvinesdal
EU89-UTM, zone 32	635476	6584817	C21957	Ski, Rakkestad
EU89-UTM, zone 32	599317	6749511	C22123	Stor-Stav, Ringsaker
EU89-UTM, zone 32	627337	6745832	C22136	Løiten
EU89-UTM, zone 32	588971	6691915	C22231	Vennolum, Gran
EU89-UTM, zone 32	581786	6572244	C22466	Sande, Sem
EU89-UTM, zone 32	601952	6729049	C22767	Gile
EU89-UTM, zone 32	427385	6438336	C23256	Voldhaugen, Eik, Søgne
EU89-UTM, zone 32	556759	6749280	C23295	Gjevre, Nordre Land
EU89-UTM, zone 32	427385	6438336	C23928	Søgne
EU89-UTM, zone 32	511063	6582896	C25078	Gvarv, Sauherad
EU89-UTM, zone 32	395296	6467162	C26001	Snartemo
EU89-UTM, zone 32	390707	6481678	C26307	Y. Vatne, Hægebostad
EU89-UTM, zone 32	455041	6466985	C26398	Birkeland
EU89-UTM, zone 32	622619	6607986	C26791	Veinstein, Askim
EU89-UTM, zone 32	618659	6669694	C27077	Raknehauge, Ullensaker
EU89-UTM, zone 32	617741	6566553	C28980	Hunn, Borge
EU89-UTM, zone 32	617741	6566553	C28985	Hunn, Borge
EU89-UTM, zone 32	617741	6566553	C28986	Hunn, Borge
EU89-UTM, zone 32	617741	6566553	C28989	Hunn, Borge
EU89-UTM, zone 32	568179	6608234	C29261	Åshaugen, haug 1, Sande
EU89-UTM, zone 32	568609	6608020	C29262	Åshaugen, haug 3, Sande
EU89-UTM, zone 32	568179	6608234	C29265	Åshaugen, haug 5, Sande
EU89-UTM, zone 32	432042	6492732	C29610	Evje

Cont. Appendix 2 ▶

Cont. Appendix 2

Coordinate system	East	North	Catalogue no.	Site
EU89-UTM, zone 32	611872	6568957	C29853	Ula
EU89-UTM, zone 32	611872	6568957	C29859	Ula
EU89-UTM, zone 32	611872	6568957	C29860	Ula
EU89-UTM, zone 32	366392	6443470	C29861	Vanse, Lista
EU89-UTM, zone 32	567266	6551420	C30149	Istrehågan, Larvik
EU89-UTM, zone 32	481181	6471550	C30491	Trålum, Fjære
EU89-UTM, zone 32	616286	6573944	C31072	Opstad
EU89-UTM, zone 32	274588	6579365	C31072	Opstad
EU89-UTM, zone 32	6820130	655821	C32693	Eltedalen nordre, Trysil
EU89-UTM, zone 33	618239	6574479	C32757	Store Tune, Tune
EU89-UTM, zone 32	556224	6577531	C34549	Lardal
EU89-UTM, zone 32	558155	6540666	C34692	Kathrineborg, Brunlanes
EU89-UTM, zone 32	574095	6694451	C34814	Tingelstad, Gran
EU89-UTM, zone 32	411850	6441834	C34884	Fuskeland, Mandal
EU89-UTM, zone 32	553049	6540346	C35042	Manvik, Brunlanes
EU89-UTM, zone 32	587001	6693524	35109	Skjervum øvre, Gran
EU89-UTM, zone 32	554644	6542121	C35237	Berg, Brunlanes
EU89-UTM, zone 32	604788	6741921	C35805	Gaalaas, Ringsaker
EU89-UTM, zone 32	482168	6469853	C35844	Fjære, Fevik
EU89-UTM, zone 32	472057	6467007	C35859	Landvik, Grimstad
EU89-UTM, zone 32	432042	6492732	C36400	Evje
EU89-UTM, zone 32	381979	6433072	C36679	Revøy Midtre, Lyngdal
EU89-UTM, zone 32	535813	6562686	C37631	Bratsberg, Skien
EU89-UTM, zone 32	631716	6554525	C38506	Berg Skole, Halden
EU89-UTM, zone 32	568529	6606020	C52083	Aas Østre, Sande
EU89-UTM, zone 32	615654	6609151	C55949	Molle østre, Spydeberg
EU89-UTM, zone 32	563133	6551928	C56077	Rødbøl
EU89-UTM, zone 32	567937	6671566	C58714	Veien kulturminnepark
EU89-UTM, zone 32	622727	6646210	C60497	Fet
EU89-UTM, zone 32	596293	6585712	C61166	Dilling
EU89-UTM, zone 32	313530	6774825	B520-525	Nedre Aure, Voss
EU89-UTM, zone 32	313449	6623845	B3358	Borgund Ø
EU89-UTM, zone 32	313407	6624873	B3855	Øvrabø Søre, Kvinnherad
EU89-UTM, zone 32	293226	6726757	B4003	Sogn
EU89-UTM, zone 32	320111	6611238	B4095	Eide, Fjælberg
EU89-UTM, zone 32	349869	6684063	B4207	Jondal, Hardanger
EU89-UTM, zone 32	322777	6823213	B4259	Naustdal i Søndfjord
EU89-UTM, zone 32	306941	6611587	B4367	Alne, Vikevik
EU89-UTM, zone 32	392258	6456026	B4414	Vemmestad, Lyndal
EU89-UTM, zone 32	353503	6851773	B4590	Evebø, Gloppen
EU89-UTM, zone 32	353176	6850969	B4591	Gloppen
EU89-UTM, zone 32	353176	6850969	B4593	Evebø, Gloppen
EU89-UTM, zone 32	330994	6616115	B4866	Etne

Cont. Appendix 2 ▶

Cont. Appendix 2

Coordinate system	East	North	Catalogue no.	Site
EU89-UTM, zone 32	399449	6792315	B4877	Kvam
EU89-UTM, zone 32	310143	6673928	B4978	Vinnes, Fuse
EU89-UTM, zone 32	394520	6479271	B5060	Spangereid
EU89-UTM, zone 32	357819	6724655	B5693	Voss
EU89-UTM, zone 32	331416	6616771	B5838	Etne
EU89-UTM, zone 32	334226	6661065	B5931	Kvinnherad
EU89-UTM, zone 32	294794	6646764	B5963	Rimbereid, Fitje
EU89-UTM, zone 32	288888	6727308	B6086	Manger
EU89-UTM, zone 32	329236	6645762	B6103	Kvinnherrad
EU89-UTM, zone 32	405931	6800926	B6111	Luster, Sogn
EU89-UTM, zone 32	381627	6865823	B6197	Stryn
EU89-UTM, zone 32	334386	6622949	B6200	Etne
EU89-UTM, zone 32	353538	6723923	B6227	Voss
EU89-UTM, zone 32	455118	6946107	C6233	Grindheim, Etne,
EU89-UTM, zone 32	455118	6946107	C6233	Grindheim, Etne
EU89-UTM, zone 32	353358	6725790	B6474	Vangen, Voss
EU89-UTM, zone 32	353358	6725790	B6475	Vangen, Voss
EU89-UTM, zone 32	384143	6867426	B6539	Stryn
EU89-UTM, zone 32	370982	6690019	B6597	Ullensvang
EU89-UTM, zone 32	369748	6775051	B6691	Vik
EU89-UTM, zone 32	310776	6658142	B6700	Tysnes
EU89-UTM, zone 32	328340	6673724	B6756	Mundheim, Strandebarm
EU89-UTM, zone 32	328340	6673724	B6756	Mundheim, Strandebarm
EU89-UTM, zone 32	373564	6716993	B6763	Ulvik, Hardanger
EU89-UTM, zone 32	339289	6702238	B6807	Norheim, Kvam
EU89-UTM, zone 32	352659	6727940	B7608	Vangen, Voss
EU89-UTM, zone 32	328988	6620876	B7908	Støle, Etne
EU89-UTM, zone 32	331714	6620387	B7956	Støe, Etne
EU89-UTM, zone 32	365076	6697137	B8200	Kinsarvik, Ullensvang
EU89-UTM, zone 32	305595	6723855	B8697	Eknes, Hosanger
EU89-UTM, zone 32	313530	6774825	B8853	Oppedal, Lavik
EU89-UTM, zone 32	346900	6939438	B8871	Haram, Sunnmøre
EU89-UTM, zone 32	350908	6938963	B8872	Haram, Sunnmøre
EU89-UTM, zone 32	381080	6788322	B8983	Hamre, Leikanger
EU89-UTM, zone 32	304585	6654216	B10042	Uggdal, Tysnes
EU89-UTM, zone 32	349999	6856316	B10097	Indre Vereide, Gloppen
EU89-UTM, zone 32	352517	6933251	B10790	Giske, Borgund
EU89-UTM, zone 32	346941	6929699	B11546	Giske, Summøre
EU89-UTM, zone 32	357732	6950357	B12048	Haram, Sunnmøre
EU89-UTM, zone 32	368504	6684365	B13367	Ullensvang
EU89-UTM, zone 32	396701	6790381	B13955	Kvåle, Sogndal
EU89-UTM, zone 32	396701	6790381	B13956	Kvåle, Sogndal
EU89-UTM, zone 32	394520	6479271	B4234	Lundevågen, Vanse
EU89-UTM, zone 32	354056	6725855	B12046	Gjerstad, Voss

Bear skin burials revisited: Norway and Sweden, mainly Migration Period

By Oliver Grimm

Keywords: Norway, Sweden, burial archaeology, bear skin, Migration Period

Abstract: In the present paper, emphasis is laid upon burials with actual or assumed bear skins from northern Europe. This relates to mainly Migration Period inhumation burials in Norway and Sweden (AD 375/400 to 550/575), whereas other bear-related furnishings (frequently claws [or rather third phalanges], rarely teeth) found in northern and central European Iron Age graves will not be considered. All in all, there is a maximum of only 15 burials with bear skin remains known from Norway and Sweden (a few more from Finland notwithstanding). This is statistically meaningless, and each new find may change the entire picture (see postscriptum). Furthermore, most finds came to light in the second half of the 19th and the early 20th century – that is, in the early days of archaeology, using rather rough excavation techniques when compared to present-day standards. Against all odds, however, a closer look at the burials with bear skins leads to somewhat surprising insights into social and gender aspects. Some graves are among the richest of their time, in cases of both male individuals ("warriors") and one female. The oldest burial (160/180–325), one from Gotland, is also that of a female. Among the earliest of these graves in Norway (325–375/400), there is yet another one of a woman that is, quite remarkably, located in the far north of the country.

Introduction

The past decades have seen numerous scholarly attempts by archaeologists to draw upon burials with bear-related furnishings, mainly those with bear claws (or rather third phalanges: see below), in an attempt to discuss the "world view" expressed by those finds (see Mansrud, this volume, on different interpretations; see also Grimm 2013, 278; Kirkinen 2017, 5). Suffice it to say that the interpretation has a wide range. On the one hand, such skins are considered as the belongings of a "berserk" (e.g. Bender Jørgensen 2001, 7; 2003, 70–71; Hedeager 2011, 91–98; see also Wamers 2009 and Sundqvist, this volume), or the reflection of a heroic deed (the killing of a bear; cf. Oehrl 2013). On the other hand, skins may have served as trading goods (e.g. Petré 1980; Iregren 1988; Lindholm/Ljungkvist 2016) with a use as, *inter alia*, bedding or shrouds (e.g. Munksgaard 1959; Sigvallius 1994, 76; Henriksen 2001, 10). In this respect, it is also worth remembering that, in large parts of northern Europe (Denmark notwithstanding), bears had a "shared history" with human beings that surely had an influence on people's minds and, possibly, beliefs, whereas on Gotland bears were foreign, but claws are found so frequently in burials of that island that a trade in skins has been suggested (see Jordahl et al., this volume).

Bear and Human: Facets of a Multi-Layered Relationship from Past to Recent Times, with Emphasis on Northern Europe, ed. by Oliver Grimm (Turnhout: Brepols, 2023), pp. 533–546 BREPOLS PUBLISHERS 10.1484.M.TANE-EB.5.134348

There is a tendency in research to equate bear claws in burials uncritically with former bear skins. Single claws may very well represent former skins, but any such assumption remains hypothetical, because they cannot tell us anything, but preserved skins (hairs) and claw groups found in correct anatomical order are able to do so. The present paper, which to some extent revisits an older one (Grimm 2013), addresses only burials with actual or assumed bear skins, as known from Norway and Sweden, from a purely archaeological point of view. The burials in question, mainly Migration Period inhumations (375/400–550/575), were mostly excavated in the early days of archaeology in the 19th and early 20th century, using rather rough excavation techniques compared to present-day standards. The 15 graves with skins are statistically meaningless, since each new find may change the entire picture (Table 1; Fig. 1; cf. postscriptum). The situation is quite different for burials with bear claws; Gotland stands out with almost 150 (!) mainly Migration and Early Vendel Period interments to that effect (see Jordahl et al., this volume). However, other areas in Norway and Sweden have also yielded many graves with bear claws of Migration Period date, which allows a well-founded archaeological discussion and conclusion (see different papers, this volume), in contrast to speculation on the basis of only a few graves with skin remains. Important findings from Finland will not be considered in this paper – that is, some burials with bear hairs that date to the Late Iron Age and medieval times, i.e. the period from the 9th to the 16th century (see Mannermaa et al., this volume). However, not all those hair finds can be associated with former bear skins as shrouds or rugs for the deceased, as the skins might also have been used for garments.

In the following, the burials are introduced first, and thereafter the actual or assumed skins are described. The text itself is kept short (cf. Table 1 for standardised information about burials and skins).

The burials with bear skins

The Norwegian and Swedish burials with bear skins amount to a number of only 15, but it cannot be ruled out that more such cases might be discovered when excavation reports from the early days of archaeology, i.e. the 19th and early 20th century, were studied again. Although there are six finds from southwestern Norway (Rogaland and the western part of Agder), two from Gotland and two from Högom (Medelpad, northern Sweden), they do not form any well-defined subgroups. Thus, the material is limited, unrepresentative and statistically meaningless. Each new find may change the entire picture (see postscriptum).

Among the 15 burials with bear skins, there are eleven inhumation and four cremation burials. This does not come as a surprise, since it is only the former context in which one would expect skins to have survived, but only under very fortunate circumstances (see below). Nine burials of the total amount of 15 belong to the Migration Period (375/400–550/575) and three more to the immediately preceeding Late Roman Iron Age ("C3" = 325–375/400), while one is even older; the burial at Smiss on Gotland reaches as far back as 160/180–325 ("C1 to C2" of the Late Roman Iron Age). In turn, one burial belongs to the Vendel Era (550/575–750/800), which followed after the Migration Period, and one defies a dating (Nedre Aure, Voss, Vestland/western Norway). Caution is due because of the small amount of graves, but the Migration Period is perhaps less dominant than has previously been thought when it comes to burials with bear skins; the given chronological distribution has to be related to that of the interments with bear claws known from some Scandinavian regions (see various contributions, this volume).

It is quite remarkable that five out of 15 burials with bear skins are among the richest of their time. This is accentuated not only by their exquisite furnishings but also the substantial grave mounds, the construction of which was energy- and time-consuming, and their oversized grave chambers/cists (however, the diameter of the mound for inhumation grave V at Snartemo is unknown, whereas the cremation burial at Haram had only a small chamber for the urn).

The oldest wealthy burial to be mentioned here is that of the "Lord of the Rings" at Haram (western Norway), a cremation dated to the timespan of *c.* 325–375/400 and furnished with a gold ring of almost 600 g, which by its weight might be the heaviest of all such rings found in Norwegian, or rather northern European Iron Age graves (HAGEN 1983, 323–328; SOLBERG 1984, 100–101; RINGSTAD 1986, cat. no. 260; SOLBERG 2000, 121). There are also lesser gold rings in the burial, meant – at least for two out of three rings – to be worn by a male person, as has been suggested. According to the burial rite in the given time span and region, weapons were only in rare instances, if at all, part of the grave furnishings in inhumation burials (BEMMANN/HAHNE 1994, fig. 33). However, it all depends on the burial rites – the man interred at Haram might have led a life comparable to those of other high-ranking male individuals who were granted magnificent inhumations with exquisite weapon furnishings.

The Migration Period "Master Warriors" in Norway (west: Evebø; south: Snartemo, grave V) and Sweden (north: Högom, mound 2) have in common exquisite weapons with mounts of precious metals (gold, silver), and their methods of manufacture reflect different branches of advanced handicraft (more generally BENDER JØRGENSEN 2001; 2003; RAMQVIST 2011). There are more common traits, too, such as the presence of colourful clothing. These burials have few, if any, counterparts in the northern European Migration Period.

In turn, the Krosshaug grave at Hauge in southwestern Norway, that of the "Petty Queen on the Bear Skin", is among the richest of all the female burials in Migration Period Norway and probably beyond (MAGNUS 1975; KRISTOFFERSEN 2000, cat. no. 47). In her case, the main dress pin – fibula – of gilded silver is an object of the highest advancement in handicraft, but there are also other extraordinary finds, such as a large bronze hanging bowl.

The mentioned burials are markers for petty kings and a petty queen at the top of the social pyramid in a regional sense. High rank is also indicated by the remains of an actual petty king's seat with an "architecture of power", excavated only 200 m away from the Evebø weapon burial in western Norway and contemporaneous with it (HATLING 2009).

In contrast, the remaining ten burials with bear skins defy any comparable interpretation. However, they mostly rank "above average" because of their grave furnishings, which include rings of gold and silver and foreign goods such as glass or bronze vessels. These burials indicate a local upper class under the control of the above-mentioned petty kings / petty queen.

Exceptions from the rule are the burials in Nedre Aure, Voss (Vestland), for which there is no proper record of find circumstances and finds, and an interment at Døsen in Os, likewise in western Norway/Vestland. The latter, the secondary burial of a man, which accompanies the primary burial of a child, has only meagre furnishings (mound IV for both). The man had with him, amongst other things, a knife, a pair of scissors, and a pair of bronze tweezers (SHETELIG 1912, 139–150). However, when seen against the background of six burial mounds, which probably represent different generations of a Migration Period farm population, the area is in part "above average" again, due to one female grave that was furnished with, among other things, a gilded silver relief fibula (mound II; cf. SHETELIG 1912, 132–139; KRISTOFFERSEN 2000, 344–345 cat. no. 74). Furthermore, among the six grave mounds, two have a diameter of almost 20 m (mounds I and IV), which places them close to the lower end of western Norway's largest burial mounds. These are thus named since their construction demanded considerable time and energy (RINGSTAD 1986).

As already mentioned, the Krosshaug petty queen's grave in southwestern Norwegian Hauge belongs to the richest of its kind in Migration Period Norway (and probably northern Europe), and, in social respects, it stands alongside the contemporaneous burials of "Master Warriors". Remarkably, a female burial on Gotland (Smiss; *c.* AD 160/180–325; cf. ALMGREN/NERMAN 1923, 89 fig. 156a; PETRÉ 1980, 6) is the oldest one with a bear skin in northern Europe. Another female burial to that effect, the earliest one in Norway (Føre, burial 2, *c.* 325–375/400; cf. SJØVOLD 1962, 77), has only two counterparts (male in gender). The find from Føre deserves particular attention due to its location in the far north (Fig. 3). However, the glass beaker found in that burial, a product of Roman origin, indicates supra-regional contacts (LUND HANSEN 1987, 442). In addition, among the Norwegian material, there is one burial of several women (Sletten; cf. KRISTOFFERSEN 2000, cat. no. 26), and at least one double burial (male and female, Vemestad; SHETELIG 1912, 158 footnote 1; BEMMANN/HAHNE 1994, cat. no. 201).

To sum up: caution is needed because of the limited number of grave finds, but a closer look reveals surprising results. Five burials – four male and one female – which are among the richest of their kind in the periods in question, against the background of a total body of 15, if at all, must not be neglected. In addition, there are other remarkable female graves; a Swedish one is the oldest of all burials with bear skins from northern Europe. Furthermore, a woman's interment, in fact from the far north of Norway, is among the oldest to that effect from this country, with two contemporaneous male graves.

THE SKINS

Eleven inhumation and four cremation burials have yielded evidence for bear skins, which does not come as a surprise. Under very fortunate circumstances – the sealing of the grave by the exclusion of oxygen in a non-acidic environment – the inhumation burial rite can lead to the preservation of the bear skin (hair) and the paws with their claws. In the case of the latter, third claw-like phalanges (bones) are most often preserved, whereas the claws themselves, which consist of keratin and cover most of the phalanges, would have vanished. In the case of cremation, however, the burning on the funeral pyre would lead to the decomposition of the skin, with only deformed third phalanges able to survive, and the remains from the funeral pyre were then either covered directly by a mound, or were taken away completely, or sorted out *pars pro toto* and buried at some other place.

Strictly speaking, there are differences between burials with actual bear skins, those that have yielded only small bear skin remains with claws still attached or missing, and burials with unspecified animal skin, found together with bear claws. It would definitely be beneficial to analyse all bear skin/hair finds preserved in archives, using modern nomenclature consistently, and this should also include the more dubious cases. In four instances, unspecified animal skin has been found together with bear claws (Eik, Haram, Krosshaug [Hauge], Vestre Skogsfjord), but are these skin remains really from bear? Likewise, all preserved remains of unspecified animal skins found together with unspecified claws, or without any of these, should also be studied to see if there are more cases with bear skins. This relates to the unspecified skin and claw find at Sletten in Norway (see also below on the weapon burial at Bø, Steigen, in northern Norway).

In three instances – inhumation burials from northern Norway (Føre, burial 2) and Gotland (Smiss and Hallvede) – groups of bear claws have been found in correct anatomical order which indicates a bear skin that has since detoriated (Føre: SJØVOLD 1962, 77; Smiss and Hallvede: PETRÉ 1980, 6). The Føre burial is somehow special since there was an organic layer (the remains of a bear skin?) found beneath the deceased. It is somewhat surprising that only seven claws were found in four groups, whereas both sites on Gotland have yielded close to twenty claws. At Smiss, there might really have been 20 claws, with three groups of five and one group of three, the latter one from a disturbed part of the cist, which might be the reason for the two missing claws; at Hallvede, three groups of five and one single claw have been recorded. There is yet another interesting case on Gotland: the weapon burial at Broa (550/575–750/800) in Halla parish (see JORDAHL et al., this volume). However, this burial has yielded only two groups with altogether eight claws (one group of five, one group of three), placed right above the head and below the feet of the deceased. So, has really a skin been placed in that burial, with two separate paws and the body of the deceased situated on top of it?

The primary inhumation burial in mound 2 at Högom deserves particular attention since many animal hairs were found, mostly from bear, when the chamber grave of 5 x 2 m was excavated under laboratory conditions after it had been transported, encased, to Statens historiska museum/The Swedish History Museum in Stockholm (Nockert 1991, 31, 36). In fact, large parts or the whole floor of the grave chamber might have been covered with bear skins, and so was the deceased, judging from bear hairs situated on the belt and sword. Also, the secondary burial in mound 4 in Högom is worth mentioning (Ramqvist 1992, 194–198). This burial's rectangular layer, 1.85 x 0.9 x 0.2 m in size, has been described as a "patch of a settlement layer". As, however, bones and charcoal occurred in the entire layer, the question must be posed whether it represents in fact the remains of the actual funeral pyre that included a bear skin from which 13 bear claws were found scattered (but not, it would seem, in groups)? A shallow pit at the eastern edge of the mentioned layer contained human and animal bones, the former of an adult person whose gender cannot be identified and the latter belonging to different animals (sheep/goat, dog, horse). Among the finds there were pieces of bronze objects, a composite bone comb and a case for it, and the head of a bone pin. No gender identification is possible (personal communication, John Ljungkvist, Sweden). There may be a female sphere (bone pin) as well as a male one (the horse, the size of the funeral pyre), but this is inconclusive.

In the case of the western Norwegian inhumation burial at Døsen in Os (Vestland), the actual bear skin had decomposed, but its dark brown hairs were preserved and so were its claws, the latter in the form of keratin, whereas the bones themselves were gone (Fig. 2). This particular preservation may have been the result of acidic soil. The bear hairs at Døsen covered most of the grave bottom, 2.5 m in length, except for the southern end (0.20–0.30 m long). This length may reflect the complete skin of an adult male bear, but it will have to remain open as to how the width of such a skin, almost two meters, was coped with – was the skin cut so that it would fit the grave's internal width of only 0.70 m or had the skin been placed in several layers? (size of bear skin: personal communication with Andreas Zedrosser, Norway). And how many claw groups were really found in that burial, besides the one known from the photograph?

For a number of burials excavated in earlier days there is very little information; bear skin (hair) was found, sometimes together with claws (cf. Table 1). In this respect, pre-modern skinning needs to be considered (for the following, personal communication with Tuija Kirkinen, Finland, is acknowledged; see also Kirkinen 2017, 5–8). The easiest way of skinning would have been to remove the paws from the carcass. However, the handling was different in pre-modern times; the paws were in fact left in the skin for some reason we do not know, be it aesthetic or symbolic. In that case, the paw needed to be skinned carefully, otherwise the claws would loosen and drop off. Bones (phalanges) would survive under fortunate circumstances, whereas the claws themselves, which consist of keratin and cover most of the third phalanx, would vanish in inhumation burials over time or would be destroyed on a funeral pyre. If only third phalanges are found in burials, they may indicate that a whole skin with paws had been placed in an inhumation or on a funeral pyre, with the deceased placed on top. Alternatively, just claws with or without attached skin remains could have been deposited. This is frequently found in archaeological contexts: third claw-like phalanges as remnants of the claws. This use of only the claws, however, would leave the question open as to what happened to the actual skin.

In the majority of inhumations that have been recorded, the deceased seems to have been placed on a bear skin under which there was often a layer of bark. However, the deceased in Högom mound 2 might have been *covered by* a skin, but the question is open as to whether the deceased also *lay on* a skin, since no hairs seem to have been found beneath the deceased (see above).

Remarkably, each of the four cremation burials with assumed bear skins represents a case of its own, and three out of four represent evidence for unspecified animal skin found together with bear claws (see above). Skin was either 1) placed on the funeral pyre and, after the fire, directly covered by a mound (see mound 4 at Högöm, the interpretation of which, however, is not beyond doubt), 2) wrapped around an urn in which no less than 15 bear claws were found (Haram; Solberg 1984, 100–101), 3) wrapped around human bones and then placed in a metal urn (Eik; cf. Holck 1986, 177), or 4) placed in a metal urn, alongside other furnishings (Vestre Skogsfjord; cf. Holck 1986, 247).

To sum up, it is only for about one half of all the 15 burials that the existence of real bear skins can actually be attested, five of them being inhumations (Døsen, Føre grave 2, Hallvede, Högom mound 2, Smiss) and two cremations (Haram, mound 4 in Högom). For the other half of the burials, the available information is rather limited; perhaps there were more cases with real bear skins, or only third phalanges with the remains of attached bear skin were placed in the burials with no connection to the skin itself.

Interestingly, it has been argued on a medical-anthropological basis that the wrapping of a body in a bear skin would have hindered the cremation of that person or made it impossible (Holck 1986, 173). This assumption, however, is contradicted by the archaeological finds themselves, which show that even horses were burnt on funeral pyres in parts of Iron Age Sweden (personal communication, John Ljungkvist, Sweden), and even more so by experimental archaeology (personal communication, Mogens Bo Henriksen, Denmark, who has studied this topic for decades; see most recently Henriksen 2019). If the cremation pyre was constructed in the right way and the fire well maintained, any corpse, human or non-human, was cremated in a couple of hours, with less than two cubic meters of firewood needed. Thus, a bear skin would not affect the cremation process.

Final remarks

The maximum number of only 15 mainly Migration Period burials in Norway and Sweden with actual or assumed bear skins is too small to generalise in any way. Actually, as it turns out, only for around one half of the burials can a real skin be documented, whereas the other cases remain dubious. Although there are differences in evidential value – actual vs. assumed bear skin – all burials shall be taken into account in the following. It also has to be kept in mind that each new find might change the entire picture (see postscriptum). Despite all source problems, however, some findings are pretty surprising.

Five out of the 15 burials are outstanding in their construction and furnishings. These people were at the top of the social pyramid during their lifetime, as petty kings and "warriors" (which would allow here to introduce the idea of berserks), but there was also one petty queen laid to rest in one of these burials. The ten other finds, mainly of men, usually indicate people of some rank, but only at a more local level.

Likewise surprisingly, the earliest of these burials, which dates to the pre-Migration Period and contains an undoubted skin (Smiss; 160/180–325), is that of a woman, and was found on Gotland, whereas the oldest female burial of this kind in Norway (Føre grave 2; 325–375/400), with two contemporaneous male counterparts, was found in the very far north of the country which makes it even more interesting.

Finally, it is no wonder that the burials with actual or assumed bear skins are mostly inhumations. In these cases, the deceased was placed on a skin which in turn had been placed on a bark layer, and it is only due to very fortunate preservation conditions that the skin (hair) has survived. Again surprisingly, each of the four cremation burials seems to represent a case of its own in the use of the bear skin; in the first case as a rug for the deceased person on the funeral pyre (?), in the second for wrapping up the urn, in the third for wrapping up human bones that were then placed into the urn, and in the fourth as partial skin, placed in the urn, alongside other furnishings.

Acknowledgements

The author would like to thank Dr. Tuija Kirkinen (University of Helsinki, Finland) for sharing her knowledge on bear skinning and bear skin/bear hair, Prof. Andreas Zedrosser (University of South-Eastern Norway) for information about bear skin size, Assoc. Prof. John Ljungkvist (University of Uppsala, Sweden) for all his expertise on Sweden, and Dr. Mogens Bo Henriksen (Odense City Museums, Denmark) for consultation about the eventualities of the cremation burial rite. Remaining errors are the author's responsibility.

Postscriptum

It was in the final phase of manuscript completion that the author became aware of important findings from northern Norway, which will be only briefly introduced here. So far, the only burial with an assumed bear skin in that part of the country is grave 2 at Føre, Bø (Nordland) mentioned above, the burial of a woman which dates back to the period of *c.* 325–375/400, with bear claws, yet only seven of them, in the four corners of the cist and an organic layer beneath the interred person, considered as the remnants of the fur. New knowledge about bear skins in burials can be gained from both the reconsideration of older finds and the excavation of new ones (Fig. 3).

As it turns out, in the case of grave 1 at Føre, the burial of a man with weapons, one has to return to the original excavation report from the 1950s; it is only now that information from that report has been published. As in the case of grave 2, an organic layer was observed beneath the interred person in grave 1, accompanied by one group of four claws in one of the corners of the burial (Klokkervoll 2015, 66–67; Roth Niemi 2018, 48). The find can be dated to the late second or the first part of the 3rd century AD (Ilkjær 1990, cat. no. 182; Bemmann/Hahne 1994, cat. no. 410).

Another case is represented by the weapon grave in mound I at Bø, Steigen (Nordland), found together with a female burial and excavated in the early 1950s (Slomann 1959). Both burials have the same dating as the aforementioned grave 1 at Føre (ibid, 3–4; cf. Ilkjær 1990, cat. no. 78; Bemmann/Hahne 1994, cat. no. 409). The find context of the weapon burial needs further clarification; in the original publication, the remains of skin are mentioned (Slomann 1959, 3–4), and the find catalogue from Tromsø Museum lists the remains of animal hair (Ts 5401 I), whereas in recent publications reference is made to "remains of fur with thick hairs, presumably from bear" (Bakke 2012, 46; cf. Roth Niemi 2018, 48–49). Thus, two questions emerge: What is really stated in the original excavation report about skin and hair, and what would be the results of modern scientific analysis of these organic materials? In the present situation, with the absence of bear claws, the burial at Bø cannot be listed among those with actual or presumed bear skin.

Remarkably, there is also one most recent find which came to light during a rescue excavation in 2017; that at Hillesøy (Kvaløy) to the west of Tromsø (Troms; cf. Roth Niemi 2018). Half of the burial had already been destroyed before the archaeologists were informed, but what was left is a weapon burial in a 5-m-long boat, dated to the transition from the Merovingian to the Viking Age, around AD 800. One group of five claws was found in the intact part of the grave; five more were discovered, however, no longer *in situ*. Beneath the deceased there was a dark layer, interpreted as the remnants of a bear skin.

The mentioned burials at Føre (grave 1), Hillesøy, and, potentially, Bø are quite similar as they are "above average" in their furnishings and belong to small groups of mounds that may reflect different generations of local farm populations (see mentioned literature). Worth highlighting are two burials at Føre, one of a male (grave 1), the other of a female (grave 2) individual; the male grave is the earliest one in Norway with an assumed bear skin, and so may be the more dubious find from Bø that would require more analysis. In turn, the interment at Hillesøy is the latest of all the burials with a bear skin known in Norway and Sweden (however, one find from western Norway – that from Nedre Aure, Voss [Vestland] – defies a dating).

Bibliography

Almgren/Nerman 1923: O. Almgren/B. Nerman, Die ältere Eisenzeit Gotlands 2 (Stockholm 1923).

Bakke 2012: S. Bakke, Etnisitet i jernaldergraver på Engeløya i Steigen. Unpubl. M.A. thesis, Tromsø University (Tromsø 2012).

Bemmann/Hahne 1994: J. Bemmann/G. Hahne, Waffenführende Grabinventare der jüngeren römischen Kaiserzeit und Völkerwanderungszeit in Skandinavien. Studie zur zeitlichen Ordnung anhand der norwegischen Funde. Bericht der Römisch-Germanischen Kommission 75, 1994, 283–640.

Bender Jørgensen 2001: L. Bender Jørgensen, Bjornekrigere. Spor – nytt fra fortiden 1/2001, 4–8.

Bender Jørgensen 2003: L. Bender Jørgensen, Krigerdragten i folkevandringstiden. In: P. Rolfsen/F.-A. Stylegar (eds.), Snartemofunnene i nytt lys. Universitetets Kulturhistoriske Museer Skrifter nr. 2 (Oslo 2003) 53–80.

Gjessing 1925: G. Gjessing, Vestagder i forhistorisk tid. In: H. Aall/A. W. Brøgger (eds.), Norske bygder II. Vest-Agder I (Bergen 1925) 33–76

Grimm 2013: O. Grimm, Bear-skins in northern European burials and some remarks on other bear-related furnishings in the north and middle of Europe in the first millenium AD. In: O. Grimm/U. Schmölcke (eds.), Hunting in northern Europe until 1500 AD – Old traditions and regional developments, continental sources and continental influences. Papers presented at a workshop organised by the Centre for Baltic and Scandinavian Archaeology, Schleswig, June 16th and 17th, 2011. Schriften des Archäologischen Landesmuseums, Ergänzungsreihe 7 (Schleswig 2013) 277–296.

Gustavson 1889: G. Gustavson, Evebøfundet og nogle andre nye gravfund fra Gloppen. Bergens Museums Aarberetning for 1889: 1. Bergen (Bergen 1889).

Hagen 1983: A. Hagen, Norges Oldtid (Oslo 31983).

Hatling 2009: S. H. Hatling, Gloppen i Folkevandringstiden. En sosial analyse av Evebøhøvdingen. Masteroppgave i Arkeologi. Institutt for AHKR, Universitetet i Bergen (Bergen 2009).

Hauken 2005: Å. D. Hauken, The Westland cauldrons in Norway. Arkaeologisk museum i Stavanger monographs 19 (Stavanger 2005).

Hedeager 2011: L. Hedeager, Iron Age Myth and Materiality. An Archaeology of Scandinavia AD 400–1000 (London 2011).

Henriksen 2001: M. M. Henriksen, Bjørnen – fruktbarhetssymbol i elder jernalder? Spor – nytt fra fortiden 1/2001, 10–13.

Henriksen 2009: M. B. Henriksen, Brudager Mark – en romertidsgravplads nar Gudme pa Sydostfyn. Two vol. Fynske jernaldergrave 6,1–2 (Odense 2009).

Henriksen 2019: M. B. Henriksen, Experimental cremations – can they help us to understand prehistoric cremation graves? In: A. Cieśliński/B. Kontny (eds.), Interacting Barbarians. Contacts, Exchange and Migrations in the First Millennium AD. Neue Studien zur Sachsenforschung 9 (Warszawa, Braunschweig, Schleswig 2019) 289–296.

Holck 1986: P. Holck, Cremated bones. A medical-anthropological study of an archaeological material on cremation burials. Antropologiske Skr. 1 (Oslo 1986).

Hougen 1935: B. Hougen, Snartemofunnene. Studier i folksvandringstidens ornamentikk og tekstilhistorie. Norske Oldfunn VII (Oslo 1935).

Ilkjær 1990: J. Ilkjær, Illerup Ådal. Die Lanzen und Speere. Jysk Arkæologisk Selskabs Skrifter 25, 1.2 (Århus 1990).

Iregren 1988: E. Iregren, Finds of Brown Bear (Ursus arctos) in Southern Scandinavia – Indications of Local Hunting or Trade? In: B. Hardh/L. Larsson/D. Olausson/B. Petré (eds.), Trade and exchange in prehistory. Studies in honour of Berta Stjernquist. Acta Arch. Lundensia. Ser. in 8° (Lund 1988) 295–308.

Kirkinen 2017: T. Kirkinen, "Burning pelts" — brown bear skins in the Iron Age and Early Medieval (0–1300 AD) burials in South-East Fennoscandia. Estonian Journal of Archaeology 21(1), 2017, 3–29.

Klokkervoll 2015: A. Klokkervoll, Dyrenes role i germansk gravskikk. En komparativ analyse av dyreneddleggelsene i nordnorske jernaldergraver. Unpubl. M.A. thesis, Tromsø University (Tromsø 2015).

Kristoffersen 2000: S. Kristoffersen, Sverd og spenne. Dyreornamentikk og sosial kontekst. Studia Humanitatis Bergensia (Bergen 2000).

Lindholm/Ljungkvist 2016: K.-J. Lindholm/J. Ljungkvist, The bear in the grave: exploitation of top predator and herbivore resources in first millennium Sweden – first trends from a long-term research project. European Journal of Archaeology 19(1), 2016, 3–27.

Lund Hansen 1987: U. Lund Hansen, Römischer Import im Norden. Warenaustausch zwischen dem Römischen Reich und dem freien Germanien während der Kaiserzeit unter besonderer Berücksichtigung Nordeuropas. Nordiske Fortidsminder B10 (Copenhagen 1987).

Magnus 1975: B. Magnus, Krosshaugfunnet. Et forsok pa kronologisk og stilhistorisk plassering i 5. arh. Stavanger Mus. Skr. 9. (Stavanger 1975).

Magnus 2014: B. Magnus, Kvinnene fa Krosshaug I Klepp og Hol på Inderøy. In: S. Kristoffersen (ed.), Et akropolis på Jæren? Tinghaugplatået gjennom jernalderen. Arkeologisk museum i Stavanger Varia 55 (Stavanger 2014) 71–87.

Munksgaard 1959: E. Munksgaard, Soveskind – eller sort magi. Skalk 1959(3), 1959, 18.

Nockert 1991: M. Nockert, The Hogom find and other Migration period textiles and costumes in Scandinavia. Archaeology and Environment 9. Hogom II (Uddevalla 1991).

Oehrl 2013: S. Oehrl, Bear hunting and its ideological context (as a background for the interpretation of bear claws and other remains of bears in Germanic graves of the 1st millennium AD). In: O. Grimm/U. Schmölcke (eds.), Hunting in northern Europe until 1500 AD – Old traditions and regional developments, continental sources and continental influences. Papers presented at a workshop organised by the Centre for Baltic and Scandinavian Archaeology, Schleswig, June 16th and 17th, 2011. Schriften des Archäologischen Landesmuseums, Ergänzungsreihe 7 (Schleswig 2013) 297–332.

Petré 1980: B. Petré, Bjornfallen i begravningsritualen – statusobjekt speglande regional skinnhandel? Fornvännen 75, 1980, 5–14.

Ramqvist 1992: P. Ramqvist, Högom. The excavations 1949–1984. Archaeology and environment of Högom I (Neumünster 1992).

Ramqvist 2011: P. Ramqvist, Folkvandringstida kammargravar – Nagra omtolkningsforslag av Evebo- och Snartemogravarna. Viking LXXIV, 2011, 103–118.

Ringstad 1986: B. Ringstad, Vestlandets storste gravminner. Et forsok pa lokalisering av forhistoriske maktsentra. Unpubl. M.A. thesis, Universitetet i Bergen (Bergen 1986).

Roth Niemi 2018: A. Roth Niemi, Nordvegen på Hillesøy. Utgravning av båtgrav fra yngre jernalder. Tromura/ Tromsø Museums Rapportserie 48 (Tromsø 2018).

Shetelig 1912: H. Shetelig, Vestlandske graver fra jernalderen (Bergen 1912).

Sigvallius 1994: B. Sigvallius, Funeral Pyres. Iron Age cremations in Northern Spanga. Theses and Papers in Osteology 1 (Stockholm 1994).

Sjøvold 1962: T. Sjøvold, The Iron Age Settlement of Arctic Norway. A study in the expansion of European Iron Age Culture within the Artic Circle I. Early Iron Age (Roman and Migration periods) (Tromso, Oslo 1962).

Slomann 1959: W. Slomann, Et nytt romertids gravfunn fra Nord-Norge. Viking 23, 1959, 1–28.

Solberg 1984: B. Solberg, Haramfunnet. In: S. Indrelid/ S. Ugelvik Larsen (eds.), – fra de forste fotefar: Sunnmores forhistorie (Alesund 1984) 100–101.

Solberg 2000: B. Solberg, Jernalderen I Norge. 500 før Kristus til 1030 etter Kristus (Oslo 2000).

Wamers 2009: E. Wamers, Von Bären und Männern. Berserker, Bärenkämpfer und Bärenführer im frühen Mittelalter. Zeitschrift für Archäologie des Mittelalters 37, 2009, 1–46.

Dr. Oliver Grimm
Centre for Baltic and Scandinavian Archaeology (ZBSA)
Stiftung Schleswig-Holsteinische Landesmuseen, Schloss Gottorf
Schleswig
Germany
oliver.grimm@zbsa.eu

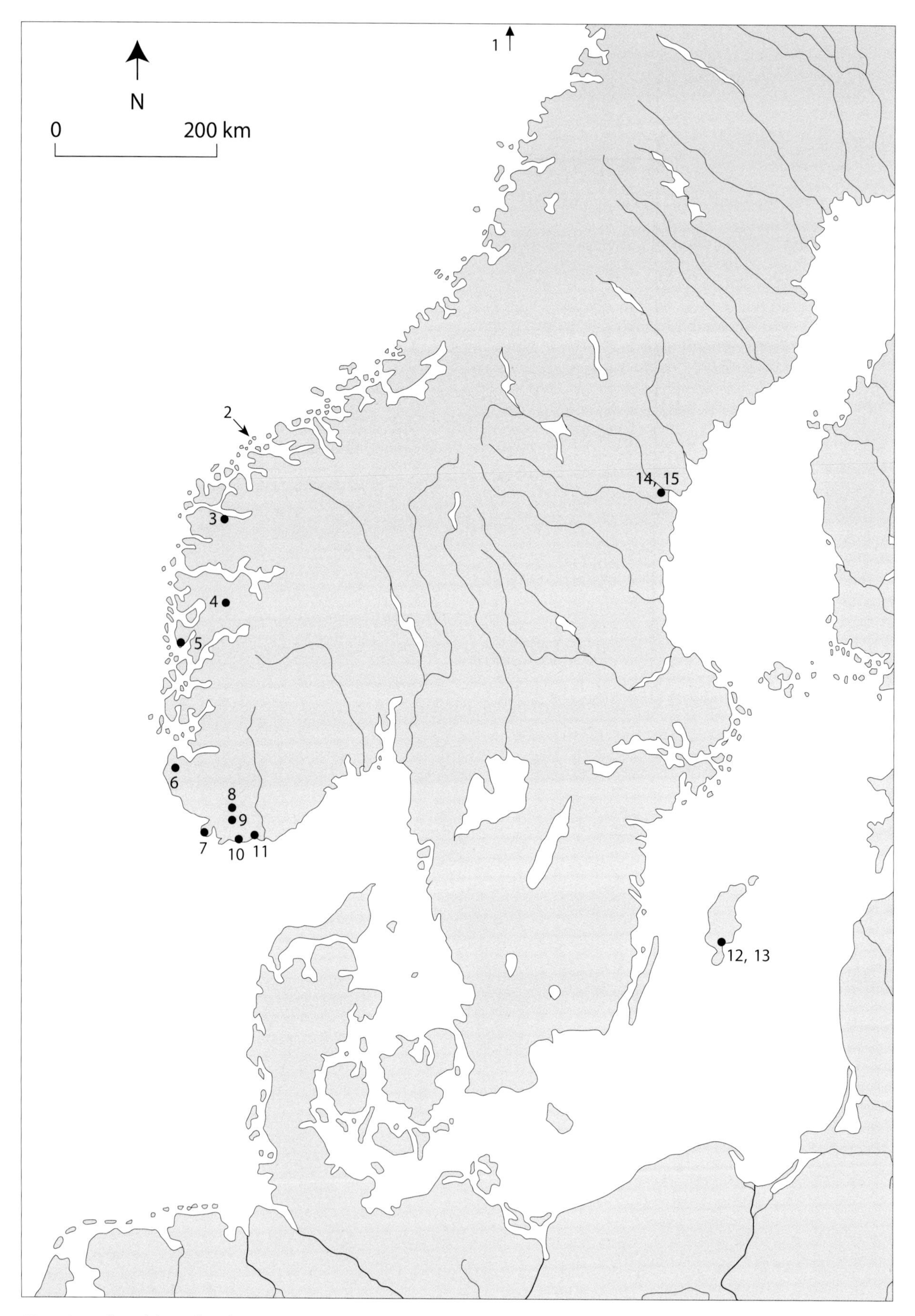

Fig. 1. Burials with bear skins from Norway and Sweden, mainly Migration Period (1–11: Norway; 12–15: Sweden. Cf. Table 1).

Fig. 2. Bear skin (hair) found in the Migration Period male grave (mound IV) in Døsen, Os, Hordaland in western Norway (after Shetelig *1912, fig. 346).*

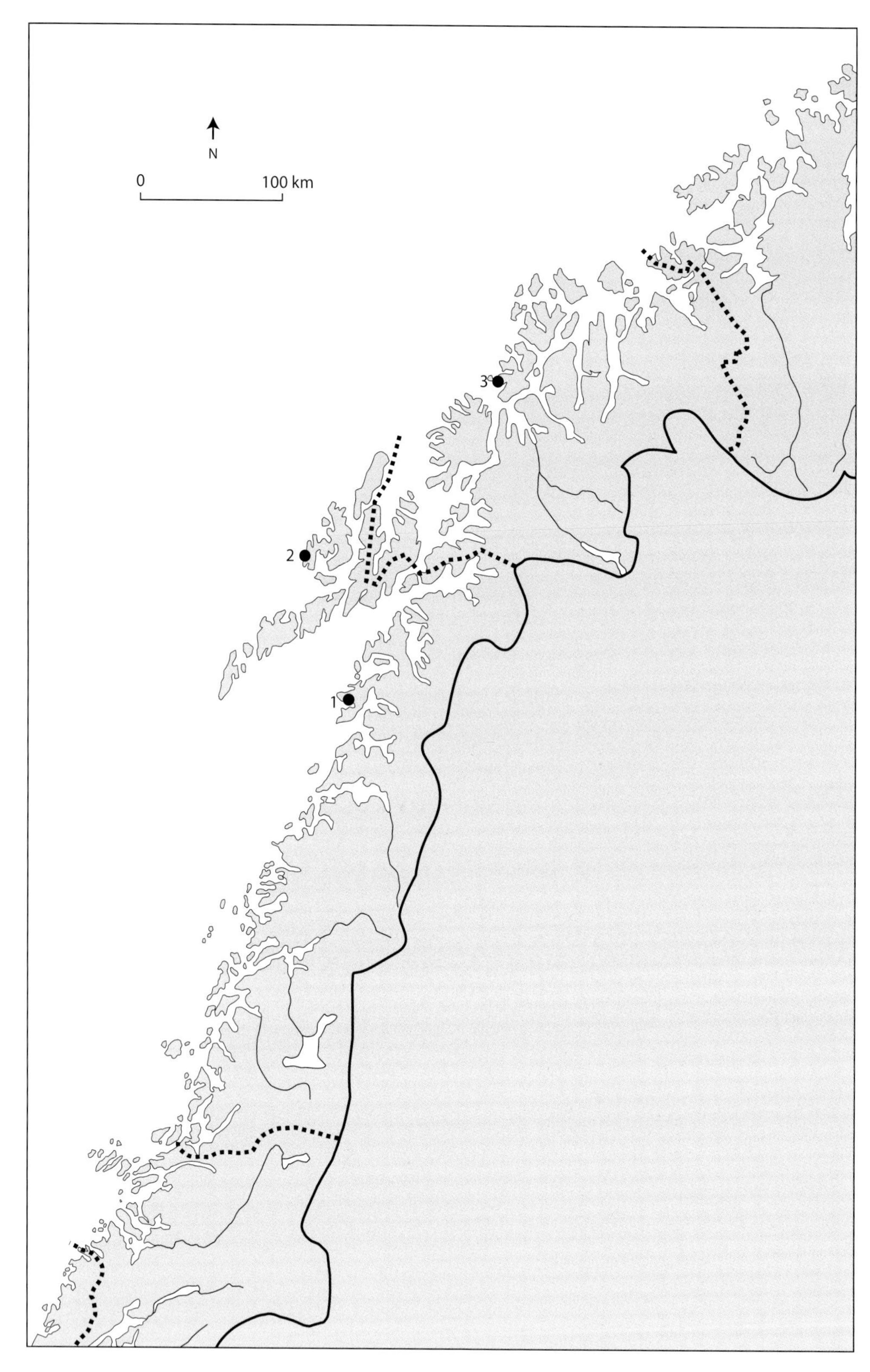

Fig. 3. Burials with actual or assumed bear skins from northern Norway. 1: Bø, Steigen (Nordland); 2: Føre, Bø (Nordland); 3: Hillesøy (Kvaløy) to the west of Tromsø (Troms).

Table 1. Burials with actual or assumed bear skins from Norway and Sweden, mostly from the Migration Period (nos. 1–11: Norway; 12–15: Sweden). B = Bergen Museum, Norway; C = Oldsaksamlingen, Oslo, Norway; SHM = Statens historiska museer, Stockholm, Sweden; Ts = Tromsø Museum, Norway.

No./Location	Mus. find no.	Excavation year	Dating	Burial type	Mound	Chamber/cist	Gender	Description of bear skin (hair)	Social status of the deceased	Selected bibliography (bear remains, find catalogues)
1. Føre, grave 2 (Bø, Nordland)	Ts 5338	1950s	*c.* 325–375/400	Inhumation	"Quite large"	2.5 x 1 m	Female	Skin as rug (preserved bear skin, plus seven claws in four groups)	Above average	Sjøvold 1962, 77; Lund Hansen 1987, 442
2. Haram (Haram, Sunnmøre)	B 12048	1960s	*c.* 325–375/400	Cremation	30 x 1.5 m (diameter)	Small pit	Male	Skin wrapped around the urn (preserved animal hair, plus 15 bear claws)	Petty king	Solberg 1984, 100–101; 2001, 121; cf. Lund Hansen 1987, 440
3. Evebø (Gloppen, Vestland)	B 4590	1880s	Late 4th/5th century	Inhumation	25 x 4 m (diameter)	4 x 1 x 1 m	Male	Skin as rug (preserved bear skin and hair)	Petty king	Gustavsson 1889; Shetelig 1912, 111–117; Ilkjaer 1990, cat. no. 157; Bemmann/Hahne 1994, cat. no 308; Kristoffersen 2000, cat. no. 92
4. Nedre Aure (Voss, Vestland)	B 520–525	1835	Not datable	Inhumation	Unknown	4.5 x 0.9 x 0.8 m	Double grave; unknown gender	Remains of a bear skin	Average?	Description of B 520–525 in the find catalogue (Bergen Museum)
5. Døsen (Os, Vestland)	B 6091	Early 1900s	*c.* 425–475	Inhumation (secondary burial)	18 m (diameter)	2.7 x 0.7 x 1.20 m	Male	Skin as rug (preserved hair and at least five claws from one bear paw)	Below average	Shetelig 1912, 141–148
6. Krosshaug (Hauge, Klepp, Rogaland)	B 2269-82, 2288-92, 2294-99	1860s	*c.* 450	Inhumation	30 x 5 m (diameter)	5 x 1 x 1 m	Female	Skin as rug (four bear claws and animal hair)	Petty queen	Magnus 1975, 19–20, 106; 2014, 74; cf. Kristoffersen 2000, cat. no. 47
7. Sletten (Vanse, western Agder)	B 4234	1880s	Late 5th or early 6th century	Inhumation	Remains of a mound	Unknown	Females (several)	Skin as rug (preserved animal skin and one claw)	Above average	Description of B 4234 in the find catalogue (Bergen Museum); Kristoffersen 2000, cat. no. 26
8. Snartemo, grave V (Hægebostad, western Agder)	C 26001	1930s	Late 5th or 6th century	Inhumation	Unknown	5 x 1 m	Male	Skin as rug (three bear claws, hair, and possibly skin)	Petty king	Hougen 1935, 8–9; Bemmann/Hahne 1994, cat. no. 199; Kristoffersen 2000, cat. no 19

Cont. Tab. 1. ▶

Cont. Tab.1

No./Location	Mus. find no.	Excavation year	Dating	Burial type	Mound	Chamber/cist	Gender	Description of bear skin (hair)	Social status of the deceased	Selected bibliography (bear remains, find catalogues)
9. Vemestad (Lyngdal, western Agder)	B 4414	1880s	Mid-5th century	Inhumation	"Substantial"	3.3 x 0.9 x 0.8 m	Male and female	Skin as rug (preserved bear skin)	Above average	SHETELIG 1912, 158, footnote 1; BEMMANN/HAHNE 1994, cat. no. 201
10. Vestre Skogsfjord (Mandal, western Agder)	C 2701-2707	Pre-1860s	Late 5th/6th century	Cremation	Unknown	Unknown	Male	Skin in a metal urn (three bear claws and animal skin)	Above average	GJESSING 1925, 44; HOLCK 1986, 247; HAUKEN 2005, cat. no 21
11. Eik (Søgne, western Agder)	C 23256	Pre-1925	*c.* 325–375/400	Cremation	Mound of unknown size	Small chamber	Male and second person	Bones wrapped up in skin (preserved animal skin and nine bear claws)	Above average	GJESSING 1925, 44; HOLCK 1986, 277; LUND HANSEN 1987, 436; HAUKEN 2005, cat. no. 25
12. Smiss (Eke, Gotland)	SHM 16113	1910s	*c.* 160/180–325	Inhumation	"Shallow"	3.4 x 0.6 m	Female	Skin as rug (four groups with 18 claws; 3 x 5 plus 1 x 3)	Above average	ALMGREN/NERMAN 1923, 89 fig. 156a; PETRÉ 1980, 6; LUND HANSEN 1987, 446
13. Hallvede (Eke, Gotland)	SHM 25133	1950s	*c.* 550/575–750/800	Inhumation	Flat grave	2.2 x 1.5 m	Male	Skin as rug (four groups with 16 claws; 3 x 5 plus 1 x 1)	Above average	Description of SHM 25133 in the find catalogue (Statens historiska museum); PETRÉ 1980, 6 fig. 2
14. Högom, mound 2, (Selånger, Medelpad)	Sundsvall Museum (no.?)	1950s	*c.* 350–450	Inhumation	40 m (diameter)	5 x 2 x 1 m	Male	Skin as cover and on the floor of the grave chamber (preserved hair)	Petty king	NOCKERT 1991, 31, 36; RAMQVIST 1992, 46–47; RAMQVIST 2000
15. Högom, mound 4, (Selånger, Medelpad)	Sundsvall Museum (no.?)	1950s	*c.* 375/400–550/575	Cremation (secondary burial)	40 m (diameter)	Man-sized layer	Unknown	Skin and 13 claws on the funeral pyre (?)	Above average	RAMQVIST 1992, 194–198

Sámi bear graves – results from archaeological and zooarchaeological excavations and analyses in the Swedish part of Sápmi

By Elisabeth Iregren

Keywords: Sámi culture, bear grave, bear hunting, zooarchaeology, archaeology

Abstract: This contribution deals with a specific ritual expression among the Sámi, the bear grave. Brown bears were sacred in the Sámi society. They were hunted, killed, and then consumed during a feast. All bones of the killed bear were afterwards collected and carefully buried. At present, in the Swedish part of Sápmi, we have evidence of definite Sámi bear graves going back in time for one thousand years. This article mainly deals with the knowledge about bear graves from archaeology and zooarchaeology. A typical Sámi bear grave consists of all the bones of one buried bear individual, but of no other animal species. The bones have been split for marrow. The burial was erected close to a newly built hut, and the construction of the grave was elaborate. Artefacts were rarely deposited in the grave. New radiocarbon dates are presented here. The bone material is used to find out more about the traditions surrounding the burying of the bear. Weapons and hunting injuries are described. Further, the location of bear graves in relation to other human constructions is discussed. To some extent historical sources are also dealt with. Ideas for future work are suggested.

Background

The Sámi culture is of course complex and varied, and there are temporal variations. As a cultural historian one could deal with, for example, settlements, burial grounds, and ritual expressions of the society. Human culture, beliefs and rituals are visible today through offering sites, depositions of bones or objects, the ceremonial drums, and in the bear and reindeer ceremonies. Here, I will deal with the Sámi bear graves.

The brown bear is a typical species of the taiga in the northern hemisphere. Around the North Pole many groups of people living on hunting and fishing have had special and ritual bonds to the bear in the past. Examples of these are the Sámi, Finns, Karelians, Shantis, Udmurts, Ostyaks as well as the Ainu in Japan, and Algonkin-speaking Indians in North America (see different contributions in the present volume; see also Zachrisson/Iregren 1974, chap. 6–7; Bäckman 1981, 44; Edsman 1994, 42–49; Wamers 2015, 41–52). Thus, as many hunters reverenced the bear, Sámi and Finnish bear rituals are parts of a much larger sphere of hunters' rituals. Bear remains have been found in recent years in Siberia, close to Lake Baikal, in human burials at very early ritual sites. These graves were created 6,000–5,000 and 3,400–2,000 years BC (Losey et al. 2013, 65 table 4.1). Many bear bones were excavated from these early graves. The most frequent remains of the bear found are the cranium and baculum (*Os penis*).

Bear and Human: Facets of a Multi-Layered Relationship from Past to Recent Times, with Emphasis on Northern Europe, ed. by Oliver Grimm (Turnhout: Brepols, 2023), pp. 547–586 BREPOLS PUBLISHERS 10.1484.M.TANE-EB.5.134349

As regards the Sámi traditions, there are written sources, ethnographical investigations and archaeological documentations that form an excellent basis for our studies. Among the Sámi the bear belonged to a divine sphere, but it was not considered a god. Thus, no offerings were made to it (Bäckman 1981, 45, 48). On this point, however, Edsman (1994, 50) seems to disagree. The bear belonged to the supernatural worlds and was highly respected. It was regarded as a living, feeling, and thinking creature (Bäckman 1981, 47). Through the ceremonies the bear was supposed to come back to its own world while it was honoured. It was further looked upon as the most sacred of all animals (Bäckman 1981, 48; Quigstad 1903, 27; citation after Schanche 1997, 262).

As in most hunting societies, many rituals concern the violent and bloody relation between the hunter and his prey. Thus, there were rules to follow so that the delicate balance between Man, the supernatural powers and the beasts could continue and remain in balance (Edsman 1970, 48). Honouring the bear during the feast and erecting a bear grave are just a few of these demands to be fulfilled (Schefferus 1956, 269).

The first written notation about the Sámi and their bear ceremonialism has been found in a Norwegian royal chronicle from 1606, *Cronicon Regiae Norvegiae* (Edsman 1994, 51). The chronicle was written by Halvard Gunnarsson who was working as a clergyman in Oslo.

Sámi beliefs and traditions, in particular testimonies on Sámi religion, have been recorded since the 17th century. However, there are substantial problems in interpreting the historical sources on Sámi rites, beliefs, and religion. The documents are few and they frequently quote one another. Further, many of the authors were priests and missionaries with limited understanding of non-Christian believes and the Sámi culture in general (Hultcrantz 1983, 23; Fossum 2006, 16). Fossum (2006, 7–14) gives a detailed account of the geographical and temporal origin of historical sources in Sweden and Norway.

In this article I will mainly rely on and cite Pehr Fjellström's book from 1755 (Fjellström 1981). I will follow the facsimile text published in 1981, including valuable comments by Professor emerita Louise Bäckman. Fjellström was a clergyman in Lycksele parish, Lapland. The bear graves that we know most about derive from this region (Fig. 1).

The first bone finds in Sweden to be identified as a Sámi bear grave came to light around 1914 in Jämtland, close to Lake Värjaren (Manker 1957, 279; Zachrisson 1983). In Norway, a bear grave was recorded at Salsfjället in Namdalen in 1940 (Petersen 1940). It is situated not far from Värjaren. During the 1950s scholars investigated the Nedre Vapstsjön (Manker 1957, 254) and Sörviken bear burials in Sweden (Janson/Hvarfner 1960; 1966).

Bears and bear graves in the Nordic area

There were several ways to establish and maintain close links between man and bear in the old cultures in the Fenno-Scandic area. These bonds might be expressed as bone depositions, elaborated artefacts, or depictions of the species.

In settlements in northern Sweden finds of bear bones are, however, very scarce (cf. Zachrisson/Iregren 1974, tables 3, 5; Ekman/Iregren 1984, table A; see also Magnell on bears and humans in Sweden, this volume). At one of these sites, however, three bear bones were found, all deriving from the paws. Most interesting is also that this is where one of the extremely few bear-shaped artefacts was recovered (Fig. 2). It is a whetstone made of sandstone, shaped as a bear's head. The settlement is dated to 4100–1700 BC.

In northernmost Norway a great number of rock carvings with bears, their dens, bear tracks, and hunts have been depicted, for example, in Alta. Knut Helskog has dealt with these in detail (e.g. Helskog 1999; 2012). The rock carvings are dated to the Late Mesolithic and the Neolithic periods (ibid.), and thus indicate a very long tradition of hunting brown bears at the den. Spears, and bow and arrow seem to have been the weapons most frequently used, judging from the rock panels. These rock carvings thus reveal the hunters' intimate knowledge of and their bonds to the species.

In contrast to Norway, rock carvings of bears are rare on the Swedish side of the Scandinavian Mountain Ridge. At the site Nämforsen, Ådalsliden parish, Ångermanland, Hallström (1960, 291) early on recorded a small number of figures as possible bear representations among more than 2,000 carvings. The datings of a nearby settlement and this enormous site with rock carvings cover 1,500 years; they cannot be younger than 2,000 BC (Baudou 1993, 81–91). Lindqvist (1994, table 9.5.), in his thesis on rock art, reports 19 (?) possible single rock engravings of bear (and dog?) at the very large Nämforsen site.

Bears have also in the past been painted in red ochre on rocks in northern Sweden, for example, on the cliff sides at the site Flatruet, Härjedalen (Fig. 3; cf. Hallström 1960, 95). The bear figure is considered to be the oldest painting at the site, more than 4,000 years old (Sandell 2010, 24). Baudou (1993, fig. 59) dates the site to the same period and somewhat later.

Sámi bear graves in the Swedish part of Sápmi

In this contribution, I will focus on Sámi bear graves, definite and possible ones, in the Swedish area of Sápmi, the cultural region raditionally defined as inhabited by the Sámi people. Many burials have also been found in Norway (see Sommerseth, this volume; Petersen 1940; Myrstad 1996; Schanche 1997, 243–245; Fossum 2006, 100–107; Sommerseth 2021). My new investigations of bone material in 2020 mainly concerned the bear bones from Sörviken (site no. 2), Nedre Vapstsjön (3), Tiirivara (4), Mjösjö (5), Avaträsk (9), Långbäck (14), Gråtanån (15) and Öhn (16–17; see Table 1).

Table 1 contains a list of irrefutable and possible bear graves; site, location, ownership, and inventory numbers are included. I would assume that bear rituals were also performed and burials constructed in Finland and in the Kola peninsula, as these are known Sámi territories. Such burials have not yet been found, but then this is the late distribution area of Sámi groups. A bear spear from northern Finland that has been located in the County Museum of Västerbotten at least indicates bear hunts (cf. Table 5 for inventory no. 80481). We would also expect bear graves to be found in many more southern areas in Sweden, too, e.g. the counties of Dalecarlia (Iregren 2022, 61–68), Västmanland, Västernorrland and Gävleborg.

In Sweden we have performed many professional excavations and detailed analyses, and now many burials have been dated (Zachrisson/Iregren 1974; Melander 1980; Zachrisson 1983; Mulk/Iregren 1995; Broadbent 2005; 2010). Archaeological and zooarchaeological observations and data form the basis of this study. Thus, details of the bear graves can be found in the mentioned publications as well as in a report from the Norwegian Sámi Council (Iregren et al. 2017).

Description of an irrefutable bear grave

I will start by introducing a typical bear grave in the taiga, based on the excavations and analyses of the bear graves in Gällholmen, Sörviken, Gråtanån, and Karats. I took part in the work with all of them. The first three constructions were complete when found.

A Sámi bear grave ideally consists of all the bones (unburnt) from one killed and consumed brown bear individual. Among the bones, the skull, the lower jaws, and the shoulder blades are never damaged during parting, fileting, cooking, and consuming the bear. Accordingly, these skeletal parts always hold prominent positions in the construction (Zachrisson/Iregren 1974, fig. 21). All bones containing bone marrow have been cleaved in all the bear graves revealed so far.

Burial construction and bone depositions

The heap of bones is built either on the ground surface, in a pit, under a cairn, or in a crevice or similar place. Thus, we see a variation. Mulk/Iregren (1995, 24) divide the bear graves into three variants – one with wooden supports, a second one under a cairn, and a third one in a crevice. At present, we have more information from the burials in woodlands than in mountainous areas. Further, regarding Gråtanån in the Vilhelmina parish, an archaeological excavation immediately followed the discovery of the bear grave and thereby information on more fragile details was not spoilt. Logs and other wooden remains in the Gråtanån bear grave have been examined and determined as pine (Melander 1980).

The stratigraphy of the Gråtanån bear grave is described as follows in the report (Melander 1980, translation and summary by E. Iregren): The turf had been removed from the ground surface. In this area people had laid thin poles and birch bark. Above it, cut wooden slabs were placed crosswise in two or three layers, perhaps even a woven container or mat, a wickerwork of laths or likewise. A larger stem was also found here, and above it a further layer of birch bark was laid.

The Karats bear grave reveals other details of preparation. Initially the ground surface was burned – just to melt the snow? Then birch bark formed the base of the bed of bones. On top of the bone concentration stones sized 0.10–0.45 m had been put, presumably to cover and protect this bear grave (Mulk/Iregren 1995, 13–14).

The bones in the Gråtanån bear grave were then organised as described below. It is evident that these burials were built with great care and devotion. On top of the heap of bones, birch-bark and logs of pine covered and protected the grave from scavenging animals, as is also the case in Sörviken and Gällholmen. In Karats the bear grave was sealed by several rounded stones of different size (Mulk/Iregren 1995, 13–14).

Still more bone elements than skull and shoulder blades may have been arranged intentionally, e.g. the vertebrae of the neck and bones from the paws in the Sörviken and Gråtanån burials (Fig. 4; Zachrisson/Iregren 1974, fig. 26–27; Melander 1980, figs. 12, 15). The vertebrae of the neck were deposited close to the skull. In the Sörviken and Gråtanån bear graves many bones of the paws lay together (Fig. 5). In the Gällholmen burial just a few of these bone elements were placed close to one another. The paws had obviously been split lengthwise along the five digits. Sinews and/or meat connected many of these small paw bones in neat rows when they were put into the burial.

Further, other vertebrae and bones might create a mid-line in the burial (Zachrisson/Iregren 1974, fig. 26; Melander 1980, fig. 13) and may supposedly be an intentional depiction of the backbone of the bear. The Sámi seem to have striven meticulously to re-shape the anatomy of the bear in the burial, although all long bones had been split to get at the marrow. Accordingly, this was a difficult task. According to historical sources, this aim is clearly referred to and must be fulfilled (e.g. Fjellström 1981, 28, see Zachrisson/Iregren 1974, 92–93 for a more thorough discussion).

Terminal phalanges of a fur animal may have the status of buried bone parts, or they can belong to a pelt, or both. In the latter case they might be regarded as an artefact (see below). These phalanges are only rarely found. In the Gråtanån bear grave (Melander 1980) seven terminal/third phalanges were recovered, and in the Nåttinäset and Aspnäset burials one single third phalanx was found in each of them. In all other burials in Sweden these elements were missing (Table 2).

Few and rare artefacts

Archaeological finds from bear graves have rarely been documented. Artefacts or constructions of organic material might have existed, as the "wicker" in the Gråtanån burial indicates, but may now often have disappeared.

On two occasions, lead bullets have been found in graves (Sörviken: one bullet; Gråtanån: four bullets; Fig. 6; cf. Zachrisson/Iregren 1974, 22, 27; Melander 1980). The weapons belonging to these bullets are judged to be muzzle-loading guns. During the second half of the 17th century these were becoming more frequent among the Sámi. The bullets in the two graves do not seem to have been fired (Zachrisson/Iregren 1974, 22; Melander 1980).

The cause of the depositions of these bullets has not been much discussed earlier. K.-Å. Aronsson (pers. comm.) points to the work by Högström 1980, 210). He refers to a tradition of putting not only the bear bones in the grave, but also objects such as skies, a knife, a wooden planer, or a piece of brass. Aronsson (pers. comm.) suggests that in these cases the objects seem to be offerings. Högström (1980, 210) mentions the depositions might have been intended to be of use in the bear's other world. The tradition of burial gifts in pre-Christian *human* burials was widespread among the Sámi and in other cultures. Many researchers within Sámi archaeology further stress the many similarities between bear graves and human burials (Zachrisson/Iregren 1974, 88; Schanche 1997, 261–264; Fossum 2006, 101–107).

Another artefact type has been exposed in one single bear burial in Sweden – Värjaren (Fig. 7) – and another one in Norway – Salsfjället in Namdalen: A brass chain had been hung at each cranium (Zachrisson 1983, fig. 6; Petersen 1940). Schanche (1997, 243, citing Bäckman/Kjellström 1979, 181) notes that rings of brass could be given as burial gifts to deceased humans. Högström's information (1980, 210) should also be kept in mind here. In this case, the brass might have been positioned as a kind of adornment or as a burial gift.

Discussion and clarifications

In this section I will not repeat for discussion features in the bear graves that are constant or *not* disputable. The following section is instead an attempt to discuss the variations noted and their implications for our interpretations. It should also be mentioned that bears of all ages (Table 2) and most likely of both sexes (Zachrisson/Iregren 1974, chapter 5.2) were buried.

When the Sámi reconstructed the bear in its grave, the skull, including the two lower jaws, was put in front of a heap of bones or on top of some bones (Gråtanån, Karats). Almost every single bone of the animal was carefully collected after cooking and consumption (Tables 2–3; Zachrisson/Iregren 1974; Melander 1980; Mulk/Iregren 1995). Only *one* bear individual is buried in a single spot. With this description as a basis – how should we look upon bear finds that do not demonstrate all the mentioned characteristics of a bear grave hitherto proven?

We must of course acknowledge the risk of lacking information or parts of the find, when local people have found the bear bones, or else when no proper excavation has been performed. Other circumstances might also have influenced the data. As the Gällholmen and Sörviken burials had been kept indoors for one or many years before investigation, possibly existing delicate wooden details or other organic constructions may have dried to dust (Zachrisson 1983, 90).

I look upon a bear bone collection where elements are missing in the following way: If the skull is missing, this is to me not an ultimate argument against the find being a Sámi bear grave, when many other bones of different body parts were found. A skull is an attractive trophy and might easily be picked up and taken away from a site by passers-by. As bear graves were often located close to the ground surface or in crevices or clefts, they can be detected relatively easy and contents can be removed.

Schanche (1997, 263), however, sees another possibility of intentional depositions of parts of bears. She stresses that the historical sources are dated to the same time span as the late bear graves, which are represented by many examples of complete burials. Thus, she mentions the possibility of depositions of only few skeletal parts/single crania, etc. in earlier rites. This is important to keep in mind.

Of course, such variations in rituals and behaviour might occur, although they are not yet proven. Schanche (1997, 264) also discusses another possible interpretation. One may look at bear graves as new expressions of ritual behaviour in a changing society. Stress from church, state and settlers in Sámi vital territories might have created a political and cultural situation where religion, traditions, and rituals were more important to keep, maintain and develop.

However, I here want to remind readers that in Norway many bear graves are not properly documented or excavated. Further, the tradition of complete burial is at least 1,000 years old in Sweden (see below, on dating). So, for the moment, my assessment is that it is important that many bones from the skeleton, preferably from all parts of the body, can be collected, as this is in firm agreement with hitherto irrefutable bear graves found and also the written sources. However, FJELLSTRÖM (1981, 31) mentions a situation when parts of the bear might be missing already at the establishment of the grave. He records a situation when hunters other than Sámi were members of the hunting party. Different meaty parts of the bear's body could then be given away to these people and thus these bones would not be included in the grave.

One find of bear bones that is difficult to interpret under this aspect stems from the Aspnäset site (ZACHRISSON/IREGEN 1974, 31; MULK/IREGREN 1995, 11), in particular because of its ancient date. No cranium has been found, but bones from many body parts were collected (Table 2). The site has not been excavated, but the bones were gathered during a cultural inventory performed by the Swedish National Heritage Board (Table 1). At present its very old date complicates the interpretation as a bear grave. The calibrated date shows that it is more than 4,000 years old (cf. Table 4 and Fig. 8).

At the site of Öhn, Ström parish in Jämtland, two bear mandibles were found by locals in 1950. They were discovered during peat digging for soil improvement at a depth of 1.2 m. We do not know more about this site. The mandibles have different sizes, and they both carry cut marks. Evidently, their origin is anthropogenic. Both mandibles yielded old dates, as did the find from Aspnäset. The calibrated results are different, the bones from Öhn are thus about 2,600 and 3,100 years old. At present the type of deposition is uncertain. A short archaeological investigation was performed in 2022, but no more bones were found (A. Hansson, pers. comm.; cf. Table 4 and Fig. 8).

Another situation to discuss is when a bear find includes bones from more than one bear individual. Such sites were identified at Nedre Vapstsjön and Långbäck. At Nedre Vapstsjön single bones from three bears were collected but none of the skeletons is complete. Only one skull now remains in a museum collection (Tables 2–3).

We know, however, from the many bear finds at Skagedalen, Spildra, Kvænangen, all Northern Norway, that one small area might be used over and over again by the Sámi. In this area four bear graves have been constructed (MYRSTAD 1996, 32–34; IREGREN et al. 2017). At Spildra, however, the graves were deposited in a small bay as separate burials at separate locations (cf. Fig. 14). Further, FJELLSTRÖM (1981, 27) mentions that the knowledge of a bear killing and of the bear's burial is narrated later. This, of course, makes it easy to return to the same area for a new feast and ceremonies. Regarding Nedre Vapstsjön, most bones seem to have been deposited in the same cleft in a split boulder (MANKER 1957, 254; ZACHRISSON/IREGREN 1974, 28–29, 39). Still in a cist, close to the boulder, a bear canine has been documented, and MANKER (1957) also mentions an existing cultural layer there. So, these constructions might be separate bear graves. Two radiocarbon datings were made on bones of two bears and they gave similar dates, younger than AD 1533 (cf. Table 4 and Fig. 8; ZACHRISSON/IREGREN 1974, 29).

Långbäck is another site that contains bones that indicate that bear graves were located close to one another (ZACHRISSON 1975b, 6–8; MULK/IREGEN 1995, 32–33). The site(s) were obviously heavily destroyed by water level changes due to the damming of the River Umeälv. Locals have retrieved scattered bear bones on several occasions when the water level was low. Three mandibles prove the existence of two bear individuals. Otherwise, the number of bones found is low (Table 2). Långbäck is likely to be a site where the local *siida* (a Sámi village, but an extended area) constructed several bear graves. An important source in this respect is the account of the traveller Daniel von Hogguér, who in 1828 (VON HOGGUÉR 1841, 78) witnessed how two bears, killed from the same lair, were buried in two separate bear graves.

The bones from Nedre Vapstsjön demonstrate another feature that disturbs the interpretation of (a) bear grave(s). A wolverine cranium *(Gulo gulo)* and another 32 bone elements of the individual were found in the bone complex. I do not know of any ritual parallel to the occurrence of this species. Above, I argued that this could be a site with several bear graves. Or is it perhaps a Sámi sacrificial site or another type of bone cache not properly investigated, described, and understood earlier? In this context, we might consider the variety of reindeer *(Rangifer tarandus)* bone depositions that have been recognised and documented (Manker 1957; Zachrisson 1975a; b; Iregen 1983; Kjellström 1983; Zachrisson 1983; Mulk 1994, 170–176, and passim; Boëthius 2010).

Another site that further complicates our understanding of bear graves is the find at Grundskatan (Broadbent/Storå 2003; Broadbent 2005; 2010). Several bear bones, presumably from the same individual, have been identified by Jan Storå (Table 2). Broadbent/Storå (2003) point out many features at Grundskatan that coincide with the bear grave at Karats and other burials (Broadbent 2005, 25). However, the bones were found in-doors, a location which has not been acknowledged earlier regarding bear finds (cf. Fig. 21). In Norway, however, the archaeologist S. E. Grydeland has identified different ritual structures with wild and/or domestic animal species as depositions inside huts on the island of Spildra and elsewhere (Grydeland 2001, zooarchaeological notes therein by E. Iregren). The Grundskatan hut is dated to the Early Medieval Period. Furthermore, the bear bones there have partly been burned, but burnt animal bones are otherwise rare in Sámi ritual contexts in Sweden.

Dating of Sámi bear graves

In accordance with what we at present know for certain, a Sámi bear grave in Sweden might have been constructed any time during the last millennium (Table 4; Fig. 8). The dating results show that many of the excavated and now dated burials are relatively recent, later than AD 1533. The written evidence indicates that bear rituals were still performed as late as the 19th century. Zachrisson/Iregren (1974, 14, 94) refer to late sources on bear rituals. A traveller from Germany witnessed the construction of bear burials in Arjeplog as late as 1828 (Von Hogguér 1841, 78), but also even later recordings are known.

These late radiocarbon dates, further, coincide with the artefacts deposited, as for instance the lead bullets. They also generally correspond to the typological dating of the brass chain in the bear grave from Värjaren (see below; Fig. 7). On the other hand, artefacts, stones, and coal in the vicinity of a bear grave have not regularly been dated or closely examined. At the site of Nåttinäset a human burial from the Late Iron Age was also found (Zachrisson/Iregren 1974, 30). It is, however, to be noted that at present most Sámi bear graves in Sweden are dated younger than their counterparts in the Sámi areas of Norway (Myrstad 1996, 46–47; see Sommerseth, this volume).

The oldest Sámi bear graves in Sweden found so far are the unquestioned burials from Karats and at Nåttinäset, together with the supposed burial at Grundskatan (Table 4; Fig. 8). The oldest one is Karats, newly calibrated to AD 607–879 (1 sigma). Grundskatan seems at present a rather untypical bear grave (Broadbent 2010, tables 16–17; also see above). These bear finds are dated up to the 10th to 11th centuries (Mulk/Iregren 1995, 20; Broadbent 2005, 25–26, 34–35). These are also the northernmost bear graves hitherto found in the Swedish part of Sápmi. It is somewhat unexpected as they are not located in the areas in Sweden with the best possibilities for preservation of bones. In Norway, one bear grave – Kjærfjorden – is dated as early as the Early Iron Age, AD 220–331 (1 sigma, new calibration; T-12023; cf. Myrstad 1996, 39, 46; Sommerseth 2021, table 1). It is the oldest typical bear grave known in Scandinavia so far (cf. Sommerseth, this volume).

There are two bear finds from Middle Sweden that have similar dates as the Kjærfjorden bear grave in Norway. They have been recovered from two cremation graves in a burial ground in Härjedalen,

I have earlier incorrectly described this injury as possibly caused by a gunshot (ZACHRISSON/IREGREN 1974, 82). However, this is not likely as the holes are much larger than the lead bullets found in contemporary bear graves. As already mentioned, bullets have been excavated from two bear graves. The four ones in Gråtanån are round and have a diameter of 9.45 to 9.77 mm. The single one in the Sörviken grave is 10–11 mm and hemispherical (ZACHRISSON/IREGREN 1974, 22, 27; MELANDER 1980). According to K.-Å. Aronsson (pers. comm.) the diameters of bullets could vary between 6 and 16 mm, and the lead of the bullets could be recast and used again (cf. Fig. 6).

Peter Krantz argues that the bullets of those days did not inflict much larger wounds than their diameters, as the velocity was low. Further, the openings of the cranial bone walls are unsymmetrical, and their sizes also coincide better with the dimensions of bear spears.

Nedre Vapstsjön, Tärna parish, Lapland

The stone constructions at Nedre Vapstsjön form complicated structures, which are not understood in all their details (ZACHRISSON/IREGREN 1974, 28–29, 54–55). There is only one cranium among the many bones.

The bear cranium from Nedre Vapstsjön shows an injury in the right frontal bone (Fig. 13a–b). The attack came from above, from the back and from the left side in case of a right-handed huntsman. The skull vault was damaged, and an opening of 22 x 13 mm was created. This kind of damage is called a buttonhole fracture, and fissures due to the reduction of induced tension run forwards and backwards.

If a spear had been used, it would have first penetrated the sinuses, then the soft brain tissue and at last damaged the relatively weak bony structures of the skull base. No damages are noted in the skull base, however. Thus, we judge this injury to be a result of a hard stroke by a club or another blunt and heavy weapon. This injury alone would not have killed the bear, as it caused damages no deeper than the sinuses of the skull. However, this was a rather powerful blow that fractured the bone walls. It may briefly have paralysed the bear, making the kill by other hunters somewhat easier.

Spildra, Kvænangen, Troms, Norway (bear grave C)

The Sámi buried at least four bears in a small valley (Fig. 14) on the island of Spildra, Norway (IREGREN et al. 2017). Ingrid Sommerseth reports that two of these burials were dated by Audhild Schanche in 1994. Their radiocarbon dates are recalibrated to AD 1030–1220 (T-11214) and AD 1132–1328 (T-11215; SOMMERSETH 2021, table 1).

In the bear graves Spildra A (formerly I) and B (formerly II; IREGREN et al. 2017, table 1) no crania were found during the investigation in 2013. From bear grave C, however, a bear skull with a severe injury was examined (Fig. 15a–c). A heavy crack surrounds the braincase of this bear on almost all sides. The fissure passes through the parietals and runs to the right temporal bone. From there the fissure runs backwards almost to the occipital crest. On the left side a small fissure approaches the sagittal crest. On the ventral side of the skull a massive crack is seen. The maxillary bones seem to have been separated from one another, and a suture closed by age has been split open again due to the force applied. An impact mark, about 15 mm long, is visible close to the sagittal crest. The symmetric damages clearly demonstrate the application of blunt force, P. Krantz notes.

The following interpretations were made: The blow was directed from above, most likely when the bear was about to leave the den. A heavy object, with a rather smooth surface, inflicted the injuries. It might have been the neck of an axe, as it caused the entire braincase to crack. The force probably induced a concussion of the brain, so the bear might have been unconscious. A spear (or a knife) probably ended the bear's life. These heavy damages indicate a strong hunter with good knowledge of anatomy and vulnerability of the bear, as P. Krantz argues.

In all these instances, we notice that the hunters' attacks most likely occurred at the den. One hunter seems to have been positioned above it. He had the important task to strike as hard as possible, when the head of the bear was first visible and in reach, as it started to leave the den. Axes, spears and perhaps clubs seem to have been utilised. Most likely both the edge and the neck of an axe were used. It should be noted that the island of Spildra is too small (30 km^2) to hold a bear population. Female home ranges amount to 100–300 km^2 (only 100 km^2 in the case of females with their cubs of the year) whereas in the case of male bears, it is 300–1,000 km^2 (A. Zedrosser, pers. comm.).

This small study of five hunting parties reveals the following results: one bear dead through the first attack, one bear immediately unconscious, and three occasions when the hunt must have continued longer. The Sámi must, in all ways, have tried to avoid a hazardous struggle in the open space although such scenes are often depicted in Sámi art. Of course, this has happened, and the happy outcome of the event would have been retold. In a painting by Wallander we can observe a hunting party with many weapons in use, such as guns, axes, and spears, i.e. all the kinds of weapons discussed here (Fig. 11). In this picture there is also a brave dog, as sometimes shown in Sámi art. The presence of dogs in the bear hunting parties might, however, be questioned, if we trust some experiences from the early 20th century (Toulja 2007).

Hindering the bear from leaving the den

The injuries found all give the impression of the bear having been struck from above. It also seems as if Sámi hunters in all cases have tried to hit the bear as soon as it was starting to move from its hibernating position. We suppose that they tried to incapacitate the bear as efficiently as possible, before it broke out of the den. This strategy was essential for the success of the hunt and to avoid human casualties. In a couple of complementary sources, we have evidence of supporting tools during the hunt, which might have been very important to reach this goal.

An intriguing photo was first published in a book on bears by H. Zetterberg (1879–1961) (Fig. 16; cf. Zetterberg 1951; Zachrisson/Iregren 1974, 82 fig. 86). The photo shows the beginning of a hunt at a den in southern Lapland in 1910. The hunters use a couple of long rods blocking the exit of the den, thus for some short moments hindering the bear from leaving it. This would be one way to slow down the motions of the bear and leave some more time for the hunters better to hit and incapacitate their prey.

Bear spears in museum collections in Sweden

Regarding weapons, I have mapped bear spears in the collections of some museums digitally: Ájtte (Swedish Mountain and Sámi Museum), the County Museums of Norrbotten, Västerbotten, Jämtland and Dalecarlia, the Skellefteå Museum, and Silvermuseet in Arjeplog (Table 5).

There are more county museums in Sweden owning Sámi spears, but I have concentrated upon museums in the counties were bear graves have already been proven. I must stress that I have received important help from many museum curators (see Table 5 for the size of the spears and spearheads). Note that I have only included those spears where measurements and/or other interesting details were available to me. A more detailed analysis of the spears is available in Swedish on the web in the regional journal "Västerbotten" (Iregren 2021).

The bear spears are quite long, longer than a man's body height (Table 5; Fig. 17a, c). The total length does not seem defined as it varies from 168 to 266 cm. Many spears exceed 200 cm in length. The variation is spectacular. The length of the spears most likely depends on the bears' general size, and furthermore every hunter probably adapted the spear to his own comfort and preferred mode of use. In today's sports (javelin throw), however, the javelin's length is defined to 260–270 cm,[1] so some bear spears hold a substantial length.

However, the long spears seem to us to have been rather difficult to manoeuvre when one is close to a prey. On the other hand, the length made it possible to reach the animal from a slightly safer distance. Heidi Henriksson argues that a bear spear was a shock weapon for use at close distance and not intended for throwing (Henriksson 1978, 29–31). Leif Braseth, however, claims that the spear could be used both for throwing and at a close distance (Braseth 2014, 99). The material of the spear shaft is always wood. The catalogues inform that the shafts were often made of birch wood. The spearheads were made of iron.

Also, the length of the spearhead varies, from 15 to 48 cm, and many of them are slender and pointy (Table 5; Fig. 17e; cf. Iregren 2021, 12). Their cross section is frequently rhomboid (Fig. 17d, f) and the diameter 11 to about 55 mm (Table 5). The hunters have rarely marked their initials, the year of use or time of manufacturing (Fig. 17g). Decorations might occur at the spearhead (Fig. 17f), or at the piece of copper or brass linking the head to the shaft (Fig. 17e; cf. Iregren 2021; see also the former spear from Inari [later tansformed into an ice axe], northern Finland in Fig. 17h). There might be chronological or geographical variations within the bear spears, even if this is not clear today.

The injuries recorded here seem to fit the thinner types of spearheads. The bear bones in Sörviken, Gråtanån, and Nedre Vapstsjön all demonstrate traces of hunts that took place later than AD 1533 (Table 4). Spears and axes were very important in hunting predators (bear, wolf, lynx; Björklöf 2010). They all seem to have been hunted with the same kind of spear, but other methods were also used (cf. Henriksson 1978, regarding bear hunting pp. 37–40).

Hunting seasons

There are scientific possibilities to evaluate the period when hunting took place. This question is also pinpointed by the knowledge that many bears were hunted at the den. The act of ringing (locating by searching in circles) the bear is treated in many sources; a Sámi hunter may track a bear down during several days, when it is searching for a suitable place to den. Finally, this place is found and marked for the forthcoming hunt (Schefferus 1956, 261–262). The search for a suitable bear, before its going into hibernation, often took place in October, it is written. Later in the winter a group of huntsmen returned to kill the bear.

Reports on the time of hunting exist in historical sources. Schefferus (1956, 262) mentions that among the Sámi the bear hunt took place in March or April, when skiing could be at best carried out in middle and northern Sweden. The frozen crust of the deep snow cover should bear the weight of a human, but not of the bear (Fjellström 1981, 8). Zachrisson/Iregren (1974, 79–81) mention more sources on this topic.

Turning to game biology of today, Evans et al. (2016) have investigated denning in bears. In conclusion they state: "Hibernation in brown bears seems to be initiated based on environmental cues and terminated due to physiological cues" (ibid., 12). Evidently, there would have been an extremely short period, perhaps only a few days, when the tracking could have taken place – just when it starts snowing, when snow is still scarce, and the bears get ready to den. All bears enter their dens when the snow begins to fall, the pregnant females even earlier.

My study of the bear graves demonstrates that in many cases the hunt took place at the den and during the winter season. Thus, the human behaviour and the season mentioned in the sources are confirmed. Table 6 presents the results of a study of the bears' age. Bears of all ages were hunted, consumed, and buried.

1 For men; cf. Javelin Throw: www.en.wikipedia.org (accessed 14.08.2020).

A good possibility to find out more about the time of death is to investigate the teeth of the bear. Game biologists, working in the northern hemisphere, have for decades made thin sections of teeth of for example wolf, bear, and elk to understand age-composition and changes in game populations (Lieberman 1994). The physiology behind the studied structures of the teeth is the following: Animal species live, behave and eat differently during the seasons of the year. Increment bands are regularly formed in the cementum at the roots, as summer and winter bands. Due to differences in growth, the orientation of collagen fibres shift, and these bands can be recognised. When the outermost layer is identified under the microscope, one can decide upon the season of death (Lieberman/ Meadow 1993).

In 1988 I received support to cut and evaluate bear teeth from Enid A. Goodwin in the Alaska Dept. of Fish and Game, Anchorage, United States. Four bear teeth from small bone samples collected at a Sámi sacrificial site in Sweden (Unna Saiva) were successfully treated and proved informative. Some teeth could not be evaluated. On the individual level, the investigation gave two important results – the bear's age at the time of death and the season of death.

My data are based on thin sections of roots. The lower fourth premolar of the bear's mandible was used. The results enlighten the lives of four bear individuals who were of different age (4–10 years) when killed. The kills might all have been executed at the bear's den. As demonstrated, the results indicate a certain variation in hunting periods – both winter and late winter to spring.

Evidently, the archaeological and zooarchaeological evidence well agree with the written sources. The hunts were induced by climate, weather, and bear biology. But it would certainly be valuable to follow up this topic further and investigate a much larger number of bear remains. Also, individuals from bear graves ought to be included.

Bear feast and ceremonies performed

During archaeological investigations light has been shed upon how the bear carcass was treated, how the bones were used in reconstructing the bear's body, and how the grave was built in every detail. I have here assembled data on how the bears of Gällholmen, Sörviken, Gråtanån, and Karats were dealt with.

Skinning, parting, cooking, and consuming the bear

The skinning and parting process of a prey is always adapted to the species killed and the rules of culture and believes. Deer are flayed in a way that suits their anatomy, and in predators the skins are also taken care of in an appropriate way.

In the typical bear burials from Gällholmen, Sörviken (Zachrisson/Iregren 1974, table 1) and Karats (not complete when found; Mulk/Iregren 1995, table 4) no terminal phalanges have been included in the burial. I therefore interpret the bears to have been flayed and that the pelt, with the third phalanges in their places, was utilised separately. In historical sources, ceremonies during the bear feast also included the use of the skin. For example, when the eating was over, the women, shooting blindfolded with a bow and arrow or throwing alder sticks, should try to hit the skin (Fjellström 1981, 32).

However, in Nåttinäset and in Aspnäset one single third phalanx was recovered, but these bears were not found complete. In the Gråtanån bear grave seven of these phalanges were recorded. One possible explanation for the presence of the many terminal phalanges in Gråtanån might be found in the historical records presented by Fjellström (1981, 15–17). He refers to an old tale of a Sámi family where a woman marries a male bear. Later she returns to her mean brothers, and they go bear hunting. After the killing of the "bear husband", the son of the human-bear couple collects all the bones, wraps them in the bear skin and buries all. Evidently, questions remain regarding the presence of terminal phalanges in bear graves. Their presence might also relate to the custom of consuming the edible parts of the bear paw.

Among circumpolar peoples many sources mention rituals connected to the head in general as well as the tip of the bear's nose (see Zachrisson/Iregren 1974, 85 with references). I have thus investigated the bones of the crania surrounding the olfactory organ carefully, the *Maxillae* and *Ossa incisivii* left and right. In the skulls studied in 2020, marks related to ceremonies involving the tip of the nose have been observed only in one skull, Sörviken (Fig. 9c). A cut mark from a knife is seen on the right maxilla. Fjellström (1971, 24, 30) described rituals concerning the nose.

In historical sources regarding Lapland, it is mentioned (Fjellström 1981, 27, 30) that particular Sámi rituals involved the bear's tail. This tradition has so far (Zachrisson/Iregren 1974, 85, 86) not been documented in Sámi bear graves.

Use of the bear's head

In the historical sources it is firmly decreed that the bear skeleton should not be harmed during cooking and feasting. This agrees with the archaeological material regarding the bear's head, shoulder blades, and a few other bone elements, as some vertebrae and bones of the paws. Sometimes we can, nevertheless, trace injuries from the hunt (see above).

However, in some burials there are clear marks of removing the mandibles from the skull. Cut marks can be observed in the upper ascending part (*Processus coronoideus*) of the lower jaw in the bear graves from Sörviken (left side), Gråtanån (left side; Fig. 18), and in the two individuals from Öhn (Fig. 19).

Use of the bear's tongue, lower jaw, brain, and spinal column

The tongue consists of bundles of muscular fibres and proves a good piece of meat. The practice of eating the tongue has been recognised in prehistoric sites, as for example at the Mesolithic settlement Bökeberg in Scania (Eriksson/Magnell 2001, 69). When cutting loose the tongue it is difficult not to touch the jawbones with your knife, as the cutting of the strong chewing muscles requires some effort. I have observed cut marks that might be related to such a procedure in one single mandible in the Gråtanån bear grave. Breaking or splitting of the jaw to consume its fatty tissue was not allowed due to the traditions (Zachrisson/Iregren 1974, photos pp. 50–65). This is otherwise a common behaviour in early societies (cf. Magnell/Iregren 2000; see Magnell, this volume, on Frösö).

In many societies and cultural groups, the fatty tissues of the brain have been of important use. It may be utilised during fur preparation or else consumed. I have looked for tiny cut marks in the finds, but damages related to removal of the brain cannot be identified in any of them. On the other hand, many vertebrae, except atlas and axis, were split longitudinally, which shows that the spinal cord was regularly consumed.

Traditions surrounding the paws

In two bear graves, Sörviken (Zachrisson/Iregren 1974, 25, 27) and Gråtanån (Melander 1980), many small bones from the paws have been deposited together. It is also evident that some metapodials and phalanges must have been connected by for example sinews at disposal (cf. Fig. 5). In the historical sources, Zachrisson and myself did not find any mention of acts involving the bear paws. The fact that such bone collections were found in two bear graves indicates that the depositions did not occur by chance. Levi Johansson, dealing with old food practices among the Sámi in the county of Jämtland (Johansson 1947, 95), mentions that reindeer feet were cooked and eaten as "pig feet". Probably the paws of the bear were eaten in a similar way and the bones were afterwards deposited together.

In the historical sources different possibilities are mentioned regarding how to choose the location for cooking, eating, and burying the bear. Huts must be built, or otherwise huts in the dwelling were chosen. According to Schefferus (1956, 264–265) a special hut is built for the men, where the bear should be skinned, dismembered, and cooked. Another hut was erected for the women, as they were to be separated from the dangerous power of the bear and the huntsmen. Fjellström (1981, 30) further specifies that the bear was prepared, boiled, consumed, and then buried, in one and the same place.

Turning to archaeology, one aspect that has often been overlooked during the excavation of bear graves is searching for traces of the bear feast and for possibly related constructions. Many have read about this lavish party in the sources (Schefferus 1956, 264–265; Fjellström 1981, 28–29), and we have noticed sites with clear archaeological indications of the event. At five sites it seems as if the bear has been cooked and consumed close to the place of its burial, and most likely relatively close to the Sámi *siida*.

Traces of cooking, a coal layer and a hearth, were found under the bear grave at Gällholmen (Fig. 20; cf. Zachrisson/Iregren 1974, 18 fig. 6). As one bone, evidently belonging to the buried bear, was found in the coal layer, the connection in time is proven. Unfortunately, the coal layer has not been dated.

At Karats, a hearth was excavated about two meters from the bear grave (Mulk/Iregren 1995, 18–20). The hearth was clearly a part of a hut. The radiocarbon dates indicate possible simultaneous construction of the two structures (Table 4). The Sámi may have used this small island repeatedly for bear feasting. A number of hearths and pits prove a long-term use. The corresponding dwelling site was very likely located at the nearby shores of Lake Karats (cf. Mulk/Iregren 1995, fig. 2).

The third example is a well investigated bear grave with surrounding areas excavated at Gråtanån. Close to the burial, an earth oven was found as well as a possible hearth. Other results prove human activities on several occasions (Table 4). The supposed hearth was located about seven to eight meters from the bear grave. It was most likely situated in a hut but could not be dated. Ernst Manker registered a Sámi dwelling site located about 400 m northeast of the bear grave in 1950 (RAÄ[2] 574, confirmed during field work in 1975; see Melander 1980).

Also, at Grundskatan a larger area with several huts and other constructions was excavated. The bear grave was located inside a hut, covered by a cairn, and placed very close to a hearth (Fig. 21; cf. Broadbent 2005, 25–31). The hearth and the burial yielded similar dates (Table 4). At the site of Nåttinäset patches of coal and burnt sand were found together with the remnants of the bear burial (Zachrisson/Iregren 1974, 30).

Another possible proof of these traditions might have been found in Sörviken. A hearth was found close to the bear grave (Zachrisson/Iregen 1974, 20). To my knowledge, the hearth has not been dated, but both structures were located at a Sámi dwelling site, however, overlaying a find spot of Stone Age character.

As pointed out it is claimed that the bear, if possible, should be brought to the Sámi village for consumption and feasting. In Karats this does not seem to have been the case, on the contrary. The bear has been transported over the ice to a close inlet. A similar situation is represented by the Gällholmen bear grave. It was also established on an island (Zachrisson/Iregren 1974, 15–16). It must be remembered that islands held particular connotations among the Sámi. Human burials in Sápmi were for a long period located on islands, capes, and isthmuses (Fossum 2006, 91–93). The closeness to water is striking.

2 Official numbering of ancient monuments in Sweden (found in official digital registers of Riksantikvarieämbetet).

Here I would like to add a complementary remark. I believe that water and the bear were closely related in traditional beliefs, and this ought to be further researched. Knut Helskog has discussed this theme thoroughly in connection with the studies of bears in rock art at Alta, Norway, and he also included the cliff structures in his arguments. He mentions water as essential and stresses the possibility of bears to reach all other worlds and to travel through these (Helskog 2012, 220–223). Some of the bear graves in Sweden have obviously been deliberately located at inlets or other natural structures in lakes, although we can understand that the dwelling site lay elsewhere but near (Gällholmen, Karats, Nåttinäset). Furthermore, most bear graves are located close to water, often lakes, as for example, the ones from Sörviken, Värjaren, Långbäck, and Grundskatan (seashore).

The feast and the burial can also take place closer to the site of killing, according to Edsman (1994, 65). Fjellström (1981, 19, 21) instead stresses that the ceremonies should be performed nearer to the habitation. The above-mentioned archaeological examples definitely demonstrate that the preparation of the bear and the feasting took place very close to the burial of the bear bones.

From the island Spildra in Northern Norway, we also gained the new and important knowledge that different social units might have used specific areas for bear burials and other bone depositions. In this small island of about 30 km², there are two areas where bears have been buried (Iregren et al. 2017; Sommerseth 2021). This island is, however, much bigger than Vägvisarholmen in Lake Karats and Gällholmen in Lake Storuman.

Thoughts about future work on Sámi bear bone finds

Bear graves have, so far, been found in the counties *(län)* of Norrbotten, Västerbotten, Jämtland, and possibly ritual remains of bear in Ångermanland. We should, however, keep our eyes open to identify these remains in other counties of Sweden, as for instance Dalecarlia, but also closer to the coasts of the Baltic Sea. Conspicuous unburnt bear finds from Dalecarlia ought to be studied further (Iregren 2022).

We should further search for more graves in the vicinity of an earlier find, as the investigations at Spildra encourage us to do (Iregren et al. 2017). Similar locations in Sweden might be Nedre Vapstsjön, Långbäck, Värjaren, and Öhn.

In many important respects the Sámi, also during recent centuries, followed the ceremonial rules and hunting instructions, mentioned in the preserved written sources from the 17th and 18th centuries. These concern the burial of the whole bear, which means the burial of all its bones, including the skull, the tail, and the tip of the nose. During fieldwork we have identified the intention to mimic the anatomy of the bear in the burials. The neck vertebrae have been positioned close to the head, and other vertebrae and bones have been laid out to give the impression of a backbone. Furthermore, the many bones of the paws may be placed together. The prohibition to injure the bones resulted in a complete skull, as well as complete jaws and shoulder blades.

When searching for constructions linked to settlements, this has been discerned through traces of hearths and huts found, also in accordance with historical records. However, this topic needs closer investigation. Also, as already stated, we ought to search for more bear graves in an area where one has already been found. Was this really the result of one single event? The fact that bear graves have always come to light unexpectedly and by chance may have fooled us. There is also a gap in our knowledge about the use of organic containers to collect and bury the bones.

We have rather little knowledge on bear graves in mountainous areas in Sweden. If a burial is placed in a crevice or among boulders, we must reflect about these constructions and their details. We should, for instance, check the arrangement of bones in crevices or the like. Fjellström (1981, 31) notes that the Sámi could also choose to reconstruct the bear in a vertical position. In this case, the deposition should start with the hind parts of the animal, he writes.

On the other hand, there are indications for a freer attitude to the traditions, too. The most conspicuous trait is that all bones with marrow contents have been cleaved, though carefully deposited. The degree of fragmentation varies, but the occurrence of splitting correlates neither to the dating nor to the Sámi geographical area, as far as I can see. Here I want to clarify the information by ZACHRISSON (1983, 88) in her article on "New South Samish Finds": The bones in the bear grave at Värjaren *were cleaved* to get at the marrow, but the bone fragments seemed bigger than in other burials.

Some of the statements from the sources which still need to be investigated further concern the ceremonies around the nose and the tail of the bear. Regarding zooarchaeology, questions such as the skinning and general dismembering of the bear remain to be enlightened. According to the sources, women and men were allowed to eat different parts of the bear only. Can this tradition be spotted on bones or in the depositions within the burial? A detailed inspection for hunting injuries on the bones of the bear's body might also give new results. Chemical analyses to identify, for example, remains of fat in the bear graves might be useful. Isotopic investigations of bear bones might enlighten the size of the area used for bear hunting by the *siida*. Further, determining the tree species of the twigs, logs/planks in the burials would add more information.

There is also a need to go deeper into some aspects of behaviour during the ceremonies. Why have brass chains been fastened to bears' crania? These chains deserve a closer look, not least because brass was an important and holy metal among the Sámi (FJELLSTRÖM 1981, 21–22, 30–33; BÄCKMAN 1981, 52). Brass rings were used on several occasions in connection with the bear feast, too, for instance during the preparations for the feast, in ceremonies directed to the bear, and as a protection of the women but also of the reindeer involved in the transportation of the bear. It was further applied for protection in connection with human burials (SCHANCHE 1997, 242–243). Can the dating of the chains from Värjaren and Namdalen further be made more precise? ZACHRISSON (1983, 88, 95) remarks that the design is ancient. Further, she notes that brass rings in a Sámi context are not known earlier than from the 16th century.

Further on, why were the bullets put in the bear graves? Bullets are found not only in such burials, but also in settlements as Jättens grotta (Giant's cave) at Lake Kakirjaure in Sweden (MULK 1994, 90–91). Could these depositions in graves be regarded as offerings (cf. HÖGSTRÖM 1980, 210) or, more likely, burial gifts? So far, we do not know much of the Sámi casting procedures and the composition of the lead alloy.

Another difficult question that we ought to think much more about is: How can we perform a meaningful separation between bear bone depositions (if there are any), bear graves and bear finds from sacrificial sites respectively? And how can we learn to understand this different behaviour? MULK (2005) made an important contribution to enlighten this when she found out that the objects and the bones/antlers, respectively, were not deposited during identical time periods in Sámi sacrificial sites (see also SPANGEN et al., this volume, on bear remains from Sámi offering sites).

In Sweden we have been slow in dating the bear graves. This has, however, partly been dealt with in this contribution. Since the investigations by MULK/IREGREN (1995) and BROADBENT (2005) a much larger time span of their occurrence has been documented than before. The new date of the Nåttinäset bear grave (Ua-67288: 1023–1150 cal AD, 1-Sigma) further strengthens the evidence for this long tradition. It might expand even further back in time. Thus, we need to date the bone depositions more swiftly in the future, not to delay research or the possibility to ask new questions.

ACKNOWLEDGEMENTS

The costs of the important dating of bear finds have been covered by financial support from the County Museum of Norrbotten (Nåttinäset), the Department of Archaeology and Ancient History,

Lund University (Gråtanån, Långbäck), the Swedish Museum of Natural History through a bear project run by Erik Ersmark (Öhn) and, finally, Öhns byalag (the village community of Öhn) and Ströms Hembygdsförening (the Ström Local History Society). Markus Fjellström kindly performed some extractions of collagen to ease the dating procedures.

Employees working in a great number of museums have given me important help in answering questions, providing me with samples, bone material for study and interesting photos. I here want to mention Johnny Karlsson, Swedish History Museum, who gave me the opportunity to study many of the bear finds in their collections. He also kindly took some of the photographs included in this article. Furthermore, I have received most valuable help from the County Museums of Jämtland, Västerbotten, and Norrbotten. Other museums in Stockholm, Lund, and in Northern Sweden also supported me during my work, such as the Swedish Museum of Natural History, Lund University Biological Museum, Ájtte – Swedish Mountain and Sámi Museum, Silvermuseet in Arjeplog, and Skellefteå Museum. My colleague Jan Storå, Osteoarchaeological Research Laboratory, Stockholm University, helpfully provided me with most useful data on the bear find from Grundskatan.

I have further received valuable information from Kjell-Åke Aronsson (pers. comm.), senior researcher at Ájtte. My former colleague Anders Ödman helpfully shared with me his experience and expertise in smithing and of guns. Ingvild Larsen, senior advisor at the Sámi Parliament, Norway, has supported me in many ways during this writing and kindly shared her knowledge with me. Högne Jungner, former leader of the Laboratory of Chronology, University of Helsinki, has been an important partner during the intriguing discussions of the datings. He also performed all calibrations and created Figure 8.

Further, important discussion partners have been Erik Ersmark, Natural History Museum, an aDNA specialist, and Markus Fjällström, Department of Archaeology and Ancient History, Lund University, Sweden, a specialist in isotope studies. Both have lately been working on bears from different aspects. I also want to thank Enid A. Goodwin, who worked in the Alaska Dept. of Fish and Game, Anchorage, US. She kindly performed the thin section analyses of bear teeth from the sacrificial site of Unna Saiva, Sweden.

Last but not least, I have received very important knowledge and support from two scholars in Lund. Associate professor Peter Krantz, former forensic pathologist, gave me invaluable help in studying the bears' injuries from hunting (from photos). Peter is also an interested hunter, and his skills and experiences were essential in evaluating and understanding the bear hunt. Finally, I want to mention my close colleague Ola Magnell, archaeozoologist at Archaeologists/Lund, the Swedish History Museum. Ola helped me understand the traces of cuts and blows, to enlighten the skinning and consumption of the bear, and also in finding examples to compare with. Boel Billgren scrutinised my English. My warmest thanks to all of you for your interest and invaluable support!

Bibliography

Ambrosiani et al. 1984: B. Ambrosiani/E. Iregren/P. Lahtiperä, Gravfält i fångstmarken. Undersökningar av gravfälten på Smalnäset och Krankmårtenhögen, Härjedalen. RAÄ-SHMM Rapport 1984:6 (Stockholm 1984).

Baudou 1993: E. Baudou, Norrlands forntid – ett historiskt perspektiv (Höganäs 1993).

Björklöf 2010: S. Björklöf, Björnen i markerna och kulturen (Möklinta 2010).

Boëthius 2010: A. Boëthius, Osteologisk analys av rendeponeringen vid Gransjön – Frostvikens socken, Jämtlands län. Reports in osteologi 2010:11, Institutionen för Arkeologi och Antikens historia, Lunds universitet (Lund 2010).

Braseth 2014: L. Braseth, Samer sør for midnattssola, Sørsamenes historie, kultur og levemåte (Bergen 2014).

Broadbent 1987: N. D. Broadbent, Iron Age and Medieval Seal Hunting Sites. Center for Arctic Cultural Reerach, Research Reports 5 (Umeå 1987).

Broadbent 2005: N. D. Broadbent, Excavation report Grundskatan, Lappsandberget, Jungfruhamn Raä 70, Stora Fjäderägg, Snöan, Lövångers kyrkstad, Västerbottens län, Sweden. Arctic Studies Center, Smithsonian Institution (Washington D.C. 2005).

Broadbent 2010: N. D. Broadbent, Lapps and Labyrinths. Saami Prehistory, Colonization and Culture Resilience (Washington D.C. 2010).

Broadbent/Storå 2003: N. D. Broadbent/J. Storå, Björngraven i Grundskatan. Populär arkeologi 1, 2003, 3–6.

Bäckman 1981: L. Bäckman, Kommentar. In: P. Fjellström, Kort berättelse om lapparnas björna-fänge, Samt deras der wid brukade widskeppelser (Umeå 1981) 43–63.

Bäckman/Kjellström 1979: L. Bäckman/R. Kjellström (eds.), Kristoffer Sjulssons minnen om Vapstenlapparna i början af 1800-talet upptecknade af O. P. Pettersson. Acta Lapponica 20 (Lund 1979).

Edsman 1970: C.-M. Edsman, Jägaren, villebrådet och makterna. Norrbotten 1970, 37–60.

Edsman 1994: C.-M. Edsman, Jägaren och makterna. Samiska och finska björnceremonier. Publications of the Institute of Dialect and Folklore Research C: 6 (Uppsala 1994).

Ekman/Iregren 1984: J. Ekman/E. Iregren, Archaeozoological investigations in Northern Sweden. Early Norrland 8, The Royal Swedish Academy of Letters, History and Antiquities (Stockholm 1984).

Eriksson/Magnell 2001: M. Eriksson/O. Magnell, Jakt och slakt. In: P. Karsten (ed.), Dansarna från Bökeberg. Om jakt, ritualer och inlandsbosättning vid jägarstenålderns slut. Riksantikvarieämbetet Arkeologiska undersökningar 37 (Stockholm 2001) 48–77.

Evans et al. 2016: A. L. Evans/N. J. Singh/A. Friebe/J. M. Arnemo/T. G. Laske/J. E. Swenson/S. Blane, Drivers of hibernation in the brown bear. Frontiers in Zoology 13, 2016, 1–13.

Fjellström 1981: P. Fjellström, Kort berättelse om lapparnas björna-fänge, Samt deras der wid brukade widskeppelser (Umeå 1981; originally published in 1755).

Fossum 2006: B. Fossum, Förfädernas land. En arkeologisk studie av rituella lämningar i Sápmi, 300 f. Kr. – 1600 e. Kr. Studia Archaeologica Universitatis Umensis 22 (Umeå 2005).

Grydeland 2001: S. E. Grydeland, De sjøsamiske siidasamfunn, En studie med utgangspunkt i Kvaenangen, Nordtroms. Nord-Troms Museums skrifter 1 (Sørkjosen 2001).

Hallström 1960: G. Hallström, Monumental art of Sweden from the Stone Age. Nämforsen and other localities (Stockholm 1960).

Helskog 1999: K. Helskog, The Shore Connection. Cognitive Landscape and Communication with Rock Carvings in Northernmost Europe. Norwegian Archaeological Review 32(2), 1999, 73–94.

Helskog 2012: K. Helskog, Bears and meanings among hunter-fisher-gatherers in northern Fennoscandia 9000–2500 B.C. Cambridge Archaeological Journal 32, 2012, 209–236.

Henriksson 1978: H. Henriksson, Popular hunting and trapping in Norrland. Early Norrland 6, The Royal Swedish Academy of Letters, History and Antiquities (Stockholm 1978).

Von Hogguér 1841: D. von Hogguér, Reise nach Lappland und dem nördlichen Schweden (1828) (Berlin 1841).

Hultcrantz 1983: Å. Hultcrantz, Reindeer nomadism and the religion of the Saamis. ARV Scandinavian Yearbook of Folklore 1983, 11–28.

Höglund 1958: N. Höglund, Skall vildrenen återinföras i Sverige? In: H. Hamilton (ed.), Sveriges hjortdjur I–II (Stockholm 1958) 945–972.

Högström 1980: P. Högström, Beskrifvning öfwer de till Sweriges Krone lydande Lapmarker. Norrländska skrifter 3. Ed. R. Jacobsson (Umeå 1980; originally published in 1747).

Iregren 1973: E. Iregren, 3000-årigt björnfynd Ströms socken. Jakten i Jämtland och Härjedalen 1973, 19–20.

Iregren 1983: E. Iregren, Osteological evaluation of reindeer bone finds from the territory of the Southern Saamis. ARV Scandinavian Yearbook of Folklore 1983, 101–113.

Iregren 2021: E. Iregren, Samisk björnjakt och jaktlycka – björngravar och björnspjut. Västerbotten förr & nu 2021, 1–15. https://nattidskriftenvasterbotten.se/2021/06/12/samisk-bjornjakt-och-jaktlycka-bjorngravar-och-bjornspjut/.

Iregren 2022: E. Iregren, Samiska björngravar. In: J. Wehlin/S. Cassé/M. Andersson (eds.), Samiska spår. Dalarna, Gävleborg, Västmanland. Dalarnas Fornminnes- och Hembygdsförbund, Dalarnas museum, Länsmuseet Gävleborg, Västmanlands Hembygdsförbund och Fornminnesförening och Västmanlands läns museum (Västerås 2022).

Iregren et al. 2017: E. Iregren/I. Larsen/C. Olofsson, Osteologisk rapport fra Spildra i Kvænangen. Untatt offentlighet. Sámediggi, Sametinget 2017, 1–87.

Janson/Hvarfner 1960: S. Jansson/H. Hvarfner, Från Norrlandsälvar och fjällsjöar (Stockholm 1960).

Janson/Hvarfner 1966: S. Jansson/H. Hvarfner, Ancient Hunters and Settlements in the Mountains of Sweden (Stockholm 1966).

Johansson 1947: L. Johansson, Frostvikenlapparnas födoämnen och maträtter fordomdags. Jämten 1947, 69–95.

Kjellström 1983: R. Kjellström, Piles of bones, cult-places or something else? ARV Scandinavian Yearbook of Folklore 1983, 115–120.

Lieberman 1994: D. E. Lieberman, The biological basis for seasonal increments in dental cementum and their application to archaeological research. Journal of Archaeological Science 21, 1994, 525–539.

Lieberman/Meadow 1992: D. E. Lieberman/R. H. Meadow, The biology of cementum increments (with an archaeological application). Mammal Review 22, 1992, 57–77.

Lindqvist 1994: C. Lindqvist, Fångstfolkets bilder. En studie av de nordfennoskandiska kustanknutna jägarhällristningarna. Theses and Papers in Archaeology, New Series A5 (Stockholm 1994).

Losey et al. 2013: R. J. Losey/V. I. Bazaliiakii/A. R. Lieverse/A. Waters-Rist/K. Faccia/A. W. Weber, Shamanka II cemetery, Lake Baikal, Siberia. In: C. Watts (ed.), Relational Archaeologies. Human, Animal, Things (London, New York 2013) 65–96.

Fig. 2. A bear's head made of sandstone, used as whetstone, found at the dwelling site Lemnäset, Bodum parish, Ångermanland, Sweden. It dates to the period 4100–1700 BC (Swedish History Museum [SHM] inv. no. 27962, artefact on exhibition in the SHM; photo J. Karlsson, SHM).

Fig. 3. Rock painting of a bear at Ruändan, Storsjö parish, Härjedalen county, Sweden. The painting in red ochre is regarded to be c. *4,000 years old. This and other paintings are located in a mountainous area (photo A. Hansson, County Museum of Jämtland).*

Fig. 4. The bear grave at Gråtanån (site no. 15), Vilhelmina parish, Lapland, Sweden, in an early stage of excavation. Vertebrae and ribs have been placed behind and close to the cranium (SHM inv. no. 31310; after Melander *1980, fig. 12).*

Fig. 5. The bear grave at Gråtanån (site no. 15), Vilhelmina parish, Lapland, Sweden, during excavation. Many bones from the paws of the bear were located together in layer 3, and close to the cranium (after Melander *1980, fig. 15).*

Fig. 6. Four lead bullets from a muzzle-loading gun were found in the bear grave at Gråtanån (site no. 15), Vilhelmina parish, Lapland, Sweden. They were located close to the cranium. Their diameters vary from 9.45 to 9.77 mm. These bullets have not been fired (photo J. Karlsson, Swedish History Museum).

Fig. 7. Cranium from the bear grave at Värjaren (site no. 13), Frostviken parish, Jämtland, Sweden. Note the brass chain and ring fixed to the zygomatic bone. The find now belongs to the Museum of Ethnology, Hamburg, Germany, inv. no. 1795:1 (photo G. Jansson, Antiquarian-Topographical Archives [ATA]).

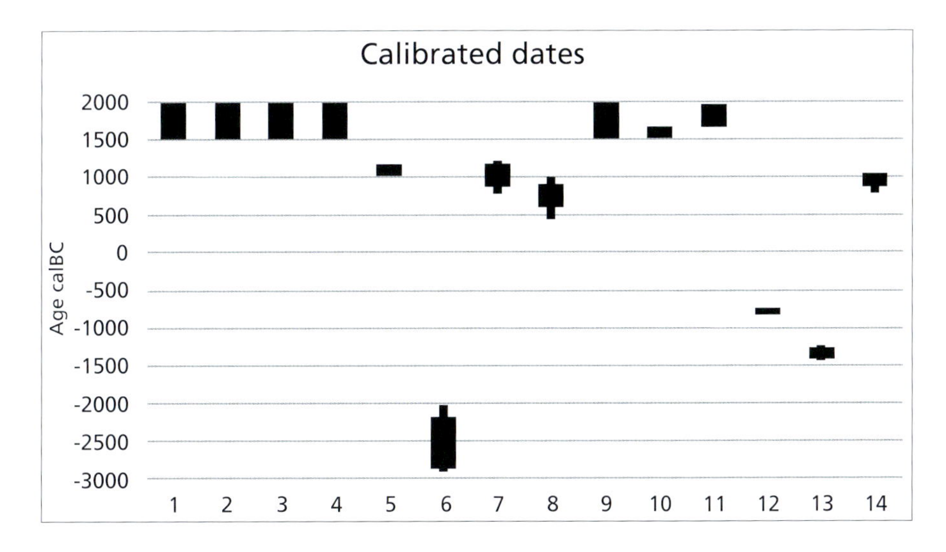

Fig. 8. Calibrated ^{14}C-dates of irrefutable bear graves, possible bear graves, and a couple of bear finds of unknown origin in Sweden. For numbers, see Table 4 (graphics H. Jungner, Helsinki University, Finland; modified by M. Bolte, ZBSA).

Fig. 9a–c. Cranium from the bear grave Sörviken (site no. 2), Stensele parish, Lapland, Sweden; SHM inv. no. 25465. The bear has been hit by a sharp axe across the forehead, presumably three times from the right side. The damages are at their deepest there. These injuries were lethal. In Fig. 9c the nose of the bear is visible from the right side. The bear has been cut by a knife in the right maxilla. This cut is most likely linked to ritual behaviour in connection with the tip of the nose (photos E. Iregren).

Fig. 10a–b. The cranium from Avaträsk, Dorotea parish, Lapland, Sweden (site no. 9). Find circumstances are not known; Swedish Museum of National History, inv. no. 583087 (formerly 3087). The bear was most likely hit by a spear, which resulted in an injury with a symmetrical opening located close to the bear's left eye. The injury was not lethal. Note that the cranium has later been damaged during storing (photos E. Iregren).

Fig. 11. A bear hunting party, painted by the artist J. Wilhelm Wallander (1821–1888). It is interesting to see that all the weapons discussed in this paper are used by the hunters in the picture – spear, axe, and gun. Both Sámi men and settlers participate. The picture seems relatively realistic, as one person is injured and lies in the snow. The hunter close to him has abandoned his gun and switched to the spear. However, the spear seems somewhat short. The picture was painted in 1858 and first published in "Svenska Folket sådant det ännu lefver wid Elfwom, på Berg och i Dalom" by Carl Anton Wetterbergh (Stockholm 1865; see Ahrland/Magnusson*, this volume). The reproduction is kept in the Royal Library in Stockholm. The oil painting is owned by Nordiska Museet, Stockholm.*

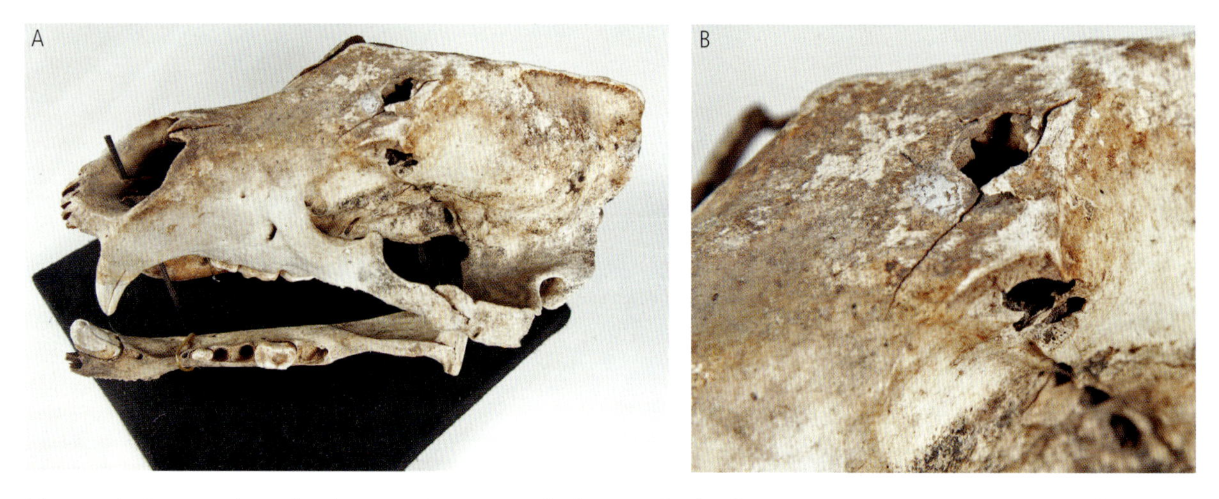

Fig. 12a–b. Bear cranium of unknown origin, originally from a school collection in Umeå (site no. 7), now in the County Museum of Västerbotten, Sweden (inv. no. Vbm 14070). The left frontal bone has been struck once, most likely by a spear. Note the fractures running from the entrance opening. The injury was not lethal (photos K. Stenman, County Museum of Västerbotten).

Fig. 13a–b. Irregular opening of the forehead of a bear from Nedre Vapstsjön, Tärna parish, Lapland, Sweden (site no. 3), probably caused by a strike with a heavy and blunt weapon. A club or the neck of an axe might have been used. Fractures run both upwards and downwards on the cranium. This kind of injury is called a buttonhole fracture, and fissures due to reduction of induced tension run in different directions. This injury did not cause the death of the bear (SHM inv. no. 30 691; photos E. Iregren).

Fig. 14. Four bear graves have been documented by Norwegian archaeologists in this small valley (Skagedalen) at the west coast of the island of Spildra, Kvænangen, Troms, in Northern Norway. Ingvild Larsen from the Sámediggi and responsible for the investigations in 2013 is visiting the sites (photo M. Chruickshank, Sámediggi).

Fig. 15a–c. The bear in the grave Spildra C (III), Kvænangen, Troms, in Northern Norway, was struck during a fierce attack, supposedly at its den. Cracks created by blunt force run through the bone walls surrounding the brain on all sides. The neck of an axe might have been used as the entire braincase is cracked. The bear became unconscious and was then easy to kill (photos E. Iregren and I. Larsen, Sámediggi).

Fig. 16. Photo from Hilmer Zetterberg's publication on bears from 1951. Possible Sámi hunters at the den of a bear in southern Lapland in 1910. Guns and spears, but also long rods are used to prevent the bear from leaving the den quickly (after Zetterberg *1951;* Zachrisson/Iregren *1974, fig. 86).*

Fig. 17. Bear spears from Sweden.

Fig. 17a. Bear spear in the collections of Ájtte Swedish Mountain and Sámi Museum, Jokkmokk, inv. no. JM 1439. Note the long shaft (photo Ájtte, Swedish Mountain and Sámi Museum, Jokkmokk).

Fig. 17b. The same spear as in Fig. 17a. Note the slim blade of the spearhead and its sheath. A piece of copper further connects spearhead and shaft, protecting the sharp blade. The decoration is commented upon in Iregren *2021, 13 (photo Ájtte, Swedish Mountain and Sámi Museum, Jokkmokk).*

Fig. 17c. Spear from the ethnographical collections of Ájtte, Swedish Mountain and Sámi Museum, Jokkmokk, inv. no. 03.1.30. Note the part of brass between shaft and head and the protected part at the rear end of the spear, the ferrule (photo Ájtte, Swedish Mountain and Sámi Museum, Jokkmokk).

Fig. 17d. The same spear as in Fig. 17c. Note the (often occurring) rhomboid shape of the spearhead (photo Ájtte, Swedish Mountain and Sámi Museum, Jokkmokk).

Fig. 17e. Head of a spear, also used as a ski pole, in the collections of the County Museum of Västerbotten, Umeå, Sweden (Svenska skidmuseet); a relatively short and pointy head combined with a long, decorated brass part (photo County Museum of Västerbotten).

Fig. 17f. Spear head in the exhibitions of the County Museum of Västerbotten, Umeå, Sweden (Svenska skidmuseet), inv. no. SSM 80275; a rhomboid head with decorations, rivets and a part of brass (photo H. Forsberg, County Museum of Västerbotten).

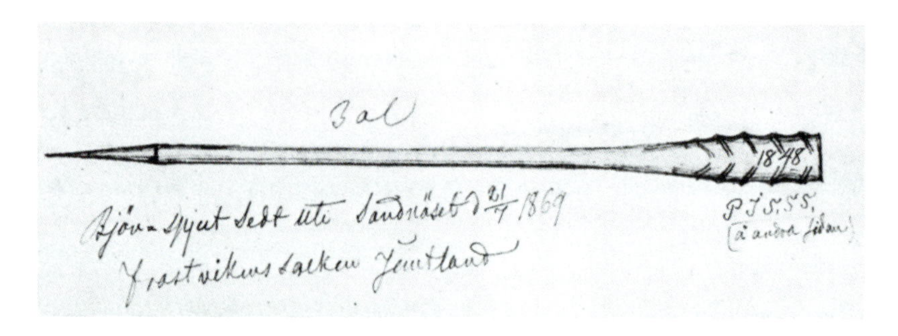

Fig. 17g. Drawing by the famous ethnographer Nils Månsson Mandelgren, 1869, showing a bear spear (County Museum of Jämtland; inv. no. JLM 77s1213 40–41). The spear was observed at Sandnäset, Frostviken parish, Jämtland, Sweden. Note the two markings on the shaft of the spear: the year "1848" and the owner's/owners' initials "P I S. S S.". The length of the spear is about 178 cm. See Table 5. The drawing was first published in the journal "Jämten" (1978, 107).

Fig. 17h. A former bear spear from the area of Lake Inari in Northern Finland, transformed into an ice axe; now kept in the County Museum of Västerbotten, Umeå, Sweden (Svenska skidmuseet), inv. no. SSM 80481 (photo L. Trygg, County Museum of Västerbotten).

Fig. 18. Left lower jaw of the bear from Gråtanån (site no. 15), Vilhelmina parish, Sweden, SHM, inv. no. 25465. The inside of the ascending part of the jaw (processus coronoideus) *displays a number of fine cuts from a knife. Apparently, the jaw has been separated from the skull before the cooking of the bear (photo E. Iregren).*

Fig. 19. Mandible of the bear from Öhn, Strömsund parish, Jämtland, Sweden (site no. 16), kept in Lund University Biological Museum, inv. no. Lzzz/5006. The inside of the lower jaw displays a number of fine cuts from a knife. Evidently, this find has an anthropogenic origin (photo E. Iregren).

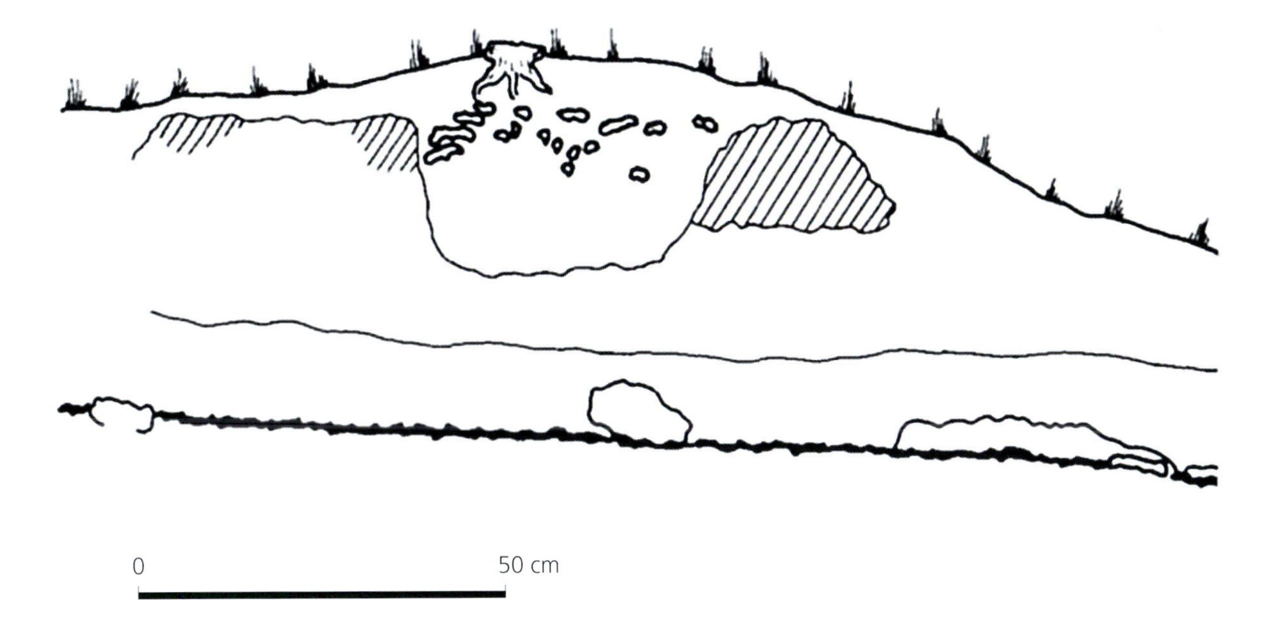

Fig. 20. The profile through the Gällholmen bear grave (site no. 1), Stensele parish, Lapland, Sweden. A layer of coal was found c. 20 cm below the pit of the grave. Presumably this is the rest of a cooking fire during the bear feast. Also, one bear bone missing in the burial was recovered there (after Zachrisson/Iregren *1974, fig. 6).*

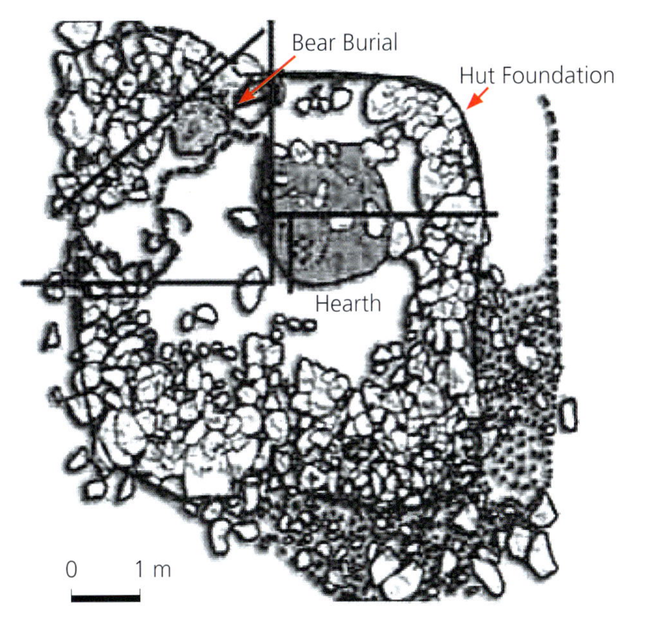

Fig. 21. Drawing of structures at site no. 18 (Feature 4) at Grundskatan, Lövånger parish, Lapland, Sweden. The hut foundations, the bear grave, and the hearth are all visible (after Broadbent *1987; 2005, fig. 16; modified by M. Bolte, ZBSA).*

Table 1. Sites with proven or supposed Sámi bear graves and a few bear finds of unknown provenance in the Swedish part of Sápmi (information on site nos. 1–11, 13–14, 16–17 after Zachrisson/Iregren 1974 and references therein; 12–14 after Mulk/Iregren 1995; 13 after Zachrisson 1981, 85–86; 1983, 88, 95; 15 after Melander 1980; 1981, 78–81; Zachrisson 1983, 89–90; 16–17 after Iregren 1973; 18 after Broadbent/Storå 2002; Broadbent 2005, 24–25; Broadbent 2010).

No.	Site	Locality	Coordinates	Museum, inv. no.	Further information
1	Gällholmen	Stensele parish, Lapland Lat. 65°16' N., Long. 16°47' E	N 7238730 E 582404	County Museum of Västerbotten Vbm 21564	Exhibited in the museum
2	Sörviken	Stensele parish, Lapland Lat. 65°06' N., Long. 17°06' E		The Swedish History Museum SHM 25465	
3	Nedre Vapstsjön	Tärna parish, Lapland Lat. 65°25' N., Long. 14°60' E	N 7255404 E 488939	The Swedish History Museum SHM 30691	
4	Tiirivara	Överkalix parish, Lapland Lat. 66°46' N., Long. 22°23' E		The Swedish History Museum SHM 30706	
5	Mjösjö	Junsele parish, Ångermanland Lat. 65°16' N., Long. 16°47' E		The Swedish History Museum SHM 34813	
6	Nåttinäset	Arjeplog parish, Lapland Lat. 66°04' N., Long. 17°59' E		County Museum of Norrbotten Nbm 40011:41	Exhibited in Silvermuseet, Arjeplog
7	Östra gymnasiet	Umeå		County Museum of Västerbotten Vbm 14070	
8	Zetterberg			Jakobsberg	Private owner
9	Avaträsk	Dorotea parish, Lapland		Swedish Museum of Natural History 583 087 (formerly 3087)	
10	Aspnäset	Tåsjö parish, Ångermanland Lat. 64°07' N., Long. 16°07' E		The Swedish History Museum Dnr 7434/71	
11	Linköping				Private owner
12	Karats	Jokkmokk parish, Lapland Lat. 66°41' N., Long. 18°48' E	N 7401975 E 666697	Ajtte, Swedish Mountain and Sámi Museum AJA:000174	Exhibited in the museum
13	Lake Värjaren	Frostviken parish, Jämtland Lat. 64°47' N., Long. 14°40' E		Museum of Ethnology, Hamburg 1795:1	
14	Långbäck 1:3	Stensele parish, Lapland	N 7241442 E 574052	The Swedish History Museum SHM 30951	

Cont. Tab. 1. ▼

Cont. Tab. 1.

No.	Site	Locality	Coordinates	Museum, inv. no.	Further information
15	Gråtanån, Vojmsjön	Vilhelmina parish, Lapland	N 7195227	The Swedish History Museum	
		Lat. 62°52' N, Long. 16°47' E	E 584777	SHM 31310	
16	Öhn	Strömsund parish, Jämtland	N 7085587	Lund University Biological Museum	
		Lat. 63°90' N, Long. 15°46' E	E 522447	Lzzz/5006 Mandibula sin.	
17	Öhn	Strömsund parish, Jämtland	N 7085587	Lund University Biological Museum	
		Lat. 63°90' N, Long. 15°46' E	E 522447	Lzzz/5007 Mandibula dx.	
18	Grundskatan, Bjuröklubb	Lövånger parish, Lapland	N 7162882	Skellefteå Museum	
	RAÄ 78, A 4		E 817609		

*Table 2. Number of identified bone fragments (NISP) of bear (*Ursus arctos*), divided into body parts in bone collections (information on site nos. 1–10 after* Zachrisson/Iregren *1974, table 1; 12–13 after* Mulk/Iregren *1995, Bilaga 7; 14–15 new data presented here; 18 after* Broadbent *2010, table 17). Bear finds represented by only parts of the head are not included here (cf. Table 3 and Table 1 for site nos.).*

Bone elements	**Site nos. and sites (see Table 1)**									
	1 (Gä)	2 (Sö)	3 (NV)	6 (Nå)	10 (Asp)	12 (Ka)	13 (Vä)	14 (Lå)	15 (Grå)	18 (Gru)
Calvarium	2	1	1	4		1	1	4	1	
Mandibula	2	2	2	2		3	4	3	2	5
Dens	1		1			1		2		50
Os hyoides		4								
Total	**5**	**7**	**4**	**6**	0	**5**	**5**	**8**	**3**	**55**
Vertebra	101	133	79	10		22	25	4	88	
Costa	113	73	74	10	2	19		1	58	
Sternum	5	2	1			1			6	
Scapula	3	2	8	1		2	1	1	1	
Humerus	24	23	18	1	3	2	3	1	11	1
Radius	11	13	11	1		4	1		11	1
Ulna	14	10	13	3	1	1	1		3	1
Coxa	19	7	16				4		12	
Femur	22	24	15				4		8	
Tibia	14	12	13	1	1	1	1		19	2
Fibula	2	3	3			5	1		3	
Extremity	1	4				5		2	24	6
Total	**326**	**306**	**251**	**27**	7	**62**	**41**	9	**244**	**11**
Os metacarpus	9	10	9	2	1	2				
Os metatarsus	8	10	9	3	2	3				
Metapodial bones	2			1		6			17	
Carpus	12	14	6		1	9				
Patella	2	1	1						1	
Astragalus	2	2	2			1			2	
Calcaneus	3	2	2			3			3	
Tarsus	10	10	2			5				
Carpus/Tarsus									23	
Phalanx I	15	17	4	2	3	3	1	1	16	
Phalanx II	10	17	2		1	5			14	
Phalanx III				1	1				7	
Os sesamoideum	18	46				8			18	
Os penis	1									
Total	**92**	**129**	**37**	9	9	**45**	**1**	**1**	**101**	0
Total identified	423	442	292	42	16	112	47	19	348	66
Unidentified	219	592	71	31	74	122		4	209	6
Number of individuals (MNI)	**1**	**1**	**3**	**1**	**1**	**1**	**1**	**2**	**1**	**1**

*Table 3. Number of identified bone elements (NISP) of bear (*Ursus arctos*) from sites where only parts of the head have been identified (cf. Table 1 for site nos.).*

Bone elements	**Site nos. and sites (see Table 1)**							
	4 (Ti)	5 (Mjö)	7 (Ös)	8 (Ze)	9 (Av)	11 (Li)	16 (Öh)	17 (Öh)
Calvarium	1	1	1	1	1	1		
Mandibula			2	2	2	1	1	1
Dentes				1				
Os hyoides								
Total	1	1	3	4	3	2	1	1

Table 4. Dating of irrefutable and possible bear graves. Dates on other close archaeological structures are also mentioned.

Site no.	**Site, locality**	**Find circumstances**	**Sample no.**	**Lab. code**	**Result BP**	**+/-**	**Result cal BC / AD (1-Sigma)**
1	Gällholmen Stensele parish, Lapland	Discovered in 1969 by locals, saved in 1970, excavated indoors in 1971 by archaeologist and zooarchaeologist	1	St-3770	<250		AD 1533–1950
2	Sörviken Stensele parish, Lapland	Found during archaeological fieldwork in 1955, excavated indoors in 1972 by archaeologist and zooacharchaeologist	2	St-207	215	60	AD 1533–1950
3	Nedre Vapstsjön Tärna parish, Lapland	Excavated in 1952 by ethnographer E. Manker	3 4	St-3993 St-3994	<250 <250		AD 1533–1950 AD 1533–1950
6	Nåttinäset Arjeplog parish, Lapland	Found in 1970, damaged, excavated by archaeologists	5	Ua-67288	982	33	AD 1023–1150
10	Aspnäset Tåsjö parish, Ångermanland	Found in 1971 during inventory by Swedish National Heritage Board; bones collected	6	St-10192	3965	165	2843–2206 BC
12	Karats Jokkmokk parish, Lapland	Found in 1983 in connection with archaeological fieldwork, damaged, excavated indoors in 1986 by archaeologists and zooarchaeologist	7 8	Ua-507 (tooth) St-11213 (birch bark from bear grave)	1035 1315	90 130	AD 893–1153 AD 607–879
13	Lake Värjaren Frostviken parish, Jämtland	Found by locals about 1914. Later sold.	9	St-7870	<250		AD 1533–1950

Cont. Tab. 4. ▶

Cont. Tab. 5.

Inv. no.	Spear total length (cm)	Spear total diameter of shaft (mm)	Spearhead length (cm)	Spearhead diameter (mm)	shape	Parish, notes
County Museum of Jämtland						
JLM 77S1214 40-41	178					Jä, Frostviken, Sandnäset; notes; shaft "1848", "P I S. S S."; Fig. 17g
JLM 1053			50	42	1+2 edges, knife-shaped	Jä, Sundsjö
JLM 21536			28	35		
Skellefteå Museum						
SM650			20	24	rhomboid, iron	La, Jokkmokk, Murjek, for wolf hunting?
SM651	266	20				La, Jokkmokk; decorated
SM655	208		36.5	35		Vb, Skellefteå
SM656	212	37	41	55		La, Skellefteå; note; "1714", "NIS", "brand"
SM657	213	40	26	23	rhomboid	La, Skellefteå, for wolf
SM3024			47.5	55	rhomboid	La, Skellefteå
SM7724			38	24		La, Arvidsjaur, for wolf
County Museum of Västerbotten						
1429	168.5	36	27?			La, Sorsele
3484			24			La, Karesuando
4076			47	38		La, Sävar, Botsmark
SSM80132	169	35				Jä, Åre, Duved
SSM80216	205	37				Jä, Tåsjö, Brattbäcken; notes; cover: "Sven Wikström Brattsbäcken 1906", head: "NAGL" (SW 1860–1931)
SSM80259	225	60				La, Jokkmokk, Njunjes
SSM80260	260					La, Jokkmokk, Aktse; marks from bear claws
SSM80275	207		29			La, Arvidsjaur/Boden Fig. 17f; decorated; cf. Iregren 2021, fig. 13
SSM80460	>144		14.5			La, Jokkmokk, marks from wolf's teeth
SSM80481	112.5	45	47	48		Finland, Lake Inari; re-shaped into ice axe; Fig. 17h

Cont. Tab. 5. ▼

Cont. Tab. 5.

	Spear total		Spearhead			
Inv. no.	**length (cm)**	**diameter of shaft (mm)**	**length (cm)**	**diameter (mm)**	**shape**	**Parish, notes**
SSM80510	169	35				La, Arvidsjaur, Pjesker
Silvermuseet Arjeplog						
Sma 320	230		30/24			La, Arjeplog
Sma 916			20	20	rhomboid	La, Arjeplog
Sma 2572			20	25		La, Arjeplog
Sma 5167			18	20		La, Arjeplog
Sma 5810			26	32	rhomboid	La, Arjeplog; note; “P. Lahrson Sakkavara 1880” (PL 1830–1913)
Sma 7912			17	20	rhomboid	La, Arvidsjaur, Aggnäs
County Museum of Dalecarlia						
DM 00574			62.5	50		Gä, Torsås, Fagersta
Kulturen i Lund, Scania						
KM13629						Orrarnäsfjäll, Jä, Frostviken, KM13629; decorated; notes; cover: “M. Andersson Orranäsfjäll 22/4 1884”, “1887”, ferrule “1884”; Mattias Andersson born in 1865. Cf. IREGREN 2021, 13–14, for comments on name and decoration

Table 6. Epiphyseal growth of long bones and disc fusion in vertebrae in well-documented Sámi bear graves found in Sweden and at Spildra, Kvænangen (northern Norway). Age estimation after Weinstock *(2009) for Ursus arctos horribilis, the brown bear of North America. - = open epiphysis, ± = epiphysis in fusion with the main element, + = epiphysis fused, px = proximal epiphysis, di = distal epiphysis. Regarding the bear graves at Spildra, all burials have been damaged and further, only part of the remaining bones were investigated due to ethical considerations.*

Site/	Nåttinäset		Karats		Gråtanån		Nedre Vapstsjön		Sörviken		Gällholmen	
Element	px	di	px	di	px	di	px	di	px	di	px	di
Vertebra		-			-		-					
Humerus			-		-	±	-	+	-	+	+	+
Radius		-	+	-	+	-	+	±			+	+
Ulna	-			-	-		+	-	+	-	+	+
Pelvis, acet.												
Femur			-		-	-	-	-	-	±	+	+
Tibia			-		-	-	-	±	-	+	+	+
Fibula				-	-	-	-	+	+	+	+	+
Calcaneus					-		+		+		+	
Metapodial		-		-		+		+		+		+
Phalanx I	-		-±+		+		+		+		+	
Phalanx II			-±+		+		+		+			
Age (years)	<2–3		3–4		4–6		6–7		6–7		> 6–9	

Site/	Spildra A		Spildra B		Spildra C	
Element	px	di	px	di	px	di
Vertebra	±/-		-		+	
Humerus	+	+	-	+	+	+
Radius	+	+	+	+		
Ulna		+	+		+	
Pelvis, acet.	+		+			
Femur	+	+	-			
Tibia				-		
Calcaneus	+					
Metapodial		+				
Phalanx I	+					
Age (years)	*c.* 6–8		*c.* 6–8		> 9	

Sámi bear graves in Norway – hidden sites and rituals

By Ingrid Sommerseth

Keywords: Sámi, brown bear, ritual, burial, graves, Sámi drums

Abstract: This paper presents a research project on bear-human relationships, focusing on the Sámi bear grave rituals and burial customs in Norway. The background to this project is a study of 30 bear burial sites, of which several were archaeologically recorded. Based on this information, a typical feature of these sites are burials in natural rock cavities, caves and in screes and under large boulders. In general, these date to between c. *AD 300–1800. This makes the Sámi bear burial custom one of the longest surviving burial customs of any kind known in northern Europe.*

Introduction

The brown bear (*Ursus arctos*) was feared, and at the same time appreciated by and sacred to the Sámi, and it was hunted for several needs and purposes. It was normal for the Sámi to worship and praise the animal, although it could be scary, in order to succeed in hunting and to ensure the bear's return and its rightful place in nature. Sámi is an extremely rich language in terms of concepts for nature, animals, hunting and fishing (Gaski 2020). The way the bear was worshipped and understood is strongly attached to the Sámi term and concept of *lihkku* (happiness and prosperity). This term, *lihkku*, possesses an aesthetic dimension with certain rules and rituals, and success in hunting was an important matter in the old Sámi hunting culture. The recognition and understanding of the term, *lihkku*, are evident and continue in the Sámi society today, especially in the reindeer herding society (in North Sámi), where they call it *boazolihkku* "reindeer happiness" (Sommerseth 2009b, 158). According to the Sámi philosopher, Nils Anders Oskal, reindeer or hunting prosperity lasts from the cradle to the grave but may change over time (Oskal 1995). You can influence your own reindeer or hunting prosperity through actions and thoughts, and to a certain degree you can improve your own happiness, but you can also spoil it. How your *lihkku* develops depends on how you live as a person, not only on how you treat animals.

The traditions and hunting rites around the Sámi bear hunt in Scandinavia are well documented in ethnographical written sources, through stories of the bear hunt and other human-bear relations that are still preserved in living traditions and in place names to this day. This article aims to present the archaeological sites and ethnographical sources of the Sámi bear hunt in parts of northern Norway. In general, brown bear bones are found inside natural caves and beneath large boulders, and the bear grave sites are often located along the coast, on islands and in the fjords. These graves are therefore interpreted as a specific practice that relates to pre-Christian Sámi belief and cultural context (Petersen 1940; Myrstad 1996; Schanche 2000; Dunfjeld-Aagård 2005; Svestad 2018; Sommerseth 2021). The practice around the Sámi bear hunt is often recorded in the older ethnographic and

Bear and Human: Facets of a Multi-Layered Relationship from Past to Recent Times, with Emphasis on Northern Europe, ed. by Oliver Grimm (Turnhout: Brepols, 2023), pp. 587–602 BREPOLS PUBLISHERS 10.1484.M.TANE-EB.5.134350

written records that describe the hunting traditions and rituals. Evidence is found in some of the missionary accounts from the 17th and 18th centuries, from priests who sought to convert the Sámi to Christianity (Niurenius 1905 [orignally published 1645]; Rheen 1897 [originally published 1671]; Schefferus 1956 [originally published 1673]; Dass 1992 [originally published 1739]; Fjellström 1981 [originally published 1755]; Leem 1975 [originally published 1767]; Friis 1871). Later, throughout the 20th century, living traditions and old stories told by the Sámi themselves were recorded, adding new knowledge of the historical context of the bear cult (Turi 1911; Bjørnson 1916; Edsman 1994; Ryd 2007; Borgos 2013; 2020).

The discovery and dating of bear graves in Norway

A total of 30 bear burial sites are known in Norway. Many graves have been archaeologically recorded where the bones of brown bears, along with traces of ritual burial, were found in specific places in the landscape (Fig. 1; Table 1; cf. Myrstad 1996; Dunfjeld-Aagård 2005; Svestad 2018; Sommerseth 2021). A common feature of the graves is that they are discovered in natural cavities and screes, where the bones have been placed in dry and airy places directly on the surface. Only five out of 30 burials sites appear to have constructed elements, such as stone slab covers and other traces that show that the cave has been rearranged by humans. Only one grave, the Salfjella grave, which is the southernmost bear grave in Norway, was arranged as a burial in the ground covered with stone slabs and gravel. This grave, which is dated to between *c.* AD 1471–1681, also contains the only artefact found in connection to a bear – a brass chain attached to the right cheekbone of the cranium (Petersen 1940, 158; Dunfjeld-Aagård 2005).

The sparse grave material from the northernmost part of Finnmark, such as that from Nesseby, has a similar appearance to the Salfjella grave material. This bear grave was found on the top of a small hillside in a scree area in 1985; it was covered with slabs and gravel and was dated to between *c.* AD 860–1169 (Myrstad 1996). Another important feature of the bear graves is that very few have been found in the high mountains. Only four bear burial sites, mainly located in the south of Nordland County, have been found far away from the coast with no visibility of the sea. Nevertheless, all bear burial sites have some typical attributes in that they have been found in natural caves, under boulders and in screes, which are normal landscape features along the coast of northern Norway (Fig. 2).

Only seven bear graves along the coast are intact, with bear bones, hidden in natural caves in the landscape. One site is situated in the Vesterålen region, four sites are on the island of Spildra, and two more sites are in the western part of Finnmark. These sites are very vulnerable, and they are automatically protected through the Cultural Heritage Act, supervised by the Sámi cultural heritage management. The bear graves on the large island of Spildra in the northern part of Troms County are the only ones that have information signs and posters. These posters are there to inform visitors to take care and not to disturb the unique sites. Three of the bears lying *in situ* at Spildra are dated to between *c.* AD 1030–1285 (Myrstad 1996).

Many bear graves in Norway have been discovered by chance, through the construction of roads and housing or power line routes. Only a few bear graves have been discovered through research and local knowledge. Between 1911 and 1980, most of the bear graves from the coast were incorporated into the collection of the Arctic University Museum. Some of the bear bones from areas that were exploited around the turn of the century are kept in other museum collections, such as the one at the Institute of Basic Medical Sciences, University of Oslo (The Schreiner Collection).

The very first bear grave was found on the island of Senja in 1911, during road works. The bear cranium was discovered inside a small cave that was later destroyed (Fig. 3). A bone sample from this large cranium is dated to the modern period between *c.* AD 1694–1917, the latest definite date that can be linked to the Sámi hunting tradition and ritual burial. The bear was probably deposited in the cave according to old Sámi burial practice but was most likely hunted down using new hunting methods, such as firearms, which became more common throughout the 18th century (Sommerseth 2009a, 262).

In total, we know of 44 bears from the grave sites, and there is a great variation in the number of bones and teeth that have been preserved and counted. Few sites have finds of bear crania, which are mostly fragmented, and we know of only two sites where almost the entire skeleton has been found – on the island of Spildra and in the Røykenes locality, which have been dated to between *c.* AD 1645–1800 (Fig. 4). For most of the sites, we have only a few intact skeletal parts documented, which seems to be normal for such finds. This situation can reflect the conservation conditions at the site, and bones may also have been carried away by other animals and later damaged by natural decomposition. Another factor that may have played a role could be the circumstances around the discovery. Information from the museum's archives indicates that bear crania and bones had often been stored for a long time locally in the village before they were collected and sent to the museum. In this way, the information on the site and the conditions around the discovery were missing, and further archaeological investigation became impossible. From the 44 known bear individuals, there are 28 bear crania kept in the museum's collection; only a few are intact while most are fragmented.

In eight of the 30 bear burial sites, more than one bear has been documented, and at two of these sites we have solid documentation of bear cubs (Myrstad 1996, 97–100; Sommerseth 2021). One of the cubs was found on the island of Ringvassøya in the northern regions of Troms County, during road works in 1985. The site has now been destroyed, but a new study of the bone material revealed a bear cub in addition to an adult bear. The adult one is probably a female, and a bone sample from its humerus is dated to between *c.* AD 1027–1153 (Sommerseth 2021).

Seines, a spectacular bear grave site with eight bears, one of them a cub, was found on an islet near the town of Narvik in Nordland County in 1970. Five bone samples from different jawbones have been dated to between *c.* AD 1052–1424 (cf. Table 1: Seines), which indicates that the Seines locality was used intensively over a specific period. For several years, many bones were collected from a more or less ruined site, but the central location and appearance is similar to nearby bear graves in the region, indicating that the place must have been most suitable for ritual bear burials. In addition to the many bears, bone fragments from reindeer, cows, and sheep were also documented (Myrstad 1996, 98). This indicates that the grave might also be interpreted as an offering site, with other purposes than that of a bear grave (see Spangen et al., this volume).

This assumption, though, needs a larger empirical basis through an archaeological investigation. Still, burials and sacrifices can be complementary and may have been performed in the same favourable place in the same period, similar to the bear graves on the island of Spildra further north.

One of the goals of the project was to initiate new datings of some of the material, since the first radiocarbon datings were performed more than 25 years ago. The aim was to test older data and compare them to new ones, to see if the sites' timelines match with more accuracy. A total of 22 bears in Norway have been radiocarbon dated, and the results are surprising, as the new datings indicate a burial practice that was in use over a long period of time, *c.* AD 236–1917. The oldest Sámi bear grave found in Scandinavia is from a cave on the island of Tjeldøya in Nordland County. The site was discovered in 1961 and is unavailable due to its location by steep mountains and large boulders. The find consists of 11 bone fragments, including jawbones from a small cranium, which indicates that it was a young animal. A bone sample from a femur yielded a date between *c.* AD 236–385.

A newly discovered bear grave site was investigated on the island of Hovøya in 2005, in the southern part of Nordland County, which yielded a bear cranium with a sensationally old dating (Dunfjeld-Aagård 2005). The remains of two bears were found, one of which consisted of a large bear cranium that was already linked to a local story from more recent times. It is said that the bear died

that you were not killed by the bear yourself (Schefferus 1956, 255). This is supported by the fact that on one of the Sámi drums we can see the bear walking towards its protector and leader, the god *Leibolmai*, both on their way to the sacred mountain (Fig. 7; cf. Friis 1871, 40).

The bear hears and understands everything!

In recent written and oral sources from the Sámi themselves, the bear is still a noble animal who enjoys respect, but this is not totally unconditional (Turi 1911; Edsman 1994; Ryd 2007). The bear was considered to be big and strong, but not too smart and not cunning at all (see Grimm et al., on Bears – fact or fiction, this volume). In several of Qvigstad's (1928, 1–28) notes on old Sámi legends and stories from Troms and Finnmark County, it is emphasised how the fox always fooled the bear, even though the bear was the stronger of them both. According to ancient myths, the partnership and cooperation between humans and bears was about equality and respect for each other (Turi 1911). Similar stories are found in 18th-century sources, where the consideration for the bear and its place alongside humans is reflected in the language, the hunting traditions, the ceremonies, and in burial rituals.

Sámi is a rich language in terms of its concepts for nature, animals, hunting and fishing. As Gaski (2020, 14) explains: "There are several hundred distinct terms for different aspects of snow and ice, and a similar abundance of terms for different aspects of reindeer, including the animal's appearance, age, sex, and color of fur. In the past, Sámi also employed many metaphorical terms for predators like bears and wolves, because these were regarded as so intelligent that they could understand ordinary human language".

The real name of the bear in the Sámi language is *guovža* (North Sámi), and *duvrie* (South Sámi). These names are rarely used, because the bear can hear its name and could thus be unintentionally summoned (Fjellström 1981). The bear has therefore been given metaphorical names, to avoid the animal overhearing the humans' plans for hunting or trapping. Using the bear's proper name would alert the bear to a hunter's intentions, while employing a metaphor, calling the bear by one of its physical attributes or by using kinship terms, would help the hunter or the community to plan the hunt in secret (Qvigstad 1904; Gaski 2020, 14).

The Sámi names and metaphors used for naming the bear are rich and have been used differently from one region to another. Stories and attributes around the names are linked to nature or to kinship among people, for example, *áddja* (grandfather), *áhkku* (grandmother), *dárffot* (turf-like), *muodda* (the old fur), *basse-váise* (the holy, wise animal) *ruomse-gállis* (old moss man), *suohkat* (thick-fur man) (Friis 1871; Rheen 1897; Fjellström 1981).

Many of the names describe the bear as a good-natured and pleasant animal, and some names refer to an old relative, despite its fearsome strength. Most of the metaphorical names are known from 18th-century sources, and the names have at least two purposes – first and foremost, to mask the hunter's real intentions and to keep the hunter's plan secret from the bear, and second, to include and relate to the bear as vital to the society. This was perhaps especially important to appease the gods, and especially the hunting god, *Leibolmai*, who was the keeper of the bear and who, on the one hand, protects the bear, but on the other allows the Sámi to hunt it and decides the hunter's success (Mebius 1968, 128; Leem 1975, 413).

In every fjord and mountain area in northern Norway, it is normal to find bear-related place names in the landscape, most of them are mapped in the Norwegian language, but some bear names have survived in the Sámi language. Sámi place names in general are known for their mapping of the cultural landscape, and they are used as a topographic tool built around narratives related to specific landscape features and landmarks (Mathisen 1997, 120–133). The Norwegian bear names on the map often refer to local stories around the hunt or to rich hunting areas or places where the bear was seen, such as *Bjørnskaret*, *Bjørnsletta*, *Bjørnsund*, *Bjørnlikollen*, and *Bjørnknorran*. The Norwegian name, *Bjørnhellarbukta*, which means "the bear-boulder in the bay", corresponds directly to a bear grave site in Nevelsfjord, which is dated to between *c.* AD 1030–1274. This name has probably been translated from an old Sámi place name, but due to changes around the bear hunt and the consequences of Norwegianisation during the 19th and 20th centuries, the Sámi name was lost. Only some Sámi place names in northern Norway can be related to the bear and in some cases to the landscape where we have documented old bear grave sites. These places often include local knowledge and stories, such as the *Guovžabákti* (the small bear hill) in Nesseby in Finnmark, located one kilometer from a bear grave site that is dated to between *c.* AD 860–1169. It is difficult to search for new bear grave sites based on modern maps with Norwegian place names, so it is therefore crucial to study old Sámi place names along with landscape features and landmarks to understand the cultural landscape.

Traces of the Sámi bear cult in the 18th and 19th centuries

In the area between Nordland and Troms County, a cluster of 15 bear graves with a total of 19 bears have been documented. Some of the graves have been left intact in the landscape; the finds from the other graves are kept in museum collections and described in written sources. The large cluster of bear graves in this particular area is quite special and makes up 45 % of all bear grave sites found in Norway. The question is, why are there so many bear graves in this area? Was the population of brown bears larger here than elsewhere in historical times? Have the hunting traditions and burial customs remained intact over time in this area? Or have the cavities with burials been well hidden and later forgotten until today? Not all of the questions can be answered, but we can try to interpret the sites' presence and time of use.

The region of Vesterålen stretches from the outer coast fjords to the inland coastal areas. The graves are present from Tysfjord to the large islands of Hinnøya and Senja, an area between Nordland and Troms County. This region is also represented by some of the latest ^{14}C-dated burial sites, suggesting that the cultic and ritual traditions lasted longer here than elsewhere in Sápmi and long after the Sámi were Christianised. The time span between the earliest and latest site in this region is substantial, and the oldest bear grave in Tjeldøya, as mentioned earlier, dates to between *c.* AD 236–385 and, close by in the same area, is the Djupfest site, which is dated to between *c.* AD 1442–1690. The distance between these two burial sites is only five kilometers and they are situated in the same type of landscape; they have the same appearance and the physical remains of ritual burial practices. From oral and written sources, we know that ritual bear burials were practiced in this region until the beginning of the 19th century (Bjørnson 1916).

This area is a geographically limited one, as mentioned above, where we find some of the latest ^{14}C-dated bear graves in Norway. Statistics on the number of bears hunted in Norway in 1850 show that 60 brown bears were hunted down, with an estimated population of 3,000 individuals (www.skandobs.no). This suggests that there was a large population of bears in the 19th century. The brown bear must have been very numerous even before the 1800s, as it is mentioned in much older sources. In the 16th century, it appears from the missionary sources of the priest Peder Claussøn Friis (1545–1614) that there were many bears ravaging small fishing villages in the Vesterålen and Lofoten regions (Storm 1881, 375). The authorities saw it as necessary to organise a municipal bear hunt, which was paid for and carried out by skilled hunters. It does not tell us who the hunters were, but recent research has revealed the historical presence of several Sámi fishing villages in the Lofoten and Vestårelen areas (Borgos 2020).

One of the stories that is known to local people concerns a famous Sámi bear hunter named Åne Ånesen (1745–1811), who is said to have killed over a hundred bears as an authorised hunter (Borgos

2020). Throughout the 18th century, it was common in many coastal areas to have a scheme for "bear tax", where the Sámi hunters could be paid by the local farmers for their services. It is said that the payment took place when the bear hunter rowed with his boat from farm to farm with the felled bear visible in his boat. This was to show that the bear hunter was skilled and trained and ready for his reward and new assignments.

To complete the story of the bear hunter from Vesterålen, a large boulder named Ånesteinen, after the same famous hunter, Åne Ånesen, is documented (Fig. 8; cf. Borgos 2013). Today, this large boulder has many local stories and myths associated with it and its location, and the bear hunter's name is a topographic tool used to keep the narratives alive in the local community. The boulder itself and the site represent also a very strong and visual landmark. Best of all, there is a bear grave under the boulder, registered around 1979, but unfortunately many of the bear bones have been disturbed and moved inside the cave. The bear bones were therefore examined and documented by The Arctic University Museum of Norway, and they originate from one bear, from which three different bones were dated as belonging to the period between *c.* AD 1726–1815 (Sommerseth 2021). The historical sources about the bear hunter Åne and the places where he lived represent actual events. He lived in the Vesterålen region with his family at the end of the 1700s, and one story is about the bear that injured Åne's face with its claws, just before he managed to kill the bear. It is said that he carried the scars on his face with pride and great honour (Borgos 2013).

Final remarks

During the course of the 19th century, the traditional knowledge around the Sámi bear cult and the bear graves disappeared (Myrstad 1996; Schanche 2000). The ritual traditions changed to practical explanations, and bear bones found under boulders and in screes around local communities could not be explained, so alternative stories arose. Some stories about the bones are explained by and connected to strangers that have perished, or robbers who have starved to death, or bears that have died in forest fires. This was easily done because some of the bones from the brown bear are very similar to those of humans and are easily misinterpreted by non-professionals. Other stories are, for example, that the bear was buried as a food store for use in bad times. Such stories are many, especially along the coast of northern Norway. The loss of knowledge, memories and traditions around the Sámi bear hunt and grave rituals is also a consequence of Norwegianisation during the 19th and 20th centuries (Sommerseth 2021; cf. Svestad 2018).

The bear bones from Ånesteinen in the region of Vesterålen, which are dated to the end of the 18th century, were probably buried there in line with the old Sámi burial customs, to honour and respect the bear and to renew good hunting fortune (*lihkku*), despite the intense missionary work and Norwegianisation by the church and the authorities. The bear grave at Vesterålen does not stand out from the rest of the material, and this 18th-century grave represents the same ritual practices regarding localisation in the landscape, treatment of bones, and burial practices as those at the much older Iron Age and medieval sites. Burials in natural cavities demonstrate that landscape affordances had a vital impact on the religious concepts and burial practices in Sápmi, the longest surviving burial customs of any kind known in northern Europe. It is said that when Åne the famous hunter was about to die, he heard a bear scratching on the wall outside his home and said: "Now, let it be" (Borgos 2013). Today, Ånesteinen is a popular hiking destination with spectacular views, and the locals are still telling the story about Åne the hunter. This is a reminder that the bear bones in general and bear graves in particular are still visible narratives in the northern landscape. They have an important role for us all and are a strong reminder of ancient stories that are still alive in Sámi history.

Bjørnson 1916: I. Bjørnson, Dundor-Heikka og flere lappers historier fortalt av dem selv. Optegnelser fra Ofoten (Kristiania 1916).
Borgos 2013: J. Borgos, Bjørnejegeren Åne Ånesen. Nordnorsk Magasin 3. https://nordnorsk-magasin.no/om/.
Borgos 2020: J. Borgos, Samer ved storhavet (Stamsund 2020).

Dass 1992: P. Dass, Nordlands Trompet eller Beskrivelse over Nordlands Amt (Oslo 1992).
Dunfjeld-Aagård 2005: L. Dunfjeld-Aagård, Sørsamiske kystområder. Tolkning av samisk tilstedeværelse i Ytre Namdal. Unpublished Master's thesis in Archaeology; UiT, The Arctic University of Norway (Tromsø 2005).

Edsman 1994: C. M. Edsmann, Jägaren och makterna. Samiske och finska björnceremonier. Dialekt- och folkeminnearkivet i Uppsala (Uppsala 1994).

Fjellström 1981: P. Fjellström, Kort berättelse om Lapparnas björna-fänge Samt deras der wid brukade widskeppelser. Norrlänska Skrifter nr.5, Facsimilieutgåva (Umeå 1981).
Friis 1871: J. A. Friis, Lappisk Mythologi. Eventyr og Folkesagn (Christiania 1871).

Gaski 2020: H. Gaski, Myths, Tales and Poetry. From Four Centuries of Sámi Literature In: H. Gaski (ed.), An Anthology (Kárášjohka-Karasjok 2020).

Hansen/Olsen 2014: L. I. Hansen/B. Olsen, Hunters in transition: an outline of early Sámi history. The Northern World 63 (Leiden 2014).
Helskog 2014: K. Helskog, Communicating with the World of Beings. The World Heritage Rock Art Sites in Alta, Arctic Norway (Oxford 2014). https://doi.org/10.2307/j.ctvh1dmmh.3.
Hultkrantz 1985: Å. Hultkrantz, Reindeer nomadism and the religion of the Saamis. In: L. Bäckman/Å. Hultkrantz (eds.), Saami Pre-Christian Religion. Studies on the oldest traces of religion among the Saamis (Stockholm 1985) 10–28.

Leem 1975: K. Leem, Beskrivelse over Finnmarkens Lapper, deres tungemål, Levemaade og forrige Afgudsdyrkelse. Efterord av Asbjørn Nesheim (København 1975).

Mathisen 1997: H. R. Mathisen, Tanker om kart. In: A. Greve/S. Nesset (eds.), Filosofi I et nordlig landskap. Jakob Meløe 70 år. Ravnetrykk 12 (Tromsø 1997) 120–133.
Mebius 1968: H. Mebius, Värro. Studier i samernas förkristna offerriter (Uppsala 1968).
Myrstad 1996: R. Mystad, Bjørnegraver i Nord-Norge. Spor etter den samiske bjørnekulten. Unpublished Master's thesis in Archaeology; UiT, The Arctic University of Norway (Tromsø 1996).

Niurenius 1905: O. P. Niurenius, Lapland eller beskrivning över den nordiska trakt, som lapparne bebo i de avlagsnaste delarna av Skandien eller Sverige. Bidrag til känndom om de svenske landsmålen ock svensk folkeliv. In: K. B. Wiklund (ed.), Bidrag till kännedom om de svenska landsmålen och svenskt folkliv XVII:4 (Uppsala 1905).

Oskal 1995: N. Oskal, Det rette, det gode og reinlykke. Unpublished PhD thesis in Philosophy, UiT, The Arctic University of Norway (Tromsø 1995).

Petersen 1940: T. Petersen, En bjørnegrav. Trekk av lappisk folketro. Viking. Tidsskrift for norrøn arkeologi IV, 1940, 153–167.

Qvigstad 1904: J. Qvigstad, Kildeskrifter til den lappiske mythologi. Det kongelige Norske vitenskabers Selskabs Skrifter 1903(1), 1904, 1–90.
Qvigstad 1928: J. Qvigstad, Lappiske eventyr og sagn fra Troms og Finnmark. Institutt for sammenlignende kulturforskning. Serie B: skrifter X (Oslo 1928).

Rheen 1897: S. Rheen, En kort relation om Lapparnes Lefwarne och Seder, wijd-Skiepellsser, sampt i många Stycken Grofwe wildfarellsser. De Svenska landsmålen och Svens Folkeliv. In: K. B. Wiklund (ed.), Nyare bidrag till kännedom om de svenska landsmålen ock svenskt folkliv XVII:1 (Uppsala 1897).
Ryd 2007: Y. Ryd, Ren och Varg. Samer berättar. Natur och Kultur (Stockholm 2007).

Schanche 2000: A. Schanche, Graver i ur og berg. Samisk gravskikk og religion fra historisk tid til nyere tid (Karasjok 2000).
Schefferus 1956: J. Schefferus, Lappland. Översättning från latin av H. Sundin. Granskad og bearbetad av J. Granlund/B. Löw och J. Bernström. Acta Lapponica VIII. Nordiska Museet (Stockholm 1956).
Sommerseth 2009a: I. Sommerseth, Villreinfangst og tamreindrift i Indre Troms: belyst ved samiske boplasser mellom 650–1923. Unpublished PhD thesis in Archaeology, UiT, The Arctic University of Norway (Tromsø 2009).
Sommerseth 2009b: I. Sommerseth, Boazulihkku (Reindeer Luck) as a link, between past and present reindeer landscapes – reflected in archaeological remains and Sámi placenames. In: T. Äikäs (ed.), Máttut – Máddagat. The Roots of Sámi Ethnicities, Societies and Spaces / Places. Publication of Giellagas Institute 12 (Oulu 2009) 150–164.
Sommerseth 2021: I. Sommerseth, Den samiske bjørnekulten; arkeologiske spor til samisk historie og religion. META – Historiskarkeologisk tidskrift: special issue Sápmi och samisk historisk arkeologi 2021, 9–31.
Storm 1881: G. Storm, Om Peder Claussøn Friis og hans skrifter: Indledning til den norske historiske Forenings Udgave af hans Skrifter (Kristiania 1881).
Storm/Fonneland 2022: D. Storm/T. Fonneland, Indigenous Religions in the Sixth Missionary District: The case of the Hillsá Drum. In: H. Rydving (ed.), Religions around the Arctic (Stockholm 2022) 113–138.
Svestad 2018: A. Svestad, Entering other realms: Sámi burials in natural rock cavities and caves in northern Fenno-Scandinavia between 900 BC and AD 1700. In:

K. A. Bergsvik/M. Dowd (eds.), Caves and Ritual in Medieval Europe AD 500–1500. Part I: Northwestern Europe (Oxford 2018) 13–31.

Turi 1911: J. Turi, Muittalus Samid Birra. En bok om Lapparnas Liv 1. Udg. med dansk overs. af Emilie Demant. Paa Foranstaltning og med forord af Hjalmar Lundholm (Stockholm 1911).

Zachrisson/Iregren 1974: I. Zachrisson/E. Iregren, Lappish bear graves in Northern Sweden. Early Norrland 5 (Stockholm 1974).

Dr. Ingrid Sommerseth
The Arctic University Museum
UIT – The Arctic University of Norway
Tromsø
Norway
ingrid.sommerseth@uit.no

Fig. 1. Location of the Sámi bear graves mentioned in the text, including a picture of the coastal bear grave site in Nevelsfjord (graphics I. Sommerseth, The Arctic University Museum, UIT).

Fig. 2. A bear grave site in the southern end of the Lofoten archipelago, Moskenes Island. The grave is in the scree area along the shoreline (photo I. Sommerseth, The Arctic University Museum, UIT).

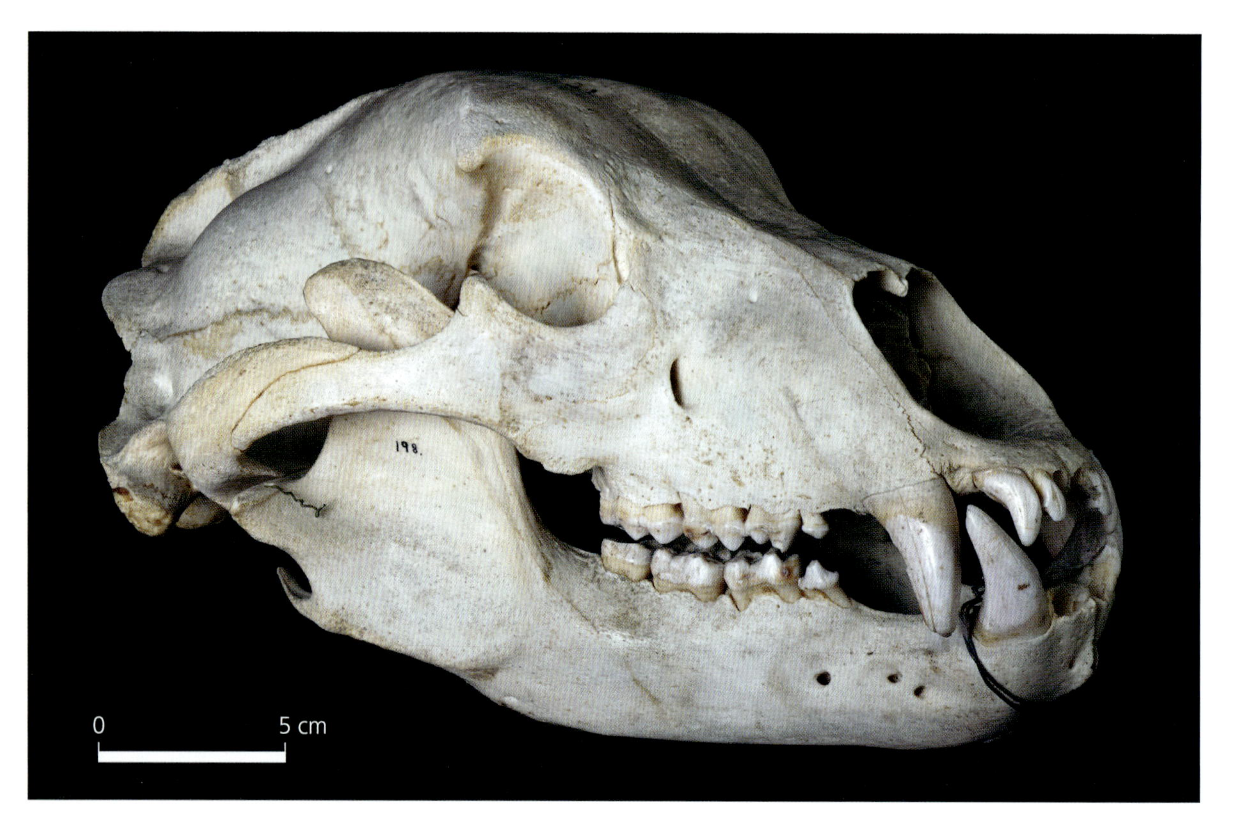

Fig. 3. Bear cranium found in a small cave near the village of Lekangen on the island of Senja in 1911 (photo G. E. Lien, The Arctic University Museum, UIT).

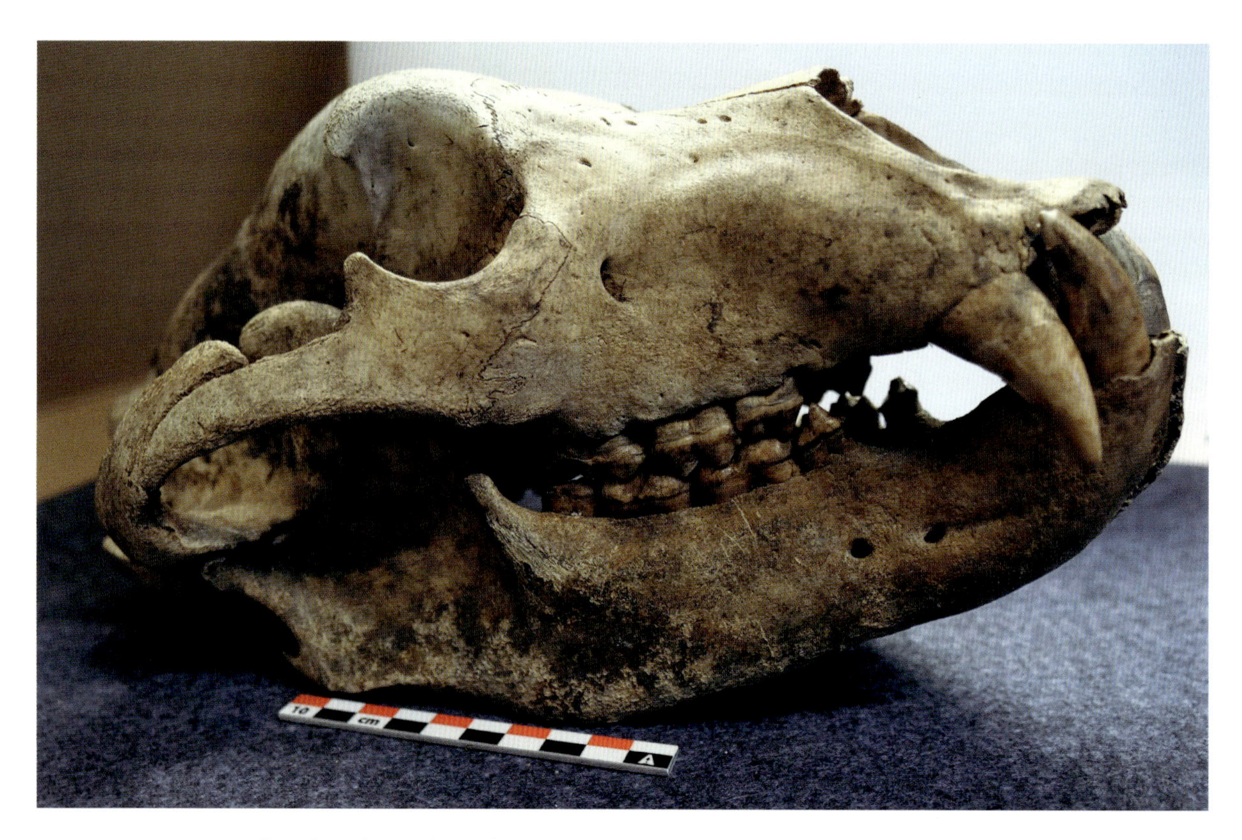

Fig. 4. Bear cranium found at the Røykenes farm in Troms County, belonging to a complete skeleton, dated to between c. AD 1645–1800 (photo I. Sommerseth, The Arctic University Museum, UIT).

Fig. 5. A 6000-year-old hunting scene – a wounded bear and the hunter with his spear. Motif from the panels at the World Heritage Rock Art Centre at Alta, Alta Museum in Troms and Finnmark County (photo K. Helskog, The Arctic University Museum, UIT).

Fig. 6. A typical bear grave site in the Vesterålen area. The bear bones are still in situ *inside the boulder, the site is kept secret (photo I. Sommerseth, The Arctic University Museum, UIT).*

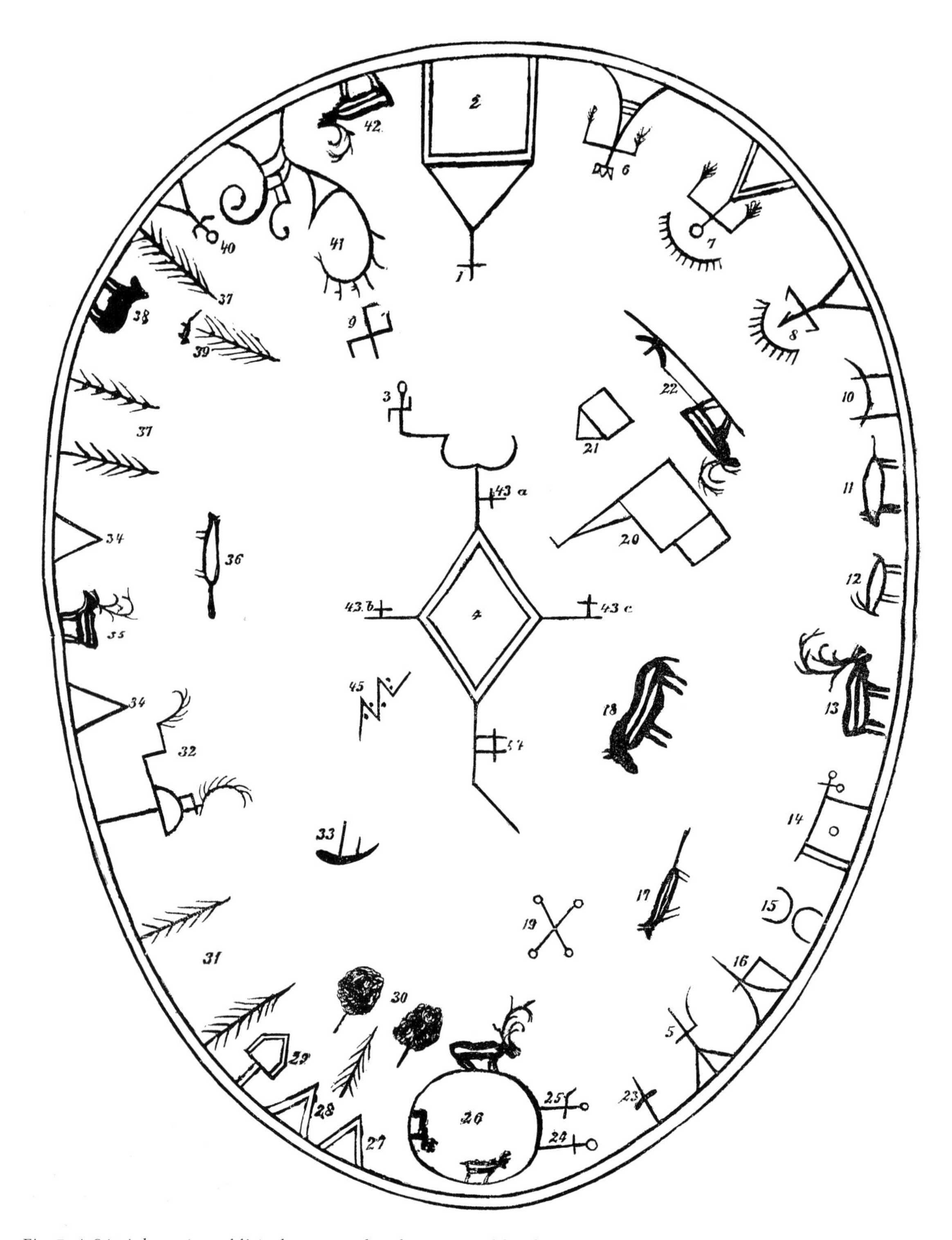

Fig. 7. A Sámi drum (goavddis)*, documented and interpreted by the missionary Knut Leem in 1767. According to him, the bear (no. 38) is walking to the thick spruce forest (no. 37) and to his keeper, the god Leibolmai (no. 40), both near the sacred sacrificial mountain (no. 41; graphics E. Kjellman, The Arctic University Museum, UIT).*

Fig. 8. The famous bear grave site "Ånesteinen" in Vesterålen (photo I. Sommerseth, The Arctic University Museum, UIT).

Table 1. Bear graves in Norway (n = 22) that have been radiocarbon dated.

Locality	County	Year of dating	Lab. no. (result BP)	OxCal v4.4.2 (2020) cal. 95.4 %
Hovøya	Nordland	2005	TUa-5026 (2960±45)	1294–1021 BC
Tjeldøya	Nordland	2019	Beta-538923 (1740±30)	AD 236–385
Bunkholmen	Troms	1996	T-12020 (1330±75)	AD 583–881
Nesseby	Finnmark	1996	T-12025 (1030±75)	AD 860–1169
Ringvassøya	Troms	2020	Tra-15653 (970±15)	AD 1027–1153
Skagedalen I, Spildra	Troms	1994	T-11214 (910±60)	AD 1030–1220
Nevelsfjord	Nordland	1996	T-12022 (855±75)	AD 1030–1274
Seines 5	Nordland	2020	Tra-15650 (890±15)	AD 1052–1215
Seines 1	Nordland	1996	T-12021 (755±90)	AD 1120–1399
Skagevågen II, Spildra	Troms	1994	T-11215 (750±70)	AD 1132–1328
Seines 4	Nordland	2020	Tra-15649 (850±15)	AD 1164–1225
Fjellnes, Spildra	Troms	1994	T-11213 (815±70)	AD 1170–1285
Seines 2	Nordland	2019	Beta-538924 (560±30)	AD 1310–1425
Seines 3	Nordland	2020	Tra-15648 (550±15)	AD 1327–1424
Rørmark	Nordland	2005	TUa-4991 (530±45)	AD 1385–1447
Djupfest	Nordland	1996	T-12024 (290±75)	AD 1442–1690
Salfjella	Trøndelag	2005	TUa-5051 (270± 50)	AD 1471–1681
Skrolsvik	Troms	1996	T-12026 (245±55)	AD 1478–1814
Juvika	Troms	2019	Beta-538922 (320±30)	AD 1482–1646
Røykenes	Troms	2020	Tra-15652 (970±15)	AD 1645–1800
Lekangen	Troms	2020	Tra-15651 (100±15)	AD 1694–1917
Ånesteinen	Troms	2016	Beta-450244 (180±30)	AD 1726–1815

Bear bones at Saami offering sites

By Marte Spangen, Anna-Kaisa Salmi, Tiina Äikäs and Markus Fjellström

Keywords: Saami, bear offering, Fennoscandia, ritual, territoriality

Abstract: Saami traditions related to bear hunting and bear burials are quite well known, both from written and archaeological sources. However, the Saami also included bears in their repeated rituals at offering sites, which has been less explored. In this article, we present the archaeological sources for this offering tradition. Further, we discuss the chronology and geography as well as the content and context of such archaeological finds. As with bear burials, the deposition of bear bones at offering sites has not been a uniform tradition in all Saami communities, which gives an interesting insight into how rituals can both bind a community together and create boundaries with other groups.

Introduction

There are hundreds of recorded Saami offering sites in Sápmi, the Saami areas of Fennoscandia and northwest Russia. The sites have been identified through oral traditions, ethnographic and historical sources, place names and archaeological surveys, and they testify to a cohesive Saami ritual tradition and animistic world view (Helander-Renvall 2010). The sites are usually related to natural features such as cliff formations, boulders or peculiar rocks, but, in the past, offering sites could also be related to trees, rivers and lakes or "altars" built of wood with roughly shaped wooden idols (Jessen-Schardebøll 1767; Friis 1871; Olsen 1910; Qvigstad 1926; Hallström 1932; Manker 1957). Contemporary reports on 17th and 18th century Saami offering traditions describe offerings of mostly reindeer but also birds, fish, wild animals such as bears, and domesticated animals such as cows, sheep, goats, roosters, cats, and dogs (Mebius 1968). However, it is crucial to acknowledge that these sources are time-specific, that the information they convey is drawn from particular regions, even if the sources do not always specify this, and that the authors (usually Christian priests and missionaries) mix local knowledge with generalised accounts from other authors (Rydving 1995). Before the 17th century, there are few written sources with reliable descriptions of the culture and social life of Saami groups; thus we have relatively little knowledge about changes in the offering traditions over time. It is therefore interesting to see that the written sources we have do not always coincide with the archaeological material available (Manker 1957, 40–45).

Very few known offering sites have been investigated through archaeological excavations, but we have information about observed and collected offering matter from quite a few. A series of recent studies of previously collected animal bones from offering sites in Sweden and Finland have discussed species variation and chronology and suggest that the earliest offerings were of wild animals such as bear and swan, with radiocarbon dates of bear bones from the famous offering site of Unna Saiva in northern Sweden (Fig. 1) stretching back to the 6th century AD. Only in the late 12th century do the

Bear and Human: Facets of a Multi-Layered Relationship from Past to Recent Times, with Emphasis on Northern Europe, ed. by Oliver Grimm (Turnhout: Brepols, 2023), pp. 603–618 BREPOLS PUBLISHERS 10.1484.M.TANE-EB.5.134351

first reindeer bones occur (Salmi et al. 2015; 2018). It is difficult to deduce whether these were bones from wild or domesticated reindeer. However, there is a marked increase in the amount of reindeer bones at the offering sites between the 15th and 17th centuries. This coincides with the first confirmed offerings of reindeer belonging to the genetic lineages of present-day domesticated reindeer, which may indicate an increased economic, and thus cultural, importance of this animal, related to more extensive domestication (Salmi et al. 2018; Heino et al. 2021). In the same time period, ovicaprid bones were introduced at the offering sites, supporting the notion that domesticated animals became more important in the Saami economy. In the 17th century, the offering of animals was drastically reduced, probably because of the intensified Christianisation and severe punishments for maintaining pre-Christian rituals of this sort (Salmi et al. 2015; 2018). The deposition of shed reindeer antlers, foodstuff, and minor objects such as antler spoons and jewellery at known Saami offering sites has continued throughout the centuries up until today. In some places, this has also included the occasional deposition of animal parts, particularly of reindeer (Qvigstad 1926; Mebius 1972; Äikäs/Salmi 2013; Äikäs/Spangen 2016; Spangen/Äikäs 2020). Generally, one may say that the offering matter found at Saami offering sites more or less consciously represents what was available and was of economic and cultural importance at any given time and place.

In large parts of the Saami area, bear hunting has been of importance until modern times for cultural and economic reasons and to decimate the population of this feared and respected animal. Intricate rituals related to the hunt are described in several sources, as are the deposition of the bear bones in "bear graves" after eating the meat (e.g. Petersen 1940, 159; Fjellström 1981). There is less mention in known historical or ethnographic sources of bears being deposited at specific offering sites (see Paulaharju 1941, 7). Nevertheless, the archaeological material tells a different story.

Finds of bear bones in Saami landscapes may of course sometimes be from bears that have died of natural causes, but finds of bears that have clearly been slaughtered and eaten, for instance assemblages of gathered bear bones, often with some elements missing, are usually suggested to be bear graves (see Sommerseth; Iregren, this volume). However, the distinction between such individual depositions and offerings is not straightforward, since the sources we have for the bear hunting rituals and burials are time- and place-specific and may not cover all the alternative practices that were acted out in the past (Myrstad 1996, 4). Even if offerings were often placed on the ground surface, on or near a focal point of the offering site, such as a rock formation, offerings of animals could also be buried in the ground (Mebius 1968, 19). Furthermore, the division between bear burials and bear offerings may have been entirely irrelevant to the Saami groups performing ritual depositions in the past. One example of the difficulty of defining finds as one or the other is a bear skull that was discovered tucked away, but not buried, under a large rock near Mjösjö village in Junsele parish, Ångermanland, Sweden. Since no postcranial bones were collected, it has been suggested that this possibly represents an offering rather than a burial (Zachrisson/Iregren 1974, 34, 96). The case demonstrates how finds of bear skeletal remains within Saami contexts will have to be evaluated on a case-to-case basis.

The same is true when considering so-called bone deposits. In one case, informants in the South Saami area describe that they have observed bear bones placed on platforms in trees. This is explained as an alternative to the traditional hiding of reindeer and bear bones in the ground or under rocks, a feature called *daktsie*, i.e. "bone deposits". The informants' interpretation was that, in this instance, placing the bones in a tree was a more practical solution in winter because of the snow (Petterson 1946, 148, cited in Manker 1957, 268).[1] This suggests that, at least in some times and places, the deposition of bear bones was conceptualised in a similar way as *daktsie* depositions of reindeer

1 The site in question is Trettondagsberget (Idvatnet) in the Saami village/area of South Vilhelmina, Sweden; Manker 1957, cat. no. 453.

bones. It has been debated whether the deposition of reindeer bones is to be considered as a kind of offering practice, or whether it had more practical reasons, in terms of disposing of meal remains and keeping dogs from eating sharp pieces of marrow-split bones. Recent studies conclude that practical and ritual intentions are difficult to separate based on the archaeological material, as these are not separate spheres but tend to be interwoven within Saami culture. A ritual aspect of this practice is probably also time- and place-specific (e.g. Rydving/Kristoffersen 1993; Andersen 2009; Ljungdahl 2012). Burials of entire individual reindeer, which occur both in South Saami areas in Sweden (Zachrisson 1985, 84–86), and possibly in the scree burial field in Mortensnes, northern Norway (Schanche 2000, 297), should probably be interpreted as more certain ritual depositions. The same ambivalence may of course be discussed concerning the described practice of depositing bear bones, and thus even the tradition of bear burials. When placed by an offering site, the religious connotations of a deposition are more obvious.

In this article, and as an operational category, we will define "bear offerings" as bear bones found on sites that are known from other sources to be offering sites, finds of bear bones from more than one individual gathered at a defined site, or bear bones found together with bones from other animals in contexts that indicate repeated ritual depositions at a site. Thus, we exclude most bear graves and other bone deposits, whether or not these have been religiously motivated or are related to rituals.

Geography and chronology

In the present study, we have looked at the reported and collected archaeological materials from offering sites that include bear bones. The investigation results suggest a regional variation that is quite interesting considering the ubiquitous bear and its importance in Saami culture. There is no doubt that traces of bears at offering places are most frequently found in the inland areas of today's northern Sweden (Fig. 2). Ethnographer Ernst Manker reports on the finds of bear bones at 15 of the Swedish locations listed in his seminal work on Saami offering sites, though only 13 are actually counted in the overview table (Table 1; cf. Manker 1957, 45 and table 3).[2] For one additional offering site, at Akkavare, there is recorded information about the observations of bear bones (ibid., 165).[3] In comparison, the neighbouring areas in today's Finland have only one example of bear bones at an offering site at Näkkälä, Enontekiö (Äikäs 2015, 294), apart from uncertain information about one previously recorded find on the island of Äijihsuálui (Ukonsaari) in lake Inari (Okkonen 2007, 30). In today's Norway the examples are also very few, and the ones we include here have previously been defined as bear burials. One is a find of bones and a bear cranium in a cave near a large boulder traditionally said to be an offering site on the island of Årøya in the Alta fjord, Finnmark county (Qvigstad 1926, 340; Myrstad 1996, 31). Another accidental find was made in 1970 on the headland in Seines in the Herjangsfjorden, Nordland county. This was an assemblage of bear bones, especially cranial bones, and teeth from eight bears. These were found together with cow, sheep/goat and reindeer teeth near a conspicuous rock formation, suggesting that this was an offering site. The rock is called "Dead Man's

2 There seem to be some inconsistencies in the information given by Manker concerning offering sites with bear bones. In his table 3, he notes finds of bear bones in the Saami village/area of Jåkkåkaska, but going through his descriptions of the offering sites in this area, we have not been able to deduce what site he means. He also mentions five sites with bear and bird bones: Vieksa, Paddustieva, Haltenjarka, Vierronjarka, and Abelvattnet (Manker 1957, 45 cat. nos. 28, 89, 110, 137, 429). However, bear bones have neither been recorded in the further description of these sites nor in the osteological study performed on the material he discusses (Gejvall 1956).

3 Other finds of singular bear skeletons in Auttejaure, Vesken and Gammgårdshobben in Vapsten, and Värkaren in Frostviken, all in Sweden, are called bear graves by Manker 1957 (cat. nos. 427, 434, 435, 482), and are also defined as such by us.

Rock", because bones were found by the rock. It is not unusual for bear bones to be confused with human bones, and most likely the myth related to the rock was based on this confusion (Myrstad 1996, 9, 38–39). Very little is known about Saami offerings in Russia, and to our knowledge there are no accounts about offerings of bears here, even if we know about burial rituals related to the hunting and slaughter of bears (see below). In Table 1, we have included offering sites from which bear bones are either preserved or where the information about finds of bear bones is specific enough to validate this (Table 1).

Contrary to the geographical concentration of bear offerings in one area, there are records in most Saami areas, from Østerdalen in southern Norway to the Kola peninsula in Russia, of rituals related to the bear hunt and the burial of the bear bones after slaughtering and skinning. Sources describe burials of the bear skeleton in its anatomically correct order (Randulf 1723; Hallström 1922; Petersen 1940, 159; Kildal 1945; Leem 1975), but archaeological investigations have shown that the mode of deposition varied significantly. In some contexts, less attention was paid to anatomical order, and different amounts of individuals, body parts, and other objects and animals are included in the burials (Zachrisson/Iregren 1974; Myrstad 1996; Sommerseth 2021). As noted, it may be discussed whether or not these should be redefined as offerings. An ethnographic example, from Bjälaja guba, Imandra, on the Kola peninsula, shows a specific variation for this region; archaeologist and ethnographer Gustaf Hallström reports in the early 20th century that the Saami here had rituals related to bear hunting, but they never ate the meat. They only preserved the skin, claws, and teeth, while the rest was buried. If it was not possible to bury everything, they cut off the head with the neck and chest and buried that, sometimes marking the place with a stick (Hallström 1922, 176). In both Norway and Sweden, there are accounts that testify to the continued ritual burial of bear bones at least into the 19th century (Zachrisson/Iregren 1974, 13; Myrstad 1996, 20).

The variation in bear rituals such as burials and offerings might potentially be due to chronological variation. Contemporary written sources that describe bear graves range from the 17th to the 20th centuries. In the archaeological material, however, the earliest radiocarbon-dated bear grave is located in Kjærfjorden on the island of Tjeldøya in Tjeldsund, Nordland, Norway, and has been radiocarbon-dated to AD 235–385 (Beta-538923, Sommerseth 2021, 15), while another on Bunkholmen in Lyngen, Troms, Norway, dates to AD 583–881 (T-12020, Myrstad 1996; new 2-sigma-calibration in Sommerseth 2021). In Sweden, one bear grave in Karats, Jokkmokk, northern Sweden, was AMS-dated to AD 775–1035, based on a bear tooth (Ua-507), while radiometric dating of birch bark from the grave yielded an even older date, AD 437–1014 (St. 11213, Mulk/Iregren 1995). Bear bones from another alleged bear grave at Grundskatan on the island of Bjurön, along the Bothnian coast of Sweden, were dated to AD 709–1160 (Ua-18930, Edvinger/Broadbent 2006, 37, with 2-sigma-calibrations performed for this article using OxCal v4.4.4, IncCal20, Bronk Ramsey 2009; Reimer et al. 2020). However, it is contested whether or not the latter does in fact represent a bear grave (Liedgren/Ramqvist 2012). Concerning bear bones found at known offering sites, only two radiocarbon dates have been acquired from this specific material so far. A bear bone from the offering site by lake Unna Saiva, northern Sweden, was dated to AD 557–774 (Ua-48702, cf. Salmi et al. 2015, 12, new calibration). A bear tooth from the offering site at Näkkälä, Finland, was dated to AD 1174–1267 (Hela-1885/1133, Äikäs 2015, 294, new calibration). This indicates that wild predators continued to have a role in rituals at offering sites into the 13th century, when reindeer became, in general, more and more common as offering gifts. It should, however, be noted that the Finnish Saami areas are distinctive in several ways. For instance, the more extensive reindeer herding known from other Saami contexts was not a widespread enterprise here until modern times (e.g. Tegengren 1952; Harlin et al. 2019; Salmi et al. 2021). A complete lack of bear graves is another peculiarity that sets this region of Sápmi aside.

If the bear bones from Dead Man's Rock on the Seines headland in Norway, which were found together with teeth from cows, sheep/goat and fish, are to be counted as depositions at an offering site, and not a bear burial as previously suggested, their chronology is also relevant. One bear bone from the site has been radiocarbon-dated to 755 BP ± 90 (T-12021). This was previously calibrated to AD 1220–1300 (Myrstad 1996, 9, 38–39), while a new calibration suggests a wider time span and dating to AD 1120–1399 (Sommerseth 2021, table 1: Seines 5). Samples from several other bear individuals from the site were also radiocarbon-dated. The time spans of the three oldest bear bones overlap in the 11th–13th centuries (Seines 1, 4, 5), while the dating results of the two youngest bones do so closely in the 14th–15th centuries (Seines 2 and 3, cf. Sommerseth 2021). Consequently, depositions must have taken place here on at least two separate occasions.

The chronological range of dated bear graves and bear offerings fits well within the general expected deposition period for animal bones at Saami offering sites, but the dates available do not suggest a specific time period in which bears in particular were deposited more frequently. This is of course in part due to sample size, and possibly several source-critical factors, such as the purely coincidental finding of bear graves, the preservation conditions at individual sites, and the selection of species and bone elements various visitors to offering sites have chosen to collect and render to museums. This may affect our current knowledge about the Saami bear rituals, including its chronology. Based on the fact that the oldest bear deposition's 3rd–4th-century radiocarbon date is from the bear grave in Tjeldsund, and the oldest radiocarbon date in Sweden is the 6th–8th-century dating result from the offering site Unna Saiva, we can perhaps hypothesise that the earliest bear rituals among the Saami took place along the Norwegian coast and focused on individual bear graves, while Saami groups in inland Sweden initially placed bear bones at common offering sites such as Unna Saiva, while also taking up the tradition of bear burials from the 8th century. In general, however, the depositions of bear bones in graves and at offering sites overlap chronologically.

The content and context of Saami bear offerings

The reported numbers of bear bones found at offering sites vary from a few fragments to 42 identified specimens at the offering site of Unna Saiva, while the minimum numbers of individuals vary from one to 14 at Unna Saiva (Table 1). The percentages of bear bones are *c.* 1–2 % of the number of identified specimens and *c.* 14–20 % of the minimum numbers of individuals in assemblages from Näkkälä and Unna Saiva, where precise counts are available (Äikäs et al. 2009; Salmi et al. 2015). The finds of bear bones at offering sites consist mainly of cranial bones and teeth (Fig. 3; cf. Salmi et al. 2018, 476). Due to the lack of precise zooarchaeological analysis of the faunal assemblages from many of the sites, the exact numbers of bear bones and their skeletal frequencies at each site are impossible to catalogue at the moment. At Näkkälä, the left and right upper molars of a bear (Fig. 4) were found with decomposed bone material, probably deriving from the maxilla, in an anatomically correct order, suggesting that a complete bear skull was probably deposited at the site. The occlusal surface of the molars was facing up, which means that the skull was probably deposited upside down (Äikäs et al. 2009). At some sites, postcranial bones are also deposited, but the sources often only mention "bones" in addition to teeth and cranial bones, with no further distinction of skeletal elements. A bear vertebral fragment was reported to have been found at Haltenjarka (Gejvall 1956). At Seines, some marrow-split postcranial bones were found, though the majority were cranial bones (Myrstad 1996, 38). The marrow-split bones have later been redetermined as reindeer bones (Sommerseth 2021, 19).

Compared to the hundreds of offering sites described by ethnographer Ernst Manker in northern Sweden, the frequency of bear skulls and teeth at these sites is not very high, even within the region where they are most common. Despite being concentrated in the area of northern Sweden, there are also great distances between the known locations of bear offerings. The fact that only two relevant finds have come to light along the extensive Norwegian Atlantic coast could, in theory, be related to the bears' habitat; in the 19th century, bears were numerous in Norway, but mostly so in the forested inland regions, while a few larger islands did not have bears at all (Stensli 1993, 40; Myrstad 1996, 7). However, estimates suggest that the bear population has been as large as, or even larger, in Norway than in Sweden during historical periods, with about 65 % of the bears in Scandinavia found in Norway (*c.* 3,100 in Norway vs. 1,650 in Sweden) in the mid-1800s (Swenson et al. 1995). The Prefect's Office in Norrbotten, northern Sweden, reported that 257 bears were killed in the area during the period 1855–1865 (Von Düben 1873, 26, 80), attesting to the great availability of this prey, which should have been at least as frequently encountered on the Norwegian side of the border. Bear burials are also relatively frequent along the Atlantic coast, which suggests that bears were not hunted less here than in the inland Saami areas of today's Sweden, but perhaps that there was a difference in deposition practices in different areas and possibly in different contexts. A similar explanation is likely for the lack of bear burials in the Saami areas of today's Finland and the low frequency of bear bones at offering sites here. Bear bones, teeth and pelts are known from ritual contexts, such as graves and foundation deposits, in Finland from the Iron Age to the 17th century (Leppäaho 1937; Puputti 2010; Kirkinen 2019; see Mannermaa et al., this volume), and bears were probably hunted in the boreal forests of Finland throughout prehistory (Ukkonen/Mannermaa 2017). The knowledge about bear bones at offering sites could be biased due to differences in the thoroughness and methods of recording this tradition. However, it is striking that Norway and Finland not only have few archaeological finds, but there is also little information in ethnographic sources about bear bones at known offering sites (e.g. Qvigstad 1926, who mentions only one instance in Norway, the location on Årøya, Alta, from where bear bones have indeed been retrieved). This further supports that bears as offering matter were less frequent in these areas.

Bear heads and placemaking

It is difficult to know whether the differences in the body parts of the bear that are found at offering sites are related to a specific significance of various elements or to taphonomic processes or sampling strategies. For instance, perhaps only teeth were left at smaller offering sites, while entire skulls and other bones were left at larger offering places. However, cranial bones are present in seven out of nine sites where such details have been recorded, suggesting a special significance related to the bear head. Similarly, it seems to have been common in northern Sweden to place the heads of male reindeer of considerable size at offering sites, while smaller female individuals are mostly represented by long bones (Salmi et al. 2018, 476). Consequently, and not entirely surprisingly, the most impressive and communicative parts of the animal bodies were used to adorn and honour offering sites, as well as to interact with both in- and out-group human actors who encountered the site. These offerings would provide a variety of information concerning, among other things, the status of the site and the people present in the area, including their religious beliefs and contact with non-human powers. While possibly related to what was seen as a valid and valuable offering (Äikäs et al. 2009, 117), the attention to skulls also reflects a persistent cross-cultural fascination with "head-objects", where human and animal heads are transformed into ritual objects that are deposited in ways that seem related to placemaking (Eriksen 2020), i.e. processes that transform nondescript space into recognised places of particular meaning (e.g. Smith 1998, 32–33, 45; see also Tuan 1977). It is possible that the offerings of bear crania or cranial elements were associated with conceptions about bear personhood and its transformations in offering rituals, particularly because the Saami perceived animals as persons, with animal personhood coming into play relationally in various contexts (HELANDER-RENVALL 2010). The offerings of bear heads and cranial elements may be related to an idea of the head as the locus of personhood, but also to more complex ideas about personhood and its entwinements with places, actions, and events. The ritual head deposition potentially manipulated the identity and personhood of the bear by transforming it into a head-thing, a thing with potency in the ritual place and context, but no longer a social personhood (ERIKSEN 2020).

In other contexts, depositions of skulls, or head-objects, are not always visible but may be retrieved from subterraneous contexts such as graves and underground building features (ERIKSEN 2020). Thus, bear burials might perhaps be interpreted in the same way, where the bear person is given transformed significance through rituals. However, since more of the skeleton is usually found in bear graves, and written sources emphasise the necessity of preserving the entire skeleton, these contexts appear more akin to human burials in Saami contexts. It also seems significant that bear burials would involve hiding or putting away the skull and bones, while depositions at offering sites could be seen as display. Consequently, offering sites with impressive skulls cannot be seen only as reflections of local economic adaptations or ritual practices within a broader tradition of Saami offerings: They should probably also be seen as possible instruments of power, where successful hunters or patrons could attain recognition both in this world and the other for their achievements and ability to procure impressive offerings.[4] Furthermore, such offerings would not only have made the place but maintained and enforced it as a vibrant meeting-place in the landscape, in the sense that the (shifting) combination of topography, impressive depositions, and human-animal-thing encounters would have made such sites affect those who came across them, regardless of their prior knowledge of the offering rituals performed there. Other offering sites that were made into distinct places by their users may have later disappeared into oblivion, because they were not the subject of similar attention and impressive depositions during their use.

This underlines that offering sites are not static places. In previous studies, we have called this "site biographies" (ÄIKÄS/SPANGEN 2016), but we would like here to emphasise the palimpsestic qualities[5] this creates in the sites. The sites are not relics but multi-layered and complex actors in the landscape that are transformed over time and simultaneously transform how people interact with them. One result is that some offering sites may be forgotten due to little or changed use of the sites and of the landscape. This is especially true in areas where Saami groups in the Middle Ages went from mainly hunting and fishing to taking up reindeer husbandry, such as in inland northern Sweden and Norway, or fishing-farming, such as along the Atlantic coast of Norway, as important parts of their livelihood. In the process, they would to some extent be leaving previously familiar landscapes less used (e.g. MULK 1994; SOMMERSETH 2011; ANDERSEN 2019).

These reflections are relevant when considering that most of the offering sites known today were in continued use until at least the 19th century. They often have quite conspicuous features or large amounts of offering matter (ÄIKÄS 2015; SALMI et al. 2015; 2018). Among these are the well-known offering sites with preserved metal objects from the Middle Ages, such as coins, jewellery, and arrowheads (SERNING 1956). At six out of 15 known offering sites where bear bones have been collected, there are also finds of medieval metal objects and/or coins (Table 1). These sites have often been well

4 This is an argument similar to that formerly made concerning the offering of valuable metal objects at, in part, the same offering sites (MULK 1996; see also below).

5 The term "palimpsest" originally refers to the practice of writing, erasing and rewriting on the same surface, particularly to describe the re-use of medieval manuscripts on parchment. The term has been used in archaeology to describe traces of various processes in historical landscapes and to describe the archaeological record as such. It is here used to describe how material remains of past activities are not only part of the past but remain present and active in subsequent time periods up until today (see e.g. BAILEY 2007).

Helander-Renvall 2010: E. Helander-Renvall, Animism, personhood and the nature of reality: Sami perspectives. Polar Record 46(1), 2010, 44–56.

Jessen-Schardebøll 1767: E. J. Jessen-Schardebøll, Afhandling om de norske finners og lappers hedenske religion. In: K. Leem, Beskrivelse over Finmarkens lapper, deres Tungemaal, Levemaade og forrige Afgudsdyrkelse, oplyst ved mange Kaabberstykker = Canuti Leemii De Lapponibus Finmarchiae, eorumqve lingva, vita et religione pristina commentatio, multis tabulis aeneis illustrata / med J. E. Gunneri Anmærkninger. Og E. J. Jessens Afhandling om de norske Finners og Lappers hedenske Religion (Copenhagen 1767).

Kirkinen 2019: T. Kirkinen, Between skins. Animal skins in the Iron Age and historical burials in Eastern Fennoscandia (Helsinki 2019).

Kildal 1945: J. Kildal, Afguderiets Dempelse. In: O. Solberg/M. K. Johannessen (eds.), Finner, Nordlands og Troms i eldre handskrifter 2 (Oslo 1945) 135–144.

Leem 1975: K. Leem, Beskrivelse over Finmarkens lapper. Edited by A. Nesheim (Copenhagen 1975; originally published 1767).

Leppäaho 1937: J. Leppäaho, Savukosken Mukkalan lappalaiskalmisto. Kotiseutu 3–4, 1937, 134–144.

Liedgren/Ramqvist 2012: L. Liedgren/P. Ramqvist, Noel Broadbents lappar och labyrinter. Fornvännen 107, 2012, 212–217.

Ljungdahl 2012: E. Ljungdahl, Om bengömmor och hornsamlingar – vanliga kulturlämningar i det samiska kulturlandskapet. In: E. Ljungdahl/E. Norberg (eds.), Ett steg till på vägen. Resultat och reflexioner kring ett dokumentationsprojekt på sydsamiskt område under åren 2008–2011 (Östersund 2021) 93–113.

Lund 2015: J. Lund, Living Places or Animated Objects? Sámi Sacrificial Places with Metal Objects and Their South Scandinavian Parallels. Acta Borealia 32(1), 2015, 20–39.

Manker 1957: E. Manker, Lapparnas heliga ställen: kultplatser och offerkult i belysning av Nordiska museets och landsantikvariernas fältundersökningar (Stockholm 1957).

Mebius 1968: H. Mebius, Värro. Studier i samernas förkristna offerriter (Stockholm 1968).

Mebius 1972: H. Mebius, Sjiele. Samiska traditioner om offer (Uppsala 1972).

Mulk 1994: I.-M. Mulk, Sirkas. Ett samiskt fångstsamhälle i förändring Kr. f. –1600 e. Kr. (Umeå 1994).

Mulk 1996: I.-M. Mulk, The role of the Sámi in fur trading during the Late Iron Age and Nordic Medieval Period in the light of the Sámi sacrificial sites in Lapland, Northern Sweden. Acta borealia 13(1), 1996, 47–80.

Mulk 2009: I.-M. Mulk, From Metal to Meat. Continuity and Change in Ritual Practices at a Saami Sacrificial Site, Viddjavárri, Lapland, Northern Sweden. In: T. Äikäs (ed.), Máttut – Máddagat. The Roots of Saami Ethnicities, Societies and Spaces/Places (Oulu 2009) 116–133.

Mulk/Iregren 1995: I.-M. Mulk/E. Iregren, Björngraven i Karats: Arkeologisk undersökning. Duoddaris, Rapportserie 9 (Ájtte 1995).

Myrstad 1996: R. Myrstad, Bjørnegraver i Nord-Norge. Spor etter den samiske bjørnekulten. Unpubl. Master's thesis in archaeology, University of Tromsø (Tromsø 1996).

Okkonen 2007: J. Okkonen, Archaeological investigations at the Sami sacrificial site of Ukonsaari in lake Inari. Fennoscandia Archaeologica XXIV, 2007, 29–38.

Olsen 1910: I. Olsen, Om lappernes vildfarelser og overtro (Trondheim 1910).

Paulaharju 1941: S. Paulaharju, Sodankylän seitoja. Ylipainos Pohjois-Pohjanmaan maakuntaliitonvuosikirjasta 1939–1940 (Oulu 1941).

Petersen 1940: T. Petersen, En bjørnegrav. Trekk av lappisk folketro. Viking 4, 1940, 153–166.

Petterson 1946: O. P. Petterson, Gamla byar i Vilhelmina III. Skogstrakterna i öster och Volgsjötrakten (Stockholm 1946).

Puputti 2010: A.-K. Puputti, Bones in pits and ditches. A contextual approach to animal bone distribution in Early Modern Tornio. Journal of Nordic Archaeological Science 17, 2010, 53–64.

Qvigstad 1926: J. A. Qvigstad, Lappische Opfersteine und heilige Berge in Norwegen (Oslo 1926).

Randulf 1723: J. Randulf, Relation angaaende Finnernes saa vel i Finmarken som Nordlandene deres Afguderie og Sathans Dyrkelse, som af Guds Naade ved Lector von Westen og de af hanem samestæds beskikkede Missionairer, tiid efter anden ere blevne udforskede. Manuscript, Gunnerus library, NTNU Trondheim (1723). https://ntnu.tind.io/record/107936.

Reimer et al. 2020: P. J. Reimer/W. E. N. Austin/E. Bard/A. Bayliss/P. G. Blackwell/C. Bronk Ramsey/M. Butzin/H. Cheng/R. Lawrence Edwards/M. Friedrich/P. M. Grootes/T. P. Guilderson/I. Hajdas/T. H. Heaton/A. G. Hogg/K. A. Hughen/B. Kromer/S. W. Manning/R. Muscheler/J. G. Palmer/C. Pearson/J. van der Plicht/R. W. Reimer/D. A. Richards/E. Marian Scott/J. R. Southon/C. S. M. Turney/L. Wacker/F. Adolphi/U. Büntgen/M. Capano/S. M. Harni/A. Fogtmann-Schulz/R. Friedrich/P. Köhler/S. Kudsk/F. Miyake/J. Olsen/F. Reinig/M. Sakamoto/A. Sookdeo/S. Talamo, The IntCal20 Northern Hemisphere Radiocarbon Age Calibration Curve (0–55 cal kBP). Radiocarbon 62(4), 2020, 725–757.

Rydving 1995: H. Rydving, The end of drum-time. Religious change among the Lule Saami, 1670s–1740s (Uppsala 1995).

Rydving/Kristoffersson 1993: H. Rydving/R. Kristoffersson. Några samiska offerplatser. Fornvännen 88, 1993, 195–210.

Salmi et al. 2015: A.-K. Salmi/T. Äikäs/M. Fjellström/M. Spangen, Animal offerings at the Sámi offering site of Unna Saiva – Changing religious practices and human–animal relationships. Journal of Anthropological Archaeology 40, 2015, 10–22.

Salmi et al. 2018: A.-K. Salmi/T. Äikäs/M. Spangen/M. Fjellström/I.-M. Mulk, Tradition and transformation in Sámi animal-offering practices. Antiquity 92, 2018, 472–489.

Salmi et al. 2021: A. Salmi/M. van den Berg/S. Niinimäki/M. Pelletier, Earliest archaeological evidence for domesticated reindeer economy among the Sámi of Northeastern Fennoscandia AD 1300 onwards. Journal of Anthropological Archaeology 62(1–2), 2021. doi.org/10.1016/j.jaa.2021.101303.

Schanche 2000: A. Schanche, Graver i ur og berg. Samisk gravskikk og religion fra forhistorisk til nyere tid (Karasjok 2000).

Serning 1956: I. Serning, Lapska offerplatsfynd från järnålder och medeltid i de svenska lappmarkerna (Stockholm 1956).

Smith 1998: J. Z. Smith, Å finne sted. Rommets dimensjon i religiøse ritualer (Oslo 1998).

Sommerseth 2011: I. Sommerseth, Archaeology and the debate on the transition from reindeer hunting to pastoralism. Rangifer 31(1), 2011, 111–127.

Sommerseth 2021: I. Sommerseth, Den samiske bjørnekulten – arkeologiske spor til samisk historie og religion. META – historiskarkeologisk tidskrift 2021, 9–29.

Spangen 2005: M. Spangen, Edelmetalldepotene i Nord-Norge. Komplekse identiteter i vikingtid og tidlig middelalder. Unpubl. Master's thesis, University of Tromsø (Tromsø 2005).

Spangen 2010: M. Spangen, Silver Hoards in Sámi Areas. In: P: Halinen/M. Lavento/M. Suhonen (eds.), Recent perspectives on Sámi archaeology in Fennoscandia and North-West Russia. Proceedings of the First International Conference on Sámi archaeology, Rovaniemi, 19–22 October 2006. Iskos 17, 2010, 94–106.

Spangen/Äikäs 2020: M. Spangen/T. Äikäs, Sacred Nature. Diverging Use and Understanding of Old Sámi Offering Sites in Alta, Northern Norway. Religions 11(7), 2020, 317. doi.org/10.3390/rel11070317.

Stensli 1993: O. M. Stensli, Forvaltning av store rovdyr. Ottar 196, 1993, 40–50.

Swenson et al. 1995: J. E. Swenson/P. Wabakken/F. Sandegren/ A. Bjärvall/R. Franzén/A. Söderberg, The near extinction and recovery of brown bears in Scandinavia in relation to the bear management policies of Norway and Sweden. Wildlife Biology 1(1), 1995, 11–25.

Tegengren 1952: H. Tegengren, En utdöd Lappkultur i Kemi Lappmark: Studier i Nordfinlands kolonisationshistoria (Åbo 1952).

Tuan 1977: Y.-F. Tuan, Space and Place. The Perspective of Experience (Minneapolis 1977).

Ukkonen/Mannermaa 2017: P. Ukkonen/K. Mannermaa, Jääkauden jälkeläiset. Suomen lintujen ja nisäkkäiden varhainen historia (Helsinki 2017).

Zachrisson 1984: I. Zachrisson, De samiska metalldepåerna år 1000-1350 i ljuset av fyndet från Mörtträsket, Lappland. The Saami metal deposits A.D. 1000–1350 in the light of the find from Mörtträsket, Lappland. Archaeology and environment 3 (Umeå 1984).

Zachrisson 1985: I. Zachrisson, New archaeological finds from the territory of the Southern Saamis. In: Å. Hultkrantz/L. Bäckman (eds.), Saami Pre-Christian Religion. Studies on the oldest traces of religion among the Saamis (Stockholm 1985) 83–99.

Zachrisson/Iregren 1974: I. Zachrisson/E. Iregren, Lappish bear graves in northern Sweden – an archaeological and osteological study (Stockholm 1974).

Associate Prof. Marte Spangen
UiT – the Arctic University of Norway
Tromsø
Norway
marte.spangen@uit.no

Associate Prof. Anna-Kaisa Salmi
University of Oulu
Oulu
Finland
anna-kaisa.puputti@oulu.fi

Dr. Tiina Äikäs
University of Oulu
Oulu
Finland
tiina.aikas@oulu.fi

Dr. Markus Fjellström
University of Oulu/Stockholm University,
Silvermuseet/INSARC
Oulu/Stockholm
Finland/Sweden
markus.fjellstrom@arklab.su.se

Fig. 1. This rock served as the focal point of the offering site by lake Unna Saiva, Gällivare, Sweden (photo E. Manker, Nordiska Museet archives).

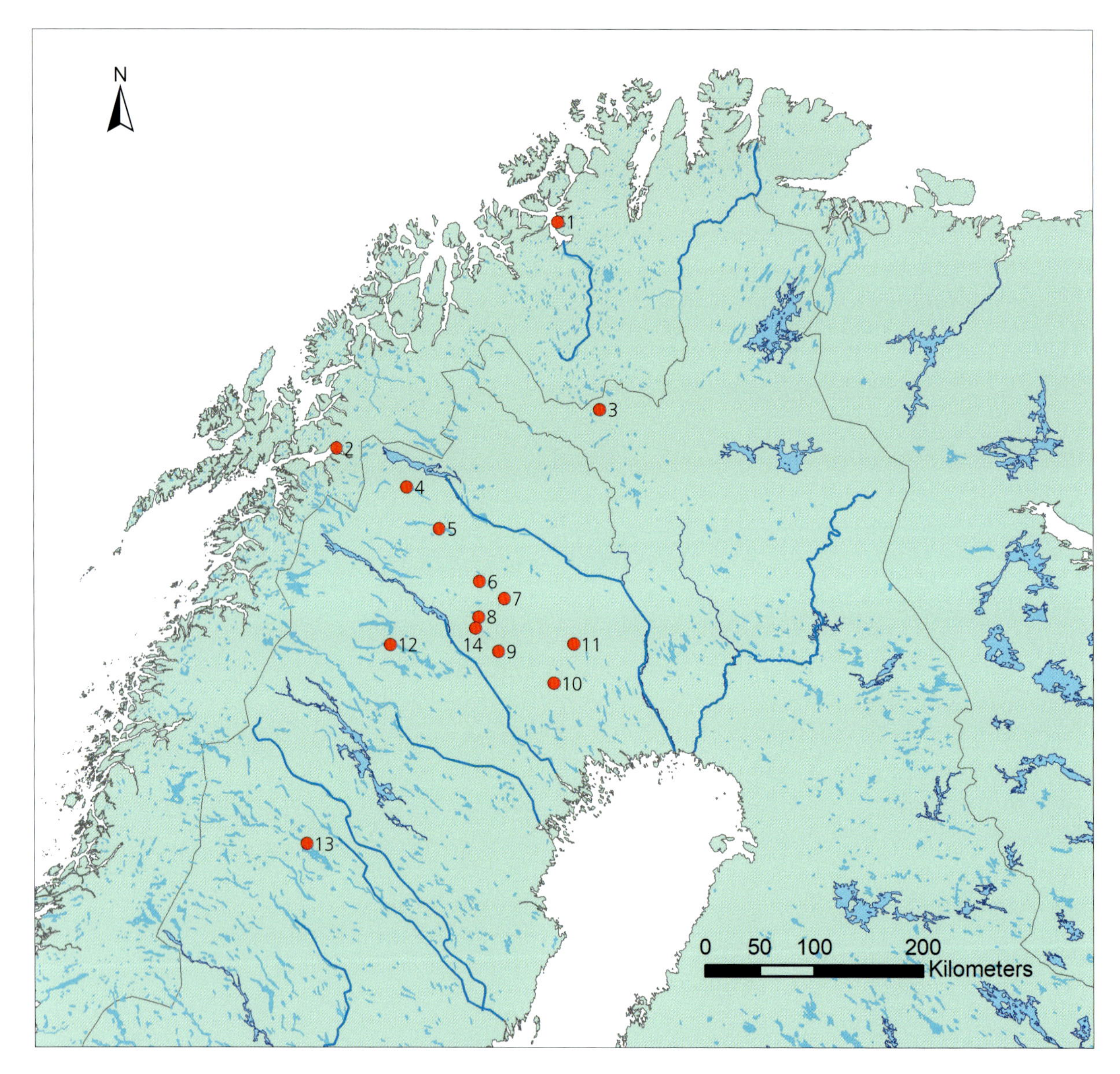

Fig. 2. Map of Saami offering sites with finds of bear bones or information about such bone finds. The numbers in the map correspond with Table 1. Äijihsuálui (Ukonsaari), Inari, Finland, and Aktse, Sirkas, Sweden, are not included, as the information about bear bones/offerings related to these sites is uncertain (graphics M. Spangen).

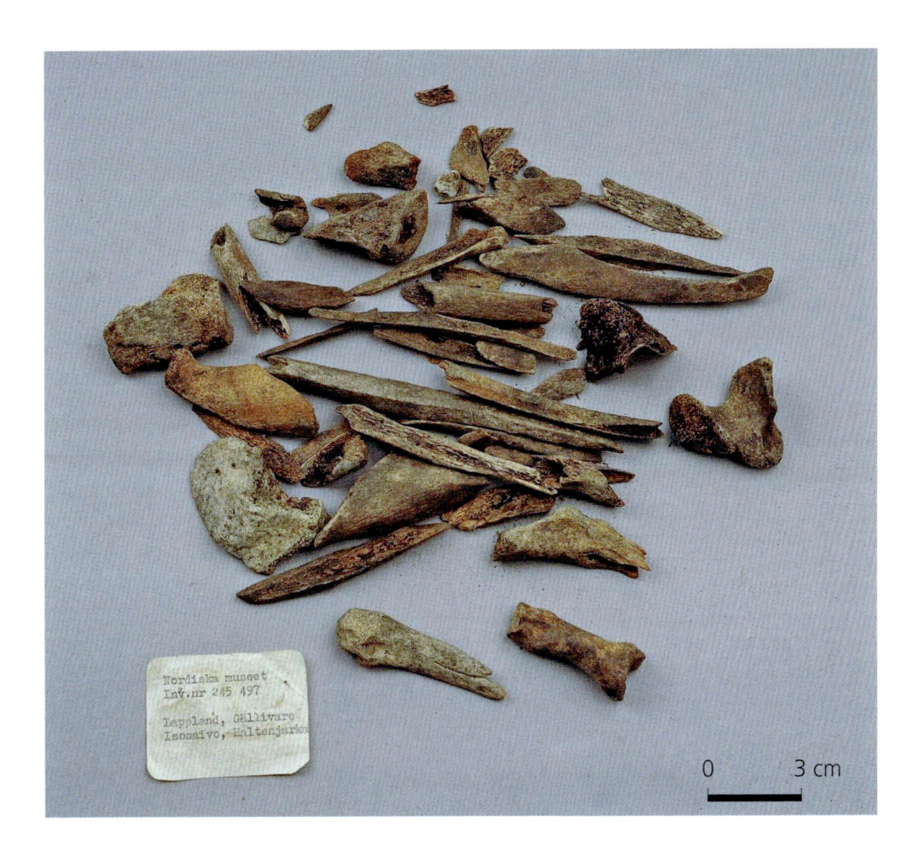

Fig. 3. Reindeer and bear bones from the offering site of Haltenjárka, Gällivare, Sweden (photo J. Karlsson, Historiska museet/SHM [CC BY]).

Fig. 4. One of the bear molars from the Näkkäla offering site of Enontekiö, Finland (photo A.-K. Salmi).

*Table 1. Saami offering sites with finds of bear bones or information about finds of bear bones. MNI = Minimum number of individuals. *MNI is set to two where the sources only state there were bones from "several" individuals. N/D = no data.*

No.	Site name	Area	Country	Bone elements	MNI	14C-date	Cal. date 2σ	Lab. no.
1	Årøya	Alta	Norway	Cranium, bones	1			
2	Seines	Narvik	Norway	Teeth, cranial, bones	8	245 ± 55	AD 1487–(1950)	T-12026
3	Näkkälä	Enontekiö	Finland	Teeth, cranial fragments	1	830 ± 25	AD 1174–1267	Hela-1885
4	Rautasjaure/ Vidjjavárri	Rautasvuoma	Sweden		N/D			
5	Haltenjarka (Haltiosaari)	Norrkaitum	Sweden	Teeth, bone fragments, vertebra	1			
6	Veiki	Mellanbyn	Sweden	Teeth	1			
7	Saggakuoika (Haltiasuando)	Mellanbyn	Sweden	Cranial bones	2*			
8	Saivo/Saiva	Sörkaitum	Sweden	Teeth, cranial and postcranial bones	2*			
9	Atjekåive	Sörkaitum	Sweden	Not preserved	N/D			
10	Unna Saiva	Sörkaitum	Sweden	Cranial bones	14	1398 ± 56	AD 557–774	Ua-48702
11	Tiirivaara	Sörkaitum	Sweden	Cranium	1			
12	Anakkats	Jokkmokk	Sweden	Not preserved	1			
13	Strömsund-skavan	Umbyn	Sweden	Crania, bones	5			
14	Akkavare	Sörkaitum	Sweden	Crania	2			

Other bone finds	Metal finds	Site ID	Info type	Sources
			Bones finds	Qvigstad 1926, 340; Myrstad 1996
Teeth of cow, sheep/goat, reindeer			Bones finds	Myrstad 1996
Reindeer, antlers, cloven-hoofed animal, mammal, fish	Metal objects, coins	Äikäs 2015, no. 9	Bones finds	Äikäs 2015
Reindeer, antlers, cattle, capercaillie, ptarmigan, lemming	Metal objects	Manker 1957, no. 113	Bones finds	Manker 1957, 134–138; Eriksson 1996; Mulk 2009
Reindeer, goat, cattle, swan, loon, wolverine	Coins	Manker 1957, no. 137	Bones finds	Gejvall 1956; Manker 1957, 146–150
Reindeer, antlers		Manker 1957, no. 160	Bones finds	Manker 1957, 157
Reindeer, birds, antlers		Manker 1957, no 164	Bones finds	Manker 1957, 157–158
Reindeer, antlers, possibly goat, horse tooth	Metal objects	Manker 1957, no. 180	Bones finds	Serning 1956; Manker 1957, 162–165
Horse tooth	Metal objects	Manker 1957, no. 182	Bones finds	Serning 1956; Manker 1957, 165–166; Zachrisson/Iregren 1974, 36
Reindeer, elk, sheep, goat, cattle, beaver, swan, duck, scoter, loon, wood grouse, black grouse, ptarmigan	Metal objects, coins	Manker 1957, no. 186	Bones finds	Serning 1956; Manker 1957, 167–168; Salmi et al. 2015
Reindeer, antlers		Manker 1957, no. 188	Bones finds	Manker 1957, 169; Zachrisson/Iregren 1974, 34
Swan		Manker 1957, no. 246	Bones finds	Manker 1957, 192; Zachrisson/Iregren 1974, 35
		Manker 1957, no. 422	Bones finds	Hallström 1924, 803; Manker 1957, 246
Reindeer bones, ox crania, other animal bones, *sieidi* stone		Manker 1957, no. 181	Written documentation	Manker 1957

The bear skin trade in Russia has never been considered a separate entity of the fur trade. Thus, the scarce materials on it are included in the broader accounts of Russia's fur trade. These accounts are rare and may be classified into the following groups: Islamic literature (travel accounts by Arab merchants), western European travel and diplomatic accounts, Baltic Sea trade documents, and Russian diplomatic documents (Martin 2004). Birchbark documents from Novgorod the Great can also be added to this group. As the materials above cover mostly the early 2nd millennium AD, the only source for the last centuries of the 1st millennium AD is the Islamic literature. This designates northern European Russia and the northwestern corner of Siberia, from which the thickest and softest pelts came, as the "land of darkness". Islamic authors of the last part of the 1st millennium AD had an idea of the fur trade from visiting the market in Bulgar (Fig. 3). Although they do not specifically report bear skins, they mention the fur-exporting region – the lands around three lakes – Ladoga, Onega, and Beloe (Fig. 3) –, populated then by the Finnish tribe of the Ves'. The excavations of Ves' settlements revealed, among other items, claw phalanges of bears. Ibn Rusta, a 10th-century Persian explorer and geographer, testified: "The Rus' bring their goods to them [the Bulgars]. All of them [the Rus'] who live on both sides of the aforementioned river [the Volga] bring their goods, such as furs of sable, ermine, squirrel and other [animals] to them [the Bulgars]" (Khvolson 1869). Although sable, ermine, black fox, and marten were traditionally imported into Bulgar (Martin 2004), some bear pelts may also have reached the Bulgar market.

By the end of the 10th century AD, the Viking tribe of the Rus' had become a prominent exporter of luxury furs (Martin 2004). They took control over both the (older) Black Sea-Don-Volga route (Bulgar and Byzantium markets) and the younger one – the southern Baltic coast route (Baltic market), which started in Hedeby or Oldenburg, and ended up at Novgorod the Great, with the most important marketplace in Birka. We know that bear skins were a commodity on the Baltic route from the fact that Ottar, a Norwegian traveller to the court of King Alfred of Wessex (known as Alfred the Great, king of the Anglo-Saxons towards the end of the 9th century AD), counted his wealth in terms of "the tribute which the Fynnes pay [...], which was all in skinnes of wilde beastes [...]. The richest pay ordinarily [...] one Beare, [...] a coat of a Beares skinnes" (Martin 2004, 41, after Jones 2001).

During the 11th and 12th centuries, Novgorod the Great became the major fur-exporting city in three directions: to Bulgar in the east, Kiev in the south, and the Baltic market in the west (Martin 2004). Although skins of smaller fur animals, such as squirrel, sable, marten, and ermine, were the main items exported, bear skins were also somehow used in an exchange – at least three birchbark documents found at Novgorod the Great mention bear skins. The earliest document, №722, is dated to 1200–1220 AD (Fig. 4). It contains a merchant's property inventory with its value: "Money of 12 *grivnas*[1] in squirrel skins and silver. Sable skin – 4 *grivnas*. Nets, cloths, and canvases – 3 *grivnas*. **Bear skin** – 2 *grivnas*" (Birchbark Document №722). Another document, №65 from 1300–1320, contains business orders from Matphei to Esif Davidov: "Compliments from Matphei [Matthew] to Esif Davidov. Bring me **two bear skins** [...]" (Birchbark Document №65). Finally, birchbark document №354 from 1340–1360, which lists household orders from Ontsifor to his mother, gives the order to "send two tagans, karakul, labels, spindles, sacks and **a bear skin**" (Birchbark Document №354).

1 The *grivna* was a currency and a measure of weight used in the Kievan Rus' and other East Slavic countries since the 11th century. 1 *grivna* of silver (204 g) = 4 *grivnas* of *kuna* (51 g) = 80 *nogatas* (coins) = 100 *kunas* (marten skins) = 200 *rezans* (coins) = 400–600 *vekshas* (squirrel skins).

One more mention of bear skins can be found in the report of Plano Carpini,[2] who travelled through southwestern Russia in 1246. He reported that a Saracen (a term for a Muslim merchant acting as a tax collector in the region), sent by Güyük Khan[3] of the Golden Horde, asked the local people to pay the tribute, among other items, of **one white bear skin** (Rockhill 1998).

Zooarchaeological records

Bear bones are very rare in Russia's zooarchaeological published records,[4] not only for the time period described here, but also before and after that. This conclusion is based on a number of sources (Lapshin 2009; Zinoviev 2009; 2014; 2021 in print; Maltby 2013; Hamilton-Dyer et al. 2017; Brisbane et al. 2020; Maltby et al. 2020). Many of the known sites yielded single third phalanges of bears (Table 1: excavation sites at Novgorod the Great of Gorodishche [1 item], Troitsky X–XI [9], and Desyatinny-I [1]; see Figs. 5–6). Phalanges may have been attached to imported bear skins, although these bones were sometimes also used as pendants (Maltby 2013). Two phalanges from the Troitsky XI excavation site at Novgorod the Great (Table 1), for instance, show disarticulation marks near their proximal end, indicating that they had been separated from the rest of the foot and the skin. Therefore, it may only have been the third phalanges themselves that were brought to the properties rather than the skins (Maltby et al. 2020). Also, while in Gorodishche and Georgii (Fig. 5) bear bones are associated with count's seats, Novgorod's sites that yielded bear bones are well outside of Detinets – the place of power at Novgorod the Great. So, bear bones are associated here with ordinary houses. There are no indications of bears and/or bear skins used as grave goods, as was the practice in some parts of central Europe and Scandinavia from the last centuries BC until Christianisation. Evidence of actual skins is rare, whereas claws are much more numerous (Wamers 2009; Grimm 2013; Beermann 2016; Schmölcke et al. 2017; see also different contributions in this volume). Although archaeological wet layers of medieval Russian cities, including those at Novgorod the Great, favour the preservation of organic matter, no preserved parts of bear skins are known or have been reported so far.

Discussion

The fur trade was a vital economic factor in Russia's principalities (then known as Rus') and the mid-Volga lands (Martin 2004). Based primarily on smaller species, such as squirrel, marten, and sable, the fur trade in the modern European part of Russia included pelts of other fur animals, such as brown bear, too. The scarce written sources do not provide references to the volume of the bear skin trade; skins of brown bear were exported along the southern Baltic coast route around the turn of the 2nd millennium AD. Bear skins were also used to pay tribute to the authorities of the Golden Horde. However, most of the bear skins were not exported (domestic trade). The predominance of the third phalanges (claw phalanges) in archaeological finds from medieval Novgorod the Great and its vicinities suggests that skins were used by the inhabitants of particular properties, often the places of power (see Tianina, this volume). Bear skins, therefore, might have been brought there as trophies –

2 Plano Carpini, or Giovanni da Pian del Carpine (*c.* 1185 – August 1, 1252), was a medieval Italian diplomat, archbishop and explorer, and one of the first Europeans to enter the court of the Great Khan of the Mongol Empire (Montalbano 2015).

3 Güyük Khan (*c.* March 19, 1206 – April 20, 1248) was the third Great Khan of the Mongol Empire.

4 Some bear bones are, without doubt, hidden in archives, still waiting for zooarchaeological analysis.

evidence of the bravery of the owner of the property (Oehrl 2013). The bear-hunting tradition with just a spear among the Russian nobility persisted up to the Great October Revolution in 1917 (Shirinsky-Shikhmatov 1900; see Fig. 7).

There is no doubt that at least some of the details of the domestic bear skin trade in Russia at the turn of the 2nd millennium AD can be reconstructed in the future by a careful analysis of bones and bear-related furnishings in early settlements' layers and graves (Smith 2012; see Syrovatko et al., this volume).

Acknowledgments

I am grateful to Dr. Oleg Oleinikov (Institute of Archaeology of the Russian Academy of Sciences, Moscow, Russia), who kindly provided osteological material from archaeological excavations at Novgorod the Great for study. Special thanks to artist Vadim Gorbatov (Moscow, Russia), who allowed the use of his splendid painting. Dr. Mark Maltby (Bournemouth University, UK) provided valuable information and comments.

Bibliography

Beermann 2016: S. Beermann, Bärenkrallen und Bärenfelle in Brand- und Körpergräbern der vorrömischen Eisenzeit bis Völkerwanderungszeit in Mittel- und Nordeuropa. Universitätsforschungen zur Prähistorischen Archäologie 279 (Bonn 2016).

Birchbark document №65 1300–1320: Novgorod. From Matphei to Esif Davidov (business orders). In: Old Russian Birchbark Documents: Handwritten Monuments of the Ancient Rus'. http://gramoty.ru/birchbark/document/search/?number=65 [accessed: 31.12.2020] (in Russian).

Birchbark document №354 1340–1360: Novgorod. Letter from the posadnik [mayor] Ontsifor Lukinich – From Ontsifor to the mother (household orders). In: Old Russian Birchbark Documents: Handwritten Monuments of the Ancient Rus'. http://gramoty.ru/birchbark/document/search/?number=354 [accessed: 31.12.2020] (in Russian).

Birchbark document №722 1200–1220: Novgorod. Inventory of the merchant's property. In: Old Russian Birchbark Documents: Handwritten Monuments of the Ancient Rus'. http://gramoty.ru/birchbark/document/search/?number=722 [accessed: 31.12.2020] (in Russian).

Brisbane et al. 2020: M. Brisbane/E. Hambleton/S. Hamilton-Dyer/M. Maltby, Using the natural environment: wild animals in medieval Novgorod and Novgorod land. Archaeological News 28, 2020, 203–222 (in Russian, English abstract).

Grimm 2013: O. Grimm, Bear-skins in northern European burials and some remarks on other bear-related furnishings in the north and middle Europe in the 1st millennium AD. In: O. Grimm/U. Schmölcke (eds.), Hunting in northern Europe until 1500 AD. Old traditions and regional developments, continental sources and continental influences. Papers presented at a workshop organised by the Centre of Baltic and Scandinavian Archaeology, June 16th and 17th, 2011. Schriften des Archäologischen Landesmuseums, Ergänzungsreihe 7 (Schleswig 2013) 277–296.

Hamilton-Dyer et al. 2017: S. Hamilton-Dyer/M. Brisbane/M. Maltby, Fish, feather, fur and forest: Exploitation of wild animals in medieval Novgorod and its territory. Quaternary International 460, 2017, 97–107.

Herberstein 1550: S. Herberstein, Comentari della Moscovia (Venetia 1550).

Jones 2001: G. Jones, A history of the Vikings (Oxford 2001).

Khvolson 1869: D. A. Khvolson, News of the Khozars, Burtases, Bulgarians, Magyars, Slavs and Russians of Abu-Ali Ahmed bin Omar Ibn-Dast, a hitherto unknown Arab writer of the early 10th century from a manuscript of the British Museum (St. Petersburg 1869) (in Russian).

Kosarev 2009: M. F. Kosarev, On the exchange trade in ancient and traditional societies (based on the Ural-Siberian archaeological and ethnographic materials). Ural Historical Bulletin 23(2), 2009, 60–66 (in Russian).

Kutepov 1896: N. I. Kutepov, The Hunt of the Grand Dukes and Tsars in Russia from the 10th to the 16th Century (St. Petersburg 1896) (in Russian).

Lapshin 2009: V. A. Lapshin, Tver in the XIII–XV centuries (based on materials from excavations in 1993–1997) (St. Petersburg 2009) (in Russian).

Maltby 2013: M. Maltby, The exploitation of animals in towns in the medieval Baltic trading network: a case study from Novgorod. In: N. Makarov/A. Mastykova/A. Khokhlov (eds.), Archaeology of the Baltic Region (Moscow, Saint Petersburg 2013) 229–244.

MALTBY et al. 2020: M. MALTBY/M. BRISBANE/E. HAMBLETON/S. HAMILTON-DYER/O. A. KRYLOVICH/M. V. SABLIN/A. B. SAVINETSKY/A. V. ZINOVIEV, The exploitation of wild mammals in Novgorod and sites in its immediate hinterland. In: M. Maltby/M. Brisbane (eds.), Animals and archaeology in Northern Medieval Russia: Zooarchaeological studies in Novgorod and its region (Oxford 2020) 223–237.

MARTIN 2004: J. MARTIN, Treasure of the land of darkness. The fur trade and its significance for medieval Russia (Cambridge 2004).

MONTALBANO 2015: K. A. MONTALBANO, Misunderstanding the Mongols: Intercultural communication in three thirteenth-century Franciscan travel accounts. Information & Culture 50(4), 2015, 588–610.

OEHRL 2013: S. OEHRL, Bear hunting and its geological context (as a background for the interpretation of bear claws and other remains of bears in Germanic graves of the 1st millennium AD). In: O. Grimm/U. Schmölcke (eds.), Hunting in northern Europe until 1500 AD. Old traditions and regional developments, continental sources and continental influences. Papers presented at a workshop organised by the Centre of Baltic and Scandinavian Archaeology, June 16th and 17th, 2011. Schriften des Archäologischen Landesmuseums, Ergänzungsreihe 7 (Schleswig 2013) 297–332.

ROCKHILL 1998: W. W. ROCKHILL (ed.), The Journey of William of Rubruck to the eastern parts of the world, 1253–55: As narrated by himself with two accounts of the earlier journey of John of Pian de Carpine (New Delhi/Madras 1998).

SCHMÖLCKE et al. 2017: U. SCHMÖLCKE/D. GROSS/E. A. NIKULINA, Bears and beavers. 'The Browns' in daily life and spiritual world. In: B. V. Eriksen/A. Abegg-Wigg/R. Bleile/U. Ickerodt (eds.), Interaction without borders. Exemplary archaeological research at the beginning of the 21st century (Schleswig 2017) 901–916.

SHIRINSKY-SHIKHMATOV 1900: A. A. SHIRINSKY-SHIKHMATOV, On the bear trail: Essays (Moscow 1900) (in Russian).

SMITH 2012: R. E. F. SMITH, The bear cult among different ethnic groups of Russia (sacred Russian bear). In: G. Pungetti/G. Oviedo/D. Hooke (eds.), Sacred species and sites: Advances in biocultural conservation (Cambridge 2012) 278–290.

WAMERS 2009: E. WAMERS, Von Bären und Männern: Berserker, Bärenkämpfer und Bärenführer im frühen Mittelalter. Zeitschrift für Archäologie des Mittelalters 37, 2009, 1–46.

ZINOVIEV 2009: A. V. ZINOVIEV, Review of the archaeozoological material obtained from 'Desyatinny-1' excavation site of Novgorod the Great in 2008. In: V. L. Yanin (ed.), Novgorod and Novgorod Region. History and Archaeology 23 (Veliky Novgorod 2009) 189–207 (in Russian).

ZINOVIEV 2014: A. V. ZINOVIEV, Game mammals of the medieval Novgorod the Great (according to the archaeozoological data). Vestnik of Tver State University, Biology and Ecology Series 4, 2014, 86–94 (in Russian, English abstract).

ZINOVIEV 2021 in print: A. V. ZINOVIEV, Zooarchaeology of Tver Kremlin. International Journal of Osteoarchaeology 30, 2021.

Dr. Andrei V. Zinoviev
Head of the Zoology and Physiology Department
Tver State University
Tver
Russia
zinovev.av@tversu.ru

Prof. of the Higher School of Economics
Moscow
Russia
azinovev@hse.ru

Fig. 1. Buffoons in a Russian village. Painting by François Riss, 1857 (courtesy of the Museum of V. A. Tropinin and Contemporary Moscow Artists).

Fig. 2. Execution by bear under Tsar Ivan Vasilievich the Terrible. Watercolour by Nikolai Samokish (after Kutepov *1896).*

Fig. 3. Map of Russia and adjacent areas discussed in the article.

Fig. 4. Birchbark document №722 from Novgorod the Great, 1200–1220 AD (courtesy of the Novgorod State United Museum).

The bear cult in medieval Novgorod, based on archaeological finds

By Elena A. Tianina

Keywords: Russia, medieval Novgorod, archaeology, paganism, zoolatry, bear cult, amulets, chronology, topography

Abstract: The bear cult belongs to the oldest layers of religious beliefs in bear-human relations. It reflects the mutual influence of Finno-Ugric and Slavic cultural traditions. These religious beliefs were also present in medieval Novgorod. The article examines Novgorod's collection of amulets made of bear fangs, other teeth, and claws along with their chronology, topography and semantic significance. As the study shows, bear worship emerged in Novgorod at the turn from the 10th to the 11th century, based on interactions with the multi-ethnic rural environment, where it has been documented since ancient times. Amulets made of bear teeth played a role as apotropaic objects, connected with the cult of Veles and, probably, also with "bear feast" rituals known from the ethnography of the Finno-Ugrics and other peoples of northern Eurasia.

Zoolatrous cults belong to the oldest forms of religious beliefs. The worship of animals is the basis of totemism, a primitive religion, the essence of which is the worship of an animal, the ancestor of the tribal collective. However, zoolatry far outlived both the primitive social system and totemism as one of the variants of its religious worldview. In the system of pagan beliefs of various peoples, animals became part of the world mythology and acted as images associated with various deities, spirits, mythological characters, or epic heroes. At the same time, both in the perception of the revered animals in faunal composition as well as in the mythological plots associated with them, the habitat plays a huge role, as man's perception of the animal in mythology is directly related to the perception of it in real nature (Gura 1997, 18).

The bear cult represents one of the oldest, most widespread and most enduring forms of zoolatry. Bear worship arose and developed from the Stone Age onwards throughout practically the whole habitat of this animal. At the same time, this cult reveals a number of common features among various peoples who lived in different periods in the forest zone of northern Eurasia. Among them are the empowerment of the bear by attributing anthropomorphic features and other mediative functions to it (mediation between the worlds), connection with the cult of ancestors, representations of the bear as the forest master, the patron of all other animals, the mating symbolism of the bear, its connection with "werewolfism" (shapeshifting), shamanism, and witchcraft (Gura 1997, 160–167; Kosarev 2003, 40–59; Koškarova 2013, 7–8). In the culture of many peoples, including the eastern Slavs, there is a taboo attached to the bear's name (cf. Udolph, this volume; cf. Panagl 1984, 148–149; Kosarev 2003, 40–42; Slavic antiquities 2004, 214).

When speaking about the continuity of the tradition, we can trace the transformation of the bear cult in cultures at different stages of socio-economic development. Thus, researchers identified the

Bear and Human: Facets of a Multi-Layered Relationship from Past to Recent Times, with Emphasis on Northern Europe, ed. by Oliver Grimm (Turnhout: Brepols, 2023), pp. 629–638 BREPOLS PUBLISHERS 10.1484.M.TANE-EB.5.134353

most ancient, "classical" variant of the bear cult, characteristic for the peoples of the Far North, which is best presented and studied in the case of the Ob'-Ugrians (Kulemzin 1984, 85; Koškarova 2013, 8). The Finno-Ugric peoples of the Ural and Volga region followed a "transitional type" of bear cult, in which some of the most ancient totemistic representations had died out/disappeared, while such mythologemes as "bear – descendant of a deity", "bear – ancestor" (in which classical totemism is replaced by the cult of human ancestors), "bear – man" (i.e. shapeshifter), "bear – forest master" and "bear – epic/cultural hero" came to the fore (Koškarova 2013, 9–14). Similar features of bear worship among the Baltic-Finnish peoples are recorded in the Kalevala (Kalevala, 437–451). In the context of Slavic agrarian culture, the bear cult did not dissolve and disappear as a relic of the primitive epoch, but instead developed and acquired new features. For example, the totemistic component in the cult of the bear as a primary ancestor transformed into a general idea about the kinship of man and bear (Gura 1997, 161–163), while the bear as the "master of the forest" and the patron of all other animals then became the patron and protector of livestock (Slavic Antiquities 2004, 212–214).

Novgorod (Fig. 1) was undoubtedly part of the geographical area of bear worship. A group of apotropaic objects associated with bear worship stands out among the finds of pagan antiquities in the cultural layers of Novgorod. These are amulets made of bear fangs (canine teeth) and other teeth. This type of the Novgorod pagan amulets is the most widespread in time and space. "Bear amulets" are part of a large group of apotropaic objects, made of the fangs, other teeth, or bones of various animals (Tianina 2011, 162–164 fig. 2.1–4). The Novgorod collection of these items is unique not only in its number – 239 amulets were found so far – but also in terms of the dynamics in the distribution of amulets associated with different animals, which allows us to identify the chronology as well as the topographic pattern of their occurrence. All this is possible due to the well-developed stratigraphy of the cultural layers of Novgorod and the large-scale excavations of entire farmsteads or even farmstead complexes instead of only parts of the settlement.

So far, 89 amulets associated with the bear cult have been found in Novgorod's cultural layers. Among them, the vast majority (82 findings) is represented by objects made of bear fangs (Fig. 2). In addition to these, three amulets were made of bear molars (Fig. 3.1), three of claws (Fig. 3.3–4), and one of a bear incisor (Fig. 3.2). One more item is described by the excavators as a bear's tooth, but with no further details provided. It is obvious that such rare types as amulets made of incisors and molars were not traditional for the Novgorod population. Possibly, such objects were included in a necklace, with a fang for the enhancement of protective or benevolent effects. All amulets made out of fangs or other teeth show a hole for suspension that has been drilled in the middle section or at the base; they do not differ in appearance from similar amulets known in other cultures. Three amulet-fangs are, in fact, blanks, as their holes are not drilled all the way through, while, in case of a further amulet, the hole is replaced by small grooves in the fang. It should be noted that, in reality, there should have been many more amulets made of bear fangs and other teeth, as they could have been carried not only by hanging, but also by being placed in a leather pouch or sewn onto clothes. It is extremely difficult to identify such amulets with archaeological methods because they do not have a hole and thus do not differ in any way from common stray finds of bear teeth or claws, which belonged to hunting prey at the farmstead. The same can be said concerning the possible use of "bear amulets" in Novgorod, not only as personal apotropaic objects, but also for other magical purposes. For example, according to ethnographic data, parts of bears (claws, teeth, bones, skin) were used as amulets in stables for cattle or at children's cradles (Slavic antiquities 2004, 214). In such cases, the amulet was placed next to the protected object. It is possible that the above-mentioned amulet without a hole but with grooves for a string or rope was used in such a manner. An object fastened that way could have certainly slipped and got lost if worn by a person.

In 2011–2012, 59 Novgorod "bear amulets" were analysed osteologically by N. D. Burova, a researcher from the Laboratory of Archaeological Technologies of the Institute for the History of Material Culture at the Russian Academy of Sciences. Results show that both upper and lower bear teeth were used for amulet-making, while these originated from both old (Figs. 2.9; 3.2) and very young individuals. In all cases in which the sex of the animal could be determined, the fangs were from male bears.

Amulets made out of bear teeth were found in all five ancient town districts of medieval Novgorod. The finds came to light at 14 excavation sites (Fig. 4). The archaeological collections from the two main Novgorod excavation sites stand out: Nerevskij (27 specimens; see Figs. 2.1–4; 4.1) and Troickij (33 specimens; see Figs. 2.8–10; 4.2). At the Troickij excavation site there were found one amulet made of an incisor, three made of claws, and one more made of a bear molar with drilled holes for wearing (Fig. 3). Two further amulets made of bear molars were found at the Nerevskij and Michajlovskij excavation sites (Fig. 4.14). Another bear tooth amulet comes from the Rogatickij III excavation site in the carpenters' district of the town (Fig. 4.7). In addition to the two main excavation sites, one specimen of a bear fang amulet was found each at the Dmitrovskij (Figs. 2.6; 4.3) and Tichvinskij (Fig. 4.4) excavation sites in the Nerevskij town district, as well as at the Desjatinnyj excavation site (Fig. 4.5) in the Ljudinnyj town district. In the Slavenskij town district, such finds came to light at the Il'inskij (five specimens; Figs. 2.5; 4.9), the Slavenskij (two specimens; Fig. 4.8), the Posol'skij 1999 (one specimen; Fig. 4.13), and the Nutnyj (one specimen; Figs. 2.7; 4.10) excavation sites, and, in the Plotnickij town district, at the Ljubjanickij (two specimens; Fig. 4.11), the Fëdorovskij (four specimens; Fig. 4.12) and the Rogatickij III (three specimens; Fig. 4.7) excavation sites. One more find comes from the Michajlo-Archangelskij excavation site in the Zagorodskij town district (Fig. 4.6).

A total of 80 finds of "bear amulets" were suitable for dating. For nine further specimens dating is impossible because they either come from the upper redeposited layer or from the drainage trenches, or they were not documented properly. The dating of the amulets made out of bear teeth (Fig. 5) drastically differs from the general chronology for amulets made out of animal teeth and bones in Novgorod, the majority of which is dated to the 10th and 11th centuries (Tianina 2011, 160–161 fig. 1.1,3). On the contrary, "bear amulets" are practically absent in the 10th-century layers, i.e. in the pagan period of the city. Only two bear amulets can be dated to this period: one of these, originating from the Troickij excavation site, is made of a molar; the second one, found at the Nerevskij excavation site, is made of a fang. Moreover, based on stratigraphy, both early specimens can be dated rather to the end of the period in question (turn of the 10th to the 11th century). Below this, in the lowest 10th-century strata of the excavation sites, no "bear amulets" were found.

Around the turn of the 10th to the 11th century, amulets made out of bear fangs became very popular. Their occurrence at different excavation sites corresponds with the general population density and the number of finds from different parts of the town. Such objects that date to the beginning and the middle of the 11th century were found exclusively in the Ljudin and the Nerevskij town districts, where the most ancient centres of medieval Novgorod were located. Bear teeth dating to the second half of the 11th century have already been documented in the other town districts. This widespread distribution pattern of these amulets in the city layers has been observed as occurring up until the middle of the 14th century, while, during later times, finds of "bear amulets" became sporadic. It should also be noted that the teeth in the layers that are younger than the first half of the 14th century (including undated finds from the upper layer) come exclusively from the Nerevskij excavation site, where these layers contain rich archaeological material (in contrast to the Troickij excavation site, where medieval layers younger than the 14th century have essentially not been preserved).

The topography of finds can be traced by studying the collections of the Nerevskij and Troickij excavation sites, where entire homestead complexes were uncovered. As the study has shown, the find distribution is generally uniform across the homesteads. However, individual features can still be noted. At the Nerevskij excavation site, there are homesteads in which single bear fang amulets are found consistently in layers dating from the 11th to the 14th centuries, which may indicate a stable tradition of wearing them. At the Troickij excavation site, only single amulets come from the homestead complexes. An exception are three amulets from the first half of the 12th century, which originate from homestead U. The fourth amulet, found in the strata of the late 11th century, belongs to the same complex. Stratigraphic analysis of this part of the Troickij excavation site demonstrates that these layers belong to the same building period, during which the owners of this homestead had a high social status and were involved in trade. It is also the period of the farmsteads to which finds of coins, weights, and lead seals of goods belong. Of particular interest is the birch bark manuscript no. 952, which mentions trade goods sent to Smolensk as well as the huge amount of 400 *grivnas*. Taking into account the traditional connection of the bear cult with the Finno-Ugric peoples, objects of Baltic-Finnish origin in the contemporary archaeological complex (bronze equal-shouldered fibulae, chain carriers, etc.) should be mentioned (Stepanov et al. 2015, 271–273).

In general, the chronology and topography of "bear amulets" in Novgorod demonstrate that the bear cult, at least in the form that has been discussed, was not characteristic for the first settlers in the city. Its emergence and rather rapid spread in the urban environment from the beginning of the 11th century onwards was caused by the intensification of the socio-cultural relationship between Novgorod and its surroundings. The traditional character of the bear cult in the districts around Novgorod since the most ancient times is evidenced not only by finds in burials (Voronin 1941, 161–162), but also by most ancient toponyms connected with bears in the area of Novgorod Poozer'e. Another piece of evidence is the co-called "She-Bear Stone" (Novgorod Poozer'e, near the village of Verchovje, 20 km from Novgorod), a cult object, named after its shape, which resembles a sleeping bear (Šorin 1998, 223–224). The veneration of this cult stone persisted until modern times.

In the uniform, multicomponent culture of Novgorod and its surroundings in the period from the 10th century to the first half of the 13th century, the tradition of wearing "bear amulets" was established in the urban culture and was reflected in the worldview of the Novgorodians. The adoption of Christianity did not stop bear-worship in Novgorod. On the contrary, the tradition of wearing "bear amulets" coincided in time with the beginning of Christianisation or a little later. This observation vividly illustrates the complex processes of the spiritual interaction and intertwining of different cultural traditions in the medieval town. Religious syncretism was not limited to the incorporation of Christian beliefs into the existing pagan environment, but had a much more multifaceted and multidirectional character, which reflected the peculiarities of the region.

The question of the semantic meaning of "bear amulets" in Novgorod is a complicated and complex one. In the present case, it is difficult to assume the preservation of the bear cult in its "classical" form with a totemic basis. The semantics of these magical objects in medieval Novgorod certainly reflect a deeply "transformed" model, far removed from representations of the bear-totem and associated with more developed forms of zoolatry.

E. E. Levkievskaja, while developing a typology of the Slavic amulets, attributed the semantics of animal teeth and claws to the category of amulets with warding-off magic, which was believed to be based on inflicting a magical blow to the enemy. Such properties were attributed, first of all, to the fangs of predators, which served as basic tools for attacking and destroying an opponent (Levkievskaja 2002, 73–79). This meaning of an amulet made of a bear's fang (as well as of that of any other predator) is certainly universal for all forms of zoolatric cults. The use of bear fangs and claws as the strongest warding-off amulets against evil spirits is documented for many peoples in bear cult areas (Kosarev 2003, 41). This also corresponds with the worship of the bear in so-called "folk orthodoxy". According to north Russian and Belarussian beliefs, the bear was able to ward-off evil spirits, acting as an animal that could detect and disarm and/or remove the so-called spoilage and evil eye. The same properties were allocated to parts of the bear's body, first and foremost to its teeth and claws or paws (Slavic Antiquities 2004, 211–213). The worshipping of the bear as a guardian of cattle in herding magic also has a protective character (Voronin 1941, 169; Gura 1997, 165). The Novgorod material is thus not an exception. The vast popularity of fangs among the "bear amulets" of Novgorod stresses the warding-off function of the amulet as its basic characteristic, reflecting the desire of its owner not just to acquire the strength and agility of the beast, but also to receive its magical protection against evil powers. However, the semantic meaning of the amulet made out of a bear fang was definitely not restricted to general protective representations. The study of the chronology and topography of the Novgorod collection of amulets made out of animal teeth and bones shows that the bear cult was much more widespread and characteristic for the Novgorodians than any other zoolatrous cult with warding-off semantics. For example, amulets made out of bear fangs are three times more common than those made out of wolf fangs, and their popularity is not limited to a narrow early period, as is the case with the amulets made out of wild boar tusks.

The discussion above shows that the bear cult, introduced into the urban environment in the late 10th/early 11th centuries, superimposed its own mythological picture of the world onto the religious views of Novgorodians and organically blended into them. In this regard, we should keep in mind that, in studies of eastern Slavic paganism, the connection of the bear cult with one of the main deities of the Slavic pantheon, the god Veles, has been repeatedly mentioned. The image of this deity is multifaceted and, in many respects, overlaps with the elements of the worldview that characterise the bear cults of other peoples of northeastern Europe. In Slavic tradition, Veles was considered to be a patron of animals, including cattle, had mediatory functions, and was associated with the cult of ancestors as well as with the priestly class of the Old Rus' – the *volhvs* (Rybakov 2002, 409–410; Zasedateleva 2003, 158–160). Exactly these features were assigned to the bear in folk tradition. Also remarkable is the connection of Veles with the calendar holidays, during which dressing up or "walking" like a bear took place (Koljada, Maslenitsa), as well as the manifestation of the bear cult in the veneration of Christian saints such as St Blaise and St Nicolas, who replaced the pagan Veles in the Christian epoch (Voronin 1941, 151–155; Rybakov 2002, 407–410).

The form of bear veneration expressed in the wearing of a bear (tooth) amulet correlates with one more aspect. Amulets made out of bear fangs, claws and other body parts could not have been just accidentally found or taken from hunted animals. Their manufacture and wearing is directly related to one of the most important rituals associated with the veneration of the bear – the so-called "bear feasts". These rituals, complex in execution and semantics, were connected with bear sacrifice or bear hunting and are well known among the different peoples of northern Eurasia, including the Finno-Ugrics, who had great influence on the culture of medieval Novgorod (cf. Piludu, this volume, for Finland; cf. Voronin 1941, 175–177; Vasil'ev 1948, 78–104; Kosarev 2003, 42–48). "Bear feasts" included either the sacrifice of a bear bred in captivity (periodic feasts), or rituals performed over a bear that had been killed in a hunt (sporadic feasts). In some cases, the fangs, other teeth, and claws of the bear killed during these "bear feasts" were preserved and used by the participants as apotropaic objects (Kosarev 2003, 42–44). Therefore, it cannot be excluded that Novgorodians were participants in such rituals and subsequently kept the worshipped objects near or on themselves, or in their homesteads. This assumption is well established concerning, on the one hand, the stability of the tradition of wearing the bear' amulets and, on the other hand, on single finds in homesteads.

Bear claw amulets need to be addressed as such. The rarity of such finds in Novgorod can be explained by the significance of this type of amulet, which reflects the mythological beliefs of the population in the north of eastern Europe. At the end of 19th century, researchers mentioned bear bones in Slavic burial mounds (*kurgans*) of northwestern Rus', dating to the period from the 8th to the 10th centuries, as well as in the *kurgans* of the southern Ladoga region (Voronin 1941, 162). That being said, terminal phalanges are the only bear bones found in this context (Peredolskij 1898, 175). Thus, we can say that bear paws with claws were added to the pyre of a human as a funerary offering and had a special mythological meaning. In this context, bear claw amulets form a semantic unity with another category of objects – clay "bear claws", which are found both in Finno-Ugric and Slavic burials of the 9th to 11th centuries in northeastern Rus' (cf. also Gustavsson/Ljungkvist, this

Fig. 1. The Nowgorod area.

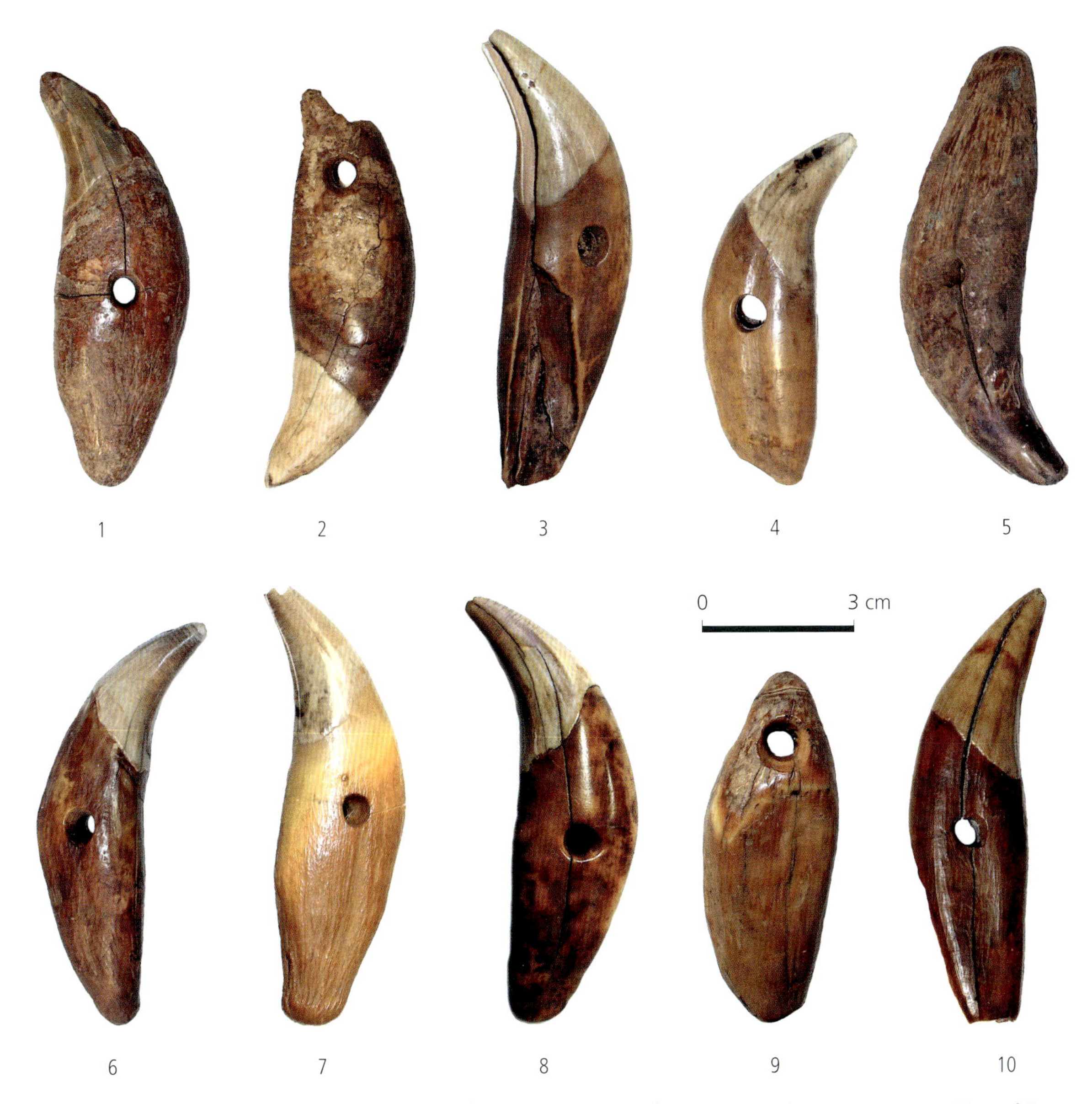

Fig. 2. Amulets made of bear fangs (excavation site, [layer]-[square]-[number]). 1: Nerevskij, 3-489-1880; 2: Nerevskij, 22-1435-28; 3: Nerevskij, 20-1695-6; 4: Nerevskij, 27-835-1262; 5: Il'inskij, 20-111-38; 6: Dmitrievskij, 15-70-12; 7: Nutnyj, 30-51-11; 8: Troickij, 18-748-59; 9: Troickij, western trench-24; 10: Troickij, 15-417-23 (photos E. A. Tianina).

Fig. 3. Amulets made of bear teeth and claws. Troickij excavation site ([layer]-[square]-[number]). 1: 16-1038-20; 2: 14-879-28; 3: 10-b/sq.-40; 4: 2-833-155 (photos E. A. Tianina).

Fig. 4. Topography and excavation sites of the amulets made of bear fangs and other teeth. 1: Nerevskij; 2: Troickij; 3: Dmitrievskij; 4: Tichvinskij; 5: Desjatinnyj; 6: Michajlo-Archangel'skij; 7: Rogatickij III; 8: Slavenskij; 9: Il'inskij; 10: Nutnyj; 11: Ljubjanickij; 12: Fëdorovskij; 13: Posol'skij (1999); 14: Michajlovskij (graphics E. A. Tianina, based on excavation plan from the archives of the Novgorod archaeological expedition).

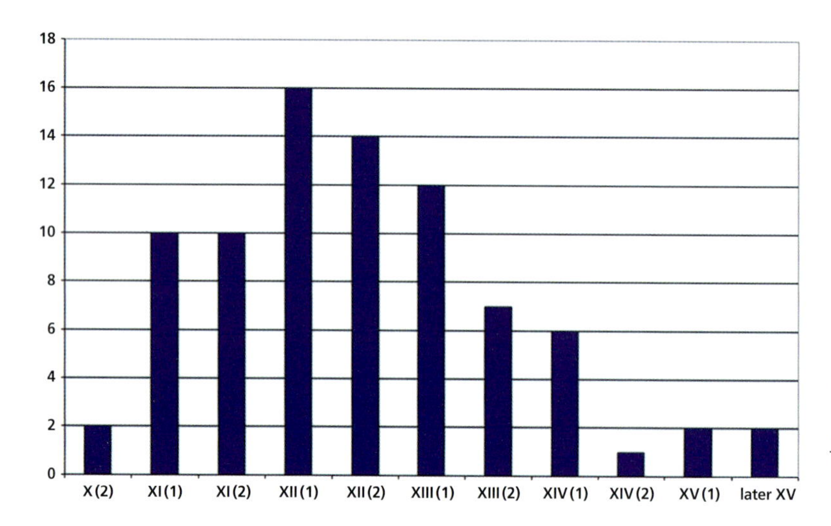

Fig. 5. Chronology of bear amulets in the cultural layers of medieval Novgorod from the 10th to 15th centuries (1 = first half of century; 2 = second half of century).

Evidence of bear remains in a cremation burial in the Moscow region (Burial 5, Kremenye burial ground on the upper river Oka, 12th century)

By Alexander S. Syrovatko, Natalia Svirkina and Liudmila Plekhanova

Keywords: Russia, Middle Ages, Viatichi, cremation, bear claws, analysis of keratinolytics

Abstract: The article presents the materials from Burial 5 at the Kremenye burial ground (Russia, Moscow Region, on the left bank of the river Oka). The burial ground consists of two parts: mounds with inhumation burials and cremation burials near the old surface level. All burials date back to the 12th century. Burial 5 is distinguished by an unusual ring structure and bear claws found among the human bones. The discovery of bear remains is exceptional for the central European part of Russia. An analysis of keratinolytic microorganisms revealed traces of skin, fur, and/or wool in different parts of the burial's ring structure.

Introduction

The burial that we present in this article is not typical for central European Russia. Everything about it is unusual: the burial rite used, its age, geographical origin and archaeological context, and even the burial construction itself.

Why is the burial rite unusual? It is a cremation burial. Such burial sites have only been found in central European Russia within the last 30 years, and there are still less than ten such sites known to us (Syrovatko et al. 2012; Syrovatko 2014). At present, all known cemeteries have been dated back to the interval between the 5th and 12th centuries.

Why is the dating unusual? The burial has, along with the whole cemetery, been dated back to the 12th century, or probably even later, and this period has been comprehensively studied archaeologically. It is considered that the Moscow region was, at that time, inhabited by the Slavic tribe of the Viatichi (modern Moscow, Kaluga, Ryazan, and Tula Oblast), whose burial rite was inhumation (Sedov 1982). The study of these monuments began in the middle of the 19th century, but, by the middle of the 20th century, interest had almost been lost. The finding of the cemetery, with its cremations of such a late date, was so totally unexpected that the discoverers had to make a considerable effort to present their views convincingly to the rest of the scientific society.

Why is the archaeological context unusual? The cremation burials (15 of them had been studied by 2019) were surrounded by the usual Viatichi inhumation mounds. So, the unusual cemetery was integrated within the grounds of the more conventional type, and both cemeteries were used at the same time. For the Old Rus epoch (in other words, the Pre-Mongolian Rus), such a combination is extremely unusual.

Bear and Human: Facets of a Multi-Layered Relationship from Past to Recent Times, with Emphasis on Northern Europe, ed. by Oliver Grimm (Turnhout: Brepols, 2023), pp. 639–650 BREPOLS PUBLISHERS 10.1484.M.TANE-EB.5.134354

What is the peculiarity of the burial construction? Each recently found cemetery containing cremation burials is unique in its own way. The common feature is that the funeral pyre was located somewhere else, and the remains were brought to the cemetery after the cremation. At the same time, the burial rite inside each cemetery was variable, and cremated remains were located near the ancient surface in shallow pits, which are barely discernable; urns were used only rarely. In one case – Burial 5, which is considered here – a ring structure has been recorded, and it is likely that such construction remains had previously been overlooked (see below). One more peculiarity of this burial is represented by the cremated bear claws, which were found together with human bones. Bear claws in a cremation burial are also unique and have not been found before in the Moscow region of Russia, or in Russia as a whole.

Materials and methods

The burial was found in the Kremenye cemetery in 2017. The cemetery is situated on the left bank of the river Oka in Moscow Oblast, not far from the town of Stupino (Fig. 1). The cemetery (Fig. 2) has long been known in archaeological literature for the excavation of six mounds with inhumation burials by V. A. Gorodtsov in the summer of 1927 (Gorodtsov 1928, 20–22). The excavation materials were also presented in the first student publication of the later academician B. A. Rybakov (Rybakov 1928; for more details on the history of research: Syrovatko et al. 2019). During the Soviet period, the mounds were robbed. In 2010, there was a fierce fire, which destroyed the forest that covered the cemetery. Furthermore, the area was ploughed in order to make way for a pine plantation (Fig. 3). During ploughing, cremations close to the surface were damaged, and the museum employee of Stupino town, E. Fomchenko, identified calcinated bones in the spoil heaps. When we arrived on location, we found artefacts from the 12th century together with the remains of cremation burials, which caused some consternation. From 2015, the site has been properly excavated, and, over the following four years, 15 cremation burials situated close to the surface and one barrow with an inhumation grave were found (Syrovatko/Fomchenko 2015; Syrovatko et al. 2019). The majority of inhumation burials were situated either close to the ancient surface, or in shallow pits right beneath the turf. In some cases, along with the main cluster of bones, there was a small pit with a black coal-like filling and a small number of bones (Fig. 3). Burial 7 stood out from the others with its location in the ditch of a burial mound; Burial 5 differed with its ring structure found beneath the bones.

Burial 5: archaeology

The burial is a flat deposit of bones in a layer of the black coal-like sand of approximately 1.9 x 1.7 m in size, but the core of the cluster was approximately 1 m in diameter (Fig. 4). In the 3D-model of the bone "cloud", built with data measured with a total station, it can be seen that some bones are situated in the upper part of the ring structure – a groove (Fig. 5). It became visible only after the bones had been removed and the surface of the subsoil had been cleaned out.

This groove had, in plan, a shape close to a circle, and it was approximately 0.9 x 1.0 m in size. The width of the groove ranged from 15 to 25 cm, and the depth from the surface ranged from 22 to 25 cm (Fig. 6a–c). The top of its filling is the same black coal-like sand into which the calcinated bones probably sank after the decomposition of the construction, which was made from an organic material (?). A part of the filling, at the bottom and along the external rim in the eastern side, was yellow-brown sand that was almost indistinguishable from the surrounding earth (Fig. 6b–d). This part of the construction was less reliable in its details. It hardly differed in colour from the surrounding area, and thus it could not be shown precisely in the drawing. In addition to the core burials in this cemetery, a small "additional" pit with a dark coal-like filling, 33 x 38 cm in size and 11–12 cm (Fig. 2) deep, was identified in the immediate vicinity of the "main" cluster.

There is a considerable number of finds in this burial when compared with the others (Fig. 6d). However, as with Burial 1, the majority of finds are indistinguishable fragments of copper alloy (15 pieces: Fig. 6d № 12–15, 18–19, 21–22, 24, 29, 30); one item can be interpreted as a small bell that was badly damaged by fire (Fig. 6d № 4). There were only two complete objects, a small cast bell with a slot (Fig. 6d № 20) and a bimetal buckle (Fig. 6d № 16). Four more objects were found in close vicinity to the bone cluster (up to two metres apart) and might be connected with it: fragments of a twisted bracelet, a neck-ring, and a small cast bell, plus indistinguishable items made of iron and copper alloy.

Thus, Burial 5 does not include artefacts with a narrow dating span. However, these artefacts do not contradict the dating of the cemetery near ground level in general. The most typical items that date the period of the cemetery's use are bipyramidal cornelian beads and bronze jewellery, characteristic for the 12th and the first half of the 13th century. These finds from the mounds indicate the simultaneity of both parts of the cemetery (inhumation burials in mounds, and cremated ones near ground level). The mound that contained Burial 7 in its ditch might, according to clothes' remains from a female inhumation grave, be dated to the time span of 1125–1150. Therefore, Burial 7, next to Burial 5, is not younger than that time period (for more detail on the dating of the whole cemetery: Syrovatko/Kleshchenko 2017; Syrovatko et al. 2019).

Burial 5: osteology

The total weight of the cremated remains from the burial is 1,340 g. The size of the fragments ranges from 5 cm to less than 0.5 cm; the average size is approximately 1–2 cm. There are different types of fractures on the fragments of the bones; and the colour of the bones is not homogeneous (from white to black), a light grey colour dominates. A significant number of the fragments are covered with coal dust. For the majority of bone fragments no further identification – species or skeletal part – is possible. Among the identifiable fragments there are bones of at least one human and one animal. The human remains originate from various segments of an adult male (?) individual of more than 30 years of age. Most informative for the determination of sex and age-at-death are fragments of the frontal bone (*margo supraorbitalis*), epiphyses of the phalanges, the humerus, and the fibula. It is worth noting that among the unidentifiable bones a glass fragment has been found. The remains from the "additional" groove have a weight of 28 g; all these remains, covered with coal dust and less than 2 cm in size, are unidentifiable. The colour of these fragments is homogeneous (light grey-white).

In addition, five distal (claw) phalanges of a bear were recorded (Fig. 7). It should be stressed here that some skeletal fragments, supposedly of an animal, could not be identified any further. The animal bones make Burial 5 stand out from the others in the Kremenye cemetery, and among other central European Russian cemeteries with cremation burials in general.

It is obvious that only a small number of human and animal bones were deposited in the burial. This is a common practice in cremations as they are known along the river Oka (Syrovatko 2014). Almost the only exception is Burial 46 in the Shchurovo cemetery, in which the total weight of the bones is 9 kg (Kleshchenko/Syrovatko 2014; Svirkina/Syrovatko 2016).

Our experience in working with medieval cremation burials from the Moscow region shows that only a selection of the bones from the funeral pyre was chosen for the burial itself at a different site, and, in such cases, the bones of animals were heavier and better preserved than those of humans. In the present case, human bones prevail, but this is no proof that there was only a small part of the bear carcass on the funeral pyre.

Burial 5: soil analysis

Dealing with such an unusual phenomenon, our team made an effort to find out whether the ring grooves were part of the enclosure or the grave construction. Burial 5 was the first case in which we applied methods of soil science to find the remains that filled the ring construction. To specify the soils of the location, data was extracted from material of a kurgan in close proximity to the cremated burial (cf. Figs. 2–3). This mound is c. 800 years old, as is the case with the cremated burial, when the beginning of mound-building at "zero-point" is taken into account (Targulian/Sokolov 1978).[1] In other words, all the natural processes that took place in this area were similar both for the soil of the mound and the cremated burial. So, it is possible to make comparisons by including the characteristics of the microbial communities.

The analysis of soil samples from the neighbouring kurgan and the ring groove of Burial 5, using the usual methods (Arinushkina 1970), has led to the identification of values of soil magnetic susceptibility, granulometric composition, humidity, and complete cation exchange capacity, humus, and mobile phosphates. Total and mineral phosphates were identified after Saunders/Williams (1955), and phosphatase activity in comparison with soils of other zones (Kashirskaya et al. 2020). In addition, the characteristics of microbial communities, namely soil respiration, basal and substrate-induced (Anderson/Domsch 1980), and also the keratinolytic (Kashirskaya et al. 2018a; b) and cellulose-decomposition (Plekhanova et al. 2020) activity of the soil and subsoils were recorded.

The method of identification of decomposed animal material in old soil (from the Bronze Age to the Middle Ages), which contains keratin (from skin, fur, and/or wool), follows Kashirskaya et al. (2018a; b). It should be noted that the laboratory had not been informed beforehand of the existence of bear claws in the grave.

Discussion

As we have already noticed, this burial significantly differs from the others on site, both in methods of funerary construction and in the composition of the bones. In the other cremation burials, such ring structures have never been found. The only exception might be Burial 11 from the same cemetery (unpublished). Probably, there was a similar ring structure, although less clearly executed (the filling is light, which made it difficult to identify in the sand layer, and the size was much smaller, no more than 0.4 m). At the same time, ring enclosures are known in funeral constructions in this area: earlier "barrows with enclosures in the mound", or the "small houses of the dead",[2] dating to the 6th–7th centuries, have been found in Shchurovo cemetery (Syrovatko et al. 2012). Later, ring structures appeared in Old Russian barrows, including those of the Viatichi (Sedov 1982; Sedova 1997). The youngest example is the barrow from the Kremenye cemetery that was examined by our expedition. All these constructions were inside the mounds and were much larger in size. It is likely that the constructions at the Shchurovo cemetery originally contained the remains of the funeral pyre, along with the subsoil taken from the ditch. Built of wood, with a diameter of 5–7 m and surrounded by a ditch, they looked exactly like "small houses". In any case, the listed examples are not identical with

1 "Zero-point" is a term taken from the theory of soil formation, as suggested by Targulian/Sokolov 1978. It defines the starting point of soil formation on bare subsurface rock. The start of forming humic acids is always connected with the colonisation of the subsurface rock by the biota.

2 "The houses of the dead" is a colloquial term, widely used in Russian archaeological literature for burial structures of various archeological cultures in forest areas. It refers to small above-ground constructions that look like houses, with inhumation or cremation burials inside.

the construction from Burial 5 near the surface; the mound and the ditches around the burial were absent.

Soil analysis has been applied for the first time for the given burials; thus, there is no comparison. However, the soils can be compared to one another. According to granulometry, their composition is identical (Petrosyan et al. 2019).

Also, the characteristics of the respiration of the microbial communities in the soils and the carbon content have been recorded. Most interesting is a conclusion arrived at by comparing the respiration activity of microbial communities (active microbial biomass the respiration of which has been recorded) and the carbon content: The sample from the eastern part of the ring groove reflects a low number of microorganisms, but the largest amount of humus (organic residues) in the groove. Such a ratio might be interpreted as "exposure to stress" for the active parts of the microorganisms; it also attests to a low threshold boundary of the "stress" for the cohesive sandy soils of that zone. This conclusion calls for additional studies and better statistical evidence, but it shows that it is possible to presume the addition of this organic substance into this sector of the groove.

The second, and probably the most important, aspect is the existence of two sectors with the highest keratinolytic values in the ring groove, which might reflect the addition of materials of animal origin, namely skin and fur, and/or wool, into the soil (Petrosyan et al. 2018).

Is there an analogy in using a bear in the cremation rite? Obviously, there are a lot of them in Fennoscandia and northern Europe on a wider scale (see several contributions, this volume; Kirkinen 2017; Simniškytė 2018, 148). It would be more relevant to provide analogies from the geographically close area, the centre or the south of the European part of Russia, but none are known to the authors at present. We can only comment on the discussion in Russian archaeology on the use and origin of the clay amulets in the shape of bear claws (or beavers?) in the barrows at Yaroslavl (282 km to the northeast from Moscow) and Vladimir (176 km to the east of Moscow) and on the Åland Islands (Golubeva 1987, 77; for the latter see also Gustavsson/Ljungkvist, this volume).

Conclusion

The Kremenye cemetery represents a unique site for the whole of central European Russia. In particular, this relates to Burial 5 containing the remains of a bear. Such circumstances may raise the self-esteem of the researcher(s) of this site, but obviously they have little from which to draw any conclusions.

Firstly, it is still unclear how to interpret this cemetery with its archaic 12th century rite. It might be the continuation of the traditions of the autochthonous population (as opposed to the Slavs), who left behind earlier cemeteries with cremations in this region: Sokolova Pustyn' 1–2 (5th–6th and 11th centuries: cf. Potemkina et al. 2013; Syrovatko et al. 2015b), Shchurovo (6th–7th and 9th–10th centuries; cf. Syrovatko et al. 2015a), Luzhki E (11th century; cf. Syrovatko et al. 2013). Any cultural and ethnic attribution of this population is impossible (it also might have been heterogeneous). On the one hand, the Kremenye cemetery and the others share the funeral rite and geographical proximity. On the other hand, clear differences in funeral rite are noticeable. For example, cremated animal bones are absent in Kremenye, apart from Burial 5; all the graves contain the remains of only one human individual, whereas in other cemeteries there might be more. There is no real evidence for continuity; the population who used the cemetery might not have been related to previous communities. Moreover, the cemetery looks unusual only for central European Russia: to the north, in the Baltic region, the cremation ritual continued to exist for a long period of time (Velius 2016; Simniškytė 2018).

Secondly, the fragmented state of burials all in all does not allow the argument that there were no bear bones and claws in other cremations. Even if they did not exist in all 14 burials – apart from

Burial 5 – in Kremenye, one cannot exclude it for earlier cemeteries. In this respect, the availability of professional archaeozoologists who are capable of working with cremated bones is a problem. Collections have been mostly analysed by anthropologists, but the detailed identification of (cremated) animal bones is beyond their competence.

So, Burial 5 at Kremenye might be explained in different ways, and it cannot be excluded that the quite distant analogies from Fennoscandia and the Baltic region are relevant in this case.

Acknowledgements

This research has been conducted with financial support of the RFBR (Russian Federation for Basic Research), project no. 17-06-00326: "The inhabitance of the left bank of the river Oka in the 'dark times' (8th–10th c.) in the context of the river valley landscapes' dynamics in the late Holocene".

Bibliography

Anderson/Domsch 1980: J. P. E. Anderson/K. H. Domsch, Quantities of plant nutrients in the microbial biomass of selected soils. Soil Science 130, 1980, 211–216.

Arinushkina 1970: E. V. Arinushkina, The manual on chemical analysis of soils (Moscow 1970) (in Russian).

Golubeva 1987: L. A. Golubeva, Merya. The Finno-Ugric peoples and the Balts in the Middle Ages (Moscow 1987) (in Russian).

Gorodtsov 1928: V. A. Gorodtsov, Archaeological Researches in Kolomenskoye and Kashira Uezd (Moscow 1928) (in Russian).

Kashirskaya et al. 2018a: N. N. Kashirskaya/A. A. Petrosyan/L. N. Plekhanova/A. S. Syrovatko/ T. N. Myakshina/A. V. Potapova, Cremated Middle Age burials and the attempts to identify fur substrate. Archeology of the Eurasian Steppes 4, 2018, 166–171 (in Russian).

Kashirskaya et al. 2018b: N. N. Kashirskaya/L. Plekhanova/A. Petrosyan/A. Potapova/A. Syrovatko/ A. Kleshchenko/A. Borisov, The Identification of Wool by the Number of Keratinolytic Microorganisms in the Ground of Ancient and Medieval burials. The Lower Volga Archaeological Bulletin 17(2), 2018, 95–107 (in Russian, with English abstract). doi.org/10.15688/nav.jvolsu.2018.2.8

Kashirskaya et al. 2020: N. N. Kashirskaya/L. N. Plekhanova/E. V. Chernisheva/M. V. Eltsov/S. N. Udaltsov/A. V. Borisov, Temporal and Spatial Features of Phosphatase Activity in Natural and Human-Transformed Soils. Eurasian Soil Science 53(1), 2020, 97–109. doi.org/10.1134/S1064229320010093.

Kirkinen 2017: T. Kirkinen, 'Burning pelts' – brown bear skins in the Iron Age and Early Medieval (1–1300 AD) burials in south-eastern Fennoscandia. Estonian Journal of Archaeology 21, 2017, 3–29.

Kleshchenko/Syrovatko 2014: E. Kleshchenko/A. Syrovatko, Cremation Ceremony Features of the Middle Oka-River Population in the Second Half of the First Millennium A.D. Considering Shchurovo Burial as an Example (Materials and Interpretation). In: 20th Annual Meeting of the European Association of Archaeologists. Abstracts of the Oral and Poster Presentations (Istanbul 2014) 439.

Petrosyan et al. 2018: A. A. Petrosyan/A. S. Syrovatko/L. N. Plekhanova/T. N. Myakishina/A. V. Potapova/N. N. Kashirskaya, Approaches to definition of fur substrates in the burial on amount of keratinolytic microorganisms. In: A. A. Tishkin (ed.), Modern solutions of the relevant problems of the Eurasian archaeology: Collection of scientific articles (Barnaul 2018) 102–106 (in Russian).

Petrosyan et. al. 2019: A. A. Petrosyan/T. N. Myakshina/L. N. Plekhanova/A. S. Syrovatko/S. N. Udaltsov/N. N. Kashirskaya, The development of the soil in an 800 years old barrow in Moscow Oblast. Russian Journal of Applied Ecology 1, 2019, 47–53 (in Russian).

Plekhanova et al. 2020: L. N. Plekhanova/N. N. Kashirskaya/A. S. Syrovatko, Cellulosolitic microorganisms activity as an indicator of details of funeral ceremony. The Lower Volga Archaeological Bulletin 19(1), 2020, 116–129 (in Russian, with English abstract). doi.org/10.15688/nav.jvolsu.2020.1.6.

Potemkina et al. 2013: O. Yu. Potemkina/A. S. Syrovatko/E. A. Klechenko, Sokolova Poustinia is a new burial site of the late Dyakovo time. Brief Reports of the Institute of Archaeology RUS 230, 2013, 260 (in Russian).

Rybakov 1928: B. A. Rybakov, On the excavations of Viatichi barrows in Myakinino and Kremenye in 1927. Collected papers of the Scientific and Archaeological Circle of the 1st Moscow State University (Moscow 1928) (in Russian).

Saunders/Williams 1955: W. M. Saunders/E. G. Williams, Observations on the determination of total inorganic phosphorus in soils. Journal of Soil Science 6, 1955, 254–267.

Sedov 1982: V. V. Sedov, Eastern Slavs in the 6th to 13th centuries (Moscow 1982) (in Russian).
Sedova 1997: M. V. Sedova, The jewelry from copper and alloys. In: Old Rus. Life and Culture (Moscow 1997) 63–78 (in Russian).
Simniškytė 2018: A. Simniškytė, Atypical burial rites or destruction of archaeological source? On the results of rescue excavations at Jakšiškis Barrow Cemetery (East Lithuania). Raport 13, 2018, 137–153.
Syrovatko 2014: A. S. Syrovatko, The cemeteries with cremations in the Middle Oka in the second half of the 1st millennium AD. Russian Archaeology 2014(4), 48–61 (in Russian).
Syrovatko/Fomchenko 2015: A. S. Syrovatko/E. E. Fomchenko, Barrow group in Kremenye, a new archaeological topic. In: E. E. Fomchenko (ed.), Oka's common thread: archaeology of Middle Oka region: the collection of the materials of the VII and VIII regional scientific and practice conferences at MBIC, Stupino History and Art Museum (Stupino 2015) 132–138 (in Russian).
Syrovatko/Kleshchenko 2017: A. S. Syrovatko/E. A. Kleshchenko, Subsoil burials with cremations of the 12th c.: new studies of the cemetery Kremenye. Archaeology of Moscow region: Materials of Scientific Workshop 13, 2017, 45–56 (in Russian).
Syrovatko et al. 2012: A. Syrovatko/N. Zaretskaya/A. Troshina/A. Panin, Radiocarbon chronology of the Schurovo burial mound cremation complex (the Viking Times, Middle Oka river, Russia). Radiocarbon 54(3/4), 2012, 771–781.
Syrovatko et al. 2013: A. S. Syrovatko/V. V. Sidorov/E. A. Kleshchenko/A. A. Troshina, The cemetery Luzhiki E: some observations of old collections. Archaeology of Moscow region: Materials of Scientific Workshop 9, 2013, 52–56 (in Russian).
Syrovatko et al. 2015a: A. S. Syrovatko/E. A. Kleshchenko/N. G. Svirkina/A. A. Troshina, Ground cremations of Shchurovo: on the question of the original form of the burials. Archaeology of Moscow region: Materials of Scientific Workshop 11, 2015, 147–154 (in Russian).
Syrovatko et al. 2015b: A. S. Syrovatko/O. Yu. Potemkina/A. A. Troshina/N. G. Svirkina, New data on chronology of Shchurovo type cemeteries: the burial in Sokolova Pustyn from the excavations of 2014. Brief Reports of the Institute of Archaeology RUS 241, 2015, 165–173 (in Russian).
Syrovatko et al. 2019: A. S. Syrovatko/N. G. Svirkina/E. A. Kleshchenko, Biritualism in funeral rite of the Vyatichi: paradoxes of the cemetery Kremenye. Russian Archaeology 2019(4), 102–117 (in Russian).
Svirkina/Syrovatko 2016: N. Svirkina/A. S. Syrovatko, 'The largest cremation' of the burial ground Shchurovo: typical or exclusive? In: 22nd Annual Meeting of the European Association of Archaeologists: Abstracts (Vilnius 2016) 162.

Targulian/Sokolov 1978: V. O. Targulian/I. A. Sokolov, Structural and functional approach to the soil: soil-memory and soil-moment. In: Mathematical Modeling in Ecology (Moscow 1978) 17–33 (in Russian).

Velius 2016: G. Velius, Underwater burial sites of the 14th century: Kernave cas. In: 22nd Annual Meeting of the European Association of Archaeologists: Abstracts (Vilnius 2016) 161.

Dr. Alexander Syrovatko
Director of the Kolomna Archaeological Centre
Russia
sasha.syr@rambler.ru

PhD student Natalia Svirkina
Russian Academy of Science
Institute of Archaeology
Russia
svirkina.natalia@mail.ru

Associate Prof. Dr. Liudmila Plekhanova
Laboratory of Archaeological Soil Science
Institute of Physical-Chemical and Biological Problems in Soil Science (Pushchino)
Russia
dianthus1@rambler.ru

Fig. 1. Location of the Kremenye burial ground, Moscow region (map GIS department, ZBSA).

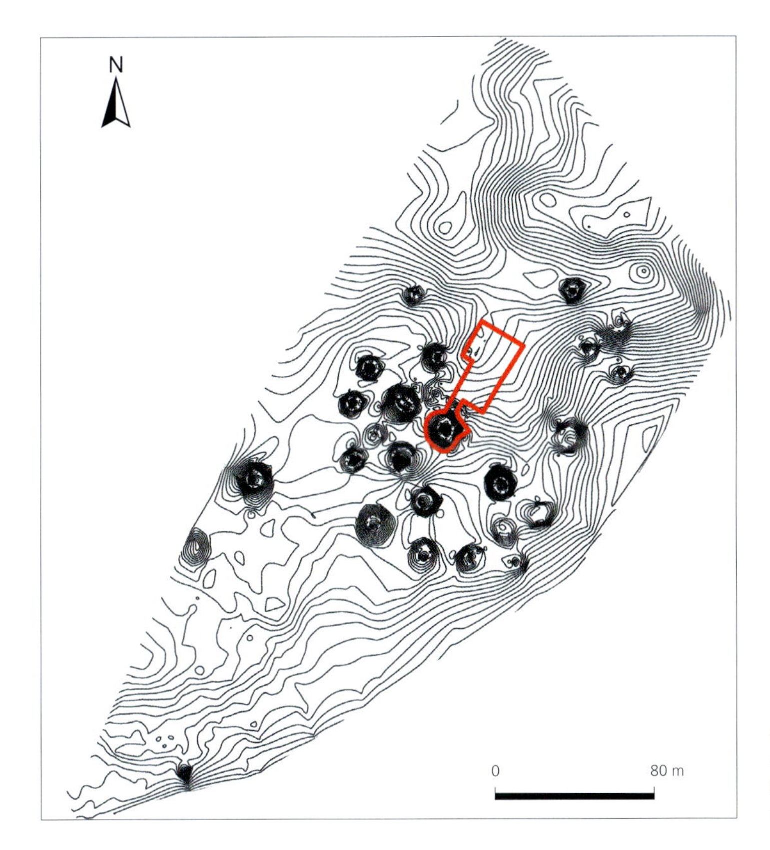

Fig. 2. Plan of Kremenye cemetery, topographic survey of 2016. The red contour line indicates the area of excavations conducted in 2015–2017 (graphics A. Syrovatko).

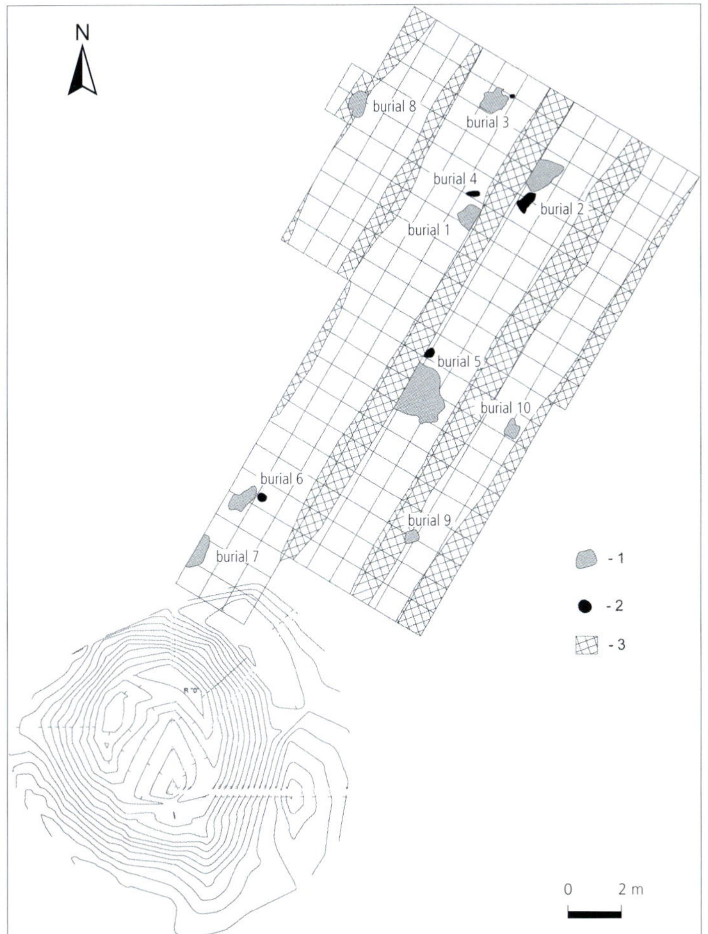

Fig. 3. Excavation plan. 1: Bone clusters; 2: Pits with dark filling near the burial; 3: Planting beds for pine plantation (graphics A. Syrovatko).

Fig. 4. General view of Burial 5 after turf removal (photo A. Syrovatko).

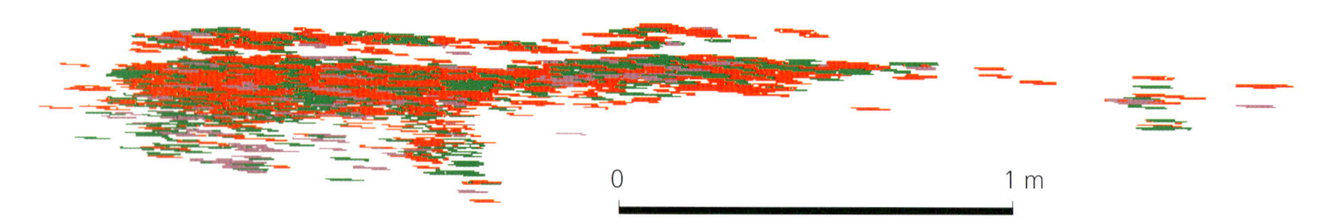

Fig. 5. 3D-model of bone "cloud", view from the southeast. Bones/bone fragments shown in red are less than 1 cm in size, bones in green are 1–2 cm in size, bones in purple are larger than 2 cm (graphics A. Syrovatko).

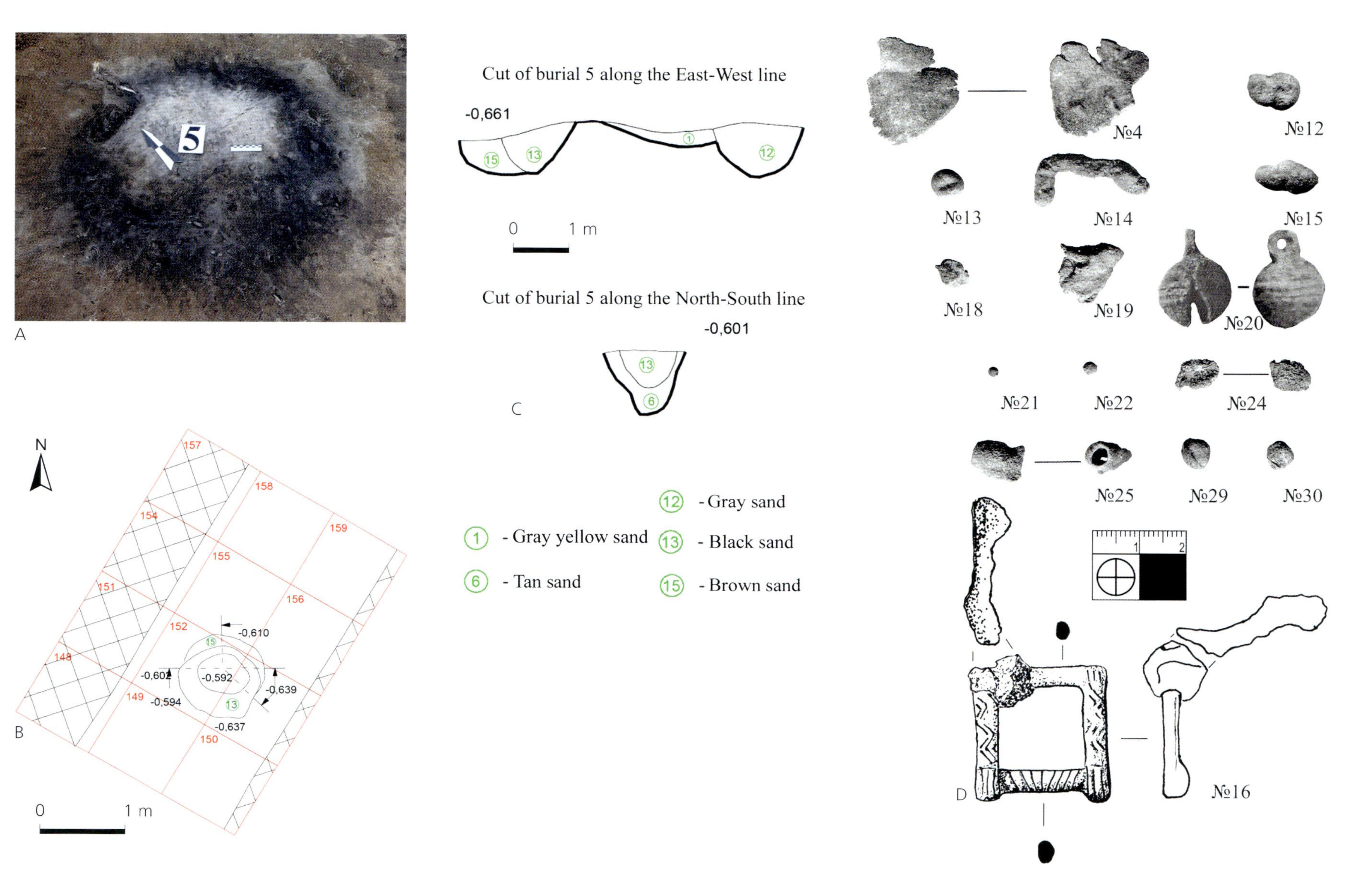

Fig. 6. Burial 5; a: Photo of the ring structure under the burial; b: Plan of the ring structure; c: Cross-section view of the ring structure; d: Burial goods. Numbers of burial goods correspond to field records (graphics A. Syrovatko).

Chapter 7

Bears in the history of religion (northern Europe)

Do these four images have an ordered thematic progression that shows the steps of an initiation sequence? The 7th-century helmet-plate dies from Björnhovda, Torslunda (Öland). Top left to bottom right: Man between bears; Man with axe holding a roped animal; Warriors carrying spears; Dancing man with horned head-dress and man with spear wearing a wolfskin (see SUNDQVIST *and other texts, this volume; image © Statens Historiska Museer, Sweden, inv. no. SHM 4325).*

Bears in Old Norse religion with specific references to the *berserkir*

By Olof Sundqvist

Keywords: berserkir, *animal-warriors, Óðinn, initiation, shapeshifting,* berserksgangr

Abstract: The notion that bear symbolism and religion played a crucial role in the warrior-band of berserkir *during the Late Iron Age (c. AD 550–1050/1100) in Scandinavia has, for a long time, been accepted in the research on Old Norse religion. However, it has recently been challenged by philologists and archaeologists. In what follows, this issue will be discussed by means of an interdisciplinary method, by which all available sources related to Old Norse religion will be taken into account. A special focus will be placed on the function of bears in relation to initiation,* berserksgangr, *and shapeshifting. Finally, the connection between bears,* berserkir, *and the worship of Óðinn will be examined. The etymology of the word* berserkir *and the factual existence of these animal-warriors will also be taken into consideration.*

Introduction

Bears are important in the religious worlds of indigenous Nordic people, for instance, those found in the Saami and Finnish traditions. However, they do not play a prominent role in the preserved Old Norse mythology. Compared to other animals, such as ravens, eagles, and wolves, they have a subordinated role here (cf. Jesch 2002). Nevertheless, previous research has regarded bears as sacred animals in Germanic religion. Jan de Vries states in his classical "Altgermanische Religionsgeschichte" (De Vries 1956/1957, §257): "Eine sehr wichtige Rolle hat der *Bär* im Kult gespielt" ("The *bear* played a very important role in cultic practice"). The sacredness of the bear, he argues, is reflected through its denominations in the Germanic languages. Cognates to the Old Norse *bjǫrn* (e.g. Old English *beorn* and the Old High German *bero*, from the Proto-Germanic **bernu*) are all derived from an old *noa*-designation, meaning "the brown one" (cf. De Vries 1961, 40–41; cf. Nedoma, Saerheim, and Udolph, all in the present volume). The Indo-European appellation of the animal – a term equivalent to the Greek *arktos* or the Latin *ursus* – is lacking in the Germanic languages. It may be no more than coincidence, but it might also be a consequence of a taboo related to the real name. In more recent research, scholars have argued that bears could have played a specific role in two religious contexts that sometimes converge; one that refers to human-bear sexual relations and genealogy, and one that is related to the religio-martial groups called *berserkir,* which probably meant "the bear-shirts" (e.g. Edsman 1994, 82–88; Tolley 2009, 563–580; DuBois 2012, 86–90; see Nedoma, this volume). For the former discourse, popular traditions with erotic undertones are often referred to, where a bear marries a woman. The offspring of such a marriage was a man-bear. For the

Bear and Human: Facets of a Multi-Layered Relationship from Past to Recent Times, with Emphasis on Northern Europe, ed. by Oliver Grimm (Turnhout: Brepols, 2023), pp. 655–676 BREPOLS PUBLISHERS 10.1484.M.TANE-EB.5.134355

latter discourse, Old Norse traditions about *berserkir* are mentioned. They were closely linked to the warriors called *ulfheðnar*, "wolf-skin-wearers", who took the role of predators in action. In *Vatnsdœla saga* (chap. 9, 24; *The Saga of the People of Vatnsdal*, chap. 9, 202) it is stated:

> *Þeir berserkir, er ulfheðnar váru kallaðir, þeir hǫfðu vargstakka fyrir brynjur ok vǫrðu framstafn á konungs skipinu [...].*
>
> Those *berserkir*, known as *ulfheðnar* – they used wolf-skin cloaks for corslets and defended the bow of the king's ship.

These elite warriors constituted some sort of warrior sodalities into which religious elements were also integrated. In what follows, I will mainly focus on the connection between bears and the *berserkir* and its significance for the Old Norse religion. The concept of Old Norse religion refers to the religion of the Germanic speaking people living in Late Iron Age Scandinavia (*c.* AD 550–1050/1100). This religion perished when Christianity became dominant around the 11th century.

Sources and source criticism

An inquiry into the significance of bears in the context of the *berserkir* reveals serious problems with the source criticism. The *berserkir* are indeed mentioned in skaldic poetry dating back to the Viking Age (*c.* AD 750–1050/1100). These poems may be considered as primary sources, since they describe these warriors and the Old Norse religion from an insider's perspective, i.e. they were composed by persons who lived with the old world-view during the Viking Age. A connection between *berserkir* and *ulfheðnar* can be found in the earliest known text about them in Old Norse, i.e. the poem *Haraldskvæði*. It was composed by the skald Þórbjǫrn hornklofi *c.* 900, but the lay is only preserved in form of quotations in medieval manuscripts. This poem, also called *Hrafnsmál*, is formed as a dialogue between a *valkyrja* (i.e. a female mythical being; literary "the one who chooses the slain"), and a raven (see stanzas 1–2), where the former poses questions and the latter replies (see the stanzas 3, 15, 18, 20, and 22). The *berserkir* and the *ulfheðnar* are here mentioned to have supported King Haraldr hárfagri at the battle of Hafrsfjǫrðr in Norway *c.* 870 (Fig. 1). It seems, according to this poem, as if these warriors were possessed by an ecstatic rage, including bestial power, during the battle as the *berserkir* were bellowing (*grenja*), and the *ulfheðnar* were howling (*emja*; st. 8; *Haraldskvæði;* text and transl. *SkP I*, 102–103):

> *Grenjuðu berserkir;* *guðr vas þeim á sinnum;*
> *emjuðu ulfheðnar* *ok ísǫrn dúðu.*
>
> Berserks (*berserkir*) bellowed; battle was under way for them; wolf-skins (*ulfheðnar*) howled and brandished iron spears.

The Old Norse verb *grenja* "to bellow"[1] is attested in connection to the sound of bears in other texts, for instance, the *Grettis saga* (74–75; author's translation):

1 Heggstad et al. 2012 translate *grenja* as "brøle, kneggje med dirrande lyd, ule" ("roar, crackle with trembling sound, howl").

Lá bjǫrninn í híðinu á daginn, en leitaði á brott jafnan, er náttaði [...] Þat var eina nótt, at Bjǫrn fór til híðsins; hann varð varr við, at dýrit var þar fyrir ok grenjaði illiga.

The bear dwelled during the day in the pit, but left it when the night fell on [...] One night Bjǫrn went to that pit; he noticed that the animal [the bear] was there, since it bellowed angrily.

However, this verb, and the Old Norse verb *emja*, are usually related to the sounds of dogs and wolves.[2] In stanza 21 of *Haraldskvæði*, the raven states that these warriors "hack at shields", which refers to the ecstatic and bestial character of them in action. The raven also mentions that the king entrusts himself to men of courage alone, indicating that the *berserkir* were perceived as his bodyguard or elite troops (text and transl. *SkP I*, 114–115):

"Ulfheðnar heita, þeir es í orrostu
blóðgar randir bera;
vigrar rjóða, es til vígs koma;
þeim es þar sist saman.
Áræðismǫnnum einum, hygg ek, þar undir felisk
skyli sá inn skilvísi, þeim es í skjǫld hǫggva."

"They are called wolf-skins, who bear bloody shields in combat; they redden spears when they come to war; there [at Haraldr's court] they are seated together. There, I believe, he, the sovereign wise in understanding, may entrust himself to men of courage alone, those who hew into a shield."

In stanza 20 the *valkyrja* asks further about *berserkir* (text and transl. *SkP I*, 113–114):

"At berserkja reiðu vil ek þik spyrja, bergir hræsævar:
hversu es fengit, þeim es í folk vaða,
vígdjǫrfum verum?"

"I want to ask you about the equipment of berserks [*berserkir*], taster of the corpse-sea [BLOOD > RAVEN]: what provision is made for war-daring men, those who surge into battle?"

It has been argued that the kenning in stanza 20, "taster of the corpse-sea [= blood]", indicates that the *berserkir* really consumed blood like a predator. The archaeologist Neil Price, for instance, translates it "the drinkers of the corpse-sea [blood]" and states: "they [the berserks] are described as drinking blood" (Price 2019, 304; cf. Dale 2014, 136). However, the kenning *bergir hræsævar* ("taster of the corpse-sea"), does not refer to the *berserkir*, which appears in plural (*berserkja* "of berserks") in this stanza. The noun *bergir* "taster" (a nomen agentis of *bergja* "to taste") is in its singular form here, and the kenning *bergir hræsævar* refers to the raven, who is the interlocutor of the *valkyrja* in this poem. She poses here a question to the raven and addresses him/her by means of this kenning. In the poem, we do not get explicit information about the assumed religious context of these warriors, or about any blood-rituals,[3] but it is clear that, in this account, they were symbolised as bears or wolves in battle.[4]

2 In *Vatnsdœla saga* (124) we meet the expression *grenja sem hundr* ("barking like dogs"). In *Grettis saga* (67) *emja* is related to wolves. Heggstad et al. 2012 translate *emja* as "skrike, hyle" ("shriek, howl"); ONP has "yell, cry ?, bellow".

3 This does not exclude that the *berserkir* ritually consumed blood in other contexts, see below.

4 The word *berserkir* is mentioned in a "free-standing" stanza, a so-called *lausavísa* created in the late 10th century by Viga-Styrr Þorgrímsson and quoted in *Eyrbyggja saga*. These *berserkir* were dangerous, according to the stanza. This

Information about the connection between bears, *berserkir*, and religion appears in Old Icelandic prose traditions, such as the Kings' Sagas (e.g. *Ynglinga saga*), the Sagas of Icelanders (e.g. *Grettis saga*, and *Víga-Glúms saga*), and the somewhat problematic Fornaldarsögur (e.g. *Hrólfs saga kraka*). These texts date back to the High Middle Ages, i.e. more than 200–300 years after the Old Norse religion had perished. These sagas are literary reflections of lost oral traditions, which also include later ideas. Hence, they must be considered as secondary sources, i.e. texts written by Christians from an outsider's perspective.

In the Sagas of Icelanders, the *berserkir* often appear as "villains" or "unwelcome suitors" (cf. Blaney 1982). This motif is probably due to later epic developments. In Snorri Sturluson's *Ynglinga saga* (*c.* AD 1230), we get the important information that the *berserkir* were regarded as Óðinn's warriors (see further below). In order to evaluate all these prose texts, other types of sources will be used, such as archaeological, runic, and onomastic materials, which are more contemporary with the Iron Age *berserkir* and also considered as primary sources. However, these types of materials also contain serious problems related to interpretation, since they are mute or fragmentary. In addition to these sources, contemporary classical and ecclesiastic texts written in Latin may contribute with important information on the connection between bears, animal-warriors, and the Old Norse religion, even if these texts must be considered as secondary sources. Important materials are also the medieval laws, written in the vernacular (see Tab. 1 for an overview).

Previous research, methods and aims

Research on *berserkir* has been extensive and it is difficult to survey. Early on, it was argued by Lily Weiser and Otto Höfler that Germanic and Old Norse sources, as well as later folklore traditions, clearly reflected the existence of a particular type of *Männerbünde* (male societies) among the Germanic peoples.[5] For Höfler, these *Männerbünde* represented more or less esoteric organisations (*Geheimbünde* [secret societies]), centred around the god Woden or Óðinn and the cult of the dead (*Totenkult*). This theory can be questioned in many respects (see e.g. Von See 1961; Meier 2001), and it has also been stated that the previously assumed evidence for this phenomenon in Tacitus' *Germania* is doubtful (Lund/Mateeva 1997). Also uncertain is the name *Winnili* (cf. German *winnig* "howling with rage") mentioned in the anonymously written text from the 7th century, *Origo Gentis Langobardorum*. It designates an elite warrior group among the Langobards, but it is not quite clear whether these warriors can be equated with the *berserkir*.[6] Equally problematic for such an interpretation is the Latin term *cynocephali* "men with dog's heads", mentioned by Paul the Deacon in his *Historia Langobardorum* (I,8, 11).[7] Notwithstanding its speculative character, the theory of Weiser and Höfler draws attention to the existence of some sort of cultic warrior groups.

Previously, scholars have thus argued that the *berserkir* were "supernaturally empowered warriors", i.e. that they used religious rites when preparing for battle and thus acted with divine force in war (see most recently Frog 2019; cf. also Nordberg/Wallenstein 2016). This position has been

word also appears in two Eddic poems, *Hárbarðsljóð* and the later *Hyndluljóð*. The former, which may have been composed during the Viking Age, tells that Þórr once fought with female *berserkir* (*brúðir berserkia barðac*) at Hlésey (Læssø). The expression "female *berserkir*" has been interpreted as a poetic expression of "giantesses", indicating that the concept of *berserkir* was regarded as something dangerous. The latter poem just mentions that Óttarr descended from a line of *berserkir*.

5 See Weiser 1927; Höfler 1934. For an overview of research: Schjødt 2008, 49–57.

6 Weiser 1927, 49, 54 interprets *Winnili* as "wütende Hunde" ("furiuos dogs"). For other explanations of the name *Winnili* (cf. **wenn-*) and its connection to the Langobards, see Nedoma 2001.

7 E.g. Kershaw 2000, 142–145. Cf. Höfler 1934, 62–63.

questioned by the philologist Margaret Clunies Ross, who thinks that "there is no unquestionable evidence in the sources to support this view" (Clunies Ross 2019, 298). Archaeologist Egon Wamers has a similar opinion regarding this and states that the relationship between bears, *berserkir*, and an Óðinn-cult has no support in the preserved sources:[8] "Aus der kritischen Analyse der antiken und der nordalpinen Bilddenkmäler mit 'Bärenszenen' und der altgermanischen Überlieferung zu Bären und Berserkern deutet bislang nichts Greifbares auf einen altgermanischen Bärenkult noch auf ein Bärenmasken-Kriegertum/Berserkertum – und erst recht nichts auf ein Odin-Bärenkriegertum. Óðinn/Wotan/Woden gewinnt in diesem Kontext keinerlei Konturen" ("According to the critical analysis of the antique and northern Alpine pictorial monuments with 'bear scenes' and the Old Germanic traditions on bears and berserks, there is nothing tangible indicating an Old Germanic bear cult or bear mask warriorhood/berserkism – and certainly no Odin-bear-warriorhood. Óðinn/Wotan/Woden does not take on any shape at all in this context"; Wamers 2009, 42).

The theory that religion and bear symbolism played a crucial role in the group of *berserkir* has thus been challenged by recent scholars. In what follows, this issue will be discussed by means of an interdisciplinary and comparative method,[9] where all available sources related to Old Norse religion will be taken into account. A special focus is going to be placed on the function of bears in relation to initiation, *berserksgangr*, and shapeshifting. Finally, the connection between bears, *berserkir*, and the worship of Óðinn will be examined. Before entering these issues, the etymology of the word *berserkir* and the factual existence of these animal-warriors will be taken into consideration.

Etymology of *berserkr* and the existence of animal-warriors in the Late Iron Age

There is little dispute about the second element of the compound *berserkr*. The Old Norse appellative *serkr* means "shirt" or "coat of mail". The first element *ber-* is, however, hotly debated, and there are two competing explanations. Some scholars argue that it refers to the adjective *berr* meaning "bar" (e.g. Noreen 1932; cf. Kuhn 1968; Speidel 2002), while others think it is related to an unattested substantive **beri* (**berr* or **ber*; cf. *bera* "she-bear") meaning "bear" (e.g. Fritzner 1954; Cleasby/Vigfusson 1957; De Vries 1961; Von See 1961; Breen 1997; Heggstad et al. 2012; Aðalheiður Guðmundsdóttir 2007; Wamers 2009; Dale 2014). Those who favour the former interpretation, usually take support from a statement in Snorri Sturluson's *Ynglinga saga* chap. 6 (1230), where it says that Óðinn's "men (the *berserkir*) went to battle without coats of mail" (*hans menn fóru brynjulausir*; *Heimskringla I*; *Snorri Sturluson; Snorri Sturluson [transl.]*). However, most scholars prefer the other interpretation. Several arguments have been adduced. The semantic structure and lexical symmetry between the concepts of *ulfheðnar*, which undoubtedly means "wolf-skin-wearers", and *berserkir* supports the idea that the latter term could be interpreted as "bear-shirts".[10] This interpretation is also in harmony with images on warrior-animal amalgamations, such as those on the

8 During the last decades, a great amount of studies has been written on *berserkir* (e.g. Blaney 1972; 1982; Breen 1997; Sundqvist/Hultgård 2004; Näsström 2006; Samson 2011; Kuusela 2012; Dale 2014; Price 2019), but mainly archaeologists have focused more exclusively on the function of bears in these contexts (see e.g. Wamers 2009; Hedeager 2011; Grimm 2013; this volume; Oehrl 2013; however: Ström 1980; Tolley 2009, 563–580; DuBois 2012).

9 By the comparative method, I am referring mainly to what Schjødt (2012, 275–280) calls "comparisons of the second level", i.e. comparisons made between source information from the Late Iron Age Scandinavian area, and its neighbours, such as Germanic tribes to the south. Comparative perspectives have been a hot topic in the discipline history of religions. Nowadays, scholars often defend such perspectives, even if they argue that such an approach should rather be described as a research design and not a single method. Cf. Stausberg 2014.

10 The compound *ulfheðinn* is composed by a first element *úlfr* "wolf" and a second element *heðinn*, meaning "kurzes Kleidungsstück ohne Ärmel, aber mit einer Kapuze von Pelz gemacht" ("a short garment without sleeves, but with a hood made of fur"); De Vries 1961, 215.

7th-century helmet plate dies from Torslunda (Öland), which suggest that warriors wore animal skins, masks, or both (Fig. 2). There we see a spear-carrying warrior with a wolf-head, wolf-skin and tail, but human feet, running or dancing with a one-eyed man to the left. It has been argued that this is a representation of Óðinn (see below). The wolf-skin-wearer is an *ulfheðinn*, who is being led by his god into battle or taking part in a ritual dance with him. This image indirectly supports the idea that the word *berserkr* should be interpreted "bear-shirt". Images from Gutenstein (Austria), Obrigheim (Germany), Sutton Hoo (England), and Källby (Sweden) do the same (Fig. 3).

The iconography and the etymology of the term *ulfheðnar*, and most likely the word *berserkr*, also indicate that this type of warrior really existed during the Iron Age, and that they wore wolf or bear skins in battle or during ritual performances. This idea can be supported by the 10th-century skaldic poem *Háleygjatal* st. 6, which mentions a wolf-skin-warrior (text and transl. *SkP I*, 204):

> *Ok sá halr*
> *at Hôars veðri*
> *hǫsvan serk*
> *Hrísgrísnis bar.*
>
> And that man wore the grey shirt of Hrísgrísnir <wolf> [WOLF-SKIN] in the storm of Hôarr <= Óðinn> [BATTLE].

It is also striking that some of the richest Migration Period weapon-burials, such as Evebø (Norway), Högom mound 2 (Sweden), and Snartemo V (Norway), have yielded bear skins. On the other hand, such skins also appear in rich female burials, such as that of Krosshaug at Tu in Rogaland (Norway). This evidence makes the connection between warriors and bear skins as something typical in rich Iron Age graves uncertain. Archaeologist Oliver Grimm states thus: "Remarkably, probably the richest of all northern European weapon burials of the Migration period (Evebø, Högom mound 2, Snartemo grave V), have yielded bear-skins that either covered or lay under the deceased, and this seems a strong argument in favour of 'bear-warriors' [...]. However, there is much counter-evidence, since the earliest northern European inhumation burials with skins were those of women. The Krosshaug 'petty queen' of the Migration period was also quite possibly among the bear-skin burials, and so were other women of that era (for example in Sletten, southern Norway [...])" (Grimm 2013, 292).

Other possible evidence for the existence of these predator-warriors are some early dithematic personal names. The theriomorphic elements in these names indicate close links to animal mimesis and costume (see Breen 1997). The personal name *Úlfheðinn* "wolf-skin-wearer" is attested in the Viking Age runic inscriptions **ulueþin** (U 799) and **ulfhiþin** (Sö 307; cf. Peterson 2007, 240). Also, the name *Biarnhǫfði* "bear-head" is evidenced in runic inscriptions dating back to the Viking Age: **biarnhufþi, biarnaffþi** and **biarnafþa** (U 1045 and U 1113; Peterson 2007, 42). In addition, the names *Bjarnheðinn* "bear-skin" and *Bjarngrímr* "bear-mask" appear in Old Norse medieval texts (*Landnámabók*, S27, H25, 66, 69; *Flateyjarbók* II, 395; cf. Lind 1905–1915, 135). One problem with these dithematic names is that the first and the second elements could have been combined by coincidence. Nevertheless, we may conclude that the evidence, when all is taken together, indicates that animal-warriors really existed in Late Iron Age society.

Initiation Rites

Certain episodes involving *berserkir* within Old Norse prose describe events that can be interpreted as literary reflections of a lost oral tradition related to memories of initiation rites (e.g. Danielli

1945; Blaney 1972; Schjødt 2008; Dale 2014; Nordberg/Wallenstein 2016). Initiation is a *rite of passage* marking entrance or acceptance into a group or society. In an extended sense, it can also signify a transformation in which the initiate is "reborn" into a new role with a new identity.[11] In the Norse prose accounts, bears and wolves occasionally play a crucial role in such initiation rituals. These high medieval narratives form a general pattern. A boy from a well-established family is placed with his foster-father, who is a warlord. Usually he is dressed in ragged clothing, such as a fur pelt, when arriving at his relative's residence. He is, at first, poorly treated and harassed by members of the retinue. The lad finally becomes the protégé of a senior member of the warband. This warrior takes the boy to a secluded place in the woods. He trains him in military skills and initiates him into secret knowledge. As a test, he has to kill a dangerous enemy, most often a ferocious animal, such as a bear. He must also drink its blood and, by means of this, in some sense incorporate the quality of the predator. He returns to his foster-father and performs a sham battle against a beast. After the ritual killing of the dummy-beast, the warriors and the warlord accept him as a full member of the warband. As a sign of this, he is given a new name and a sword as an insignia.[12]

This pattern is partly attested in the medieval Sagas of Icelanders. In *Víga-Glúms saga* chaps. 2–4, the young Icelander Eyólfr is sent to Norway. First, Eyólfr is treated poorly by his host, since he wears an old fur pelt (*loðkápa*) every day. After killing a bear and cutting off its nose as proof (chap. 9), he becomes a respected man. When he defeats the *berserkr* called Ásgautr in a *hólmganga* (f.), he also receives Ástríðr Vigfússdóttir as his wife (chaps. 11–13). A similar pattern is repeated in chapter 6, when Glúmr, Eyólfr's youngest son, travels to Norway. He is also treated poorly by his relatives until he drives the *berserkr* with the name Bjǫrn "bear" járnhauss away from the hall. He is then accepted and placed in the high-seat, together with his grandfather, Vigfúss. Later, Glúmr also receives three family heirlooms. Vigfúss says: "I will give you a fur, a spear and a sword" (*... vil ek þér gefa, feld ok spjót ok sverð*; *Víga-Glúms saga*, 16–19). According to *Grettis saga* chaps. 18–21, Grettir kills a *berserkr* and a bear in order to prove himself and become an heir to the family (*Grettis saga*, 56–78). Like Glúmr, Grettir also receives a sword after his deeds, thus indicating traces of a ritual pattern.[13]

One of the most important sources attesting to a detailed initiation of animal-warriors, is the account of the *berserkr* Bǫðvarr Bjarki and his protégé, Hǫttr. This narrative is rendered in several prose texts, such as Saxo Grammaticus' *Historia Danorum* (*c.* 1200), *Skáldskaparmál* in *Snorra Edda* (*c.* 1220), *Skjǫldunga saga* (*c.* 1200), and in Olaus Magnus' *Historia de gentibus septentrionalibus* (1555). The tradition is attested early, in the skaldic poem *Bjarkamál* (possibly 10th century; cf. M. Clunies Ross in *SkP III*, 496), and also in later poems, such as *Bjarkarímur* (15th century). The

11 Arnold van Gennep identified three successive and distinct factors in such groups of rites: separation, margin, and aggregation (Van Gennep 1960). Victor Turner proceeded from this framework, but went further and focused on an aspect of these rites that had previously been neglected, i.e. the marginal or liminal period (Turner 1967; 1969). In ritual initiations the neophytes are – during a well-marked liminal period – removed, hidden, without rank or insignia, etc. "The subject of passage ritual is, in the liminal period, structurally, if not physically, 'invisible'" (Turner 1967, 95). He/she is "betwixt and between", neither here nor there. The neophyte is thus regarded as symbolically dead. He/she is leaving his/her former being and identity, but he/she has yet not achieved what he/she will become. Only after the "exhibitions", "actions", and "instructions" (of the "knowledge"), i.e. the ritual actions that comprised the initiation, can he/she become what he/she is going to be.

12 On this pattern, see e.g. Weiser 1927, 80–82; Höfler 1934, 190–201; Danielli 1945, 229–245; Blaney 1972, 64–129; Schjødt 2008, 271–327, 352–355; Nordberg/Wallenstein 2016, 53–54.

13 Well-known is the account of Sigmundr and his son Sinfjǫtli in the late *Vǫlsunga saga* chaps. 6–8 (14th or 15th century). Scholars have argued that this story reflects an ancient initiation ritual (e.g. Weisser 1927, 70–71; Höfler 1934, 190–219; Schjødt 2008, 299–312). Sinfjǫtli is tested in several different ways. During a liminal phase, he learns the ideals of warriors. Both father and son are disguised as wolves during that period, speaking wolfish language. The boy also "dies" in a symbolic manner, after killing eleven warriors. He is rescued by his father (symbolically reborn) by supernatural means. After that he can remove his wolf dress and become a real warrior.

most extensive version is found in the fornaldarsaga *Hrólfs saga kraka*, which is a late literary product from the 14th or 15th century. The pattern of initiation is not so prominent in the account of the protagonist Bǫðvarr himself. Indeed, he is one of King Hrólfr's *berserkir* (see *Hrólfs saga kraka* 26–37).[14] He is also perceived as a man-bear, since he is son of a woman and a bear (*Hrólfs saga kraka* 26–27). This is indicated by his nickname *bjarki*, which means "the little-bear" (Falk 1924, 4–5; Heggstad et al. 2012, 73). This name reflects genealogic aspects, and possibly "initiatory" functions as well. It could have been given to him under his initiation, but it might also refer to his role as an initiatory tutor when he was teaching Hǫttr how to kill a bear. In addition, the name *bjarki* might be related to his role as a shapeshifter (see below).

Aspects of initiation are more clearly attested in connection to Bǫðvarr Bjarki's protégé, Hǫttr (cf. Schjødt 2008, 322–326). *Hrólfs saga kraka* reports that the young boy leaves his home and arrives at King Hrólfr kraki's hall. There he is ignored, mistreated, and insulted by the retainers of Hrólfr. Finally, he acquires a tutor, Bǫðvarr, who takes him into the wood and teaches him how to kill a beast. After drinking its blood and eating its heart, Hǫttr returns to Hrólfr's court, where he performs a sham battle and kills the dummy beast. He is then accepted into the group as a warrior. Since Saxo, in his much-abbreviated version of this account, states that Bǫðvarr Bjarki's (Latin *Biarco*'s) training and killing involved a bear (Latin *ursus*), it becomes possible to associate this ritual with an initiation into the band of *berserkir*. Saxo states thus (*Saxo Grammaticus*, 2.6.11):

> *Vrsum quippe eximie magnitudinis obuium sibi inter dumeta factum iaculo cinfecit comitemque suum Hialtonem, quo uiribus maioreuaderet, applicato ore egestum bellue cruorem haurire iussit.Creditum nanque erat hoc potionis genere corporei roboris incrementa prestari.*
>
> When a giant bear met him [Biarco] among the thickets he dispatched it with his javelin and then told Hialti [Hǫttr], his comrade, to apply his mouth and suck out the beast's blood so that he might achieve greater strength. It was believed that this type of drink afforded an increased bodily vigor.

Hǫttr's blood-drinking ceremony, which also is depicted in Olaus Magnus' Book 5 of his *Historia de gentibus septentrionalibus*, chap. 15 (Fig. 4), indicates an initiation. In this ritual, the bear symbolism plays a particular role, since the drinking of the bear's blood leads to physical strength and courage.[15] When Hǫttr finally, according to *Hrólfs saga kraka*, receives his new name, Hjalti, and a sword, all doubts are expelled. By means of this rite, Hjalti has included bear features in his personality, which was a prerequisite for being one of the *berserkir*.[16]

One problem when trying to reconstruct the historical background of the initiation of *berserkir* is the late dating of the extensive written sources, which are supposed to substantiate this theory. This type of initiation has, however, also been linked to the iconography of the four Torslunda helmet

14 The *berserkir* are often in this text designated as *kappar* "fighters" (m. pl).

15 In fact, blood has been recharged and surrounded by rites even in peasant communities well into the Christian era. Lid 1924 investigated slaughter customs in folkloric material from Norway and its neighboring countries. It was still customary in the 18th century to drink the blood of slaughtered animals, which was considered as nourishing and energy-providing. This was mainly done in connection with hunting, but also during the slaughter of domesticated animals.

16 The three phases of a *rite de passage* can thus be identified in this story: *a separation phase* – when Hǫttr leaves the hall and his former identity symbolically dissolves; *a liminal or transition phase* – when Hǫttr goes to the woods with his master and obtains his strength and courage by drinking blood and eating the beast's heart, where he also acquires the qualification to be a warrior, and, finally, *an incorporation phase* – when Hǫttr displays his new skills by symbolically killing the beast, he also receives a weapon and a new name. Hǫttr's social status and identity have now changed and this is irreversible. See e.g. Danielli 1945; Dumézil 1973, 69–71; Blaney 1972, 102–110; Schjødt 2008, 322–326.

plate dies. Heinrich Beck (BECK 1968) and, more recently, Roderick Thomas Duncan Dale (DALE 2014, 103–106) have identified an ordered thematic progression in them, which was related to an initiation sequence. They argue that the first two matrices show how the warrior proves himself against two bears and a monster. The third matrix is about the warrior standing in a shield-wall in a position indicating that the members in the war-band now can rely on him. In the fourth, the warrior is running or dancing as an *ulfheðinn,* together with his god Óðinn.[17] Whether Beck's and Dale's arguments can be considered valid will be left open in the present investigation. However, it should be mentioned that parts of the initiation pattern attested in the medieval prose texts also have support in much older classical sources that refer to Germanic tribes in continental Europe, especially the test of the novice's courage and manliness. Tacitus reports in his *Germania* (*c.* AD 98) that the young Chatti warriors could not cut their hair (i.e. complete their initiation) until they had slain an enemy (*Tacitus*, chap. 31):

> *Et aliis Germanorum populis usurpatum raro et privata cuiusque audentia apud Chattos in consensum vertit, ut primum adoleverint, crinem barbamque submittere, nec nisi hoste caeso exuere votivum obligatumque virtuti oris habitum. Super sanguinem et spolia revelant frontem, seque tum demum pretia nascendi rettulisse dignosque patria ac parentibus ferunt: ignavis et imbellibus manet squalor.*
>
> The ceremony, practiced by other German peoples only occasionally, and by individual hardihood, has with the Chatti become convention, to let the hair and beard grow when a youth has attained manhood, and to put off the facial garb vowed and held as due to manliness only after an enemy has been slain: standing above the sanguinary spoil, they remove the face's cover, and advertise that then and not before have they paid the price of their birth-pangs, and are worthy of their kin and country. Cowards and weaklings remain unkempt.

A. A. Lund and A. S. Mateeva (LUND/MATEEVA 1997, 213–214) argue that the description of Chatti customs in this passage is built on a literary *topos* common in the Greek-Roman world, with no source value for the reconstruction of ancient Germanic rituals in warrior contexts. Evidence from other classical authors, on the other hand, indicates that similar tests of the novice's courage and manliness were performed by different Germanic tribes. Ammianus Marcellinus (*c.* AD 330–395), for instance, reports in his *Res Gestae* that the younger warriors among the Taifali had to serve the seasoned troops until they had killed a bear or a boar by themselves (*Ammianus Marcellinus*, vol. III, book xxxi, chap. 9):

> *Taifalorum gentem turpem obscenae vitae flagitiis ita accepimus mersam, ut apud eos nefandi concubitus foedere copulentur maribus puberes, aetatis viriditatem in eorum pollutis usibus consumpturi. porro siqui iam adultus aprum exceperit solus vel interemerit ursum immanem, conluvione liberatur incesti.*
>
> We have learned that these Taifali were a shameful folk, so sunken in a life of shame and obscenity, that in their country the boys are coupled with the men in a union of unmentionable lust, to consume the flower of their youth in the polluted intercourse of those paramours. We may add that, if any grown person alone catches a boar or kills a huge bear, he is purified thereby from the shame of unchastity.

17 Based on a detailed examination of the matrix, BLANEY (1972, 66–67) argued that the horned figure is one-eyed and can be related to the one-eyed god Óðinn (cf. PRICE 2019, 308). Blaney's suggestion was based on OXENSTIERNA 1956.

Most likely, these customs among early Germanic tribes in continental Europe can be related to the initiation rituals performed by *berserkir*-groups during the Viking Age in Scandinavia, which also included the killing of a predator, such as a bear (cf. BLANEY 1972, 91–92).

OLD NORSE *BERSERKSGANGR*

As a full member, the *berserkr* performed something called *berserksgangr* in battle contexts. This compound can be translated "*berserkir* rage or frenzy", or rather "*berserkir* movement or walking", i.e. "go berserk".[18] In *Ynglinga saga* chap. 6, Snorri describes *berserksgangr* as thus (*Snorri Sturluson*, 17; *Snorri Sturluson [transl.]*, 10):

> *[...] en hans menn fóru brynjulausir ok váru galnir sem hundar eða vargar, bitu í skjǫldu sína, váru sterkir sem birnir eða griðungar. Þeir drápu mannfólkit, en hvártki eldr né járn orti á þá. Þat er kallaðr berserksgangr.*
>
> [...] his [Óðinn's] own men went to battle without coats of mail and acted like mad dogs or wolves. They bit their shields and were as strong as bears or bulls. They killed people, and neither fire nor iron affected them. This is called go beserk [sic.] (*berserksgangr*).

The expression *berserksgangr* is also attested in other Old Norse prose narratives. When, for instance, the *berserkr* Ljot the Pale in *Egils saga* chap. 64 (13th century; *Egils saga*, 202) entered the arena, the *berserksgangr* came over him and he started bellowing (*grenja*) menacingly and biting at his shield (*Ok er hann gekk fram á vǫllinn at hólmstaðnum, þá kom á hann berserksgangr, tók hann þá at grenja illiliga ok beit í skjǫld sinn*).

The Old Norse verb *grenja* is also used for the sound of a bear (see above). The Viking Age poem *Haraldskvæði* indicates that the *berserkir* and the *ulfheðnar* were bellowing (*grenja*) and howling (*emja*) menacingly like bears and wolves just before battle. They were also hewing into their shields (see above). Archaeological evidence shows that this should not be interpreted as a literary invention as, for instance, Klaus von See argued previously (VON SEE 1961). For instance, the rook from the Lewis chessmen (Scotland; 12th century), depicting a warrior biting his shield, indicates the opposite (Fig. 5). Scholars have previously interpreted this battle fury of the *berserkir* as a symptom of physical or mental illness, rabies, or a genetic predisposition to violence in certain individuals (see overview in WALLENSTEIN 2015, 31; cf. NORDBERG/WALLENSTEIN 2016, 56). Recently, it has been argued that these warriors instead used certain techniques to alter their state of consciousness as preparation before battle (DALE 2014, 96–98). This might be compared with modern athletes using techniques to prepare before events. The animal-like expressions during *berserksgangr* were not symptoms of being *berserkir*, but a means for going berserk. Dale argues that the *berserkr* imitated the predators' behaviour in order to incorporate their rage or frenzy. This was a ritual act of mimesis that would enhance their ferocity. Historians of religion, Andreas Nordberg and Frederik Wallenstein, suggest that these warriors induced their violent, ecstatic and predatory state by means of ceremonial dances, including the wearing of animal masks and other rituals: The mental state of the berserkers during the *berserkersgangr* can be considered as a kind of inspired, focused battle trance in the form of a violent, ecstatic, and instinctive animal state (NORDBERG/WALLENSTEIN 2016, 57).

18 See HEGGSTAD et al. (2012, 69), who explains the term "rassinne som stundom tok berserkane", "berserksgang".

When the predator (the bear or the wolf) takes control of the warrior during his battle trance, his human side is distanced from his violent actions. This dehumanisation results in freedom from moral responsibility and human feelings.

Shifting shape

Sources report that both the deity Óðinn and the human *berserkir* had the ritual ability of shape-shifting. In *Ynglinga saga* chapt. 7, it is described as thus (*Ynglinga saga*, 18; Snorri *Sturluson; Snorri Sturluson [transl.]*, 10):

> Óðinn *skipti hǫmum. Lá þá búkrinn sem sofinn eða dauðr, en hann var þá fugl eða dýr, fiskr eða ormr ok fór á einni svipstund á fjarlæg lǫnd at sínum ørendum eða annarra manna. Þat kunni hann en at gera með orðum einum at sløkkva eld ok kyrra sjá ok snúa vindum hverja leið.*
>
> Óðinn could shift shape. When he did so his body would lie here as if he was asleep or dead; but he himself, in an instant, in the shape of a bird or animal, a fish or a serpent, went to distant countries on his or other men's errands.

In this euhemerised context, Óðinn is not presented as a deity, but as a "shaman" with ritual skills (Lindow 2003; cf. Clunies Ross 2019, 299).[19] *Egils saga* reports that human *berserkir* also possessed a shape-shifting ability, but not in a shamanic sense. Egill Skallagrímsson was born and raised in a *berserkir*-family. His grandfather Úlfr (also called Kveld-Úlfr) was a *berserkr* and was considered "a great shapeshifter" (*miǫk hamrammr*; *Egils saga*, chap. 1, 4).[20] When fighting he shifted shape (*hamask*), but he did not act out of his body as Óðinn did. The transformation of Kveld-Úlfr took place rather on a mental level. Once, when he was fighting against King Haraldr's men on a ship with his son, Skallagrímr (Egill's father), they both shifted shape, but they did not transform into animals in a physical sense: Kveld-Úlfr had a gigantic double-bladed axe in his hand. Once he was on board, he told his men to go along the gunwale and cut the awnings from the pegs, while he stormed off back to the afterguard, where he is said to have shifted shape (*at þá hamaðisk hann*). Some other men of his shifted shape, too (*er þá hǫmuðusk*), killing everyone they saw, and so did Skallagrímr when he ran around the ship (*Egils saga*, chap. 27, 69). Afterwards, when Kveld-Úlfr's fury was past, he was so tired that he had to go to bed and eventually he died there from his exertion.

The shape-shifting-phenomenon is thus described in many different ways in the sources (see overview in Aðalheiður Guðmundsdóttir 2007).[21] Some sources report that shape-shifting is related to shamanistic phenomena where the free-soul is transported in the body of an animal, while the practitioner's body is lying as dead, as in the case of Óðinn. Bǫðvarr bjarki shifted shape and appeared in his last battle as a bear, while his body was lying in the hall of King Hrólfr. *Hrólfs saga kraka* chaps. 50–51 (*The Saga of King Hrolf Kraki*, chap. 33, 74) states thus:

19 Rydving 2011 is critical of the use of the comparative concept of shamanism in cross-cultural studies.

20 In *Egils saga* it is said that Kveld-Úlfr was *hamrammr* "able to change one's shape". This compound includes the noun *hamr* meaning "form, shape", "skin", or "shape assumed by a disembodied spirit" and the adjective *rammr* "strong, powerful, mighty". A person who had the ability to change shape was also called *hamhleypa* "ham-leaper", i.e. he had the ability to act out-of-body like a wolf or a bear, in distance from his sleeping body. Other related terms and expressions are *hamremi* "the state of being *hamrammr*", *eigi einhamr* "not one skin", *hamast* "to assume another shape". See e.g. Cleasby/Vigfusson 1957; Samson 2011, 244–260; Dale 2014, 120–127; Wallenstein 2015, 29–30.

21 There is also another type of shape-shifting, which could be designated the werewolf-transformation, where the body is believed to be transformed. See Wallenstein 2015, 27–28, and below.

> *Hugrúnar scaltu kunna, ef þú vilt hveriom vera/ geðsvinnari guma;/ þær of réð, þær of reist,/ þær um hugði Hroptr,/ ... á biarnar hrammi [...].*
>
> Mind-runes you must know if you want to be / wiser-minded than every other man; / Hroptr [Óðinn] interpreted them, / cut them, thought them out, / [...] on the bear's paw, [...].

As mentioned in the introduction of the present study, Margaret Clunies Ross has criticised the notion that the *berserkir* used rites in order to be supernaturally empowered (Clunies Ross 2019; cf. Wamers 2009). Even if I can agree with her that the *berserkir* are not set in a religious context in the oldest preserved skaldic poems, the connection between Óðinn, bears, and *berserkir* in the sources just mentioned points in another direction. If the interpretation of the one-eyed figure on the helmet plate dies from Torslunda as Óðinn is correct, we actually have first-hand evidence of the relationship between the Æsir-god and animal-warriors. In addition, in the Icelandic medieval-Christian law called *Grágás* (13th century), *berserksgangr* is listed among the magical practices that attracted a penalty of lesser outlawry. In *Kristinna laga þáttr 7* (manuscript Konungsbók, *c.* 1258–1262), it is listed in the section of forbidden (pagan) religio-magical practices. There we read thus (*Grágás*; text and transl. after Dale 2014, 314):

> *Ef maðr gengr berserks gang ok warder þat fiörbaugs Garð ok sua wardar körlum þeim er hia eru. nema þeir stöðue hann at. þa wardar eingum þeirra ef þeir geta stoðuat hann. Ef optar kemr at. ok wardar fiorbaugs garð þott stoðuat werði.*
>
> If a man goes berserk (*gengr berserks gang*) he shall be punished with lesser outlawry as shall those men that are present unless they stop him. None shall be punished if they are able to stop him. But if it happens again the penalty is lesser outlawry even if he is stopped.

This instance indicates that "to go berserk" was considered as a type of ritual practice, which was still performed when the law was formulated. It also suggests that such a practice in more ancient times included some type of magico-religious associations.

Conclusion

To conclude, we cannot rule out the notion that bears were important symbols for warrior-bands who worshipped Óðinn during the Late Iron Age. Different types of sources, such as archaeological and iconographic material, as well as Old Norse texts, support this assumption. Memories of warriors wearing bear-shirts – the *berserkir* – were narrated in the High Middle Ages, and also later. In the romantic-nationalistic paintings of the 19th century, such motifs were popular, as may be seen in O. P. Hansen Balling's large oil painting *Harald hårfagre i slaget ved Hafrsfjord* (1870; cf. Fig. 1). Still today, the expression "to go berserk" is common.

Bibliography

Primary Sources

Adam of Bremen: Adam of Bremen, Magistri Adam Bremensis Gesta Hammaburgensis Ecclesiae Pontificium. Scriptores rerum germanicarum in usum scholarum. Ex Monumentis Germaniae Historicis. Editio Tertia. Ed. B. Schmeidler (Hannover, Leipzig 1917).

Adam of Bremen (transl.): Adam of Bremen, History of the Archbishops of Hamburg-Bremen. Translated with an introduction and notes by F. J. Tschan (New York [2]2002).

Ammianus Marcellinus: Ammianus Marcellinus, Histories, Book 14–19, Book 20–26, Book 27–31, volume I–III. Ed. J. Loeb. Transl. J. C. Rolfe (London, Cambridge [Massachusetts] 1956–1963).

Edda: Edda. Die Lieder des Codex Regius nebst verwandten Denkmälern. Band 1: Text. Ed. G. Neckel, 5. verbesserte Auflage, red. von H. Kuhn (Heidelberg [2]1983).

Egils saga: Egils saga. In: Egils saga Skallagrímssonar. Íslenzk fornrit 2. Ed. Sigurdur Nordal (Reykjavík 1933).

Egil's Saga: Egil's Saga. In: The Sagas of the Icelanders. A Selection. Translated by B. Scudder (London 2001) 3–184.

Eyrbyggja saga: Eyrbyggja saga. In: Eyrbyggja saga, Grœnlendinga sǫgur: Brands þáttr ǫrva. Eiríks saga rauða. Grœnlendinga saga. Grœnlendinga þáttr. Íslenzk fornrit 4. Eds. Einar Ól. Sveinsson/Matthías Þórðarson (Reykjavík 1985).

Eyrbyggja Saga: Eyrbyggja Saga. Translated with an introduction and notes by Heimir Pálsson/P. Edwards (London 1989).

Flateyjarbók: Flateyjarbók: En samling af norske kongesagaer med inskudte mindre fortællinger, 1–3. Eds. Guðbrandur Vigfusson/C. R. Unge (Christiania 1860–1868).

Grágás: Grágás. Islandernes lovbog i fristatens tid. Udgivet efter det Kongelige Bibliotheks haandskrift. Ed. V. Finsen (Copenhagen 1852, new print: Odense 1974).

Grettis saga: Grettis saga. In: Grettis saga Ásmundarsonar, Bandmanna saga, Odds þáttur Ófeigssonar. Íslenzk fornrit 7. Ed. Guðni Jónsson (Reykjavík [2]2001).

Háleygjatal: see *SkP 1*, 195–213.

Haraldskvæði: see *SkP 1*,91–116.

Hárbarðsljóð: see *Edda,* 78–87; *The Poetic Edda,* 65–73.

Hávamál: see *Edda*, 17–44; *The Poetic Edda,* 13–35.

Historia de gentibus septentrionalibus: Olaus Magnus, Historia de gentibus septentrionalibus (Rome 1555); Olaus Magnus, Historia om de nordiska folken. New edition (Stockholm 1982).

Historia Langobardorum: Paulus Diaconus, Pauli Historia Langobardorum. Ed. L. Bethmann/G. Waits. Monumenta Germaniae Historica. Scriptores rerum Langobardicarum et Italicarum Saec VI–IX (Hanover 1878); see also: Ed. C. Clemen, Fontes Historiae Religionis gee (Berlin 1928) 48–51.

Hrólfs saga kraka: Hrólfs saga kraka ok kappa hans. In: Fornaldar saga Norðurlanda, vol. II. Ed. Guðni Jónsson (Reykjavík 1959).

Hyndluljóð: see *Edda*, 288–296; *The Poetic Edda,* 245–251.

Landnámabók: Landnámabók. In: Íslendingabók. Landnámabók. Íslenzk fornrit 1. Ed. Jakob Benediktsson (Reykjavik 1986).

Origo Gentis Langobardorum: Origo Gentis Langobardorum. In: Scriptores rerum Langobardicarum et Italicarum saec. VI–IX. Ed. G. Waitz. Monumenta Germaniae Historica. Scriptores rerum Langobardicarum et Italicarum (SS rer. Lang.) 1 (Hannover 1878); see also ed. C. Clemen, Fontes Historiae Religionis gee (Berlin 1928) 34–35.

Saxo Grammaticus: Saxo Grammaticus, Gesta Danorum I–II. Ed. K. Friis-Jensen. Transl. P. Fischer (Oxford 2015).

SkP I: SkP I = Poetry from the Kings' Sagas 1: From Mythical Times to c. 1035. Skaldic Poetry of the Scandinavian Middle Ages 1. Ed. D. Whaley (Turnhout 2012).

SkP III: SkP III = Poetry from Treaties on Poetics. Skaldic Poetry of the Scandinavian Middle Ages 3. Ed. K. E. Gade in collaboration with E. Marold (Turnhout 2017).

Snorri Sturluson: Snorri Sturluson, Heimskringla I–III. Íslenzk Fornrit 26–28. Ed. Bjarni Aðalbjarnarson (Reykjavík [4]1979).

Snorri Sturluson (transl.): Snorri Sturluson, Heimskringla. History of the Kings of Norway. Transl. L. M. Hollander (Austin 1964).

Tacitus: Tacitus, Cornelius. Germania. In: Agricola, Germania, Dialogus. Ed. J. Loeb. Transl. M. Hutton (London, Cambridge [Massachusetts] [2]1992).

The Poetic Edda: The Poetic Edda. A new translation by C. Larrington (Oxford 2014).

The Saga of King Hrolf Kraki: The Saga of King Hrolf Kraki. Translated with an introduction by J. L. Byock (London 1998).

The Saga of the People of Vatnsdal: The Saga of the People of Vatnsdal. In: The Sagas of the Icelanders. A Selection. Translated by A. Wawn (London 2001) 185–269.

Vatnsdœla saga: Vatnsdœla saga. In: Vatnsdæla saga. Hallfreðar saga. Kormáks saga. Hrómundar þáttr halta. Hrafns þáttr Guðrúnarsonar. Íslenzk fornrit 8. Ed. Einar Ól. Sveinsson (Reykjavík 1939).

Víga-Glúms saga: Víga-Glúms saga. In: Eyfirðinga sögur. Islenzk Fornrit 9. Ed. Jónas Kristjánsson (Reykjavík [2]2001).

Vǫlsunga saga: Vǫlsunga saga. In: Fornaldar saga Norðurlanda, vol. I. Ed. Guðni Jónsson (Reykjavík 1959).

Ynglinga saga: Ynglinga saga. See *Snorri Sturluson*, Heimskringla I.

Secondary Sources

Aðalheiður Guðmundsdóttir 2007: Aðalheiður Guðmundsdóttir, The Werewolf in Medieval Icelandic Literature. Journal of English and Germanic Philology 2007, 277–303.

Beck 1968: H. Beck, Die Stanzen von Torslunda und die literarische Überlieferung. Frühmittelalterliche Studien 2(1), 1968, 237–250.

Blaney 1972: B. Blaney, The Berserkr: His Origin and Development in Old Norse Literature (Ann Arbor 1972).

Blaney 1982: B. Blaney, The Berserk Suitor: The Literary Application of a Stereotyped Theme. Scandinavian Studies 54, 1982, 279–294.

Böhner 1991: K. Böhner, Die frühmittelalterlichen Silberphaleren aus Eschwege (Hessen) und die nordischen Preßblechbilder. Jahrbuch RGZM 38, 1991, 681–743.

Breen 1997: G. Breen, Personal Names and the Re-creation of berserkir and úlfheðnar. Studia anthroponymica Scandinavica 15, 1997, 5–38.

Cleasby/Vigfusson 1957: R. Cleasby/G. Vigfusson, An Icelandic-English Dictionary (Oxford [2]1957).

Clunies Ross 2019: M. Clunies Ross, Response. In: K. Wikström af Edholm/P. Jackson Rova/A. Nordberg/O. Sundqvist/T. Zachrisson (eds.), Myth, Materiality, and Lived Religion: In Merovingian and Viking Scandinavia (Stockholm 2016) 297–301.

Dale 2014: R. T. D. Dale, Berserkir: a Re-Examination of the Phenomenon in Literature and Life (Nottingham 2014).

Danielli 1945: M. Danielli, Initiation Ceremonial from Norse Literature. Folklore 56(2), 1945, 229–245.

De Vries 1956/1957: J. De Vries, Altgermanische Religionsgeschichte I–II. Grundriss der germanischen Philologie 12/1–2 (Berlin 1956/1957).

De Vries 1961: J. De Vries, Altnordisches etymologisches Wörterbuch (Leiden 1961).

DuBois 2012: T. DuBois, Diet and Deities. Contrastive Livelihoods and Animal Symbolism in Nordic Pre-Christian Religions. In: C. Raudvere/J. P. Schjødt (eds.), More than Mythology. Narratives, Ritual Practices and Regional Distribution in Pre-Christian Scandinavian Religion (Lund 2012) 65–96.

Dumézil 1973: G. Dumézil, Gods of the Ancient Northmen. Publications of the UCLA Center for the Study of Comparative Folklore and Mythology 3 (Berkeley 1973).

Edsman 1994: C.-M. Edsman, Jägaren och makterna. Samiska och finska björnceremonier. Skrifter utgivna genom dialekt- och folkminnesarkivet i Uppsala (Uppsala 1994).

Falk 1924: H. Falk, Odensheiti (Kristiania 1924).

Fritzner 1954: J. Fritzner, Ordbog over det gamle norske sprog. Omarb., foroget og forbedret (Oslo 1954).

Frog 2019: Frog, Understanding Embodiment Through Lived Religion: A Look at Vernacular Physiologies in an Old Norse Milieu. In: K. Wikström af Edholm/P. Jackson Rova/A. Nordberg/O. Sundqvist/T. Zachrisson (eds.), Myth, Materiality, and Lived Religion: In Merovingian and Viking Scandinavia (Stockholm 2019) 269–301.

Grimm 2013: O. Grimm, Bear-skins in northern European burials and some remarks on other bear-related furnishings in the north and middle of Europe in the 1st millennium AD. In: O. Grimm/U. Schmölcke (eds.), Hunting in northern Europe until 1500 AD. Old traditions and regional developments, continental sources and continental influences. Schriften des Archäologischen Landesmuseums, Ergänzungsreihe 7 (Neumünster 2013) 277–296.

Hauck 1957: K. Hauck, Alemannische Denkmäler der vorchristlichen Adelskultur. Zeitschrift für Württembergische Landesgeschichte XVI, 1957, 1–39.

Hedeager 2011: L. Hedeager, Iron Age Myth and Materiality: An Archaeology of Scandinavia AD 400–1000 (London, New York 2011).

Heggstad et al. 2012: L. Heggstad /F. Hednebø/E. Simensen, Norrøn ordbok (Oslo [5]2012).

Höfler 1934: O. Höfler, Kultische Geheimbünde der Germanen 1 (Frankfurt a. M. 1934).

Hultgård 2007: A. Hultgård, Wotan-Odin. In: Reallexikon der Germanischen Altertumskunde 35 (Berlin, New York 2007) 759–785.

Jesch 2002: J. Jesch, Eagles, Ravens and Wolves: Beasts of Battle, Symbols of Victory and Death. In J. Jesch (ed.), The Scandinavians from the Vendel Period to the Tenth Century: An Ethnographic Perspective. Studies in Historical Archaeoethnology 5 (Woodbridge 2002) 251–271.

Kershaw 2000: K. Kershaw, The One-eyed God: Odin and the (Indo-)Germanic Männerbünde. Journal of Indo-European Studies, Monograph Series 36 (Washington D. C. 2000).

Kuhn 1968: H. Kuhn, Kämpen und Berserker. Frühmittelalterliche Studien 2, 1968, 218–227.

Kuusela 2012: T. Kuusela, Varulven i fornnordisk tradition. In: E. Odstedt (ed.), Varulven i svensk folktradition (Stockholm 2012) 320–352.

Lid 1924: N. Lid, Norske slakteskikkar: med jamføringar frå naerskylde umråde: fyrste luten. Skrifter utgit av Videnskapsselskapet i Kristiania, 1923: II. historisk-filosofisk klasse (Kristiania 1924).

Lind 1905–1915: E. H. Lind, Norsk-Isländska dopnamn ock fingerade namn från medeltiden (Uppsala, Leipzig 1905–1915).

Lindow 2003: J. Lindow, Cultures in Contact. In: M. Clunies Ross (ed.), Old Norse Myths, Literature and Society. The Viking Collection 14 (Odense 2003) 89–109.

Lund/Mateeva 1997: A. A. Lund/A. S. Mateeva, Gibt es in der Taciteischen 'Germania' Beweise für kultische Männerbünde der frühen Germanen? Zeitschrift für Religions- und Geistesgeschichte 49, 1997, 208–216.

Meier 2001: M. Meier, Männerbund. In: Reallexikon der Germanischen Altertumskunde 19 (Berlin, New York 2001) 105–110.

Meulengracht Sørensen 1991: P. Meulengracht Sørensen, Om eddadigtenes alder. In: G. Steinsland/U. Drobin/J. Pentikäinen/P. Meulengracht Sørensen (eds.), Nordisk hedendom: Et symposium (Odense 1991) 217–228.

Näsström 2006: B.-M. Näsström. Bärsärkarna (Stockholm 2006).

Nedoma 2001: R. Nedoma, Langobarden I. Philologisches. § Ethnonym. In: Reallexikon der Germanischen Altertumskunde 18 (Berlin, New York 2001) 50–52.

Nordberg/Wallenstein 2016: A. Nordberg/F. Wallenstein, "Laughing I Shall Die!": The Transformations of Berserkers and *Úlfheðnar* in Old Norse Society. In:

P. Haldén/P. Jackson (eds.), Transforming Warriors: The Ritual Organization of Military Force (New York 2016) 49–65.

Noreen 1932: E. Noreen, Ordet bärsärk. Arkiv för nordisk filologi 48, 1932, 242–254.

Oehrl 2013: S. Oehrl, *Svá beitum vér björnuna á mörkinni norðr* – Bear hunting and its ideological context (as a background for the interpretation of bear claws and other remains of bears in Germanic graves of the 1st millennium AD). In: O. Grimm/U. Schmölcke (eds.), Hunting in northern Europe until 1500 AD. Old traditions and regional developments, continental sources and continental influences. Schriften des Archäologischen Landesmuseums, Ergänzungsreihe 7 (Neumünster 2013) 297–332.

ONP: ONP, Dictionary of Old Norse Prose. https://onp.ku.dk/onp/onp.php?

Oxenstierna 1956: E. Oxenstierna, Die Goldhörner von Gallehus (Lidingö 1956).

Paulsen 1967: P: Paulsen, Alamannische Adelsgräber von Niederstotzingen (Kreis Heidenheim). (Stuttgart 1967).

Peterson 2007: L. Peterson, Nordiskt runnamnslexikon (Uppsala 2007).

Price 2019: N. Price, The Viking Way. Magic and Mind in Late Iron Age Scandinavia (Oxford, Philadelphia 22019).

Rydving 2011: H. Rydving, Le chamanisme aujord'hui: Constructions et deconstructions d'une illusion scientifique. Études mongoles et sibériennes, centrasiatiques et tibétaines 42, 2011, 1–13.

Samson 2011: V. Samson, Les Berserkir. Les guerriers-fauves dans la Scandinavie ancienne, de l'âge de Vendel aux Vikings (VIe-XIe siècle) (Villeneuve d'Ascq 2011).

Schjødt 2008: J. P. Schjødt, Initiation between Two Worlds. Structure and Symbolism in Pre-Christian Scandinavian Religion. The Viking Collection 17 (Odense 2008).

Schjødt 2012: J. P. Schjødt, Reflections on aims and methods in the study of Old Norse religion. In C. Raudvere/J. P. Schjødt (eds.), More than Mythology. Narratives, ritual practices and regional distribution in pre-Christian Scandinavian religions (Lund 2012) 263–287.

Speidel 2002: M. P. Speidel, Berserks: A History of Indo-European 'Mad Warriors'. Journal of World History 13(2), 2002, 253–290.

Stausberg 2014: M. Stausberg, Comparison. In M. Stausberg/S. Engler (eds.), The Routledge Handbook of Research Methods in the Study of Religion (London, New York 2014) 21–39.

Stjerna 1903: K. Stjerna, Hjälmar och svärd i Beovulf. In: B. Salin/O. Almgren/S. Ambrosiani (eds.), Studier tillägnade Oscar Montelius 9/9 1903 af lärjungar (Stockholm 1903) 99-120.

Ström 1980: Å. V. Ström, Björnfällar och Oden-religion. Fornvännen 75, 1980, 206–210.

Sundqvist/Hultgård 2004: O. Sundqvist/A. Hultgård, The Lycophoric Names of the 6th to 7th Century Blekinge Rune Stones and the Problem of their Ideological Background. In: A. van Nahl/L. Elmevik/S. Brink (eds.), Namenwelten. Orts- und Personennamen in historischer Sicht. In: Ergänzungsbände zum Reallexikon der Germanischen Altertumskunde 44 (Berlin, New York 2004) 583–602.

Tolley 2009: C. Tolley, Shamanism in Norse Myth and Magic I–II. Folklore Fellows Communications 296 (Helsinki 2009).

Turner 1967: V. Turner, The Forest of Symbols: Aspects of Ndembu ritual (Ithaca [New York] 1967).

Turner 1969: V. Turner, The Ritual Process. Structure and Anti-Structure (New York 1969).

Van Gennep 1960: A. Van Gennep, The Rites of Passsage (Original: Les rites de passage. 1909; transl. M. B. Vizedom/G. L. Caffee) (Chicago 1960).

Von See 1961: K. Von See, Berserker. Zeitschrift für deutsche Wortforschung 17, 1961, 129–135.

Wallenstein 2015: F. Wallenstein, Varggudens raseri: extas och djurblivande hos bärsärkar och úlfheðnar. Aiolos – Tidskrift för litteratur teori och estetik 49, 2015, 25–44.

Wamers 2009: E. Wamers, Von Bären und Männern. Berserker, Bärenkämpfer und Bärenführer im frühen Mittelalter. Zeitschrift für Archäologie des Mittelalters 37, 2009, 1–46.

Weiser 1927: L. Weiser, Altgermanische Jünglingsweihen und Männerbünde. Ein Beitrag zur deutschen und nordischen Altertums- und Volkskunde (Baden 1927).

Prof. Dr. Olof Sundqvist
Department of Ethnology, History of Religions and Gender Studies
Stockholm University
Stockholm
Sweden
Olof.Sundqvist@rel.su.se

Fig. 1. O. P. Hansen Balling's large oil painting Harald hårfagre i slaget ved Hafrsfjord, *1870 (© Nasjonalmuseet, Norway).*

Fig. 2. The 7th-century helmet-plate dies from Björnhovda, Torslunda (Öland). a: Man between bears; b: Man with axe holding a roped animal; c: Warriors carrying spears; d: Dancing man with horned head-dress and man with spear wearing a wolfskin (© Statens Historiska Museer, Sweden, inv. no. SHM 4325; cf. Stjerna *1903, fig. 1).*

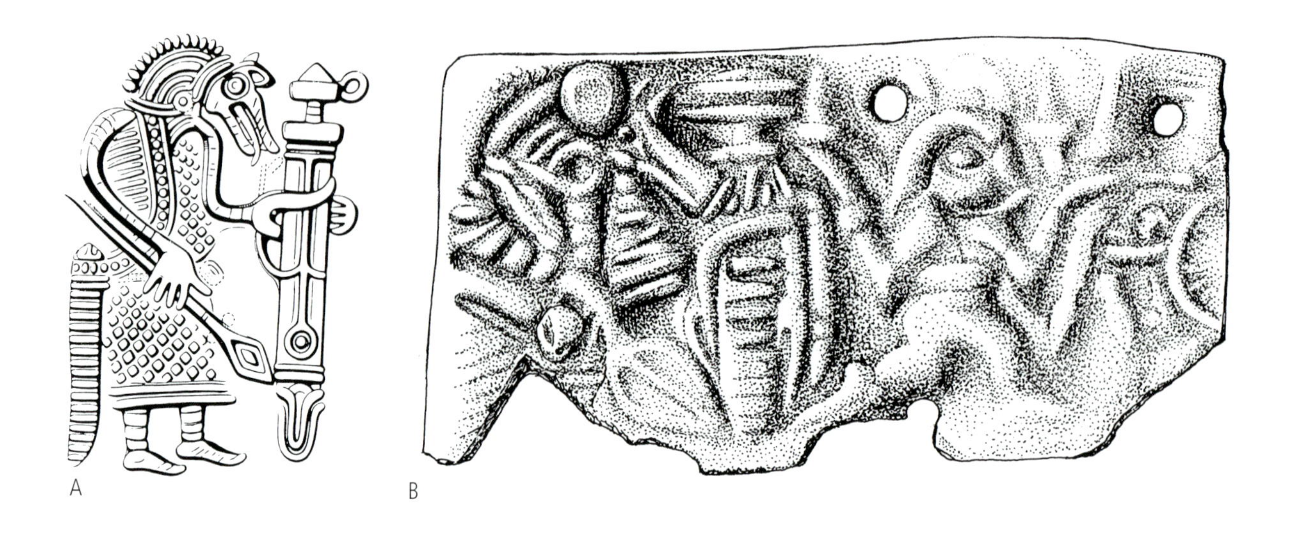

Fig. 3. Disputed berserkir *imagery. a: Scabbard mount (re-used), grave 2 from Gutenstein, Sigmaringen, Baden-Württemberg, Germany, late 7th century or around 700 (after* Hauck *1957, fig. 5); b: Sheet-metal, grave 139 from Obrigheim, Bad-Dürkheim, Rheinland-Pfalz, Germany, middle third of the 7th century (after* Böhner *1991, 717 fig. 29); c: Purse lid from the Sutton Hoo ship burial, England, 7th century (drawing L. F. Thomsen after a template); d: Image from the runic stone from Källby, Västergotland, Sweden, c. 1000 or early/first half of the 11th century (after* Paulsen *1967, vol. 1, 44 fig. 19).*

Fig. 4. Hǫttr's blood-drinking ceremony, from Hrólfs saga kraka, *as depicted in Olaus Magnus'* Historia de gentibus septentrionalibus, *Book 5, chap. 15 (after* Historia de gentibus septentrionalibus *1982).*

Fig. 5. Rook chessman depicting a warrior who bites his shield; Isle of Lewis, Scotland, 12th century (© The Trustees of the British Museum).

Table 1. Sources on berserkir *– a simplified classification model. *Skaldic and Eddic poems as composed in the Viking Age by "insiders" are direct sources, although they were not written down until the Middle Ages. Their worldview and ethics are obviously not Christian, and the medieval Icelanders regarded them as pagan. They can thus be considered as oral art forms from the past. The formalistic language in these poems, rhymes, kennings, and their metres might have allowed them to retain their original shapes for centuries. Especially skaldic poetry can be regarded as based on a firm oral tradition before being written down. Some Eddic poems are problematic for this classification, since their dating is so uncertain. Cf. MEULENGRACHT SØRENSEN 1991.*

Primary sources (produced by insiders)
Skaldic and Eddic poetry*
Archaeological and iconographic materials
Onomastic sources
Runic inscriptions

Secondary sources (produced by outsiders)
Icelandic Sagas (Kings' Sagas, The Sagas of Icelanders, Fornaldarsögur)
Medieval Laws
Classical and Medieval Latin texts

“The Bear Ceremonial” and bear rituals among the Khanty and the Sami

By Håkan Rydving

Keywords: Bear rituals, comparison, Khanty, limitative approach, Sami

Abstract: Researchers from different fields of study agree on the importance of comparison, but debate how *to compare. Rather than comparing globally, on the basis of secondary literature and looking for similarities alone, this article argues for a limitative approach that restricts itself to just a few cultures, is based on local sources, and takes both resemblances and differences into account. In contrast to the idea of a uniform and transcultural bear ceremonial in northern Eurasia, it focuses on plurality and diversity when discussing and comparing the bear rituals found among the southern Khanty (about 1900) and the southern Sami (about 1750).*

Editorial note: This text was already published in 2010 in Temenos 46(1), issued by the Finnish Society for the Study of Religion. It is reprinted here, with slight editorial changes, with the permission of the present editors of Temenos and the author.

Introduction

When the non-theological study of religions began at European universities during the late 19th and the early 20th century, analysis by means of various forms of comparison became *the* characteristic method of this new field of research; as a result, at a number of universities the subject was named “comparative religion”.

In most cases, the comparative enterprise consisted of macro-comparison and a search for similarities. Data was collected from all over the world – generally from secondary sources – and arranged according to types, based on evolutionary theories. When these theories were abandoned during the 1950s, various non-evolutionary and non-historical morphologies were developed. These were often called “phenomenologies of religion”, a confusing term, since they had little or nothing to do with philosophical phenomenology. Even in studies of singular religions, parallels and analogies drawn from one or another of the “phenomenologies of religion” were used to interpret the data, without paying any attention to time and place.

While the study of religions still adhered to comparison as its main analytic tool, anthropology had taken another direction, not least due to the seminal local studies of British anthropologists such as Bronislaw Malinowski and E. E. Evans-Pritchard, whose investigations were based on prolonged fieldwork. The anthropological focus on the local field required scholars to learn local languages and – rather than use phenomena from elsewhere to interpret the culture under consideration – to

Bear and Human: Facets of a Multi-Layered Relationship from Past to Recent Times, with Emphasis on Northern Europe, ed. by Oliver Grimm (Turnhout: Brepols, 2023), pp. 677–692 BREPOLS PUBLISHERS 10.1484.M.TANE-EB.5.134356

some 70 % of whom speak Khanty. One usually reckons with eight main dialects. Of the five different literary languages in use, the most important is the one based on the Middle Ob' dialects (Skribnik/Koshkaryova 1996). Culturally, the Khanty in the different areas are closely related to the other Ob-Ugrian people, the Mansi, but linguistically the two ethnic groups are clearly differentiated. This means for example that the northern Khanty have a culture that is more similar to the culture of the northern Mansi than to that of the southern and eastern Khanty, although their language is distinctly Khanty. A typical characteristic of the social culture of the Khanty is the division into two exogamous patrilineal phratries, the *por* and the *mosh*. The Por are linked to the bear and the Mosh to the hare (or goose; cf. Balzer 1999, 184). Since the Khanty are a small people spread over a large area, there are considerable cultural and linguistic differences between the different regions.[5] Even so, scholars have had a tendency to generalise and write as if all the Khanty (or even all the Ob-Ugrians) shared an identical (or at least very similar) culture and religion, although this tendency has not been as common in studies of the Khanty as it has in those of the Sami (see below).

The earliest information we have about Khanty bear rituals dates from the beginning of the 18th century, but it was not until the late 19th century that descriptions became more detailed. The most important information was collected and published by Russian scholars like Nikolay Gondatti (Gondatti 1888) and Serafim Patkanov (Patkanov 1897; 1900), and, at the beginning of the 20th century, by the Finnish scholar K. F. Karjalainen (Karjalainen 1914; 1927, 193–235). In the post-revolutionary era, the material collected by V. N. Chernetsov is especially valuable. Chernetsov documented bear rituals in 1936/37, succeeded in filming dances at a bear ritual in 1948, and collected new material during the 1960s (Tschernjetzow 1974, 285, n. 1). During the Soviet era the bear ceremonials did not cease, at least not in the north and east. They were so popular that the authorities even thought of declaring them secularised (Balzer 1999, 190), bringing them under the general policy of folklorisation. However, such a decision was never made. Since the fall of the Soviet Union and the renaissance of indigenous customs, some of the bear festivals have become important political manifestations of Khanty (and Ob-Ugrian) unity. Today, we might echo the Russian folklorist Olga Balalaeva in distinguishing two types of bear festival: "quite private, elder-led festivals that occur on the back rivers of Eastern Khanty camps and the larger, more popular festivals led by [members of the political Association for the Salvation of the Ugra] as well as elders" (Balzer 1999, 197). The traditional bear ceremonials differed from one Khanty region to another. In the north there were only few rituals during the actual hunt and in bringing a bear home to the settlement, with more elaborate ceremonies occurring during the bear festival itself; in the south, rituals connected to the hunt and the homeward journey were dominant; in the east, the bear ceremonials were relatively insignificant (Karjalainen 1927, 194, 200).

The example I will discuss concerns the rituals of the southern Khanty, as that area is best documented in the oldest sources. Today, the southern Khanty are Russified and totally integrated into mainstream society.

The Sami

The traditional settlement area of the Sami – the westernmost of the peoples that speak a Finno-Ugric language – stretches in an arc from the central parts of Scandinavia to the Kola peninsula (Fig. 2). There are today about 80,000 Sami, of whom approximately 40 % speak Sami. Like the Khanty, the Sami are a small group of people spread over a large area; the Sami language is therefore split into a number of different dialects and dialect groups. One usually reckons with ten main dialects, further divided into dialects and sub-dialects. Of the six Sami literary languages in use, North Sami is the most important.

5 Cf. the excellent introduction to the Khanty world-view in Jordan 2003.

Despite the great linguistic and cultural variation, most scholars who have studied Sami culture have disregarded this variation and written about the Sami as if they all shared an identical culture.

The oldest information about Sami bear ceremonials dates from the beginning of the 17th century, but the most important sources are from the late 17th and the first half of the 18th century. The first monograph was published as early as 1755 (Fjellström 1755). We know that there were regional differences, but even today no study has tried to map them. Unlike among the Khanty, the Sami bear rituals are no longer performed, since the indigenous religion perished during the 18th century. A few examples of bear ceremonials, however, are mentioned in 19th-century sources. The area that is best documented is that of the South Sami, and my example is therefore from that region. Despite several centuries of cultural and linguistic influence from Norwegian and Swedish, as well as a cultural impact, many South Sami still speak their original language. This is accounted for by several factors: the South Sami have lived isolated from Norwegians and Swedes, with an economy based almost exclusively on the reindeer; Sami is used as a means of communication within families; the language has a high status and is an important social and cultural symbol of identity (Jernsletten 1997; Rydving 2004a; b; 2008).

The rituals

How should rituals be compared? One possible way is to break them up into elements, as Anna-Leena Siikkala did in her study of indigenous ritual specialists ("shamans" in her terminology) in Siberia (Siikkala 1978). I did the same in my study of ritual aspects of the process of religious change among the Lule Sami, although I used another terminology, adopted from Melford E. Spiro, which is the terminology I will use here as well (cf. Rydving 2004a, 93). This means that I use "ritual" as "the generic term for any kind of cult behaviour, regardless of its degree of elaboration or complexity", while "rite" denotes "the minimum significant unit of ritual behaviour", "ceremony" "the smallest configuration of rites constituting a meaningful ritual whole", and "ceremonial" "the total configuration of ceremonies performed during any ritual occasion" (Spiro 1982, 199).

Bear ceremonials among the southern Khanty in the decades around 1900[6]

Among the southern Khanty, in the decades around 1900 the bear hunt was undertaken roughly as follows.[7]

A. The hunt

1) Preparations for the hunt. The bear hunt was never to be planned and it was regarded as dangerous to try to track a bear (Karjalainen 1927, 194). Those who were going to participate in the hunt first had to go through a purification ritual. Both the participants and their food were purified with incense, and they prayed to the bear that it would let them kill it without hurting any of the hunters. After the participants had made the sign of the cross in front of the icons they set out.

6 After Patkanov 1897; Karjalainen 1914; 1927, 193–235; cf. Gondatti 1888; Kharuzin 1895a; b; Kálmán 1968; Tschernjetzow 1974; Schmidt 1989; Glavatskaya 2005.

7 Since the level of analysis in this short article is restricted to the comparison between Khanty and Sami bear rituals in relation to the idea of a "North-Eurasian Bear Ceremonial", I have neither here, nor in the next section (about bear ceremonials among the southern Sami), discussed variations and changes *within* Khanty and Sami bear rituals, only between them. Discussing internal variations as well as differences between different types of sources (texts, archaeological material, participant observation) within each of these two cultures would be the next step in the investigation. Of course, I also need to discuss bear rituals among other north Eurasian groups than the two here under examination in order to test the general applicability of the results of this pilot study.

2) The hunt. Once the bear had been killed, the man who had killed it would throw snow (in the winter) or earth (in the summer) on himself before all the hunters ate the food they had brought with them. The bear, too, was regarded as participating in this meal. Then the body of the bear was laid on its back with its head towards the east, and it was skinned. During the process of removing the skin, the one doing the skinning said a short phrase and broke a few[8] short sticks that were placed beside the bear.

3) On the way home. Where the eating of bear meat was prohibited,[9] the meat was left at the site of the kill, otherwise it was taken to the village. In either case, the skin, with the head, was taken to the village. On the way to the village, those who carried or dragged the bear told it about the places they passed by.

4) Returning to the village. When the hunters and the bear arrived at the village they were greeted. The hunters fired their weapons and the people in the village answered with shots and came out to meet them. The bear skin was sprinkled with water and incense, and the one who had killed it was thrown into the water. Despite the clear division of male and female rituals, both men and women participated in these rituals.

B. The bear festival

1) Preparations for the bear festival. The skin with the head was taken through a back window into the room where the festival was to be held; there it was placed in the sacred corner, with its head resting on its front paws facing the door. A male bear was provided with a cap and a scarf, a female bear with a head cloth and a neck collar with pearl embroidery and rings on the claws. Usually the eyes were covered (in some places the nostrils as well) with coins or pieces of birch-bark. Different types of food were then placed in front of the bear (Karjalainen 1927, 203–206).

2) The bear festival (*îke-pore*). This ceremonial consisted of entertainment for the bear and the participants. It continued over several (often three) evenings. All the participants were sprinkled with water or snow, and they greeted the bear with kisses. The host (or someone else) said to the bear: "Turem's son [...]! With an arrow made by Russians you were killed, with a spear made by Russians you were killed. Don't be angry with us!" Thereafter everyone sat down in a fixed order and the festival could begin. It consisted of three elements: bear songs, dancing performances, and short plays (Karjalainen 1927, 206–224). *The bear songs* were sung by male singers without masks and they described the life of the bear.[10] *The dancing performances* took place in intervals between songs and plays. Among certain southern Khanty, for example those living along the river Konda, these dances were the most important element of the bear festival. The dances were performed by men and women in garments that differed from their ordinary clothes and with their faces covered. One of the dances, the so-called bear dance, gave an account of the life of the bear. *The plays* (which were a considerably less important part of the bear festival among the southern Khanty than among the northern) were performed by men in red masks which were often made of birch-bark. The masks that represented men had large noses, while those who performed women wore women's clothes and head-scarves. The actors distorted their voices. The themes were everyday subjects; they could be serious or humorous, and made use of many puns, and were sometimes offensive (even to persons who were present).[11] Each play was very short, but the number of plays could in some areas (but not among the southern Khanty) be very large (Karjalainen 1927, 229–230). In the breaks between the plays songs were sung. The bear festival ended with games to decide when and by whom the next bear was to be killed.

8 Five or seven, if it was a male bear, four or five, if it was a female bear (Karjalainen 1927, 197).

9 Depending on which phratrie one belonged to, the *mosh* (who could eat the meat) or the *por* (who could not eat the meat).

10 Such South Khanty bear songs are reproduced in, for example, Karjalainen/Vértes 1975; Paasonen/Vértes 1980.

11 Some plays are presented in Gondatti 1888 and Karjalainen 1927, 215–220.

3) After the bear festival, the skin was taken out through the back window near the sacred corner. If the bear meat was to be eaten, it was cooked so as to be ready when the festival was completed.

C. Afterwards
The skin was given to the host (the one who had found the bear and arranged the festival) and he could use it as he liked. Generally, it was sold to cover the expenses of the festival. However, it could not be sold until forty days had elapsed, and the host had to celebrate memorial days on the 9th, 16th, and 36th day after the festival.

Bear ceremonials among the southern Sami in the mid-18th century[12]
Unlike among the Khanty, the Sami bear festival was generally celebrated in springtime (although it could also be celebrated during autumn or winter). It thus functioned as a calendrical spring ritual. The following synthesis is based on sources describing the southern Sami festival during the first half of the 18th century.

A. The hunt
1) The bear was tracked (or encircled) during the autumn after the first snow had fallen so that it would be easy to know where it was hibernating.
2) Before the hunt. Among the southern Sami, no preparatory rituals are known from this period that would be comparable to the strict order in which the hunters approached the place where the bear was hibernating found among the Lule Sami during the 1670s. In the latter case, the person who had tracked the bear went first, followed by the others in a specific order.
3) After the hunt. Once the bear had been killed, the hunters walked over it on their skies. Then a twig was attached to the bear's mouth, and one of the hunters sang a *vuelie* (chant) and pulled the twig three times. He could also aim a spear three times at the bear. After that the bear was covered with twigs and left at the hunting ground.
4) Returning home. As the hunters approached the huts they sang a special *vuelie* in order to let the women and children know they had killed a bear. The *vuelie* also told the persons in the settlement whether it was a male or a female. Using the back door, the men entered the tent where the women were sitting with their heads covered. The women looked at the bear hunters through rings of brass, spat chewed alder bark on them, and fastened brass rings on the men's clothes (cf. Paproth 1963). Then they feasted on the best food they had, the hunters in a tent that was erected especially for the purpose of the bear rituals, the women and children in the ordinary tent. After the meal, everyone went to sleep.
5) Collecting the bear. On the second day, the bear was collected with great honours. On the way home the hunters sang different *vuelieh* (chants) and prayed to the bear to protect them from evil.

B. The feast
The bear was taken to the special tent, and the women spat red chewed alder bark at it. It was then skinned by the men, while they sang various *vuelieh*. In one of the *vuelieh* they sang to the bear that it had been killed by men from Sweden, Poland, England, and France. The other *vuelieh* were about where the bear had been taken, about the honour it should be shown, about what the women might be doing in the ordinary tent, etc. Per Fjellström, who wrote the first monograph about the Sami bear rituals, gives the following characterisation of the *vuelie* to the bear:

12 After Fjellström 1755; cf. Niurenius 1905, 18–19 (original text *c.* 1640); Rheen 1897, 43–46 (original text 1671); Lundius 1905, 18 (original text late 1670s); Högström 1747, 209–211; text by Holmberger (1770s), in Hasselbrink 1964; cf. Zachrisson/Iregren 1974; Edsman 1994; 1996; Korhonen 2008.

"[...] the so-called bear song is not the same and does not have the same contents among all [groups], nor would they decide in advance and prepare a precise order in which to sing it. Instead they probably adjusted [the song] both to existing circumstances and to the bear hunt itself, as well as to the Lapps' own conditions and nature. [...] Thus, it is believed that their bear song is performed more with voice and sound than with words. Even if their song makes use of pure words, they are such as are unusual and not used at all in the ordinary Lappish language, and therefore they cannot be understood by anyone, regardless of how skilled they might be in their language, other than those who are instructed and trained in their superstition" (Fjellström 1755, 21–22).

After the bear had been skinned, the meat was carved from the bones and boiled in a certain order. The men ate certain parts of it (which parts, depended on the sex of the bear) in the special tent, the women and children other parts in the ordinary tent. After that everyone rested. Then the hunters washed themselves in lye, ran three times around the place where the bear had been cooked and into and out of the ordinary tent through the ordinary door and the back door, while imitating the growl of the bear.

C. The burial

1) The bear's bones were buried in a precise order. It was important that no bone was broken and that all of them were buried.[13]

2) After the burial. The skin or the liver was used in a game that decided when and by whom the next bear was to be killed.

D. Afterwards

The man who had tracked the bear received the skin and sold it.

Comparison

If we now compare these two ritual complexes we have to look at both the structural level (how elements are connected, the order of the different elements) and the individual rites (the elements of each complex; cf. Tables 1–2). This is possible even if we do not know the exact meaning of all the rites performed. Earlier ritual theories regarded rituals as something scholars could use to "read" the respective culture, since rituals were regarded as communicative acts; the theories formulated by Frits Staal, Catherine Bell, Caroline Humphrey and James Laidlaw, and Roy Rappaport, in contrast, emphasise, among other things, the role of rituals as tools for enculturation and for the "disciplining of the body" even if the "meaning" (as suggested by the ritual specialists or by scholars) is not understood by all – or by any – of the participants (Staal 1975; Bell 1992; Humphrey/Laidlaw 1994; Rappaport 1999). This means that the activities (movements, sounds, etc.) are interesting objects for analysis, even if they only help us answer the question "how", but not the "why". However, for the purpose of the comparison of Khanty and Sami bear rituals, I will compare both outer form (how rites were performed) and "inner meaning", where it is known to us.

Apart from banal resemblances, such as the fact that among both the Khanty and the Sami there is first a hunt and then some kind of festivity, it is evident, even from the very brief summaries of the

13 The fact that the bones in excavated bear graves generally have been split (cf. Zachrisson/Iregren 1974, 39, 96–97) is a good example of the gap between hunting ideology and actual behaviour that Smith (1982, 53–65) called attention to in a classical article.

contents of the two rituals presented here, that the main structures are different. Most of the individual elements differ, each of them occurring in only one of the two rituals. The focus is different: for the Khanty, the most important element was the festival and its entertainments, while for the Sami it was the feast and the burial. Therefore, the principal conclusion to be drawn is that the structural differences between the bear rituals of the southern Khanty and the southern Sami are considerable.

However, there are a few elements that are strikingly similar: from the perspective of "meaning", both rituals involved a) purification rites (even if different ones) for both the hunters and the bear, and b) games to decide when and by whom the next bear was to be killed; while in terms of resemblances in outer form, both rituals involved c) prayers and songs to the bear (in one case with similar content, namely that others were to blame for the death of the bear), d) several meals (feasts), and e) the use of the back door.

It might appear that these resemblances do indeed suggest a close connection between the two ritual complexes. However, there are various types of resemblance. Purification rites, prayers, songs, and meals (feasts) are all found in various types of ritual context (not only bear rituals), and their occurrence in the two bear ceremonials thus cannot be used to support the hypothesis of a connection. What remains are three (more specific) elements: a) the fact that the killing in both contexts is blamed on someone else, b) the games to decide about the next hunt, and c) the use of the back door during the ritual. But since the first two elements are found in hunting ceremonials around the world (cf. Hutter 2001), neither of them can be used to support the hypothesis. The sacred back door is the most interesting resemblance and might indeed be a connecting element (cf. Ränk 1949). However, one or two elements do not make a ritual.

Conclusion

The comparison of the bear ceremonials among the southern Khanty and the southern Sami gives a negative result when we consider both resemblances and differences, rather than resemblances alone, as was the case in earlier versions of the comparative enterprise. It seems as if the main connecting point is the bear itself. The conclusion has to be that the two examples of bear rituals do not support the hypothesis that the different bear rituals in northern Eurasia are concrete forms, or representatives, of one common ritual. This conclusion calls into question the whole idea of a "North-Eurasian Bear Ceremonial". However, this negative result does not mean that the religions of the Finno-Ugric peoples cannot "provide an interesting test case for comparative methodology in the history of religions" (Honko 1987, 330). On the contrary: it is in my opinion evident from the case presented here that they can indeed function as exemplary sources for comparative analysis.

Acknowledgements

I am most grateful to Peter Cripps and Ellen Valle for their careful correcting of the English language, and to Veikko Anttonen and the two anonymous reviewers for their valuable comments on a preliminary version of the text.

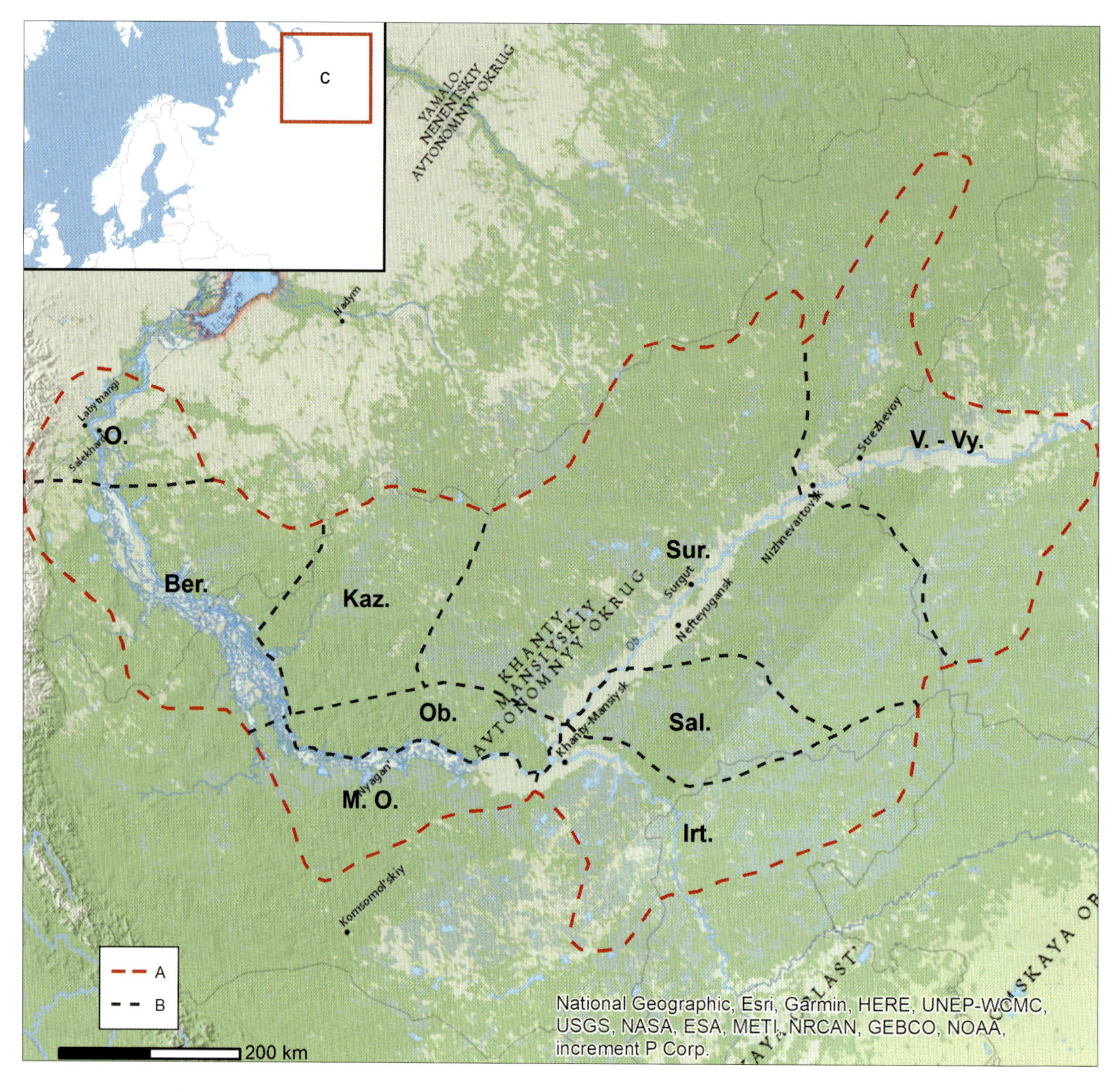

Fig. 1. The main dialect areas of Khanty. A: extent of the Khanty area of settlement; B: approximate borders of the main dialects of Khanty; C: the area on the main map. Northern Khanty: O.: Obdorsk Khanty; Ber.: Berezino Khanty; Kaz.: Kazym Khanty; M.O.: the Khanty dialects at the Middle Ob. Southern Khanty: Irt.: Irtysh Khanty (incl. Konda and Demyanka). Eastern Khanty: Sal.: Salym Khanty; Sur.: Surgut Khanty (Pim, Yugan, Trom'yugan, Agan, etc.); V.-Vy.: Vakh and Vasyugan Khanty (mainly after Schmidt *1989;* Martynova *1994; map GIS department, ZBSA).*

Fig. 2. The main dialect areas of Sami. A: extent of the Sami area of settlement (as depicted in most modern surveys, despite the fact that the South Sami language area, for example, extends to the Gulf of Bothnia); B: approximate borders of the main dialects of Sami; C: the area on the main map. Western Sami: S.: South Sami; U.: Ume Sami; Arj.: Arjeplog Sami; L.: Lule Sami, N.: North Sami. Inari Sami: I.: Inari Sami. Eastern Sami: Sk.: Skolt Sami; Akk.: Akkala Sami (extinct in 2003); Kld.: Kildin Sami; T.: Ter Sami (after Rydving *2004b, 358; map GIS department, ZBSA).*

Table 1. Bear ceremonials among the southern Khanty (decades around 1900).

Ceremonial 1: The Hunt

- Ceremony 1: ritual before the hunt
    - rite 1: purification with incense
    - rite 2: prayer to the bear
    - rite 3: the sign of the cross in front of the icons
- Ceremony 2: rituals after the hunt
    - rite 1: (purification) the throwing of snow or earth
    - rite 2: a meal
    - rite 3: the skinning (skinning + utterance of words + breaking and placing of sticks)
- Ceremony 3: rituals on the way home
    - rite 1: telling the bear about the way
- Ceremony 4: rituals on returning to the village
    - rite 1: shooting and greeting
    - rite 2: (purification) sprinkling of bear skin with water and incense
    - rite 3: the person who killed the bear was thrown into the water

---

Ceremonial 2: The Bear Festival

- Ceremony 1: preparatory rituals
    - rite 1: the bear (skin and head) was taken to the place for the ritual
    - rite 2: the bear was dressed
    - rite 3: food was offered to the bear
- Ceremony 2: the bear festival
    - rite 1: purification with water or snow
    - rite 2: the bear was greeted
    - rite 3: introductory words (the Russians killed you, not we)
    - rites 4–6: a) bear songs
        - b) dancing performances
        - c) short plays
    - rite 7: games to decide when and by whom the next bear was to be killed
- Ceremony 3: concluding the festival
    - rite 1: the bear (skin and head) was taken out
    - (rite 2: a meal in places where bear meat is eaten)

Ceremonial 3: Remembrance Rituals

- rites 1–3: remembrance of the festival on the 9th, 16th, and 38th day after the end of the festival

Table 2. Bear ceremonials among the southern Sami (mid-18th century).

Ceremonial 1: The Hunt

Ceremony 1: rituals before the hunt

(no such rituals documented)

Ceremony 2: rituals after the hunt

rite 1: skiing over the skin

rite 2: a chant was sung

allorite 3: a twig in the bears mouth was pulled three times

or

allorite 3: a spear was aimed at the bear three times

Ceremony 3: rituals on returning to the village

rite 1: a chant was sung

rite 2: (purification) the women spat chewed alder bark on the men and fastened brass rings on their clothes

rite 3: meal

Ceremonial 2: The Bear Feast

Ceremony 1: the bear was collected

rite 1: several chants were sung to the bear

rite 2: prayer to the bear

Ceremony 2: the bear feast

rite 1: the women spat chewed alder bark on the bear

rite 2: the bear was skinned and several chants were sung

rite 3: meal

rite 4: (purification) the hunters washed themselves in lye

rite 5: the men ran around the bear imitating it

rite 6: a game to decide when and by whom the next bear was to be killed

---

Ceremonial 3: The Burial

rite 1: the bear was buried

The songs and rituals of the Finno-Karelian bear hunt: Gifts, seduction and mimesis in the forest

By Vesa Matteo Piludu

Keywords: Finnish and Karelian bear ceremonialism, ritual hunt, ontology, animism, personalisation of forest and bear, mimesis, ritual seduction

Abstract: In this chapter, I analyse the first stage of Finnish and Karelian bear ceremonialism – the songs and rituals performed during the bear hunt in the forest. First, I present the whole structure of Finno-Karelian ceremonialism and its socio-economic, cultural, and religious backgrounds. I examine the conceptions of the bear and the forest in Finno-Karelian folk belief, emphasising the importance of the personalisation of the forest and the bruin – the bear. The forest was considered a mythical world with a social structure: Its inhabitants, the forest haltias *(guardian spirits) and the bear, were considered as persons of the forest, with whom the hunters tried to establish a complex ritual relationship both during and after the hunt. After this introduction, I will analyse the songs and rituals of the bear hunt, focusing: 1) on offerings to forest spirits, 2) on seductive songs for female forest spirits, 3) on songs to wake up the bear from its hibernation in its den, which were performed before the kill, and 4) on songs that portrayed the kill as an accident and denied the hunters' responsibility. The goal of the ritual communication during the hunt was to please the forest spirits and to avoid the revenge of the bears. My interpretation is based on theoretical reflections: 1) on the personhood of the bear and forest spirits, ontology, animism, and human-environmental ritual relations, and 2) on the status of the hunter as a mimetic suitor and groom of female forest spirits and bears.*

The structure, meaning and objectives of bear ceremonialism in Finland and Karelia

This chapter aims to analyse the first stage of Finnish and Karelian bear ceremonialism – the songs and rituals performed during the bear hunt in the forest. In Finland and Karelia, bear ceremonialism was performed in the winter, during hibernation, and it could be divided into three main parts: 1) songs, incantations and rituals performed in the forest during the bear hunt, 2) songs and rituals of the bear feast in the village, consisting of the ritual consumption of all the meat, fat and organs of the bear, 3) the procession with the bear's skull and bones brought to a sacred pine in the forest, the attachment of the bear's skull to a branch, and the performance of songs for the bear's skull and its soul in order to achieve the regeneration of the animal in its mythical homeland.[1]

1 For an analysis of the songs and rites of the bear feast and the bear skull rituals, see Piludu, this volume, on the Finno-Karelian bear feast and wedding, and on the Finno-Karelian bear skull rituals.

Bear and Human: Facets of a Multi-Layered Relationship from Past to Recent Times, with Emphasis on Northern Europe, ed. by Oliver Grimm (Turnhout: Brepols, 2023), pp. 693–722 BREPOLS PUBLISHERS 10.1484.M.TANE-EB.5.134357

Why did the bear hunt require the performance of complicated rituals and long songs? In Finland and Karelia, the forest was considered a mythic and sacred landscape, which was inhabited and owned by non-human persons, the forest *haltias* (guardian spirits) and their offspring and cattle, the bear and game animals. The forest was considered a sentient and perceptive environment: The spirits and the bear could see, listen to and understand human speech and the hunter's actions. Both the bruin and the forest spirits observed the hunt: If the hunter did not perform respectful rituals and *Bear Songs*, they took a harsh revenge. The hunters could be punished with bad luck in hunting, with bear attacks on cattle or humans, or with a dangerous illness (the forest *nenä* or the bear *viha*). By contrast, if the hunters performed all the rituals and the songs, they would please the forest *haltia*s and the bears, and the forest spirits would "give" them more bears and other game animals in the future. With a successful bear ceremonial, the hunters obtained good luck in all the hunting activities (not only the bear hunt) and the meaningful social status of bear killers (*karhunkaataja*).

In this article, I will introduce the topic by analysing the classic studies of Hallowell and Sarmela on bear ceremonialism, the sources of Finno-Karelian bear ceremonialism, the social and economic background of the rituals and the status of the bear and the forest in Finno-Karelian folk belief. After that, I shall analyse the songs and rituals of the bear hunt, stressing the importance of rites based on reciprocity and communication with forest spirits, in particular the offerings to forest spirits and seductive songs addressed to female forest spirits. When approaching the den, the hunters claimed to communicate with the bruin itself, and I will examine the meanings of: 1) songs to wake up the bear from its hibernation in its den, performed before the kill, and 2) songs that portrayed the kill as an accident and denied the hunters' responsibility. My interpretations are based on theoretical reflections on personhood, ontology, animism, and human-environmental ritual relationships. I will also pay attention to the status of the hunter as a mimetic suitor of female forest spirits and bears, and to the meaning of seduction and marriage in the hunt.

Irving Hallowell and *Bear Ceremonialism in the Northern Hemisphere*

In his monograph *Bear Ceremonialism in the Northern Hemisphere* (Hallowell 1926), the cultural anthropologist Alfred Irving Hallowell[2] developed the concept of bear ceremonialism, covering the whole system of rites, rituals and ceremonies connected with the hunting of the bear. According to the anthropologist Regna Darnell,[3] Hallowell's monograph is the last of the major distributional studies of the Boasian school[4] (Darnell 1977, 14). This school should not be considered a monolithic entity. Franz Boas'[5]

2 Alfred Irving Hallowell (1892–1974) was an influential American cultural anthropologist, archaeologist, and businessman. He was a professor of anthropology at the University of Pennsylvania and an expert in the ontology or animism of Native American and Ojibwa (USA and Canada) cultures. He developed the concept of ontology that is currently used in cultural anthropology as a scientific term regarding animistic beliefs among indigenous peoples. His work on bear ceremonialism was his PhD thesis, published in a special number of *American Anthropologist*.

3 Regna Darnell is an American and Canadian anthropologist who has studied the theories of the members of the Boasian school.

4 Boasian anthropology is considered one of the most influential schools of the discipline. It is named after Professor Franz Boas.

5 Franz Boas (1858–1942) was one of the most famous American anthropologists of all time. He was the founder of the so-called Boasian school of anthropology and the teacher or mentor of many influential American and Canadian anthropologists. The Boasian school is well-known for historical particularism and cultural relativism, which postulated that a culture can only be understood by its own standards and values. However, the school was also involved in diffusionist studies and in research about the Siberian origins of Native American cultural features (such as shamanism or bear ceremonialism, for example).

most renowned students — including Edward Sapir,[6] Robert Harry Lowie,[7] Clark Wissler,[8] Alfred Luis Kroeber[9] and Hallowell — "shared a disciplinary culture" (Darnell 1977, 15) and "certain assumptions about the nature of sociocultural anthropology" (Nash 1977, 7), but they "have often differed from their teacher and from each other" (Lowie 1960, 412). However, Dennison Nash[10] stressed that defining Hallowell as only a Boasian "would be to miss his special genius and his particular contribution to anthropology" (Nash 1977, 7).

Hallowell was fiercely polemical with those scholars who assumed a unilateral theory of religious evolution and considered "animal worship" as an early stage in the religious development of mankind (Hallowell 1926, 14). He openly criticised the evolutionist theories of James George Frazer and Edward Tylor (Hallowell 1926, 13–20). Hallowell joined the anthropologist Alexander Alexandrovich Goldenweiser[11] in his firm critique of the confused abuse of the evolutionist use of the term "totemism" "as a necessary stage in the development of religion" (Goldenweiser 1910/1911, 264; cited in Hallowell 1926, 14; Darnell 1977, 18). Hallowell considered animism to be valid as a general statement about attitudes towards animals, but too general to clarify the differences existing in the folk beliefs and rituals relating to animal species, such as the bear (Hallowell 1926, 15–16). By contrast, Hallowell considered bear ceremonialism to be a concept grounded on ethnographic facts and rituals: It was much more useful to grasp the native point of view (Darnell 1977, 27).

The peculiar status of the bear made a complex ceremonial necessary when it was hunted, and Hallowell remarked on the presence of common features in the bear ceremonies of different Eurasian and Amerindian peoples: 1) the season of the hunt – the end of winter – is connected with hibernation, 2) the belief that the bear sucks his paw for nourishment during hibernation, 3) the use of particular weapons for the bear hunt, 4) the custom of talking or singing to the bear, 5) the idea that the bear is protected by a guardian spirit (Hallowell 1926, 154) and the use of ritual circumlocutions and euphemisms for the bear (instead of using the generic name bear) in order to honour the bruin and avoid its revenge, 6) the rite of awakening the bear and calling it out of the den before the kill, 7) the use of conciliatory speeches, 8) the hunters' justifications or apologies for the bear's death, 9) the presence of elaborate ceremonies after the kill, including the bear feast, characterised by the "eat all (the meat and organs)" ritual, 10) the disposal of the bones – in particular, the bear skull (Hallowell 1926, 145–147, 154). Hallowell sought to find common features in almost all the bear ceremonials but remarked that each of the circumboreal peoples developed culturally specific variations of rituals to deal with these respective features, and he dedicated many sections and subsections of the book to the analysis of these variations in different geographic regions. In this respect, Hallowell was also influenced by the cultural relativism of Franz Boas.

6 Edward Sapir (1884–1939) was one of the most important linguists and anthropologists of his time and a renowned expert on Native American languages.

7 Robert Harry Lowie (1883–1957) was a professor at the University of California (Berkeley) and one of the most important scholars of the native peoples of the American Great Plains and the author of *Primitive Society* (1919).

8 Clark Wissler (1870–1947) was an anthropologist, ethnologist and archaeologist, an expert on the Northern Plains peoples and the founder of the culture area approach.

9 Alfred Luis Kroeber (1876–1960) was a professor of anthropology at the University of California (Berkeley). He was an expert on the myths and religions of the native peoples of California. He made detailed studies of Ishi (1861–1916), the last surviving member of the Yahi people (Northern California).

10 Dennison Nash (1924–2012) was a professor of anthropology, a pioneer of the anthropology of tourism and a prolific scientific writer.

11 Alexander Alexandrovich Goldenweiser (1880–1940) was a Russian-American anthropologist. He is renowned for his critical studies on totemism and primitive religions.

However, Hallowell tried to build a general historical and geographical theory to explain the presence of bear ceremonialism in the whole circumboreal region[12], including Lapland,[13] Finland and Karelia,[14] western Siberia,[15] eastern Siberia,[16] the Ainu culture (Japan)[17] and several Native American cultures of Alaska, Canada and the United States.[18] According to Hallowell, bear ceremonialism was a common feature of an ancient boreal hunting culture – associated with shamanism and the pursuit of reindeer – that originated in Eurasia and passed to North America across the Bering Strait (Hallowell 1926, 153–163). It was a product of an "ancient hunting complex" that was associated with "conservative" customs and rituals present in many different peoples (Hallowell 1926, 33; Darnell 1977, 25). If the cultures were continuous in their geographic distribution, there were possibilities of demonstrating historical connections between them. The Boasian school was strongly interested in the exploration of the cultural relations between the populations of northern Siberia and North America (Hallowell 1926, 160–161; Darnell 1977, 26; Fortescue 1998; Brightman et al. 2014, 9).

Recently, Håkan Rydving,[19] scholar of religions, contested the generality of Hallowell's theory, stressing the structural differences between the Sámi (Lapland or Samiland) and the southern Khanty (western Siberia) bear rituals (Rydving 2010, 42). Rydving suggests the existence of a plurality of bear rituals in contrast to the idea of a "uniform and transcultural" bear ceremonial in North Eurasia (cf. Rydving, this volume [reprint of Rydving 2010]). Hallowell was quite imprecise in his description of the Finnish bear ceremonial, because he used the English translation of Rune 46[20] of the epic poem *Kalevala* (1849) by Elias Lönnrot[21] as his source (Hallowell 1926, 33). The language barrier

12 There have been few comparative articles, chapters, monographs and books on bear ceremonialism after Hallowell; see Campbell 1988, 54, 147–155; Lajoux 1996; Spagna 1998; Pentikäinen 2005; 2006; 2007, Pastoureau 2011.

13 On Sámi bear ceremonialism, see Fjellström 1755; Holmberg 1915, 45–52; Itkonen 1937; 1948; Edsman 1953; 1956; 1975; 1994; Schefferus 1971; Laestadius 2002, 180–196; Pulkkinen 2005, 33–35; Rydving 2010, reprinted, this volume; Pentikäinen 2015; 2019.

14 On Finnish and Karelian bear ceremonialism, see Appelgren 1885; Sirelius 1919, 37–40; Nirvi 1944; Karhu 1947; Virtaranta 1958, 308–328; Edsman 1953; 1958; 1965; 1975; 1994; Kuusi 1963, 41–51; Vilkuna 1965; Haavio 1967, 1–41; Sarmela 1972; 1982; 1983; 1991; 2006a; 2006b, 70–94; 2009, 79–107; Ilomäki 1986; 1989; 2002; 2014a, 84–89; 2014b; Honko 1993; Tarkka 1994; 1998; 2005, 256–299; 2013, 327–381; 2014; Uusitalo 1997; Inha 1999, 349–350, 352–355; Klemettinen 2002; Miettinen 2006; Partanen 2006; Pentikäinen 2006; 2014; Piludu 2006; 2009; 2015; 2019a; 2019b; 2022; Pulkkinen/Sammelkivi 2006; Rebourcet 2006; Salo 2006; 2012, 33–73; Lehikoinen 2007, 248–259; 2009, 168–176; Partanen/Piludu 2007; Kailo 2008; Krohn 2008, 146–164; Siikala 2008, 140–144; 2016, 380–389; Meriluoto-Jaakkola 2010a; 2010b; Pulkkinen 2014, 212–229; Pulkkinen/Lindfors 2016, 106–120; Ahola 2020.

15 On Ob-Ugrian bear ceremonialism, see Ahlqvist 1881; Kannisto 1906a; 1906b; 1907; 1933; 1938a; 1938b; 1939a; 1939b; Karjalainen 1914; 1918, 512–545; Sirelius 1929; Kannisto et al. 1958; Kálmán 1968; Cushing 1977; Bartens 1986; Schmidt 1989; Lindrop 1998; Patkanov 1999; Kulemzin/Lukina 2006, 129–131; Juslin 2007; Pentikäinen 2007, 31–42; Soldatova 2008, 145–147; Rydving 2010, reprinted, this volume; Wiget/Babalaeva 2010, 133–140; 2022; Siikala/Ulyashev 2011, 91–96; Piludu 2022.

16 On eastern Siberian, Tungus and Nivkh bear ceremonialism, see Dyrenkova 1930; Zolotarev 1937; Paproth 1976; Kwon 1999; Janhunen 2003; Taksami 2006.

17 On the Ainu bear ceremonialism, see Batchelor 1901, 383–496; 1932, 37–44; Irimoto 1996; 2014; Akino 1999, 248–255; Maraini 2013.

18 On North American bear ceremonialism, see Hallowell 1926; Tanner 1979, 145–181; Nelson 1983; Rockwell 1991; Comba 1996; 2015; 2019; Spagna 1998; Hämäläinen 2011.

19 Håkan Rydving (born in 1953) is a Swedish linguist and historian of religions. He is Professor of History of Religions at the University of Bergen (Norway) and an expert on Sámi religions and languages.

20 Lönnrot composed Rune 46 of the *Kalevala* (Lönnrot 2012) using several *Bear Songs*, many of them from Viena Karelia. The *Kalevala* is a literary work strongly based on folk songs, but Lönnrot edited and changed many details of the original *Bear Songs*. The material is adapted to the epic plot; the singer is not a common hunter but a hero, Väinämöinen. See Rebourcet 2006; Piludu 2019a, 29–31.

21 Elias Lönnrot (1802–1884) is probably the most renowned Finnish scholar, ethnographer, poet, professor, and writer. His literary masterpiece is the *Kalevala* (1849), his epic poem that is based on edited versions of folk songs and incantations collected in the field. Other significant books are the *Kanteletar* (1835), a collection of edited lyric Finnic and Karelian folksongs, and the first and shorter version of his poem, the *Old Kalevala* (1835). As the *Kalevala* was the first

denied Hallowell – and other influential foreign scholars after him – access to the original *Bear Songs* in Finnish and Karelian, so they used a literary reconstruction as an ethnographic source (see PILUDU 2019a, 29–31). For this reason, Finnish scholars have rarely referred to Hallowell's book in their studies on Finnish bear ceremonials.

However, Hallowell's monograph remains a classic text on bear ceremonialism, especially if we contextualise it as a very interesting product of the American anthropologic research of the time (DARNELL 1977, 22). The Boasian school shared with the evolutionist school a deep interest in the historical development of traditions, but it studied this by "focusing on cultural variability" (WILLERSLEV 2007, 145; MILLER/MATHÉ 1997). Hallowell also wrote about the processes of modification, differentiation and assimilation of bear ceremonial features that were caused by local cultures and social organisations (HALLOWELL 1926, 162–163). Hallowell stressed that "each culture exhibits its own peculiar combination of features which cannot be deduced from any general principle of association" (HALLOWELL 1926, 18). Over time, some original traits were modified, and details and features of bear rituals could differ from one culture to another (HALLOWELL 1926, 162).

Three methods were used by Hallowell: 1) the ethnographic analysis of bear rituals that considered the point of view of a particular circumboreal culture, 2) the use of a comparative framework that brings out similarities and differences in the features of bear ceremonials of different cultures (HALLOWELL 1926, 20), and 3) the elaboration of a theory about the origins and diffusion of the bear ceremonial. Another important merit of Hallowell is that he connected bear ceremonialism with a set of important ontological problems. He considered the study of bear ceremonialism as a way of answering the theoretical problem of how human groups are related to the environment and to animals (DARNELL 1977, 21). He stressed that: a) bears and other animals are believed to have the "same sort of animating agent which man possesses" and they are supposed to have linguistic abilities, understand human speech and actions and have specific forms of family or social organisation (HALLOWELL 1926, 7), b) animals are not all equal; some beings – such as the bear – have a higher status or rank because they have stronger "powers" or they have a deep relationship with powerful spirits (HALLOWELL 1926, 8, 17), c) bears are protected by a guardian or owner spirit governing the supply of certain game animals; the bear ceremonials honoured not only the bear, but they also propitiated this supernatural agent (HALLOWELL 1926, 145). Hallowell emphasised that the bear rituals should always be connected to a larger framework of beliefs about the environment, the animals and the guardian spirits. Hallowell also elaborated these theories on animal personhood and guardian spirits later, in his influential article and book on Ojibwa[22] ontology (HALLOWELL 1960; 1974). These ontological considerations are of fundamental importance for my analysis of the Finno-Karelian bear ceremonials, as we will see in later sections of this chapter: In the Finno-Karelian *Bear Song*, communication with the forest *haltias* (guardian spirits) of the bruin was of fundamental importance.

THE SOURCES OF FINNO-KARELIAN BEAR CEREMONIALISM

In the 19th century Finnish folklore collectors transcribed a large number of *Bear Songs* and informants' descriptions of the rituals performed in all the phases of the ceremonialism. At that time, the bear ceremonial was a vanishing tradition, and the majority of the songs and information were collected

important literary publication in the Finnish language, he is considered a founder of Finnish literature (KALLIO 1921; ANTTILA 1931a; 1931b; 1962; HAAVIO 1971).

22 The Ojibwa or Ojibwe (Chippewa) are a Native American people living in Canada and on the Northern Plains of the Midwestern United States.

refused to perform the whole ceremonial. A similar process happened in Finland with the efforts of the Lutheran church to erase the bear ceremonies from the country. However, in Viena Karelian villages and parishes near the Finnish border, the folklore collectors transcribed some of the longest and most complete *Bear Songs* from skilled *runolaulajas* (singers), including the famous Arhippa Perttunen, Iivana Malinen and Jyrkki Malinen. This is not a surprise, as the singers of these Viena Karelian villages also remembered the longest and most important Kalevalaic epic songs and incantations used by Elias Lönnrot to compose his epic poem, the *Kalevala* (Lönnrot 1985; 2012; see Piludu 2019a, 29–31). For all these reasons, the material from Viena Karelia is particularly important in the three chapters I have written for this book.

The reconstruction of the Finnish bear ceremonialism eras according to Matti Sarmela

In his theorical reconstruction of the historical development of the Finnish bear ceremonials, the anthropologist Matti Sarmela[29] argues that it can be divided into three eras. According to him, the first era comprised the prehistoric Finnish culture, marked by shamanism,[30] totemism[31] and a hunter-gatherer economy. Sarmela argues that the Finnish ritual of the bear skull should relate to the natural environment of the bear; the skull and all the other bones should be returned to the forest to ensure the rebirth or regeneration of the animal in the original birthland (Sarmela 1991, 221–222). According to Sarmela, the myth of *The Birth of the Bear in the Sky* (which emphasises the regeneration or resurrection of the animal) relates to this first era.[32] In the second era the rituals evolved in a different direction: The old beliefs did not work anymore because the ecological environment changed when the Finnish people adopted agriculture (Sarmela 1991, 224). During the Iron Age, with the development of slash-and-burn agriculture, the bear became the "enemy" of the people, because it killed the cattle in the forest pastures or destroyed the farmland produced by means of the slash-and-burn technique (Sarmela 1991, 230). Sarmela stresses that the most important religious specialist of the Finnish archaic agrarian villages was no longer the shaman, but the *tietäjä*, a ritual specialist who did not travel to other worlds or dimensions in search of lost souls (like the shaman) but expelled "the sorcerer's arrows or darts" *(noidan nuolet)* or sickness from the body of an ill person. The environment in the age of the *tietäjä* was divided into two worlds, the cultivated landscape and uncultivated nature, which represented a kind of anti-world (Sarmela 1991, 229). Sarmela states that the most important goal of the *tietäjä* was to protect the cattle and the crops in the fields from

29 Matti Sarmela (born in 1937) is one of the most influential Finnish professors of cultural and social anthropology. He wrote quite intensely on Finnish traditions and bear ceremonialism. Although I criticise some of his theories in this chapter, I warmly suggest that readers interested in Finnish bear ceremonialism read all his chapters and articles on the topic (Sarmela 1982; 1983; 1991; 1994). He is also an expert on northern Thai folk cultures.

30 The use of the terms "shaman" and "shamanism" is quite problematic when they are used to define Finnish and Karelian religions in prehistory. "Shaman" is a Tungusic word used to define the local ritual trance specialist of eastern Siberian peoples. In anthropological literature, the words "shaman" and "shamanism" have been broadly used to define almost every kind of ritual specialist that uses trance technique around the world (from Siberia to Africa, from South Asia to Amazonia), regardless of the evident differences between shamanic traditions from contemporary South Korea to those of western Siberia in the 19th century. However, in recent years, there has been a tendency to use emic terminology to define local trance specialists (for example *noaiddi* in Sámi) and to emphasise the peculiarities of each trance tradition. In Finland and Karelia, the name of the trance specialist is *tietäjä* (sage, the one-who-knows) and most of the information about the traditions and incantations of the *tietäjäs* was collected in the 18th, 19th, and 20th centuries. Several Finnish scholars, including Anna-Leena Siikala, argue that the *tietäjä* tradition evolved from an older form of Nordic shamanism, more like the *noaiddi* tradition or the Siberian ones (Siikala 1992; 2002).

31 On totemism and Sarmela's totemic theory, see Piludu, this volume, on the Finno-Karelian bear feast and wedding.

32 On the *Birth of the Bear in the Sky*, see Piludu, this volume, on the Finno-Karelian bear feast and wedding; Piludu 2022.

bears.[33] The *tietäjä* asked forest *haltias* (guardian spirits) to control the bears in order to avoid a bear attack on cattle.

According to Sarmela's theory, the third era was the "age of the countryman," which developed during the Middle Ages. This period was marked by the religious dominance of the Christian faith. The Catholic and Orthodox cults of the saints influenced the rituals of the *tietäjäs*; in their incantations, the saints took the place of the earlier forest *haltias* (guardian spirits) in protecting and controlling the bear (Sarmela 1991, 236). The *Birth of the Bear* from wool thrown by a saint into (river or sea) water represents the main myth of this era; according to Sarmela, the bear was not considered sacred anymore, but as a puppet fully dominated by the saints.

Although Sarmela's reconstruction contains many useful observations and interpretations, the differences between the three eras are too sharp. Sarmela does not negate the influence of agriculture or Christian beliefs on the Finnish bear lore; however, he has a tendency to idealise the hunter-gather traditions (proposing them as a model of ecological sustainability) and to consider the influence of agriculture, Christianity and cattle herding on bear lore as a negative degeneration of the original bear ceremonial. The folklorist Lotte Tarkka has noted that for many decades Finnish scholars focused on the "ultimate origins of the song" (Tarkka 2013, 80) and that for them the "true" and "authentic" text of a song was supposed to be the most ancient or archaic part of it. Sarmela also has this tendency; in his interpretation, whatever deals with agriculture or the *tietäjäs* has a negative connotation and whatever is archaic is fully idealised.

Another problem with Sarmela's interpretation is that he considers each era as completely separated from the other, denying the possibility of communication and syncretism between the traditions of hunters and cattle herders. Agriculture, cattle herding, and the Christian faith certainly reduced the areas where bear ceremonialism was performed, but they did not completely or abruptly erase or degenerate the sacred status of the bear and its *haltias* (guardian spirits).

The folklorist Anna-Leena Siikala[34] has pointed out that, in the Finnic Kalevalaic songs and incantations, the fusion of different historical elements often acquired a significant contextual and ritual meaning; new historical layers rarely wiped away the old strata of meaning (Siikala 1994, 94). I agree with her theory, which stresses that changes in tradition were not mechanical events but complex processes in which renewing and conserving tendencies could act at the same time, influencing each other. In the Finno-Karelian folk tradition, ritual actors tended to "accumulate" supernatural helpers, mobilising quite a variegate group of beings that belonged to a historically stratified tradition. The hunters and cattle holders asked for the help of all the powerful spirits, divinities and Orthodox saints that could help them in preventing a bear attack.

The socio-economic background: The protection of the cattle and the values associated with the bear hunt

The Finnish and Karelian hunters did not perceive the forest as a "homogeneous natural environment" (Tarkka 2013, 327), as do contemporary urban dwellers. The anthropologist Rane Willers-

33 This is a simplification, if not a banalisation, of the many roles of the *tietäjä*. He or she healed sick persons and helped the hunters, the fishermen, the farmers and the cattle-herders in the case of difficulties. In Karelia, a particular *tietäjä*, the *patvaska*, protected the bridal couple and the families from the malevolent magic of *noidat* (sorcerers) and *kateet* (envious people) (Siikala 2002). The culture of the *tietäjä* deeply influenced bear ceremonialism; many famous Karelian and Savonian bear-killers were believed to be powerful *tietäjäs* and the bear, too, was believed to have the *tietäjä*'s skill.

34 Anna-Leena Siikala (1943–2016) was a Finnish professor of Folklore Studies. She is renowned for her studies on Siberian shamanism, on Finnish *tietäjäs* (sages) and on Finnic and Uralic mythologies.

lev[35] stresses that a hunter's understanding of the world is not based on an abstract contemplation of an objective "nature," but "emerges from concrete context of practical engagement" (Willerslev 2007, 94). The hunter's myths and rituals are strictly bound with practical activities carried out in certain environments (Willerslev 2007, 95). Willerslev's theoretical background is clearly influenced by the phenomenological and existential philosophy of Martin Heidegger, which deeply inspired the anthropologist Tim Ingold[36] and his school of British social anthropology. The latter scholar defines the environment of indigenous people as a "taskscape", a place to accomplish a variety of different works: "As the activities that comprise the 'taskscape' are unending, the landscape is never complete: neither 'built' nor 'unbuilt'. It is perpetually under construction" (Ingold 1993, 163). I tend to agree with them: The Finno-Karelian forest was indeed a "taskscape", but the conceptualisation of the forest in Finland and Karelia was both influenced by everyday activities (synchronic aspect) and by a complex ontological worldview based on a long and ancient tradition, which was preserved well by the *Bear Songs* (diachronic aspect).

The folklorist Lotte Tarkka[37] emphasises that, in Viena Karelia (Russia), the division of labour was gendered (Tarkka 2013, 330). This provoked a gendered separation of the taskscapes; the work in the houses, the cattle sheds and around the farm were the dominion of women and the household's mistress, while the forest and travelling outside the village were the environments of male activities, like hunting and logging (Tarkka 1998, 93). In Karelia and in Finland, there were various exceptions from the gendered taskscapes. Young women or girls often followed the cattle to the summer pastures, which were glades located in the forest or near it, or they went into the forest to pick berries and mushrooms; boys and young men could also work as cattle herders in the summertime and do several agricultural tasks (Piludu 2019a, 58). Adult men could help women in several rural activities and took care of horses.

The Karelian women did not breed cows for meat production, as the cows owned by a family were few: Dairy products were important food and a source of fat, while cattle also produced manure for the fields (Tarkka 1998, 93). In this economic context, the bear hunt was performed to protect the cattle, "since bears took a heavy toll on the cattle which were vital to the livelihood of small farms" (Tarkka 1998, 93). Tarkka's statements are also valid for the rural culture of eastern, northern and central Finland; in the summer, it was vital to protect the cows from bears. In other words, the Finno-Karelian bear ceremonialism had different historical layers; the archaic strata of the hunting rituals were influenced by the necessities of agriculture and cattle breeding.

Bears could also kill or injure horses, which were important animals for transportation, travelling and ploughing. Cattle and horses also represented the wealth of the household. According to the anthropologist Matti Sarmela, after the development of agriculture the bear became the "enemy of humankind" because it killed the cattle in the forest pastures (Sarmela 1991, 230). This statement fits well for the wolves, which were heavily demonised by rural communities for their frequent attacks on cattle (Pulliainen 1974, 16–85). However, the status of the bear, which is an omnivorous animal and not as aggressive a predator as the wolf, was much more elaborate.

35 Rane Willerslev (born in 1971) is a Danish anthropologist and a renowned expert on Yukhagir hunting rituals and culture. He is the director of the Danish National Museum. His theoretical approach includes innovative reflections about animism, ontology, and mimesis in hunting rituals.

36 Tim Ingold (born in 1948) is a British social anthropologist and an expert on the Skolt Sámi of Finland. He is interested in the historical evolution of reindeer pastoral culture and human-animal relations; his approach is strongly influenced by philosophical phenomenology, especially the thoughts of Martin Heidegger.

37 Lotte Tarkka (born in 1963) is Professor of Folklore Studies at the University of Helsinki. She is an expert on the Kalevalaic singing tradition in the Parish of Vuokkiniemi (Viena Karelia) and on bear ceremonialism and forest symbolism in Viena Karelia.

In many Karelian and Finnish villages, the bear was still considered an innocent animal of the forest, protected by powerful forest *haltias*. If the bruin attacked the cattle, an envious neighbour or a sorcerer who had bewitched the animal was held to be culpable (Piludu 2019a, 68–70). The folk believed that cattle luck *(karjan onni)*, like other forms of "luck" (for example "hunting luck") was present only in a finite quantity; someone's luck could increase only by diminishing the "luck" of their neighbours through sorcery (Stark 2006, 46). Tarkka stresses that conflict between people and bears reflected an internal struggle present in human society: The contradiction between the bear's innocence and the havoc it created was resolved by framing it in terms of "aggression within the human sphere" (Tarkka 2013, 332). Although dangerous, the bear remained sacred *(pyhä)* and innocent, "clean" (*puhdas*): Its killing was considered acceptable only if the hunters performed a huge number of ritual songs and rituals to please the bear itself and the forest spirits.

Finally, the hunt was a marginal activity in the complex economic system (fishing was more important as a source of protein), but the social values associated with it were particularly high. Tarkka stresses that the bear kill was "a proof of masculine prowess" (Tarkka 2013, 347) as well as courage, requiring successful cooperation between the male members of the hunting group and the ability to communicate with forest spirits. The hunters acquired the honourable title of "bear-killers" (*karhunkaataja)*. In the wedding songs, the ideal husband was portrayed as a manly hunter and a brave traveller (Tarkka 2013, 337): A bear-killer was a model of manhood, the groom "par excellence". In the *Bear Songs*, the hunt is deeply intertwined with wedding metaphors; the bear can be described as a bride by hunters or as a groom by the women of the villages.

The personalisation of the *haltias*, the sacred forest and the bear

The forest was not only a well-known taskscape and background of everyday activities; it was considered as an otherworld (Nadasdy 2021) or a mythic landscape, called Metsola, Tapiola or Pohjola in the *Bear Songs*. The forest and the forest *haltias* in general were defined with the adjective *pyhä* ("sacred"; Tarkka 2013, 329). The ethnologist Laura Stark[38] stresses that in folk religion the sacred "is defined by the local community rather than by a religious institution" (Stark 2002, 30). It seems that the adjective *pyhä* did not have a Christian origin, even if the word was also used in a Christian sense (Anttonen 1994, 26). The scholar of religions Veikko Anttonen[39] and the folklorist Henni Ilomäki[40] state that the *pyhä* category is generally connected with borders between "inside" and "outside" dimensions, or the "own" and "other" or "common" areas, stressing that *pyhä* could be concretely manifested in elements of landscapes (Anttonen 1992; 1994, 27; 1996, 76–151; Ilomäki 2014b, 121). The forest was a mythical world inhabited by *haltias*, but it was also the immediate and real-world hunting ground. The folklorist Frog notices that the Finnish and Karelian sacred agents, the *haltias*, did not occupy an otherworld or a mythic world completely separate from the mundane or human word (Frog 2009, 9–10), such as the Christian heaven or hell. They inhabited other parallel worlds also located in the physical environment: "The seen world was animated and affected by the unseen world and its inhabitants" (Frog 2009, 9). The folklorist Anna-Leena Siikala explains that, in

38 Laura Stark (previously Stark-Arola, born in 1966) is a Finnish-American ethnologist and Professor of Ethnology at the University of Jyväskylä. She has written several influential books on female magic rituals and incantations in Orthodox Karelia.

39 Veikko Anttonen (born in 1948) is Professor Emeritus of the Study of Religions at the University of Turku. He is an expert on Finnish folk beliefs, and he is well known for his theories about the historical evolution of the Finnish concept of *pyhä* (sacred).

40 Henni Ilomäki (born in 1941) is a Finnish folklorist and an expert on Finnish and Karelian incantations and wedding songs.

Finno-Karelian folk belief, the separation between the otherworld and "this" world was not radical (Siikala 1992, 145; Tarkka 2013, 300).

Lotte Tarkka notes that the otherworldly forest was conceived as a family, village and society that resembled the human world; it was a "sociomorphic" landscape, an approximate mirror-image of "our" world (Tarkka 2013, 330). The forest *haltias* were anthropomorphic beings and people, carrying on a life that was similar to that of humans. The forest had a basic, familiar and hierarchic structure, which resembled that of the human household. The *haltias* had epithets that indicated their gender, age and social status (Tarkka 1998, 96–97; Stark 2002, 51). The forest world was considered as another dimension that resembled the human world.

The most important male *haltia* was Tapio, the master of the forest, who had the power to provide game animals to the hunter or to deny him the quarry (Siikala 2016, 376). Tapio was both an anthropomorphic spirit and a name for the forest and the bear (SKVR I4/1266, 15; Krohn 2008, 107). This partial homology between the master and his dominion signified that the forest itself could be considered an intentional agent (Tarkka 2013, 330). Tapio was also a shapeshifter, able to transform himself into a bear or another animal and he had a wife, whose name varied in the hunting songs: Mielikki, mistress of the forest, Mielus, Hongotar, Tapiotar and Katajatar (Siikala 2016, 376). The forest mistress had many "forest girls" (*metsän tytöt*), maidens or servants, at her service: They were usually portrayed as attractive and powerful female *haltias*, and they were considered the guardians of bears. In certain *Bear Songs*, the hunters syncretised *haltias* with Christian saints (especially in Orthodox Karelia): Tapio was syncretised with St George (*Jyrki*) and Mielikki with St Anna (*Annikki*).[41]

The *haltias* were also "more-than-human" persons, in the sense that they were much more powerful than humans: They could be invisible, become as tall as the trees, appear in dreams, cause illnesses or madness, protect or bestow game animals, steal cows, make bears or wolves furious, make persons disappear or get lost in the forest.

The bear was particularly sacred because it had a deep relationship with the forest *haltias* (see Fig. 2). Belief in the existence of anthropomorphic *haltias* made the personification of the forest and the bear itself possible (Tarkka 2013, 330). The bruin was considered the offspring of a female *haltia*, and sometimes of Tapio himself (Piludu 2019a, 120–122). One of the most typical circumlocutions for the bear was "forest" (*metsä, tapio*), a metonymy or synecdoche that indicates the deep identification between the animal, its environment, and the forest master (Tapio). The bear could even share with its master, Tapio, the honourable title of "King of the Forest" (*Metän kuningas*; SKVR XII2/6481:1). As an alternative, the bear was often considered the cattle, the dog or the "pet" of the forest mistress or forest maidens (Tarkka 2005, 259). The concept of the game animals as the "cattle" or "pets" of forest spirits is present in several hunting cultures, such as the Mistassini Cree (Canada), who also performed bear ceremonials (Tanner 1979, 139). The singer Samppa Riiko emphasised that "the bear is the favourite cattle of the forest maids" (*Karhu on metsän piikojen lempi karjaa*; SKVR I4/1490, Tuhkala, Viena Karelia, Russia, 1888). The term "favourite" stressed a relationship of affection and tenderness.

What made the Finno-Karelian conception of the bear even more complex is that the animal was also considered to be a humanlike being with human origins. In Jyskyjärvi (Viena Karelia, Russia), the people believed that the similarity of the bone structures of the bear paw and the human hand demonstrated that the bear was a bewitched person transformed into a bear (SKS KRA Niemi, O. 532, 1936).[42] Some sorcerers (*noidat*) were shapeshifters, able to transform themselves into bears in

41 On popular forms of syncretism between *haltias* and saints, see Piludu 2019a, 123–129. The Orthodox and Lutheran churches fought against vernacular syncretism in Karelia and Finland for several centuries, but with limited success.

42 Sources from the SKS KRA (Archive of Folk Poetry of the Finnish Literature Society) are mentioned here, indicating: the name of the collector; the number of his manuscript and its year; the village and its parish; the name of the informant and, sometimes, basic information about him (age and profession).

order to roam the forest, sleep all winter or kill the cattle of their neighbours (Piludu 2019a, 66–68). The bear was an anomalous being; it was difficult to categorise as it was a being "in between" humanity and the world of the forest spirits.

The bear ceremonials were based on communication and reciprocity with the animals and the forest *haltias*, who were conceived of as persons (Tarkka 2013, 71). The recent anthropological debate on animism, perspectivism and ontology analyses the personhood of animals and spirits in folk and indigenous belief systems. Viveiros De Castro[43] stresses that, in Amazonia, too, non-human persons – animals and spirits – had a soul and a perceptive, appetitive, and cognitive disposition. They saw themselves as persons, and they had social relations that could be both reflexive and reciprocal (Viveiros de Castro 2009; Brightman et al. 2014, 2).

In the Finno-Karelian hunting cultures, there were similar ideas about the non-human or more-than-human persons of the forestland: bears and *haltias* were considered agents with will, intentions, and desires, and they interpreted the actions and the words of the hunters from their own point of view – their own perspective. Bruins and *haltias* were considered sentient persons with emotions, agency, a moral code, and expectations (Stark 2002, 23). The bear was supposed to have a soul (see Fig. 3), and it was considered a skilled *tietäjä* (seer). The hunters (Fig. 4) were expected to take the forest persons (*haltias* and bears) into consideration and to build a ritual relationship of reciprocity with them. The Finno-Karelian hunters developed a complex ritual relationship with the bear and the forest *haltias* as independent and competent persons (Figs. 5–6).

The bear and *haltias* were believed to be powerful beings with extraordinary senses of sight and hearing. They could see from a distance the actions of the hunters in the forest or the women in the villages, and they could hear human speech and songs from the den or the deep forest. Even when the bruin was killed, it was believed to still be able to understand the songs and the meaning of the ritual actions performed during the feast or the skull ritual. Finally, the bear was able to speak and tell the forest spirits about the ritual honours it received during the ceremonials (Piludu 2019a, 294–295).

The singer Iivana Malinen[44] said that *Bear Songs* should be performed in all the phases of the rituals "to please the forest maidens" (SKVR I4/1191, Vuonninen, Viena Karelia, Russia), and to please the master of the forest (Tapio), "so that Tapio would not get angry if a beast disappears from his cattle" (*ettei Tapio vihastuisi, jos on elukka karjasta katonnut*; SKVR I4/1244, Vuonninen, Vuokkiniemi, Viena Karelia, Russia, 1888). Angered forest *haltias* could send diseases to the human communities or bears and wolves to kill cattle. If the hunter correctly performed all the rituals and songs, the woodland denizens were "pleased": The forest spirits provided more bears or game animals, and the satisfied bear "returned" to a future village feast as the guest of honour (Piludu 2019a, 19–20).

Pleasing Tapio with offerings of ale and gold

The passage between the inner world of the village and the outer world of the forest was ritualised in many ways. The hunters sang *The Birth of the Bear*-incantations, believing that by telling the secret origins of the bruin, they gained control and power over the animal.[45] The hunters generally sang that

43 Viveiros de Castro (born in 1951) is one of the most influential Brazilian anthropologists; he developed the concept of Amerindian perspectivism.

44 Iivana Malinen or Jyrkińi Iivana was a skilled Karelian singer, *tietäjä* (seer) and *patvaska* (seer specialised in protecting people during weddings) of the Malinen family of singers. He learned his incantations and *Bear Songs* from his father, Jyrki Malinen (Ontreińi Jyrki), and his grandfather, Ontrei (Haavio 1948, 6–31).

45 On *The Birth of The Bear* in the mythical forest and in the sky, see Piludu, this volume, on the Finno-Karelian bear skull rituals; Piludu 2022. On the *The Birth of The Bear*-incantations influenced by popular and syncretic Christian faith, see Piludu 2019a, 123–129. In some incantations, a saint or St Mary created the bear by throwing wool into water.

the bear was born in the mythical forest of the *haltias* (Tapio or Mielikki): Tapiola, Metsola or Pohjola (Piludu 2019a, 144–120). When the hunters entered the forest, they sang that they were walking in Tapiola, Metsola and Pohjola; the hunting ground coincided with the mythical land of the *haltias*, the place the bear was born (Fig. 5; Piludu 2019a, 159).

In order to obtain the favour and assured help of the *haltias*, it was necessary to please them in many ways. To begin, it was mandatory to give offerings of ale to the most important male *haltia*, the master of the forest, Tapio (SKS KRA, Lilli Lilius b 181, 1888, Joutsa, central Finland, Otto Jussila). The hunters also offered small quantities of melted gold or silver, or some coins, to please Tapio (SKVR I4/1098, 36–41, Latvajärvi, Viena Karelia, Russia, 1839). Often these offerings were placed on or poured over the branches or roots of the "table of Tapio" (*Tapion pöytä*), a young spruce with an unusual shape, having a very flat top the branches of which sloped downwards so that it resembled a natural desk or altar (Piludu 2019a, 83, 154–157). If the hunters intended to kill a bear, they were expected to offer something valuable in exchange. This concept appears in many hunting cultures. Among the Mistassini Cree (Canada), the hunt was presented as a form of transaction between the hunters and the guardian of the game animals (Tanner 1979, 148). Offerings regulated the balance between "this world" and the "other side" represented by the forest spirits. If the hunters did not make offerings to the forest spirits, they could be punished with bad luck in hunting and an illness. However, when the hunters made offerings, they believed that the forest spirit was, in one way or another, "obliged" to provide the desired quarry. The rules of reciprocity were also valid for the *haltias*. The reciprocity between humans and sacred agents operated following moral principles recognised by both parties (Stark 2002, 41). The ethnologist Laura Stark stresses that "agreement upon shared 'rules of the game' and submission to a system of mutual moral obligation were expressed through collective symbols" (Stark 2002, 41). The offering of ale, one of these collective symbols, pleased the male *haltia,* Tapio, making him favourable towards the hunter. As Tapio was conceived as an anthropomorphic, male, humanlike being, he was supposed to enjoy the alcoholic beverages that humans loved. Ale was much more than an intoxicating beverage, it was a collective symbol with great relevance in almost all the rituals of the village communities, from agricultural spring rituals (*Ukon vakat*) to Christmas (Apo 2001, 369). Ale was present at all the parties and ceremonials; it was a powerful symbol of social life, sharing, "being together" and collective singing. The hunters also offered ale during the bear feast and the bear skull ritual to please the soul of the killed bruin.[46]

Seduction and mimesis in the forest: Pleasing the forest mistresses

The *Bear Songs* were characterised by a particular language with a strict ritual etiquette (Tarkka 2013, 71). The bear did not like people uttering its real names (*karhu* or *kontio*) and it got angry if someone pronounced these; the result was a bear attack against humans or cattle. The use of a multitude of honorary names involved the ritual strategy of gratification of the bear. The most common honorary names for the bear were *otso*, *ohto*, *mesikämmen* (honey-paw), *jumalanvilja* or *metsän vilja* (grain of God, grain of the forest), *kulta* (gold) or *hopea* (silver). In all the circumboreal bear ceremonials the use of ritual names, circumlocutions and euphemisms to honour the bear is present (Hallowell 1926, 43–51).

A way to please the female *haltias* protecting the bear was the "wooing or seduction strategy", which was common in the Finno-Karelian *Bear Songs*. Quite often the hunters sang to female *haltias* or to the forest, which was considered a female entity: "Become fond of my men, you forest, / fall in

46 See Piludu, this voume, on the Finno-Karelian bear feast and wedding, and also on the Finno-Karelian bear skull rituals.

love, you wild wood, with my dogs!" (*Mielly, sä metsä, miehihini, / kostu, sä korpi, koirihini!*; SKVR I4/1194, 19–22, Kieretti, Viena Karelia, Russia, 1894). The infatuation of the forest with the hunters was functional when it came to reaching their objective. Falling in love with the hunters, the forest accepted them as lovers and members of her kin or people.

What is peculiar is that the hunters requested the forest female *haltia* or the feminine forest to fall in love with them, but they never sang to the forest or a female spirit that they loved her. Here, seduction was almost the opposite of love. Rane Willerslev emphasised that the hunter/seducer remained emotionally unavailable because seduction is a game of power (Willerslev 2007, 105). Love involved self-surrender. If the forest fell in love, it gave the hunter whatever he wanted. But if the hunter fell in love, he would assume the role of the "giver", and of a person who is controlled by the forest. By being a seducer, the hunter maintained his capacity to reflect and act in a rational manner and manipulate the feelings of the forest.

Lotte Tarkka notes that, in a *Bear Song* by the Karelian singer, Arhippa Perttunen,[47] the bear hunt was portrayed as a sexual act or an erotic flirting (Tarkka 2013, 343–344) with a female forest spirit: "My mind is set, / set on visiting Metsola, / to make love to the forest maiden, / to drink the forest honey, / flesh from under the leaves, / grease from the spruce's roots" (*Mieleni minun tekisi, / mieli käyä metsolassa, / metsän nettä naiakseni, / metsän mettä juoakseni, / lihoa lehen alaista, / kuuta kuusen juurehista*; SKVR I4/1095, 1–6, Latvajärvi, Vuokkiniemi, Viena Karelia, Russia). In this case, the "gastronomic" names for the bear (honey, flesh, grease) are connected with an explicit description of sexual intercourse or a wedding proposal with a strong erotic content and a bodily fusion with the environment. However, the meaning of the verb "to make love" (*naida*) also means "to marry". The bear hunt was poetically transformed into a sexual or wedding proposal directed at the female guardians of bears. The hunters tried to appeal to and seduce the female *haltias* who could provide game (honey, flesh, grease).

With the symbols of sex and marriage, the process of identification between the hunter and the forest reached a high level of mimesis. Tarkka stresses that the ritual communication became a real blending or bodily fusion with the forest: the hunter and the forest spirit melded their "flesh, grease and honey", making love under the leaves and on the roots of the trees. The wooing hunter created a state of ritual communion, by means of which he became akin to the forest or merged with it physically (Tarkka 2013, 344).

The imitation and seduction of forest spirits and game animals are fundamental aspects of several hunting rituals around the world. The anthropologist Rane Willerslev, who studies the rituals of the Siberian Yukaghir hunters, stresses the relevance of the connections between mimesis and seduction. Willerslev observes that imitating something – the mimetic faculty – is to be "sensuously filled with that which is imitated, yielding to it, mirroring it, and hence imitating it bodily" (Willerslev 2007, 96). According to Michael Taussig,[48] mimesis has a corporeal aspect; the sensuous contact between the imitator and the original (Taussig 1993, 10; see Willerslev 2007, 12).

To seduce the forest spirits, Arhippa Perttunen also sang that "the man smells of the taste of honey" (*Mies haisee meen maulta*; SKVR I4/1095, 13). The hunter was perfumed with honey, the most sweet and sensual of the products of the woodland, the favourite food of the bear and a circumlocution for

47 Arhippa Perttunen or Arhippa Perttuńe was one of the most famous Karelian singers of all time. Elias Lönnrot transcribed his songs in 1934, and Arhippa's songs were extremely important when editing both the *Old Kalevala* (1835) and the *New Kalevala* (1949). Arhippa learned his songs from his father, the "Great Iivana". The Perttunen family came from the village of Laitasaari (Oulujoki), near Oulu in Finland, but they were fully integrated into the Karelian singing culture. Arhippa's son was Arhippaińi Miihkali or Miihkali Perttunen, another famous singer (Timonen 2008, 14–17).

48 Michael Taussig (born in 1940) is an Australian anthropologist interested in the concepts of mimesis, alterity, fetishism, and the philosophy of Walter Benjamin.

the bear itself. According to Tarkka, the sensual melding with the forest could also be interpreted as a form of camouflage (Tarkka 2013, 344). The hunter thus lost his human smell and acquired the scents of the forestland and the human scent could no longer be detected by animals.

The singer Juhana Korpelainen[49] clearly sang about a wedding proposal, declaring his desire to make love to/marry (*naida*) the forest maiden, "so I could bustle at a wedding / twirl at the table's end" (*Saisin häisä häärästellä, / pöydän pääsä pyörästellä*; SKVR VI2/4889, 30–31, Kiuruvesi, North Savo, Finland, 1819). Korpelainen also lamented: "Woe, I am indeed a poor boy, / as I was not married / with the favourite (or lovely) maidens of the forest, / with the girl with fur breast"[50] (*Voi minun polosen poijan, / kuin ei naitettu minua, / metän mieli tyttärillä / karva rinnoille kavolle*; SKVR VI2/4889, 32–35).

Korpelainen did not sing about occasional sexual intercourse but of a wedding, a ritual that created a kinship relation with forest *haltia*s that could safeguard game animals in the future. The Savonian singer Matti Waljakka asked of the forest mistress, Annika: "Marry us, our men, / regenerate our heroes / with the favourite (or lovely) daughters of the forest" (*Uuista urohitamme, / metsän mielityttärille, / karvarinta rakkahille*; SKVR VI2/4822, 31–35, Mäntyharju, South Savo, Finland, 1858). Here, the forest mistress, Annika (St Anna syncretised with the Mistress of the Forest, Mielikki), is an authoritative mother-in-law, and the brides are the forest spirits or the bears, who could be addressed with female names.

The hunter-prey relationship is conceptualised as seduction and marriage in many hunting cultures. Among the Makuna of Amazonia, the hunter explicitly declares a wish to attract and seduce his prey, and the relationship between the spirit owner and its animals is equated to that between a father and his marriageable daughters, which could be allocated to human beings if they properly performed all the rituals (Århem 1996, 192). Among the Cree of the East Coast of James Bay (Canada), the relationship between the hunter and the caribou is presented as "sexual lust", as the caribou was believed "to give itself to the hunter eagerly" (Preston 1975, 230; Tanner 1979, 138).

In Finland and Karelia, sexual intercourse or wedding proposals in the *Bear Songs* were strategies to become a relative of the forest, part of the "kin" of the forest mistress. According to Laura Stark, in Karelia the wedding was a moment in the social fabric which had the potential to alter the division of labour, power dynamics and the relations between the social classes (Stark 2006, 170–171). Sometimes farmhands and humble serving maids were able to "marry up" into the landowning class (Stark 2006, 171). Weddings could thus enrich or empower the weaker or poorer members of society. The hunter (who presented himself as a foreigner and a poor wanderer in the forest) who married a forest *haltia* was believed to be able to influence the decisions of his powerful and rich "wife": She would generously give her "gold" and "silver" (bears) to the hunters. However, the seduction or the wedding with the forest *haltia* was temporary; the hunter became a lover or husband of the forest only for the short period of the hunt. After that, he returned to the village and his wife.

If the hunter became a sexual partner or a "relative" of the forest *haltias*, he could ask of them the desired prey. In the song by Arhippa Perttunen, the seductive lines were followed by requests. He asked the Forest Mistress to send the scent of the bear towards the hound: "Slash the smell as an arc, / let it go into the dog's nostrils" (*Tuhku kaarelle sivalla / tulla koiran sieramihin*; SKVR I4/1095, 16–17). Afterwards, he asked her to set the hound to discover the location of the bear or its den (SKVR I4/1095, 21–13).

49 Juhana Korpelainen was a famous Savonian singer. However, he considered himself a *tietäjä* (seer), and the incantations and the *Bear Songs* are dominant in his repertoire (Siikala 2016, 308–309).

50 The "fur breast" of the forest maiden indicates a deep relation between forest spirits and fur animals. Although beautiful, the forest maidens were hybrid or shape-shifter beings, with some degree of animality, or with some vegetal feature. In western Finland, forest maidens were portrayed as attractive, but with backs made of bark.

The last lines reveal that the hunter did not perform a complete metamorphosis. By marrying the forest *haltia* or making love with her, he became mimetically similar to the forest, but he maintained his consciousness and identity (WILLERSLEV 2007, 107). Taussig stresses that mimesis is like "dancing between the very same and the very different" (TAUSSIG 1993, 129). While the hunter emphatically played "the role of the harmless lover" (WILLERSLEV 2007, 104), he never forgot the main goal of his hunt. If the hunter completely became a forest being, he lost his agency and capacity to hunt.

Mimesis requires profound reflexivity; the hunter, sensuously melding with the forest, should pay attention to avoid complete absorption into it. The imitator is conscious that the original is different. If there is complete homogenisation, there is nothing different to imitate. Mimesis depends on the existence of difference, and the imitator remains a being "in between" the identities (WILLERSLEV 2007, 12). The imitator voluntarily remains an imperfect copy, "a poorly executed ideogram" (TAUSSIG 1993, 17) of the original.

Mimesis and seduction are sophisticated games of power. Taussig argues that the copy or the imitator, "drawing on the character and power of the original, may even assume that character and power" (TAUSSIG 1993, 13). Willerslev stressed that hunters did not use imitation to represent something or someone but to manipulate the power relations in the environment around them (WILLERSLEV 2007, 95). In the case of Arhippa's song, the hunter's mimetic body fusion with the forest – a sexual and sensual fusion with the "grease" of forest spirits on the tree's roots – was completed to assume seductive power over the female *haltia*, who had the right to give the bear to the hunter.

The transformations of the forest spirits and the hunting ground

After the offerings of ale and the seductive songs, the hunters asked the *haltias* to bring them to the bear den. The Savonian hunter and singer Juhana Korpelainen prayed to the female forest spirits: "Bring [me] to that hill, / carry me on that mound, / where I could get the prey, / my job will bring a catch" (*Saata sillen saarexellen, / sille kummullen kuleta, / josta saalis saatasiin, / eron toimi tuotasiin*; SKVR VI2/4889, 49–52, Kiuruvesi, North Savo, Finland, 1819). The Viena Karelian hunters described the place around the hill or den as shining with gold or expensive materials: "Carry me on that mound, / where the spruces are in golden belts, / the pines are in silver cover, / the branches of the aspen in baize!" (*Tuolla kummulla kuleta, / kussa ois' kuuset kultavöissä, / petäjät hopeisiloissa, / hoavan oksat hal'l'akoissa!*; SKVR I4/1193, 23–26; Koljola, Viena Karelia, Russia, 1894). The seduced female *haltias* became the guides of the hunters. The gold and silver are the "richness" of the forest spirits – the bears. The hunters also obtained the favour of the *haltias* with other strategies; singing that they were humble, weak, starving poor people, lost in a strange land; they absolutely needed the help of the powerful and rich spirits in order to obtain prey (PILUDU 2019a, 170–174). The huntsmen avoided boasting about themselves, and they presented themselves as humble and respectful "guests" of the forestland.

The hunters asked the *haltias* to enchant the forest: "Let the groves shine like the sun, / the wild woods shine like the moon" (*Lehdot paistaa päiväsenny, / korvet paistaa kuutamannu*; SKVR VII5/3370, Salmi, Ladoga Karelia, Russia, 1884). The shining of the forest was the visual signal the hunter expected. The light, the sun and the moon represented life and birth (TARKKA 2013, 393), when the woodland was rich with game. By contrast, darkness and the cold were related to the abode of the dead (TARKKA 2013, 390) and a forest without animals.

If the request was accepted, the sombre backwoods changed into a marvellous place. The hunters asked the personalised forest to "wrap (or dress) the thickets in a broadcloth, / the deep woods in German linen" (*Viiat verkohon[!] vetele, / salot Saksan paltinoihin*; SKVR VII5/3367, 5–6). The "dressed" trees were also the "clothes" of the *haltias*. The wood should be dressed in linen clothes made in Germany; the forest's clothes were imported, coming from far away, so they were expensive, valuable. Fancy textiles and rich clothes were related to the abundance of prey. If the *haltias* wore expensive or golden clothes, the woodland would be rich with animals. If they wore poor gowns, they would not send any catch to the hunter (Tarkka 2013, 331). The hunters also asked the forest *haltias* to modify their environment, flattening the hills, for example, to make their travel in the forest easier and faster.

The songs emphasised a radical change in the behaviour and status of the *haltias*: The jealous and severe protectors of bears became lovers, brides, generous hosts, and providers of bruins; the guides who carried the hunters to the hill that shone of gold and silver (the den of the bear). The modification of the behaviour of *haltias* included a "change of their clothes" (the trees), which was a radical transformation of the forest; the spirits transformed a dark, dreadful and poor woodland into a shining, rich and welcoming environment. The richness of the trees indicates abundance of prey, as the bears were considered the "gold" and the "silver" of the *haltias*.

The bear as a maiden who has overslept: Calling the bear out of his den

When the hunters reached the bear's den, it was forbidden to kill the bruin while it still slept in hibernation. The rite of awakening the bear and calling it out of the den before the kill is present in almost all the circumboreal bear ceremonials (Hallowell 1926, 53–54).

The bear should be awakened for a fight, which was supposed to be fair. In reality, the bear was sleepy and weak, it faced a group of hunters alone, and it could be blinded by sunlight when it arose (Sarmela 1991, 211). The Karelian singer Lukkańi Huotari[51] awakened the bear as if it was a maiden: "Rise now, sooty maiden, / from the sooty fireplace" (*Nousep' pois, nokińe(n) neit'[i], / nokiselta nuotiolta*; SKVR I4/1206b, 1–2; Ponkalaksi, Viena Karelia, Russia, 1877). These lines were also a motif of the Karelian wedding songs to wake up a bride who had overslept on her wedding day (SKVR I4/2265, 1–2; Tollonjoki, Vuokkiniemi, Viena Karelia, Russia, 1918; Tarkka 2013, 334). The bruin, which in several *Bear Songs* is addressed as a male, is here a maiden. Tarkka stresses that the bear's symbolic change of gender was meaningful, as the Karelian hunter saw a similitude between the bear and the "maiden" or "bride" (Tarkka 2013, 351): Both were seen as important "catches". The bear was probably considered a catch both as sweet and desirable as a young bride sleeping in her bed. The motif is part of the seductive strategy of the hunter. First, the hunters seduced the female *haltias* in order to reach the den. After that, the hunters seduced the bruin itself, described as a maiden, to wake it up from its den. The hunt, which is a violent act, causing the death of animals, continued to be presented by the hunters as a journey to seduce or marry "maidens" of the forest village. The hunters carefully transformed the hunt, violence, and killing into sexuality, seduction and weddings – acts and ceremonies that pleased the forest spirits and game animals.

Denying responsibility for the bear kill

After the kill, the hunters continued to sing songs to the bear as if it was still alive and able to listen to their songs. At first, the responsibility for the bear kill was totally denied; in the songs, the hunters invented a false version of the facts in order to vindicate themselves (Piludu 2019a, 200–203). The

51 Lukkańi Huotari, Huotari Lukkanen or Ponkalahden Huotari was a skilled Karelian fisherman, who also knew incantations for the bear hunt. In his village, Ponkalaksi, people went often to Finland to buy and sell things, and there were Finnish influences on clothing and on the buildings.

Karelian singer Jyrki Malinen[52] portrayed the kill as an accident; the bear killed itself by falling from a tree: "You slipped by yourself from the branch, / you rolled down by yourself from the bough, / over your berry-filled belly, / broken is your golden stomach" (*Itše hairahit havolta, / itše vierit vempeleltä, / läpi marjaisem mahasi / rikki kultaisen kupusi*; SKVR I4/1207, 9–12, Vuonninen, Viena Karelia, Russia, 1872). This motif, which was very common in the Bear Songs, was the explanation furnished in Karelia and Finland for the bear's death. In this way, the bear brought about its own death and the hunter was blameless for the killing of his quarry (Tarkka 1998, 99).

The hunters' justifications or apologies for the bear's death are a fundamental phase of the circumboreal bear ceremonials. However, different explanations were possible among the various ethnic groups who performed bear ceremonials. The main strategy was to shift the responsibility to some other animal or a foreign hunter (Hallowell 1926, 55–57). The Yakut (Siberia) blamed the Russians or the Tungus, the Khanty (western Siberia) accused the Russians, the Sámi told the bear that the hunters were Russian, German, or English (Kuusi 1963, 47), the Ojibwa (Canada, USA) accused the English or Anglo-American hunters (Hallowell 1926, 55–57), and the Finno-Karelians told the bear that their knives had been made in Estonia and Germany. In North America, many native hunters simply apologised, telling the truth to the bear. The Western Abenaki hunters (Quebec, Canada, Saint Francis River) explained that they needed fur for their coats and meat to eat. The Menomini hunters (Menominee or Mamaceqtaw, United States) reminded the bear that it was intelligent and that it knew that the children of the hunters were starving and that they needed meat (Hallowell 1926, 55). The justification for the bear kill was a fundamental way of avoiding the revenge of bears and forest spirits, so the hunters paid particular attention to this delicate phase in the ritual.

Conclusions: Personhood and gifts, honour, seduction and innocence

When analysing the *Bear Songs* sung during the hunt, there is no doubt that both the bear and the forest *haltias* were personalised in quite a sophisticated way. This personalisation was a strategy to communicate with them, as one of the main characteristics of personhood is the ability to speak and understand the language. Persons interact with other persons, so the hunters created a web of social and emotional relationships with the *haltias* and the bear.

A fundamental characteristic of personhood is the capacity to feel emotions and have expectations. The main goal of the *Bear Songs* and hunting rituals was to avoid the rage of the forest *haltias* and the bear itself and to gratify them in many ways, transforming their perception of the hunt as a set of acts honouring, seducing, satisfying and pleasing them. The hunters' ritual strategies included:

- Offerings of ale and gold and silver. With these gifts, the hunters showed respect as they were foreigners, guests and visitors of the forestland owned by the *haltias.* However, the rules of reciprocity stressed that, if the *haltias* accepted the offering, they were supposed to give another kind of gift (the bear) to the hunters. The system of gifts bound the forest and human communities in a web of mutual expectations, including:
- The seductive songs for the female forest spirits. These lines pleased the forest maidens, who became infatuated with the hunters and helped them to reach the bear den;
- The respectful behaviour of the hunters, who used honorary names for the bear;
- Songs showing humbleness towards the *haltias*;
- The rite of awakening the bear and calling it out of the den: addressing the bruin as a bride.

52 Jyrki Malinen or Ontreiñi Jyrki was the son of Ontrei Malinen and the father of Iivana Malinen. The whole Malinen family had a consistent repertoire of *Bear Songs* and epic songs. Jyrki learned the *Bear Songs* from his father Ontrei (Haavio 1948, 19–20; Siikala 2016, 158).

Finally, the hunters sang songs telling the bear that they did not kill it, but that it died slipping from the branch of a tree. The rituals and the songs permitted the transformation of personhood: The bear, the *haltias* and the hunters gained different social roles in the *Bear Songs*. Several scholars have noted that, in modern Western societies, personhood means, above all, individual consciousness, while in folk or indigenous societies a person is essentially a social being (Mauss 1985; Carrithers et al. 1985; Brightman et al. 2014). A social person could be defined by a social role (foreigner, guardian, bride, groom, seducer, seduced, helper, protector), but these roles in the *Bear Songs* were far from static. With offerings and seductive songs, the forest *haltias*, who were portrayed as jealous protectors of bears, were transformed into guides who helped the hunters and gave them the prey. Before the kill, the bear was kindly awakened as a bride or a maiden that had overslept, in order to persuade the bruin to gently leave the den without attacking the hunters.

The hunters presented themselves to the forest maidens as lovers or suitors, not as the killers of their cattle or offspring (the bears). The sexual intercourse or marriage transformed the social status of the hunters, who became accepted – during the short time of the hunt – as members of the "people of the forest". Entering the forest, the hunters acted as handsome seducers of female forest *haltias* and the feminised woodland but, when they required the help of the forest spirits to find the den, they could humbly redefine themselves as orphans, poor and hungry foreigners, or young boys who absolutely needed the guidance of the *haltias*. The hunters became mimetic beings because, in the songs, they represented their sexual intercourse with the forest spirits as a bodily fusion with the forest. However, the hunters did not completely transform themselves into forest *haltias*; they maintained their human identity, as their brief seductions and weddings were aimed at convincing the forest spirits to guide them towards the bear den.[53]

The hunters' mimesis was a sensual, strategic, and voluntary manipulation of the process and degree of identification with the alterity represented by the forest *haltias*. This mimesis was a partial transformation of the hunter into the kin of the forest *haltias*. The "incomplete coping" of the forest *haltias* was characterised by a state of "in-betweenness" (see Willerslev 2007, 105–108); a liminary state between the human and the forest world. However, the hunters did not completely transform themselves into *haltias*, nor they were dominated by powers of the forest. The mimesis of the singers of the *Bear Songs* included the defence of their independent agency as hunters.

Acknowledgements

I would like to express my special thanks of gratitude to Paavo Hamunen and Susanna Lendiera for the permission to publish their photos, and to Riho Grünthal, Professor of Finnic Languages (University of Helsinki) for the permission to publish an adapted English version of his map of Finnic languages. I would like to thank Lotte Tarkka, Professor of Folklore Studies (University of Helsinki), for her valuable help in the interpretation and translation of the Viena Karelian *Bear Songs* when she was the supervisor of my PhD work on Finnish and Karelian bear ceremonialism. This chapter is dedicated to the memory of Enrico Comba (1956–2020), former Professor of Anthropology of Religions at the University of Turin, for his constant interest in my research work. I'm grateful to the Niilo Helander Foundation (*Niilo Helanderin Säätiö*) for having generously funded my research work at the University of Helsinki.

53 During the bear feast, the killed bear also participated in a mimetic wedding: In the songs, it was presented as a groom or a guest of honour at a wedding, but it did not become completely human, maintaining its alterity and dangerousness. See Piludu, this volume, on the Finno-Karelian bear feast and wedding.

Bibliography

Ahlqvist 1881: A. Ahlqvist: Unter Wogulen und Ostjaken: Reisebriefe und Ethnographische Mittheilungen. Abdruck aus den Acta Societatis Scientiarum Fennicae XIV (Helsinki 1881).

Ahola 2020: J. Ahola, Kalevalakielissä erätarinoissa kaikuja menneisyydestä. In: T. Kirkinen (ed.), Karjan eräperinne. Luovutetun Karjalan metsästyksen ja kalastuksen kulttuurihistoria (Helsinki 2020) 34–44.

Akino 1999: S. Akino, Spirit-sending ceremonies. In: W. Fitzhug/W. Dubreuil/O. Chisato (eds.), Ainu: Spirit of a Northern People (Washington 1999) 248–255.

Anttila 1931a: A. Anttila, Elias Lönnrot: elämä ja toiminta I (Helsinki 1931).

Anttila 1931b: A. Anttila, Elias Lönnrot: elämä ja toiminta II (Helsinki 1931).

Anttila 1962: A. Anttila, Elias Lönnrot (Helsinki 1962).

Anttonen 1992: V. Anttonen, The Concept of Pyhä (Sacred) in Pre-Cristian Finnish Religions. In: M. Hoppál/J. Pentikäinen (eds.), Northern Religions and Shamanism. Ehnologia Uralica 3 (Budapest, Helsinki 1992) 31–38.

Anttonen 1994: V. Anttonen, Erä- ja metsäluonnon pyhyys. In: P. Laaksonen/S.-L. Mettomäki (eds.), Metsä ja metsänviljaa. Kalevalaseuran Vuosikirja 73 (Helsinki 1994) 24–35.

Anttonen 1996: V. Anttonen, Ihmisen ja maan rajat. Pyhä kulttuurisena kategoriana (Helsinki 1996).

Apo 2001: S. Apo, Viinan voima. Näkökulma suomalaisten kansanomaiseen alkoholiajatteluun ja -kulttuuriin (Helsinki 2001).

Appelgren 1885: H. Appelgren, Karhun palveluksesta. Valvoja 1,V (Helsinki 1885).

Bartens 1986: R. Bartens (ed.), Siivekkälle jumalille, jalallisille jumalille. Mansien ja hantien runoutta (Helsinki 1986).

Batchelor 1901: J. Batchelor, The Ainu and Their Folk-Lore (London 1901).

Batchelor 1932: J. Batchelor, The Ainu Bear Festival. The Transactions of the Asiatic Society of Japan, Second Series, IX, 1932, 37–44.

Brightman et al. 2014: M. Brightman/V. E. Grotti/O. Ulturgasheva, Introduction. Animism and Invisible Worlds: The Place of Non-humans in Indigenous Ontologies. In: M. Brightman/V. E. Grotti/O. Ulturgasheva (eds.), Gender and Folklore: Perspectives on Finnish and Karelian Culture. Animals in Rainforest and Tundra: Personhood, Animals and Things in Contemporary Amazonia and Siberia (New York 2014) 1–28.

Campbell 1988: J. Campbell, Historical Atlas of World Mythology: Volume I: The Way of Animal Powers. Part 2: Mythologies of the Great Hunt (New York 1988).

Carrithers et al. 1985: M. Carrithers/S. Colling/S. Lukes, The Category of the Person: Anthropology, Philosophy, History (Cambridge 1985).

Comba 1996: E. Comba, Visioni dell'orso: ritualità e sciamanismo tra gli indiani delle pianure. In: A. Bongiovanni/E. Comba (eds.), Bestie o dei? L'animale nel simbolismo religioso (Torino 1996) 23–44.

Comba 2015: E. Comba, Tradizioni dell'orso fra i nativi nor damericani. In: E. Comba/D. Ormezzano (eds.), Uomini e orsi: morfologia del selvaggio (Torino 2015) 185–211.

Comba 2019: E. Comba, Bear Traditions in Native North America. In: E. Comba/D. Ormezzano (eds.), Men and Bears. Morphology of the Wild (Torino 2019) 185–211.

Cushing 1977: G. T. Cushing, The Bear in Ob-Ugrian Folklore. Folklore 88(2), 1977, 146–159.

Darnell 1977: R. Darnell, Hallowell's "Bear Ceremonialism" and the Emergence of Boasian Anthropology. Ethos 5(1), 1977, 13–30.

Dyrenkova 1930: N. P. Dyrenkova, Bear Worship among the Turkish Tribes of Siberia. In: Proceedings of the Twenty-Third International Congress of Americanists, held at New York, September 17–22, 1928 (New York 1930) 411–440.

Edsman 1953: C.-M. Edsman, Studier i jägarnas förkristna religion: finska björnaktsriter. In: Kirkohistorisk årskrift (Uppsala 1953) 48–106.

Edsman 1956: C.-M. Edsman, The Story of the Bear Wife in Nordic tradition. Ethnos 21, 1956, 36–56.

Edsman 1975: C.-M. Edsman, Jägaren, villebrådet och makterna. Lapska och finska björnriter. Västerbotten 1970, 1975, 37–57.

Edsman 1994: C.-M. Edsman, Jägaren och makterna: samiska och finska björnceremonier. Uppsala: Dialekt- och folkminnesarkivet (Uppsala 1994).

Eräjärvi 1963: I.-L. Eräjärvi, Kaarle Krohn. Elämä ja Toiminta. Suomi 110(2) (Helsinki 1963) 1–138.

Fjellström 1775: P. Fjellström, Kort Berättelse on Lapparnas Björnä-fänge samt deras der wid brukade widskeppelser (Stockholm 1775).

Fortescue 1998: M. Fortescue, Language Relations Across the Bering Strait: Reappraising Archaeological and Linguistic Evidence (London 1998).

Frog 2009: Frog, Do You See What I See? Mythic Landscape in the Immediate World. Folklore 43, 2009, 7–27. http://www.folklore.ee/folklore/vol43/frog.pdf.

Goldenweiser 1910/1911: A. A. Goldenweiser, Totemism, an Analytical Study. The Journal of American Folklore 23, 1910/1911, 179–294.

Grünthal/Sarhimaa 2004/2012: R. Grünthal/A. Sarhimaa (eds.), Itämerensuomalaiset kielet ja niiden päämurteet (2004/2012). https://www.sgr.fi/muutjulkaisut/ItamerensuomalaisetKieletMurteet2012.pdf

Grünthal 2020: R. Grünthal, The spread zones and contacts of medieval Finnic in the Northeastern Baltic Sea area: implication for the rate of language change. Journal of Historical Sociolingustics 6(2), 2020. https://doi.org/10.1515/jhsl-2019-0029.

Haavio 1948: M. Haavio, Viimeiset runolaulajat (Porvoo, Helsinki 1948) 1–10.

Haavio 1967: M. Haavio, Suomalainen mytologia (Porvoo, Helsinki 1967).

Haavio 1971: M. Haavio, Elias Lönnrot. In: D. Strömbäck/B. Alver/B. Holbeck/L. Virtanen (eds.), Biographica. Nordic Folklorists of the Past. Studies in Honour of Jouko Hautala (Copenhagen 1971) 1–10.

In: E. Comba/D. Ormezzano (eds.), Men and Bears. Morphology of the Wild (Torino 2019) 146–184.

Piludu 2022: V. M. Piludu, Comparative and Contextual Approaches to the Study of Finno-Karelian and Ob-Ugrian *The Birth of the Bear in The Sky* incantations and songs. In: H. Rydving/K. Kaikkonen (eds.), Religions Around the Arctic: Source Criticism and Comparisons (Stockholm 2022) 77–100.

Preston 1975: R. J. Preston, Cree Narrative: Expressing the Personal Meaning of Events (Ottawa 1975).

Pulkkinen 2005: R. Pulkkinen, Bear. In: U.-M. Kulonen/ I. Seurujärvi/R. Pulkkinen (eds.), The Saami. A Cultural Encyclopedia (Helsinki 2005) 33–36.

Pulkkinen 2014: R. Pulkkinen, Suomalaisen kansausko (Helsinki 2014).

Pulkkinen/Lindfors 2016: R. Pulkkinen/S. Lindfors, Suomalaisen kansauskon sanakirja (Helsinki 2016).

Pulkkinen/Sammelkivi 2006: R. Pulkkinen/M. Sammelkivi, Mitä paikannimet kertovat Suomen karhusta? In: C. Tolley (ed.), Karhun kannoilla. In the Footstep of the Bear. Satakunnan Museon julkaisuja 14, 2006. Turun yliopiston kulttuurituotannon ja maisemantutkimuksen laitoksen julkaisuja 9 (Pori 2006) 147–148.

Pulliainen 1974: E. Pulliainen, Suomen suur-pedot (Helsinki 1974).

Rebourcet 2006: G. Rebourcet, The Bear Cult in Finno-Ugrian Folk Poetry and in the Kalevala. In: C. Tolley (ed.), Karhun Kannoilla. In the Footsteps of the Bear. Satakunnan Museon julkaisuja 14, 2006. Turun yliopiston kulttuurituotannon ja maisemantutkimuksen laitoksen julkaisuja 9 (Pori 2006) 89–104.

Rockwell 1991: D. Rockwell, Giving Voice to Bear: North American Indian Rituals, Myths, and Images of the Bear (Niwot 1991).

Rydving 2010: H. Rydving, The 'Bear Ceremonial' and Bear Rituals among the Khanty and the Sami. Temenos 46(1). 2010, 31–52.

Salo 2006: U. Salo, Karhun kämmen ja karhun kynsi. Karhun jälkiä Suomen esihistoriassa. In: C. Tolley (ed.), Karhun kannoilla. In the Footstep of the Bear. Satakunnan Museon julkaisuja 14, 2006. Turun yliopiston kulttuurituotannon ja maisemantutkimuksen laitoksen julkaisuja 9 (Pori 2006) 167–182.

Salo 2012: U. Salo, Kalevalaiset myytit ja uskomukset arkeologian, kielihistorian ja kulttuurihistorian näkökulmasta II. Luonto ja kulttuuri. Muinaissuomalainen elämänymmärrys. (Somero 2012).

Sarmela 1972: M. Sarmela, Karhunpeijaisten arvoitus. Kotiseutu 4/5, 1972, 164–170.

Sarmela 1982: M. Sarmela, The Death of the Bear: An Old Finnish Hunting Drama. The Drama Review 26(3): Scandinavian Theatre, 1982, 57–66.

Sarmela 1983: M. Sarmela, The Finnish Bear Hunting Drama. Suomalais-Ugrilaisen Seuran Aikakauskirja 1983, 283–300.

Sarmela 1991: M. Sarmela, Karhu ihmisen ympäristössä. In: P. Laaksonen/S.-L. Mettömäki (eds.), Kolme on Kovaa Sanaa. Kirjoituksia kansanperinteestä. Kalevalaseuran vuosikirja 71 (Helsinki 1991) 209–250.

Sarmela 1994: M. Sarmela, Suomen perinneatlas. Suomen kansankulttuurin kartasto 2 (Helsinki 1994).

Sarmela 2006a: M. Sarmela. Karhun ihmisen ympäristössä. In: C. Tolley (ed.), Karhun Kannoilla. In the Footsteps of the Bear. Satakunnan Museon julkaisuja 14, 2006. Turun yliopiston kulttuurituotannon ja maisemantutkimuksen laitoksen julkaisuja 9 (Pori 2006) 42–83.

Sarmela 2006b: M. Sarmela, The Bear in the Finnish Environment: Discontinuity of Cultural Existence. Published on the author's webpage: http://www.kotikone.fi/matti.sarmela/bear.pdf.

Sarmela 2009: M. Sarmela, Finnish Folklore Atlas. Ethnic Culture of Finland 2. Published on the author's webpage: http://www.kotikone.fi/matti.sarmela/folkloreatlas.pdf.

Schefferus 1971: J. Schefferus, The History of Lapland (Stockholm 1971; originally published 1673).

Siikala 1992: A.-L. Siikala, Suomalainen samanismi (Helsinki 1992).

Siikala 1994: A.-L. Siikala, Transformation of the Kalevala Epic. In: A.-L. Siikala/S. Vakimo, Songs Beyond the Kalevala: Transformation of Oral Poetry. Studia Fennica Folkloristica 2 (Helsinki 1994) 41–55.

Siikala 2002: A.-L. Siikala, Mythic Images and Shamanism (Vammala 2002).

Siikala 2008: A.-L. Siikala, Uskomusmaailma. In: R. Räsänen (ed.), Savo ja sen kansa (Helsinki 2008) 109–220.

Siikala 2016: A.-L. Siikala, Itämerensuomalaisten mytologia (Helsinki 2016).

Siikala/Ulyashev 2011: A.-L. Siikala/O. Ulyashev, Hidden Rituals and Public Performances. Traditions and Belonging among the Post-Soviet Khanty, Komi and Udmurts (Helsinki 2011).

Sirelius 1919: U. T. Sirelius, Suomen kansanomaista kulttuuria: esineellisen kansatieteen tuloksia 1 (Helsinki 1919).

Sirelius 1929: U. T. Sirelius, Obin ugrilaisten peijaisista. Kalevalaseuran vuosikirja 9, 1929, 193–207.

Schmidt 1989: É. Schmidt, Bear Cult and Mythology of the Northern Ob-Ugrians. In: M. Hoppál/H. Miháli/J. Pentikäinen (eds.), Uralic Mythology and Folklore. Ethnologica Uralica 1 (Budapest, Helsinki 1989) 187– 232.

Soldatova 2008: G. E. Soldatova, Uj jikw (Beast Dance). In: V. Napolskikh/A.-L. Siikala/M. Hoppál (eds.), Mansi Mythology (Budapest, Helsinki 2008) 145–147.

Spagna 1998: F. Spagna, L'ospite selvaggio. Esperienze visionarie e simboli dell'orso nelle tradizioni native americane e circumboreali (Torino 1998).

Stark 2002: L. Stark, Peasant, Pilgrim and Sacred Promises: Ritual and Supernatural in Orthodox Karelian Folk Religion. Studia Fennica Folkoristica 11 (Helsinki 2002).

Stark 2006: L. Stark, The Magical Self: Body, Society and the Supernatural in Early Modern Rural Finland. Folklore Fellows Communications 138 (Helsinki 2006).

Taksami 2006: C. Taksami, Karhun merkitys Nivkhikansan perinnekulttuurissa. In: C. Tolley (ed.), Karhun kannoilla. In the Footstep of the Bear. Satakunnan Museon julkaisuja 14, 2006. Turun yliopiston kulttuurituotannon ja maisemantutkimuksen laitoksen julkaisuja 9 (Pori 2006) 31–42.

Tanner 1979: A. Tanner, Bringing Home Animals: Religious Ideology and Mode of Production of the Mistassini Cree Hunters (London 1979).

Tarkka 1994: L. Tarkka, Metsolan merkki – metsän olento ja kuva vienalaisrunostossa. In: P. Laaksonen/S.-L.

Mettömäki (eds.), Metsä ja Metsänvilja. Kalevalaseuran vuosikirja 73. Helsinki (1994) 56–102.
Tarkka 1998: L. Tarkka, The Sense of the Forest: Nature and Gender in Karelian Oral Poetry. In: S. Apo/ A. Nenola/L. Stark-Arola (eds.), Gender and Folklore: Perspectives on Finnish and Karelian Culture. Studia Fennica Folkloristica 4 (Helsinki 1998) 92–142.
Tarkka 2005: L. Tarkka, Rajarahvaan laulu. Tutkimus Vuokkiniemn kalevalamittaisesta runokulttuurista 1821–1921 (Helsinki 2005).
Tarkka 2013: L. Tarkka, Songs of the Border People: Genre: Reflexivity and Performance in Karelian Oral Poetry. Folklore Fellows Communications 152 (Helsinki 2013).
Tarkka 2014: L. Tarkka, Luonto, tuonpuolinen ja resurssien rajat. Esimodernin ympäristösuhteenkysymyksiä Vienan Karjalassa. In: S. Knuuttila/U. Piela (eds.), Ympäristömytologia. Kalevalaseuran vuosikirja 93 (Helsinki 2014) 34–58.
Taussig 1993: M. Taussig, Mimesis and Alterity: A Particular History of the Senses (London 1993).
Timonen 2008: S. Timonen, Elias Lönnrot ja Runolaulaja. In: U. Piela/S. Knuuttila/P. Laaksonen (eds.), Kalevalan kulttuurihistoria (Helsinki 2008) 3–27.

Uusitalo 1997: L. Uusitalo, Björnceremonialismen i Finland (Stockholm 1997).

Vilkuna 1965: K. Vilkuna, Björnfest. Kulturhistorisk lexicon för nordisk medeltid I (Helsinki 1965).
Virtanen/DuBois 2000: L. Virtanen/T. DuBois, Finnish Folklore. Studia Fennica Folkloristica 9 (Helsinki 2000).
Virtaranta 1958: P. Virtaranta, Vienan kansa muistelee (Porvoo, Helsinki 1958).
Viveiros de Castro 2009: E. Viveiros de Castro, Metaphisique cannibales (Paris 2009).

Wiget/Balalaeva 2010: A. Wiget/O. Balalaeva, Khanty: People of the Taiga (Fairbanks 2010).
Wiget/Balalaeva 2022: A. Wiget/O. Balalaeva, Valuing Difference: Bear Ceremonialism, the Eastern Khanty, and Cultural Variation among Ob-Ugrians. Sibirica 21(1), 2022. https://doi. 10.3167/sib.2022.210103.
Willerslev 2007: R. Willerslev, Soul Hunters: Hunting, Animism and Personhood among the Siberian Yukaghirs (Berkeley, Los Angeles, London 2007).

Zolotarev 1937: A. M. Zolotarev, The Bear Festival of the Olcha. American Anthropologist 39, 1937, 113–130.

Århem 1996: K. Århem, The Cosmic Food Web. Human-Nature Relatedness in the Northwest Amazon. In: P. Descola/G. Pálson (eds.), Nature and Society: Anthropological Perspectives (London 1996) 185–204.

Vesa Matteo Piludu
PhD, Postdoctoral researcher
University of Helsinki
Indigenous Studies,
Department of Finnish, Finno-Ugrian and Scandinavian Studies
Faculty of Arts
Helsinki
Finland
vesa.piludu@helsinki.fi

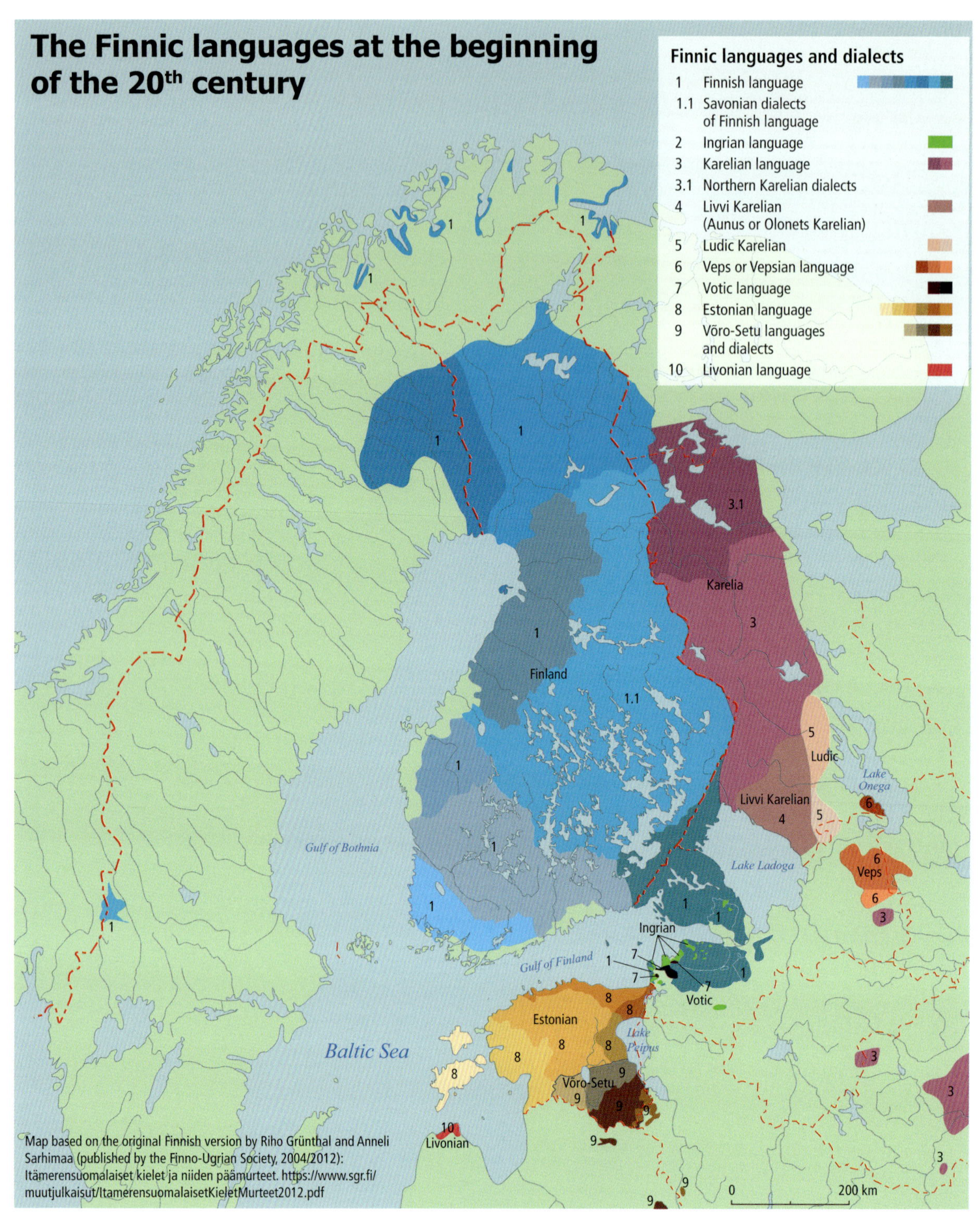

Fig. 1. Map of the Finnic languages (courtesy of Riho Grünthal, after Grünthal/Sarhimaa *2004/2012; see also* Grünthal *2020, 6). The most frequent dialects of the Bear Songs were the North Karelian, Viena Karelian or White Sea Karelian (dark purple on the map, the northernmost part of Karelia, 3.1), and the Savonian dialects (light blue or azure on the map, 1.1), which are used throughout a great part of eastern and central Finland. For details about the Finnish dialects in relation to Bear Songs, see note 25 in this chapter.*

Fig. 2. "The Bear and the Forest", photo by Paavo Hamunen, 2020. Metsä *(forest),* metsän kultaine n kuningas *(golden king of the forest) and Tapio (the most important male guardian spirit of the forest, the personalised "Forest") were ritual and honorary names for the bear in the Bear Songs. The sacred status of the bear was strictly connected with the forest and the world of the* haltias *(guardian spirits) of the woodland. Pronouncing the real names of the bear (*karhu *or* kontio*) was strictly prohibited during the bear hunt and ceremonials: The hunters believed that the bruin would be enraged and attack the gunmen. Other honorary names were* ohto, otso, mesikämmen *(honeypaw),* auvo *(honour, luck, groom),* kulta *(gold) and* hopea *(silver).*

*Fig. 3. "The Bear Soul", photo by Susanna Lendiera, 2018. According to Finnish and Karelian folk belief, the bear was sacred (*pyhä*); it had the status of a person (*ihiminen*) and a soul (*henki, sielu, haltia*). If a bear lost his soul, due to some malevolent ritual done by a sorcerer (*noita*), it became enraged and killed cattle or humans.*

Fig. 4. Karelian bear hunters from Venehjärvi, a Viena Karelian village that was famous for its bear killers and singers of Bear Songs. Photo taken by Into Kondrad Inha, during his travels in Karelia, 1894. The hunter holding the rifle was Varahvontta Lesonen, the guide of Inha (photo Museovirasto, Finnish National Board of Antiquities; Finno-Ugric Photographic Collection [Suomalais-ugrilainen kuvakokoelma] SUK5:19; public domain).

Fig. 5. "The Dark Forestland", photo by Susanna Lendiera, 2018. In Finno-Karelian folk beliefs and in the Bear Songs, the forest was portrayed as an otherworld called Tapiola, Metsola or Pohjola (The Northland). At the beginning of the hunt, the hunter sang that he was entering "the dark Pohjola" (pimeassä pohjolassa)*; the forest was described as a dangerous, deadly, otherworld that was strictly controlled by powerful* haltias *(guardian spirits). When the* haltias *accepted the hunter, they changed the aspect of the woodland, and the hunters sang that the forest shone with light, gold, and silver.*

Fig. 6. "Tellervo, Daugher of Tapio (Diana)", bronze sculpture by Yrjö Gabriel Liipola (1881–1971), 1928. Tellervo was one of several forest maidens, haltias *(guardian spirits), which protected the bears. In the Bear Songs they are presented as beautiful, and the hunters sung seductive songs to convince them to give them a bear and to guide them toward the den. Liipola, who studied sculpture in Florence, interpreted Tellervo as a northern version of the Roman goddess of the hunt, Diana (photo V. M. Piludu, Helsinki, 2022).*

eaten became a wedding and a jolly drinking and singing party. Overall, the feast transformed the hunt's violence into its exact opposite: a celebration of the continuation of life and of the good relations between the human community and the forest dwellers. The bear feast is called "post mortem ceremony" in the model of the phases of the bear ceremonialism analysed by Hallowell (HALLOWELL 1926, 61–135).[2]

In the circumboreal regions, the feasts were a fundamental part of the hunting rituals: The hunters and their communities joined together to eat the meat of large game animals (bears, elks, caribous) or sea mammals (whales). These feasts were presented as "voluntary visits" of the game animals to parties or ceremonies in which the prey was the guest of honour of the village (LOT-FALCK 1961, 180–188). There was great variation in the organisation of the feasts, depending on the hunting culture and on the species of game animal (TANNER 1979, 157–169; SCHMIDT 1989; cf. RYDVING 2010, reprinted in this volume).

Even if the bear was dead, the Finnish and Karelian hunters believed that it preserved its consciousness, and the hunters continued to sing to it, as its soul was still alive. The bear's soul was considered as a person, a guest able to understand the songs and to see what the people did during the feast. The *Bear Songs* of the bear feast resemble those sung by the hunters when leaving the village and entering the forest, but this time the main traveller and guest was the bear, and the journey was in the opposite direction. The bruin's entry into the village was considered a liminal and potentially dangerous event; small mistakes could provoke the posthumous revenge of the bear, often in the form of the bruin's attack on the cattle. Imperfection in performing the rituals could also cause a contagion that was produced by a supernatural illness – the "hate of the forest",[3] the "nose of the forest",[4] or the "hate of the bear".[5] This illness was caused by "the force of the forest" (*metsän väki*);[6] a dynamic and potentially dangerous force present in the bear meat, the *väki,* could invade human bodies.

Other hunting peoples also faced the problem of opening and closing, shifting and permeable boundaries with the external environment (the forest, the sea, the desert, the mountains). Hallowell assumes that, in Ojibwa[7] ontology, humans and animals are analogically related as human and nonhuman persons, but the Ojibwas draw some differences between the two forms of personhood (HALLOWELL 1960).[8] A fundamental problem was keeping a certain distance from the animal persons entering the village after the hunt.

The anthropologist and historian Ann Fienup-Riordan[9] states that, according to Yup'ik[10] seal hunters, both humans and animals are characterised by personhood, but there is a clear distinction between

2 For an analysis of the theories of Hallowell, see PILUDU, this volume, on the songs and rituals of the Finno-Karelian bear hunt.

3 The "hate of the forest" (*metsän viha*) related to the idea that the forest *haltias* (guardian spirits) were upset for some offence performed during the hunt or the feast.

4 The name *nenä* (nose) means something similar to a "bird beak", "something aggressive" that could penetrate the body and the flesh. The name *nenä* in Finnish related to a person who showed aggressive behaviour (*nenäkäs*) and put his/her nose in front of the face of another person.

5 The "hate of the bear" (*metsän viha*) related to the idea that the bear, or its soul, was upset and enraged about something; for example, if the bear meat was eaten without following strict ritual procedures.

6 The name *väki* means "force" (a dynamic force coming from the forest and able to invade human bodies) and "people". The two meanings are connected: The people of the forest (the *haltias* and the bear) have a force that could cause illness to humans. On the other hand, the hunters could also reinforce themselves by eating some parts of the bear and absorbing the bear force, but only if they performed the rituals. Without these rites, the bear meat was poisonous.

7 A native people living in Canada and the United States.

8 For an analysis of the importance of Hallowell's writings on bear ceremonialism and ontology, see PILUDU, this volume, on the songs and rituals of the Finno-Karelian bear hunt.

9 Ann Fienup-Riordan (born 1948) is an American cultural anthropologist and historian, known for her works on the Yup'ik culture.

10 The Yup'ik are an indigenous people living in western, central and southwestern Alaska.

human and nonhuman persons (Fienup-Riordan 1994, 48). Roy Wagner[11] and Fienup-Riordan define the human-animal relationship as a "deliberate controlled analogy" (Wagner 1977, 361; Fienup-Riordan 1994, 48), which made a carefully regulated ritual interaction between the two communities possible. Yup'ik ritual activities focused on the construction of boundaries and passages to circumscribe the flow of different persons, actions, and forces within an otherwise undifferentiated world (Fienup-Riordan 1994, 49). Interactions between human and animal persons were supposed to follow innumerable rules to create boundaries, but at the same time they enabled ceremonial exchanges between the human and animal worlds (Fienup-Riordan 1994, 48–51).

In the Finno-Karelian bear ceremonial, these "boundary problems" were extremely important. The introduction of the killed bear into the village was even presented as the bear's wedding, representing a partial unification of the forest with the human community (Tarkka 2013, 348). However, the people never forgot the potential supernatural dangerousness of the entrance of the bear into the village. For the women, who took care of the cows, the arrival of the killed bear in the village resembled its attack on cattle, so the hunters encouraged young women to protect their cows (Tarkka 2013, 354–356).

As I have already explained in this section, bringing home killed game animals, and in particular the powerful bear, could infect the humans with an illness coming from the force (*väki*) of the forest or from the bear. Eating bear meat without proper rituals also caused a "contagion" (Piludu 2019, 251–253). For these reasons, the marriage of the bear and the period of unification of the bear with the village was brief; just after the end of the feast, through the ritual of the bear skull, the bones and soul of the bear were supposed to return to their mythic homeland (Piludu 2019, 265–271).

Connecting the bruin and human communities: The bear as a guest and a groom

The hunters progressively integrated the killed bear into the human community. When leaving the forest, the Karelian hunter Moiśśeińi Kuśma[12] invited the bear to wander "to the heroic people,[13] / to the manly company" (*uroisehe väkeh, / miehisehe joukijoh*; SKVR I4/1203, 29–30, Latvajärvi, Vuokkiniemi, Viena Karelia, Russia, 1894).[14] In this song, the bruin is a male adult who is requested to join the manly hunting group. With a sort of rapid rite of passage, the killed bear was thus incorporated into the group of its killers. According to the singers, the killed bear did not cease to exist. Instead, it changed its social status, becoming a member of the human group during the bear feast. The hunters stressed that, when the bear became a member of the hunting group, it should forget its intention to take revenge on the human community. The hunters of Vuokkiniemi emphasised: "do not hate for a long time, / do not scare the Christian people"[15] (*Elkähä viikon vihoko, / kamaloiko*

11 Roy Wagner (1938–2018) was an influential American anthropologist, an expert on the cultures of Papua-New Guinea and author of the pivotal book *The Invention of Culture* (1975).

12 Moiśśeińi Kuśma, or Kuśma Ahońi, was an important *tietäjä* (seer) of the Ahonen or Ahońi family of singers from the village of Hoapovaara in Latvajärvi. He learned its incantations and *Bear Songs* from his grandfather and his father (Spiireińi Moiśśei), who were both skilled *tietäjäs*.

13 The adjective or noun *uros* means "the hero", "the man" or "the male". Here, the masculinity of the bear is underscored, whereas when the bear was awakened from the den it was feminised as a bride.

14 SKVR indicates a reference to a song or incantation from the collection *Suomen Kansan Vanhat Runot* (Ancient Songs of the Finnish People). In this chapter, the SKVR sources are indicated by the number of the volume (SKVR XII2), followed by the number of the song (/6573) and the number of the line (1–2).

15 The name *ristikansa* ("people of the cross", "Christians") means Karelians. The Karelians were Orthodox and many of them were Old Believers (*Старообрядчество*): Christian religion was a fundamental aspect of their self-identity. However, they continued to perform pre-Christian rituals and to believe in the existence of the *haltias* (guardian spirits). A similar situation could be found in Finland: The Finnish hunters were Protestants, but their folk rituals were still

ristikansa; SKVR I4/1229: 16–17, Vuokkiniemi, Viena Karelia, Russia, 1837; name of singer not mentioned in the manuscript by Elias Lönnrot). The incorporation of the bruin into the human community was functional to avoid the revenge of the bear for its death.

The entrance of the bear into the village marked a significative change in the communicative form of the *Bear Songs*. In Viena Karelia, when the hunters approached the yard of their village or home, a singing dialogue started between the hunters and the mistress of the house or village, or even a group of wives or mistresses (SKVR I4/1225 b.; I4/1227). The conversation was divided into questions (generally by the hunters) and answers (generally the women answered, but they could also ask the hunters about where the prey had come from). In the middle of this dialogue, both the hunters and the wives sang some lines to the silent "guest", the bear. A similar dialogical style was present in the wedding songs: When the groom arrived, the family of the bride and the groom started to sing a set of standard questions and answers. The folklorist Leea Virtanen[16] noticed that the songs of the bear feast have structural and meaningful affinities with the wedding songs (Virtanen 1968, 36–38).

The ritual role of the mistress of the house should not be underestimated, as she stood on the border of the household and the forest, behaving as a guardian and a mediator when the guest from the forestland came inside. This is clearly reminiscent of the role of the female forest *haltias*, the guardians of the woodland.

In Viena Karelia, the mistress welcomed the bear very warmly: "Welcome, *ohto* (the bear), upon your arrival, / honey paw, as you came by" (*Tervet, ohto, tultuosi, / mesi käm̃en, käytyösi!*; SKVR I4/1253, 1–5, Lonkka, Viena Karelia, Russia, 1834; name of singer not mentioned in the manuscript by Elias Lönnrot). The women sang to the bear as if it was a son-in-law or the groom awaited by them (Tarkka 1998, 115). In Vuonninen, the women welcomed the bear as their groom: "So I waited for you, / as a maiden for a young man, / the red-cheeked one for a spouse, / or the ski for new snow" (*Niimpä mie sinuo vuotin, / niinkuin neiti nuorta miestä, / puna-poski puolisuo, / eli suksi uutta lunta*; SKVR I4/1241, 21–24, Vuonninen, Viena Karelia, Russia, Varahvontta Lesońi,[17] 1872). In Venehjärvi, the mistress sang that she waited for the bear/groom until the ice melted under her feet and she walked until her shoes were broken (SKVR I4/1235, 8–9, Venehjärvi, Vuokkiniemi, Viena Karelia, Russia, Varahvontta Lesońi).

The welcome songs referring to the bear as a groom or son-in-law are particularly interesting because in one of the oldest documents about Finnish bear ceremonialism[18] the whole bear feast was called "The wedding of the bear", and it included a staged wedding: "When the bear had been successfully killed and flayed in the forest, and also the flesh with the skin was brought back home to the settlement, a day was settled upon when the so-called 'wedding of Couvo'[19] (the bear) would be celebrated. For this important celebration, some barley was gathered to brew beer and *viina*.[20] When the arranged day arrived, people gathered in church clothes at a house. Here, a boy was chosen in honour of the bear as a bridegroom and, following the custom of the land, a girl clad in bridal costume was chosen as bride" (SKVR IX4/1096, Viitasaari, central Finland; name of singer not

influenced by the belief in pre-Christian *haltias* (guardian spirits) and syncretic Catholic saints, which the Protestant churches abhorred. Throughout many centuries, the official Catholic, Orthodox and Protestant churches have made many efforts to erase the belief in the *haltias*, but with scarce results in the remote countryside up until the first decades of the 20th century.

16 Leea Virtanen (1935–2002) was a Finnish professor of Folkore Studies at the Univerity of Helsinki, an expert on Finnish and Estonian traditions, children's stories, and urban and paranormal legends.

17 Varahvontta Lesonen or Varahvontta Lesońi or Feropotij Terentjev Lezhev was a singer and the guide of the photographer I. K. Inha when he travelled in Viena Karelia. He learnt his Bear Songs from his father Tero Lesońi, who was an expert singer and hunter.

18 *Text of Viitasaari* (after 1750).

19 *Couvo* or *kouvo* means "the bear" or "the deceased", "the ghost".

20 *Viina* or *paloviina* (*bränvin* in Swedish) is a hard liquor or clear spirit drink made of barley.

mentioned; Pentikäinen 2007, 74; 2014, 424). However, the bridal couple was almost never mentioned in the 19th or 20th century sources. Was the presence of the bridal couple in Viitasaaari a local tradition, restricted to certain villages or areas? Or was it a once common tradition that disappeared before the 19th century?

Even though we do not have many sources that mention the presence of a bridal couple or a bride, in some 19th-century North and Viena Karelian *Bear Songs* singers clearly referred to the bear feast as a wedding of the bear, or sang to the bear as if it was a groom or bride.

The singer and hunter Juhana Kainulainen[21] invited the forest spirits to the feast by singing: "Honeyed matron of Mehtola,[22] / golden king of the forest,[23] / come now to the wedding of your ox, / to the feast of your long wool" (*Mehtolan metinen muori, / metän kultanen kuningas, / tule nyt häihiin härköisiis, / pitkän villaisi pitohon!*; SKVR VII5/3390, Villala, Humuvaara Kesälahti, North Karelia, Finland, 1828). The feast was presented to the forest spirits as the wedding of an ox, for the bears were often mentioned as the "cattle" (*karja*) of the forest spirits (Piludu 2019, 63–66). The latter song is interesting because it reveals that the spirits protecting the bear were also invited to the feast and the wedding.

The wedding was a way to humanise and personalise the bear, making the bruin a member of the village community and honouring it with its participation in the most important of the human ceremonies. However, the bear was a bride or groom only during the feast. Otherwise, the bruin's identity was mimetic: The bear was a humanlike being, but it did not become completely human, and its soul was supposed to return to the forest and regenerate there after the feast.

Mimesis depends on the existence of difference, and the imitator (in this case, the humanised bear, "imitating" the role of a human groom in the women's songs and in the bear wedding) remains a being "in between" identities (Willerslev 2007, 12). The imitator (the bear) is an imperfect copy, "a poorly executed ideogram" (Taussig 1993, 17) of the original (the human groom). The bear is an imperfect "copy", because it retained a potentially dangerous part of its alterity when it entered the village and the house: It was feared because it could possibly attack the cattle or cause an illness. The incomplete and mimetic wedding of the bear reminds us of the episode of the hunters' seduction of, or wedding to, the female forest spirits present in the songs of the bear hunt: In this case, the hunters were temporarily accepted as lovers and kin of the forest people to obtain the prey, but they maintained their identity as humans and hunters.[24]

The protection of the cattle and the women

After the welcoming songs, the hunters brought the bear skin, the bear head and the bear meat into the cabin and, more specifically, into the *tupa*; a living room with a kitchen, a brick oven, and a table. In Finland and Karelia, the *tupa* was the centre of all the indoor activities engaged in by the women and other members of the family. In eastern Finland and Karelia, it was also the main bedroom: The family slept on the wall benches, or on a bed on the top of the oven (Räsänen/Räsänen 2008, 335).

21 Juhana Kainulainen (1788–1847) was a famous *runolaulaja* (singer), *karhuntappaja* (bear hunter) and *tietäjä* (seer). Elias Lönnrot, the collector and editor of the Finnish epic poem *Kalevala* (1849) met Kainulainen in 1828 and transcribed 57 Kalevalaic songs and incantations, including the *Bear Songs* (Haavio 1948, 122–132). The pine tree where Kainulainen hung bear skulls is still alive in Kesälahti and is now a tourist attraction (called *Lönnrotin petäjä*, the "Pine of Lönnrot").

22 Mehtola or Metsola (The Forestland) was a name of the otherworldly forest governed by the forest *haltias* (guardian spirits). The matron of Mehtola was the Mistress of the Forest, the main female *haltia*, who was called Mielikki in some songs.

23 "Golden king of the forest" was one circumlocution for Tapio, the Master of the Forest, the main male forest *haltia*.

24 On the concept of mimesis, see Piludu, this volume, on the songs and rituals of the Finno-Karelian bear hunt.

In Viena Karelia and Finland, the bear killer exhorted the boys and girls to stay away from the doors when the bear was brought inside: "Go away, boys, from the porch, / maids, from the doorjambs, / when *ohto*[25] is entering the *tupa*!" (*Pois, pojat, porštuošta, / piijat, piht'ipuol'isešta, / ohon tullešša tupah!*; SKVR I4/1220a, 6–7, Kivijärvi, a Viena Karelian village in the territory of Finland, 1910, Maura Marttińi[26]). Because the bear was considered an older, male guest of honour, or a groom, the young and unmarried members of the village were expected to show their respect, leaving the places around the door free (see Fig. 2). The elders represented authority, so young people were supposed to give them precedence on the road and at the door (Siikala 2016, 101–102).

The presence of the mistress of the house (*emäntä*), a mature and authoritative woman, was allowed in the cabin and the *tupa* during the feast; she continued her singing dialogue, as she represented the household and its ownership. According to the folklorist Satu Apo, the women ensured the integrity of the farm household, and they acted as guardians and gatekeepers of the human habitat, the village, and the house (Apo 1998, 72–73; Stark-Arola 1998b, 39). This role not only included defending the household and the cattle, but also the proper welcoming of foreigners and the official acceptance of the guest, hosts and grooms, such as the bear. The women carefully controlled who and what entered and left the village.

However, the women protected the cattle when the bear passed across the threshold. The appearance of the bear corpse in the yard was considered a potential danger for the cattle, as the people feared a bear attack as a revenge for the killing of the bruin. This revenge could also be caused by other bears, informed about the kill of their "brother". The double role of the women (mistresses of the house and keepers of the cattle) could explain the apparently ambivalent positions and attitudes of the women towards the bear. The Viena Karelian hunters sang to the women, warning them: "Beware now, poor wives, / that the cattle won't vanish, / that the 'grains[27] of the mistress' (the cattle) won't disappear, / when *ohto* (the bear) is coming into the *tupa*" (*Varokaatte nyt, raukat vaimot, / ett'ei karja kaipastuisi, / viipastuisi emännän viljat, / ohon tullessa tupahan*; SKVR I4/1219b, 32–35, Niskajärvi, Viena Karelia, Russia, 1915, Hovatta Lesońi[28]).

Sometimes, the hunters exhorted their wives to guard their genitals too, when the bear entered the village. The singer Lukkańi Huotari[29] sang about the risk of a bear's sexual assault: "Beware, you poor wives, / when I move my golden one, / so that the hair (the pubic hair of the women, and the cattle) won't vanish, / the mistress' wool (the pubic hair of the women, and the cattle) won't fail, / when *ohto* (the bear) is coming to the fireplaces, / the honeyed dick (the bear) to the estate" (*Varokate vaimo raukat, / kuin ma kultańi kuletan, / jott' ei karva kaipastuise, / epeä emännän vil'l'a, / ohon tullessa tulilla, / kartanoh kalun met'isen*; SKVR I4/1206c: 9–14, Ponkalaksi, Viena Karelia, Russia, 1877).

The folklorist Lotte Tarkka stresses that in this song the words for the female sex (*karva*, "the hair" and *emännän villa*, the "wool of the mistress"; the vagina with pubic hair) could also be interpreted as a reference to the cattle (*emännän vilja*, "the grain of the mistress", the cows); therefore,

25 *Ohto* was one of the most common honorary names of the bear.

26 The family of Maura Marttińi came from Vuokkiniemi (Viena Karelia, Russia). She was a singer and *tietäjä* (seer); her son, Iivana Marttinen, was a folklore collector who has also transcribed this song. She learnt her songs and incantations from her grandmother (Outi Jyrintytär Remsunen, a *tietäjä*) and grandfather (Outokkainen).

27 The world *vilja* (grain) was very polysemic; in the songs, it means both grain (product of the field), the cattle (the "grain" of the mistress), and even the prey of the hunt (*jumalan vilja* means the "grain of the god [the *haltia*, guardian spirits]", the bear or some other forest animal).

28 Hovatta Lesońi Teronpoika was the brother of Varahvontta Lesonen or Varahvontta Lesońi and the son of Tero Lesońi, who was a skilled hunter and singer. He was born in Venehjärvi and moved to Niskajärvi. According to the folklore collector Samuli Paulaharju, Hovatta was a *tietäjä* (seer) and lived in a poor house.

29 Prokkoni Lukkańi or Ponkalahden Huotari was a fisherman from the village of Ponkalahti. The folklore collector Axel August Borenius (1846–1931) transcribed a certain number of songs and incantations from him.

the menace represented by the bear was double. In Karelia and Finland, a very intimate relationship existed between the cattle, femininity, and female genitals and sexuality (Tarkka 1998, 122). A ritual called *harakoiminen* was usually performed to protect cattle during the grazing season: A woman climbed on to the gate and the cattle passed below her legs and sex. It was a preventive protection – after the ritual the female force (*naisten väki*) present in her genitals remained attached to the cattle. This powerful, strong sexual force, and the "scent of the women", protected the cows from bears for the entire summer. The bear was "ashamed" by the female force and scent – it did not dare to touch the cows.

A woman protected what belonged to her house (cows) and what moved from the "inside" dimension of the home to the "outside" forestland (Stark-Arola 1998b, 39–40). The female force was considered particularly effective against bears. In the forest, a woman could even expose her sex and pronounce an incantation to embarrass and startle a bear (Piludu 2019, 108). In Sodankylä, this type of "exposure" was followed by a short incantation: "Look at that, and shame, rascal!" (*Katto tuohon ja häpiä, koranus!*; SKVR XII2/6512: 1, Sodänkylä, Lapland, Finland, 1920, Rantalan Joopi).

However, when the bruin itself entered the village as a guest of honour or a groom, it had a position of power, and it could be sexually aggressive with women or pose a danger to the cattle. According to Lotte Tarkka, in Lukkańi Huotari's song the arrival of the bear in the household could be described as including a potential danger of sexual penetration or harassment (Tarkka 1998, 122).

The singer Matti Karjalaińi called upon the wives to watch out for their "womb" when the bear entered the cabin: "Guard, poor wives, / protect your wombs / when *Ohto* (the bear) comes to the fireplaces / the furry muzzle forces his way!" (*Varuotos, vaimo raukat, / kokuotos kohtuon, / Ohon tullessa tulille, / karvaturvan tungetessa!*; SKVR I4/1254: 22–25, Lonkka, Viena Karelia, 1892). The advice was probably intended to protect against the menace of a bear's sexual assault. We can also suppose that this could be a humoristic version of the standard lines: "furry muzzle" (*karvaturpa*) in Karelian sounds like a funny, somewhat erotic metaphoric name for the bear.

However, a different and more dramatic explanation is also possible. The singer Anni Lehtonen[30] from Vuonninen explained: "[…] when the bear was brought into the cabin, a pregnant woman should not even be in the room. It is not so contagious to other women [or: "the bear does not seize, grip, attack other women"][31], only to pregnant ones" (*Kun karhua tuuvah pirttih, niin ei pie olla paksun naisen huonehessakaan. Ei muihin naisih niin tartu, kuin paksuh*; SKS KRA Samuli and Jenny Paulaharju 18553, 1932, Vuonninen, Viena Karelia, Russia, manuscript from 1916).[32] It seems that an illness called "forest hate" or "bear hate", coming from the meat or fur of the bear, could be particularly contagious for pregnant women. Lotte Tarkka stresses that, in Viena Karelia, the bear was dangerous for child-bearing women, and a pregnant woman was more susceptible to being infected by the bear's force (Tarkka 2013, 354). In Viena Karelia, a pregnant woman walking in the forest also could lose her unborn male child to the bear, if it identified the foetus as a potential hunter in the future (Virtaranta 1958, 309, 322; Paulaharju 1995, 29; Tarkka 2013, 354).

30 Anni Lehtonen (in Karelian Ańńi Lehtońi) was one of the most important Viena Karelian singers. Her grandfather was the famous *runolaulaja* (singer), *tietäjä* (seer) and hunter, Ontrei Malinen. The collector Samuli Paulaharju transcribed 450 songs and incantations by Anni Lehtonen. She learned her songs from the sons of Ontrei and from the great *tietäjä* Ohvo Homanen of Vuokkiniemi and from Martiska Karjalainen, the brother of the grandfather, Ontrei (Haavio 1948, 26–31; Kaukonen 1984, 24–25).

31 The verb *tarttua* is polysemic: it means "seize, attack" (in case of an animal) and also "be contagious" (in case of an illness). The context of the sentences allows both interpretations.

32 The sources from the SKS KRA (Archive of Folk Poetry of the Finnish Literature Society) are mentioned indicating: the name of the collector; the number of his manuscript and its year; the village and its parish; the name of the informant and, sometimes, some basic information about him (age and profession).

the meat was not poisonous, but like sheep meat" (SKS KRA Paulaharhu 39755; Paulaharju 1953, 205). Alder was used in many other ways to neutralise the "poison" of the bear meat (Piludu 2019, 242–246). In Finland, however, alder was also considered a bad tree and it is described as being created by the devil. In western Finland, the people believed that the cross of Jesus had been made of alder; therefore, in the popular European Christian faith the tree became a dangerous one (Krohn 1917, 50; Guenat 1994, 129). Even so, the alder was considered powerful for its red fluid, which was associated with human blood (Aaltonen 1915, 130; Guenat 1994, 129–130). The evil or dangerous forces contained in the alder could also be used for protective purposes; they could be apotropaic, and the tree was used to drive away other evil forces (Guenat 1994, 130). Aaltonen notes that, in Finland, alder was used in a great number of rituals associated with healing, agriculture, cattle breeding, hunting, and fishing: "It seems that the fundamental feature in the use of alder is that it works as a kind of fastener, something that securely locks up, ties and rules its bad spiritual power, which could be used at any given time" (Aaltonen 1915, 121). In Finland, Scandinavia and central Europe (Germany), the alder was also used to drive away bad spirits or the souls of the dead (Guenat 1994, 130). Considering the multitude of ritual uses of the alder, is not surprising that the tree was also used to expel the poison that was believed to be present in the meat of the bear.

The theory about Karelian and Finnish totemic clans and the refusal to eat bear meat in Viena Karelia

An influential theory by the anthropologist Matti Sarmela is strictly related to the ritual eating the bear meat and to some taboos about eating the bear meat in Karelia. His theory is often mentioned in scientific literature about Finnish bear ceremonials and in some books popularising these scientific debates (Hyry et al. 1995, 20–21; Pulkkinen 2014, 220–223; Pulkkinen/Lindfors 2016, 50). Sarmela stresses the importance of the discovery of Stone Age bear-headed axes and elk-headed ones (see Figs. 4–5) in a large area from Scandinavia to the Urals (Carpelan 1974; 1975), and he supposes that two totemic clans existed in Finland and Karelia (Sarmela 2009, 93; Pentikäinen 2007, 3).

Sarmela adds that the famous Latin writer Cornelius Tacitus (AD 56–120), in his *Germania* (*c.* AD 98) wrote about two mysterious peoples living near the people of the Fenni (Pulkkinen/Lindfors 2016, 38). According to Tacitus, the *Hellusios* and *Oxionas* have the faces and expressions of men but the bodies and limbs of wild beasts. Sarmela elaborates a theory by the Latinist Tuomo Pekkanen that: a) the name *Oxionas* may derive from the Finnish word *oksi* (*otso*, *ohto),* meaning "bear", and b) the name *Hellusios* may derive from the word *elg* ("elk"), found in many languages, for example, the Ancient Greek *ellós* (Sarmela 2009, 94; cf. Pekkanen 1983; 1984; Pentikäinen 2007, 23).

It is unclear if the Fenni of Tacitus were the Sámi, another Finnic people living somewhere in Fennoscandia, or another Germanic people. Tacitus himself wrote that he was not sure if the Fenni were Germans or Sarmatians (De Anna 1988, 49–50). Some modern scholars have argued that the Fenni lived in Lithuania rather than in Finland or Sámiland. In the Scandinavian languages, Finn is a common name for Sámi (Finnmark in Norway) or Finns (Finland). As for the *Hellusios* and *Oxionas*, practically no other historical source mentions them. Sarmela speculates that the *Hellusios* and *Oxionas* were two large totemic clans, the Finns and the Karelians, who wore the skins of their animal ancestors – the elk (*Hellusios,* the Finns) and the bear (*Oxionas*, the Karelians; cf. Sarmela 2009, 93).

Yet Tacitus did not write about peoples wearing skins but about monsters (animal-human hybrids): The *topos* was very common in Ancient Greek and Latin historical and geographical literature dealing with fantastic barbarian peoples living at the edge of the world (De Anna 1988, 55). He was not historically accurate and himself very doubtful about the existence of the two peoples.[46] Sarmela defines the totem as an animal or plant ancestor of a group of people. This "totemic alliance group" allegedly shared a myth about the origin of the animal and a prohibition against eating the meat of the totem animal (SARMELA 2009, 94).

At this point, as already stated, Sarmela theorises that in prehistory the Finns considered the elk as their totemic ancestor and the Karelians had the bear as their ancestor (SARMELA 2009, 93). I am sceptical about this theory: Finland and Karelia are large geographic areas, and the assumption of a totemic bear alliance covering the whole of Karelia and an elk alliance covering the whole of Finland presupposes a belief in the existence of a totemic "national consciousness" in Finnish and Karelian prehistory.

The archaeological evidence of Sarmela's theory is weak, as he does not consider the fact that the prehistoric bear-head axes have also been found in Finland, in Sweden, and in Onega Karelia (HUURRE 2001, 290–291). Archaeologists have found both bear-headed and elk-headed axes (see Figs. 4–5) in a huge geographic area from Scandinavia to the Urals, but most of these axes have been found in Finland and Onega Karelia (HUURRE 2001, 290). Another problem is that in Finland the bear-headed axes were more common than elk-headed ones (HUURRE 2010, 292). The archaeologist Matti Huurre stressed that bear-headed axes and elk-headed ones could have had different religious or cultural uses: They could be symbols of totemic clans and families, status symbols of local chieftains, or objects used in ceremonials related to hunting (HUURRE 2001, 293).

Elk-headed axes, staffs and other objects were found in the whole Scandinavian and Baltic area, Lithuania included. Such staffs or axes seem to be represented in some petroglyphs of Alta (Northern Norway) and Nämförsen (central Sweden): Human figures use them killing or sacrificing an elk or performing other rituals involving elks (IRŠĖNAS et al. 2018, 133–136). Although the elk-headed and bear-headed axes and staffs were important in the religion of the Stone Age and in a huge Nordic geographic area, the exact ritual use of these objects remains open to scientific debate and speculation.

Sarmela's "totemic alliance" concept (a totemic system in which an animal is the ancestor of large human groups living in huge geographic areas) is very uncommon in anthropological literature: More typical is the presence of two or more moieties (descent groups) with two or more animal or plant ancestors inside the same ethnic communities or clans that are numerically very small, although the dimension of the geographic area varies a lot (DESCOLA 1996).[47] The prohibitions related to totemic beliefs are most variable: Sometimes a moiety has a set of animals or plants that should not be eaten, and these species may or may not include the ancestor animal or plant. The term "totemism" is one of the most discussed in anthropological literature, and it has been thoroughly analysed and redefined over recent decades.[48] However, two of the most renowned scholars of totemism, Claude Lévi-Strauss (LÉVI-STRAUSS 1964; 1966) and Philippe Descola,[49] do not consider totemism as a way to distinguish

46 Tacitus, *Germania*, Book 1, Chapter 46: *Cetera iam fabulosa: Hellusios et Oxionas ora hominum voltusque, corpora atque artus ferarum gerere: quod ego ut incompertum in medio relinquam* (http://www.thelatinlibrary.com/tacitus/tac.ger.shtml#1). English translation by Alfred John Church and William Jackson Brodribb [1864–1877]: "All else is fabulous, as that the Hellusii and Oxiones have the faces and expressions of men, with the bodies and limbs of wild beasts. All this is unauthenticated, and I shall leave it open" (http://www.sacred-texts.com/cla/tac/g01040.htm).

47 By contrast, it is common that only one animal became the main symbol of peoples with a strong national consciousness and written history, such as the Ancient Romans, who had the wolf suckling Romulus and Remus as the coat-of-arms of their city, army, and political power (the Senate of the Roman People, SPQR). However, the she-wolf of Rome was not a totem or ancestor *strictu sensu*, because she was not the biological mother of the twins that founded the city, but she was surely a powerful animal who fed them when they were abandoned. In Rome, the wolf was related to warfare and the god Mars, so it was a symbol of the belligerent character of Rome and of its army (RISSANEN 2018).

48 See DESCOLA 1992; 1996; 2005; ÅRHEM 1996; 2016, 6–9; BRIGHTMAN et al. 2014, 16–19; VIVEIROS DE CASTRO 2009; WILLERSLEV/ULTURGASHEVA 2014.

49 Philippe Descola (born in 1949) is a French scholar renowned for his studies of the Achuar, a Jivaroan people of Amazonia, and for his contribution to anthropological theories about environment.

entire peoples from each other (such as the Karelians from the Finns), but to define moieties, clans, social groups, or other social identities (warriors, hunters, healers) inside the same – and often small – ethnic group, using two or more animals and plants to build up a web of exogamic relations. In one of his articles, Descola defines totemism as a classificatory system to create distinctions within the same human society (ethnic group) based on the differences between animal species (Descola 1996, 88; Willerslev/Ulturgasheva 2014, 49). According to Descola, totemism models human society after nature (Descola 1996, 88; Århem 2016, 7).

Sarmela states that the prehistoric Karelians had a bear ancestor because, according to him, the people from Viena and Olonets Karelia did not perform any bear rituals or *Bear Songs* (Sarmela 1991, 224), and they did not hunt the bear (Sarmela 1991, 225). This information is incorrect. In western Viena Karelia, the collectors transcribed a great number of *Bear Songs* (79 in all), and these are often longer than the ones transcribed in Finland (SKVR I4/1189–1267). Sarmela mentions that, in Viena Karelia, no information or songs about the ritual of the bear skull were collected (Sarmela 1991, 224; 2009, 79). However, Viena Karelian singers, including the famous Iivana Malinen, described the ritual of the bear skull and sang some of the most typical lines related to this ritual (SKVR I4/1245b; I4/1245c; I4/1253: 16–17, I4/1943). Sarmela states that Viena Karelians did not eat bear meat at all.[50] Yet the Karelian sources and songs contain several references to eating and cooking bear meat, as we have seen in this chapter.[51]

For the theoretical problems in the definition of totemic beliefs and the evident inaccuracy in dealing with Viena Karelian sources, I consider Sarmela's historical reconstruction very problematic. It is not easy to compare archaeological findings, Ancient Latin written sources (Tacitus) and ethnographic songs and information collected in the 19th and 20th centuries: The historical distance between the ages of the sources is huge.

However, Sarmela is right in stressing that the Stone Age bear-headed and elk-headed axes are historically meaningful because they are valuable archaeological sources demonstrating that the bear and the elk were animals with a deep meaning in the remote prehistory of Fennoscandia and North Russia (an area inhabited by other Finno-Ugric peoples). The axes are in good condition, so they were not used in battle as normal weapons. They were probably ritual objects or objects representing the social status of a chief or family (Pentikäinen 2007, 3). Unfortunately, we do not have any written source that demonstrates for certain that the bear or the elk were considered ancestors or totems in Finno-Karelian prehistory. However, the idea that both the elk and the bear could be two ancestors of two moieties of the same people is a hypothesis that could be evaluated by making comparisons between archaeological and ethnographic material. However, we should be very careful that we propose explanations based on valid sources rather than on Latin historical literature on fantastic barbarian peoples and monsters.

In Viena Karelia, there were persons – generally women, like in Finland – who avoided eating bear meat. On the one hand, eating the bear was considered a "sin", as it was part of a "pagan" or pre-Christian ritual that was abhorred by the Orthodox church or by the Old Believers, who were a very conservative Orthodox group that had very strict rules about eating (Pentikäinen 1999). On the other hand, the anthropomorphic features of the bruin's body were considered problematic, as they were like those of humans – eating the bear was compared to anthropophagy (see Fig. 6). However, eating the bear (or avoiding eating it) was generally a personal or local choice, not a general prohibition that applied to the whole territory of Viena or Orthodox Karelia. This refusal seems to relate to the anomality of the bear and its humanlike features, not to a prehistoric totemic system based on the belief that the bear was the only animal ancestor of all Karelia.

50 Sarmela 1991, 225. This idea seems to be based on the statements by Kaarle Krohn about the Orthodox Karelians' prohibition against bear meat (Krohn 2008, 162) and Virtaranta's Karelian informants (Virtaranta 1958, 313).

51 See, for example, SKVR I4/1245: 10–16; I4/1242b: 10–16.

The information about women eating bear meat or not varied from one village to another. It seems that in some places in Finland and Viena Karelia the women almost categorically refused to eat bear meat, while in other villages they ate it, albeit after the men. Pregnant women avoided contact with bears, as the bear soul could kill the unborn male child in the womb, preventing him from growing into a hunter.

The local behaviour or the women's decision to avoid eating bear meat was also based on the anomalous nature and ambiguity of the bear's identity: It was a wild animal that had human features and potentially human origins.

In 1958, Pertti Virtaranta[52] published a book with his interviews and conversations with Viena Karelian women about the prohibition against eating bear meat. Oksenia Nykänen from Akonlahti said: "About the bear, it was told that the bear is a bewitched person. The bear, when it is skinned, is like people. For this reason, bear meat was not eaten" (VIRTARANTA 1958, 313). Domna Huovinen[53] from Vuokinsalmi (Kontokki) was convinced that the bear originated from humans, because when it was skinned its body resembled a human one, and its "toes are like human toes" (VIRTARANTA 1958, 313). Domna added that in Finland the bear meat was eaten, but this was not the case in their Viena Karelian village, because it was a sin (VIRTARANTA 1958, 313).

Because the bear was considered a humanlike being or a human transformed by sorcery into a bruin, eating it was considered by these Karelian informants to be an abominable sin, akin to anthropophagy. The bear was a very anomalous creature, as it was a wild animal of the forest and it was considered an offspring of the forest spirits, but it had potential human origins according to other folk beliefs. In other words, the bear did not fit into normal categories (DOUGLAS 2002, 47). Mary Douglas[54] emphasises that what is anomalous and ambiguous is also potentially disturbing, dangerous, and polluting (DOUGLAS 2002, 48–49). In some cases, as the mentioned Domna Huovinen illustrates, the ritual answer to the bear's anomality was negative, resulting in the refusal to eat its meat. However, we must consider that Virtaranta's interviews were done quite late, in the 20th century: The memory of bear ceremonialism was fading away. According to the *Bear Songs* of the 19th century, in Viena Karelia eating the bear meat was allowed, but only when performing all the rituals and songs described in this chapter.

Eating the organs of the bear's head: Sharing the same mind and song

In Karelia and Finland, eating the organs of the muzzle was one of the most delicate moments of the whole feast. The head of the bear was the most powerful part of the bruin, full of bear force (*karhun väki*), and the eating of it was probably reserved for the men or hunters. Iivana Malinen sang: "I take from *ohto* his tongue, / to be my tongue,[55] to be my mind,[56] / for me to sing the kindred song, / [to be] joy for the sitting places,[57] / singing I eat the eye from *ohto,* / lilting, I dig out the ear, / rejoicing,

52 Pertti Virtaranta (1918–1997) was a Finnish linguist and professor of the Finnish language. He studied the Karelian and Veps languages and collected a huge number of folk stories in their original dialects from Karelia, Finland, Tver Karelia and from the Finno-American communities.

53 Domna Hillipantytär Huovien (1878–1963) was a Viena Karelian singer, born in Vuokinsalmi, who moved to Kuivajärvi (a Viena Karelian village in Finnish territory). She was famous for her laments (*itkuvirret*) sung at Karelian funerals or weddings.

54 Mary Douglas (1921–2007) was a British cultural and social anthropologist who worked in the United States for 11 years. She is famous for her theories on purity and impurity and the notion of dirt and for her classic monographs *Purity and Danger* (first edition 1966) and *Natural Symbols* (1970)

55 The word *kieli* is polysemic; it could mean "tongue" or "language".

56 The word *mieli* is polysemic; it could mean "mind", "mood", "desire", "agreement" or "pleasure".

57 The singers and the audience were sitting on the benches at the table, eating the bear meat and organs.

I dismember" (*Kielenpä otan oholta, / kielekseni, mielekseni, / laji virttä laulaakseni, / hoksi istuma sijoille, / laulellen syön silmän oholta, / koikaten korvan koverran, / ilon lyöen irtauttele*; SKVR I4/1245: 10–13, Vuonninen, Viena Karelia, Russia, 1888).

Matti Sarmela emphasises that, by eating the ears and eyes of the bear, its powerful senses were transferred to humans (Sarmela 1991, 218, 232). Lotte Tarkka stresses that these lines reveal a deep level of communion and physical fusion between the singer, the hunters and the bear (Tarkka 1998, 103). By eating the tongue, the hunters acquired a shared language expressed in the form of a "kindred song" (*laji virssi*), which was considered part of a long tradition, passing from one generation to another in the kin or family (Tarkka 1998, 103–104). If the bear tongue (*kieli*) was used in the singing of a "kindred song", the bear itself was considered a skilled singer. Learning the songs of the bear could be useful for the hunters, as these songs could be the ones used in bear ceremonialism to please the bear and the forest *haltias*. Tarkka observes that "tongue" (*kieli*) in Finnish and Karelian also means "language"; it was eaten to find an agreement with the bruin, or a common "mind", expressed by the Finno-Karelian word *mieli* (Tarkka 1998, 103). The word "mind" (*mieli*) refers to "sense" or "meaning" as well as the "mind, mood, sensibility, aspiration, desires and even memory of the person" (Tarkka 1998, 136).

These lines probably represent the maximum level of mimesis between the bear and humans, as they share the same "mind", "mood", memory, and songs. Nevertheless, the identification is not complete.

The hunters maintained their human identity and they were aware of the alterity and dangerousness of the bear because, in the following ritual phase, they detached the bear's fangs and claws from the skull, depriving it of its "weapons" and using the bear force contained in them for many rituals, including self-protection from other bruins. The teeth and claws were stored and used as powerful amulets (see Figs. 7–8). The hunters took the bear's fangs or teeth with them in their bags during the hunt, as they could be used to cure wounds caused by bears or to neutralise the "poison" in the bear meat (Piludu 2019, 260–262).

Final thanksgiving songs and the hope for the bear's future return

The bear feast often ended with some songs of thanksgiving. In these last songs of the bear feast the members of the villages hoped for a future return of the guest.

The Finnish and Karelian vernacular idea of the bear's "return" or "rebirth" on earth after its death was expressed by the hope that the bruin would again be the guest of the village in the future. The concept of the bear's return to the village is present in the Viena Karelian songs. In the feast of Venehjärvi, the mistress asked God to send another bear in the future: "Let again, God, / another time, true God, / during the life of this mistress / this *auvo* (bear, groom, luck, honour) enter!" (*Annappa on vastaki, Jumala, / toitsiki, totini Luoja, / eleässä tämän emännän, / tämän auvon astuossa!*; SKVR I4/1235: 75–79, Venehjärvi, Viena Karelia, Russia, 1894, Varahvontta Lesońi).

Similar lines were sung in the Karelian wedding songs, when the mother-in-law hoped for new suitors (a new *auvo*, a new groom) for the still unmarried daughters. The polyvalent term *auvo* in this case means "bear", but it derives from its additional meanings of "groom", "beloved" or "desired person", "relative" and "kinsman". In wedding songs or sexual magic, the word means also "luck", "honour", "desire", and "intention" (Tarkka 2013, 346), and it could refer to the sexual desire or the desired boy or girl, groom or bride. In the song, the bear was addressed as a desired groom (*auvo*) who brings "honour" to the village. The polysemic word *auvo* (desire / desired person) reveals a web of connections between the bear hunt and the weddings: The bear was the desired prey in the hunt, as a bride or a groom was the desired "catch" in weddings. The bear was a very important prey, and the hunters gained "honour" (social respect) and good "luck" in hunting if they performed the whole ritual hunt well. A good groom gave "luck" (prosperity) and "honour" (social respect) to the bride, the mother-in-law, and the family. In the bear feast, the visit of the bear to the house was an "honour" for the whole village.

The proper treatment of the bear throughout a well-organised bear feast was the prelude to the return of the satisfied guest and groom and subsequently a successful bear hunt in the future, followed by yet another joyful feast to please the bruin and the forest spirits. In this sense, the bear feast ensured good luck in hunting.

Conclusions: The multi-layered connection between the bear feast and the weddings

The connection between the bear feasts and weddings was complex and multi-layered. The wedding was much more than a sexual adventure; it was a ritual connected with the building and rebuilding of social ties, and it was a ritual of transformation (Stark 2006, 170–171). By means of a wedding it was possible to unite what was previously separate, such as two distinct families – or the bear and a member of the village. The wedding was a way to redefine social identities.

Lotte Tarkka stresses that, in Viena Karelia, the bear ceremonial recalls both rites of passage and of territorial mobility (Tarkka 2013, 338). In Karelia and eastern Finland, the wedding blended both concepts. Like the bear ceremonial, the wedding included several crossings of boundaries between the groom's village, the bride's village, and the forest in between.

The representation of the bear feast as a marriage in the *Bear Songs* is peculiar to the Finnish and Karelian bear ceremonials. Among many other hunting cultures, the marriage between a bear and a woman is present in myths but is not represented in the rituals (McClellan 1970; Spagna 1998). The most common of the myth plots concerns a girl who fell into a den and married a bear, and they had children or cubs. One day, the brothers of the girl came to the den to rescue their sister, and they killed the bear. The bruin, seeing the brothers of his wife coming, told her all the procedures to organise a proper bear feast and ritual of the bear skull. The story is the aetiological myth that explained why the bear ceremonialism should be organised by the hunters. In certain Ob-Ugrian bear ceremonials, the myth about the primordial union between the bear and a primordial mother was represented with dances and songs (Pentikäinen 2007, 31–42). A version of this myth was collected among the Sámi of Northern Fennoscandia (Edsman 1956). In Finland and Karelia, the aetiological myth was not collected in any village, but it was common to represent the bear feast as a bear marriage or the bear as a groom.

Henni Ilomäki notes that both the bear ceremonialism and the wedding were finalised with a change of social status. In the wedding, the groom and the bride switched their social positions and became husband and wife. The groom took the bride from her house; she left her family to become a member of the groom's family (Ilomäki 1986, 131). In a similar way, the hunter took the bruin from its homeland and its parents (the woodland and the forest spirits) and the bruin became a member of the community of the village during the fictive "wedding" of the bear feast. The difference is that the bride became a wife for her whole life and did not return to her previous home; thus, she passed through a definitive social metamorphosis. By contrast, the bear was married only for the brief time of the feast, and even in this situation it maintained a part of its original alterity and dangerousness. After the feast, the bruin was supposed to go back to its homeland and parents to ensure its regeneration from the bones.[58] This made a new hunt possible, with a future feast and wedding.

58 See Piludu, this volume, on the Finno-Karelian bear skull rituals.

Lévi-Strauss 1964: C. Lévi-Strauss, Totemism (London 1964).
Lévi-Strauss 1966: C. Lévi-Strauss, The Savage Mind (Chicago 1966).
Lot-Falck 1961: E. Lot-Falck, Riti di caccia dei Siberiani (Milano 1961; original French edition 1953).

McClellan 1970: C. McClellan, The Girl who Married the Bear: A Masterpiece of Indian Oral Tradition. Publications in Ethnology 2 (Ottawa 1970).

Paulaharju 1953: S. Paulaharju, Sompio. Luiron korpien vanhaa elämää (Porvoo, Helsinki 1953).
Paulaharju 1995: S. Paulaharju, Syntymä, lapsuus ja kuolema. Vienan Karjalan tapoja ja uskomuksia (Helsinki 1995).
Pekkanen 1983: T. Pekkanen, Vanhin kirjallinen tieto suomalaisista. Suomalais-Ugrilaisen Seuran Aikakauskirja 78, 1983, 173–185.
Pekkanen 1984: T. Pekkanen, The Hellusii and the Octiones of Tac. Germ. 46,4. Arctos 27, 1984, 49–60.
Pentikäinen 1999: J. Pentikäinen (ed.), Silent As Water We Live. Old Believers in Russia and Abroad (Helsinki 1999).
Pentikäinen 2007: J. Pentikäinen, Golden King of the Forest. The Lore of the Northern Bear (Helsinki 2007).
Pentikäinen 2014: J. Pentikäinen, The Songs of the Bear in Hunt Rituals. In: V. M. Piludu/Frog (eds.), Kalevala: epica, magia, arte, musica / Kalevala: Epic, Magic, Art, Music (Viterbo 2014) 415–443.
Piludu 2019: V. M. Piludu, The Forestland's Guest: Mythical Landscapes, Personhood and Gender in the Finno-Karelian Bear Ceremonialism (Helsinki 2019) https://helda.helsinki.fi/handle/10138/282220.
Pulkkinen 2014: R. Pulkkinen, Suomalainen kansausko (Helsinki 2014).
Pulkkinen/Lindfors 2016: R. Pulkkinen/S. Lindfors, Suomalaisen kansauskon sanakirja (Helsinki 2016).

Räsänen/Räsänen 2008: M. Räsänen/R. Räsänen, Kotipiiri arkena ja pyhänä. In: R. Räsänen (ed.), Savo ja sen kansa (Helsinki) 293–371.
Rissanen 2018: M. Rissanen, Rooma, suden kaupunki (Helsinki 2018).
Rydving 2010: H. Rydving, The "Bear Ceremonial" and Bear Rituals among the Khanty and the Sami. Temenos 46(1), 2010, 31–52.

Sarmela 1991: M. Sarmela, Karhu ihmisen ympäristössä. In: P. Laaksonen/S.-L. Mettömäki (eds.), Kolme on Kovaa Sanaa. Kirjoituksia kansanperinteestä. Kalevalaseuran vuosikirja 71 (Helsinki 1991) 209–250.
Sarmela 2009: M. Sarmela, Finnish Folklore Atlas. Ethnic Culture of Finland 2. Published on the author's webpage: http://www.kotikone.fi/matti.sarmela/folkloreatlas.pdf.
Schmidt 1989: E. Schmidt, Bear Cult and Mythology of the Northern Ob-Ugrians. In: M. Hoppál/J. Pentikäinen (eds.), Uralic Mythology and Folklore. Ethnologica Uralica 1 (Budapest, Helsinki 1989) 187–232.
Siikala 2016: A.-L. Siikala, Itämerensuomalaisten mytologia (Helsinki 2016; originally published 2012).
Spagna 1998: F. Spagna, L'ospite selvaggio. Esperienze visionarie e simboli dell'orso nelle tradizioni native americane e circumboreali (Torino 1998).
Stark 2006: L. Stark, The Magical Self: Body, Society and the Supernatural in Early Modern Rural Finland. Folklore Fellows Communications 138 (Helsinki 2006).
Stark-Arola 1998a: L. Stark-Arola, Magic, Body and Social Order: The Construction of Gender through Women's Private Rituals in Traditional Finland. Studia Fennica Folkloristica 5 (Helsinki 1998).
Stark-Arola 1998b: L. Stark-Arola, Gender, Magic and Social Order. In: S. Apo/A. Nenola/L. Stark-Arola (eds.), Gender and Folklore: Perspectives on Finnish and Karelian Culture. Studia Fennica Folkloristica 4 (Helsinki 1998) 31–56.

Tanner 1979: A. Tanner, Bringing Home Animals: Religious Ideology and Mode of Production of the Mistassini Cree Hunters (London 1979).
Tarkka 1998: L. Tarkka, The Sense of the Forest: Nature and Gender in Karelian Oral Poetry. In: S. Apo/A. Nenola/L. Stark-Arola (eds.), Gender and Folklore: Perspectives on Finnish and Karelian Culture. Studia Fennica Folkloristica 4 (Helsinki 1998) 92–142.
Tarkka 2013: L. Tarkka, Songs of the Border People: Genre: Reflexivity and Performance in Karelian Oral Poetry. Folklore Fellows Communications 152 (Helsinki 2013).
Taussig 1993: M. Taussig, Mimesis and Alterity: A Particular History of the Senses (London 1993).
Turner 1969: V. Turner, The Ritual Process (London 1969).
Turner 1992: V. Turner, The Anthropology of Performance (New York 1992).

Virtanen 1968: L. Virtanen, Kalevalainen laulutapa Karjalassa (Helsinki 1968).
Virtaranta 1958: P. Virtaranta, Vienan kansa muistelee (Porvoo, Helsinki 1958).
Viveiros de Castro 2009: E. Viveiros de Castro, Metaphisique cannibales (Paris 2009).

Wagner 1977: R. Wagner, Analogic Kinship: A Daribi Example. American Ethnologist 4(4), 1977, 623–42.
Wiget/Balalaeva 2022: A. Wiget/O. Balalaeva, Valuing Difference: Bear Ceremonialism, the Eastern Khanty, and Cultural Variation among Ob-Ugrians. Sibirica 21(1), 2022. https://doi. 10.3167/sib.2022.210103.
Willerslev 2007: R. Willerslev, Soul Hunters: Hunting, Animism and Personhood among the Siberian Yukaghirs (Berkeley, Los Angeles, London 2007).
Willerslev/Ulturgasheva 2014: R. Willerslev/O. Ulturgasheva, Revisiting the Animism versus Totemism Debate: Fabricating Persons among the Eveny and Chukchi of North-Eastern Siberia. In: M. Brightman/V. Grotti/O. Ulturgasheva (eds.), Animals in Rainforest and Tundra: Personhood, Animals and Things in Contemporary Amazonia and Siberia (New York 2014) 48–67.

Århem 1996: K. Århem, The Cosmic Food Web. Human-Nature Relatedness in the Northwest Amazon. In: P. Descola/G. Pálson (eds.), Nature and Society: Anthropological Perspectives (London 1996) 185–204.
Århem 2016: K. Århem, Southeast Asian Animism in Context. In: K. Århem/G. Sprenger (eds.), Animism in Southeast Asia (London 2016) 3–30.

Vesa Matteo Piludu
PhD, Postdoctoral researcher
University of Helsinki
Indigenous Studies,
Department of Finnish, Finno-Ugrian and Scandinavian Studies
Faculty of Arts Helsinki
Finland
vesa.piludu@helsinki.fi

Fig. 1. The bear feast was organised in the village, just after the bear had been killed by the hunters. The hunters continued to sing to the dead bear, considering its soul sentient and able to understand human language. The bear was treated as a guest of honour or as a groom by the mistress of the house where the feast was held. The whole feast was based on the ritual model of a wedding. Photo of Venehjärvi (Viena Karelia, Russia), a village that was famous for its bear hunters, taken by I. K. Inha in 1894 (Museovirasto [Finnish National Board of Antiquities], Finno-Ugric Photographic Collection [Suomalais-ugrilainen kuvakokoelma]: SUK: 22, public domain).

Fig. 2. The entrance of the bear into the house and into the tupa *(main room) was a delicate phase in the feast. The young and unmarried women and girls were invited to leave the yard, protect the cattle and their "wombs" or genitals. The soul of the dead bear was considered extremely dangerous for pregnant women, who did not participate in the feast. An older, married woman (the "mistress of the house") welcomed the bear into the village and the house as a guest of honour or a groom. When the bear entered the house, the children, boys and girls, should leave the outdoor area and the porch as a sign of respect for the bear. Photo of the yard and the outdoors of a house in the village of Pirttijärvi (parish of Vuokkiniemi, Viena Karelia, Russia) with women and children at a local wedding ceremony (photo of the Finnish Army [SA-Kuva 1942-06-21], 21 June 1942, public domain).*

Fig. 3. In Orthodox Karelia, bear fur and the bear head were brought to a place of honour during the bear feast: the rear or holy corner of the tupa *(main room). In this corner, there were the family icons. All the most important family rituals were carried out in this corner – the bear feast, some phases of funerals, and some of the most important parts of the wedding rituals, such as the engagement. Photo of a Karelian wedding ritual phase, the wooing or wedding proposal with a woman lighting up a candle in front of the family icons, taken by I. K. Inha in 1894 (Museovirasto [Finnish National Board of Antiquities], Finno-Ugric Photographic Collection [Suomalais-ugrilainen kuvakokoelma]: SUK: 76, public domain).*

Fig. 4. A prehistoric axe with a bear head from Paltamo (Finland), 9 x 12 cm. The presence of prehistoric axes with bear or elk heads was the basis for several theories about the existence of bear and elk totemic clans in Finland and Karelia. The age of these axes is difficult to establish; they occur in the whole Neolithic period in Finland (6000–1000 BC) (photo M. Haverinen, Museovirasto [Finnish National Board of Antiquities]: KM 13275; Photographic Collection of Archaeology [Arkeologian kuvakokoelma Diakokoelma]: AKD54353:1, public domain).

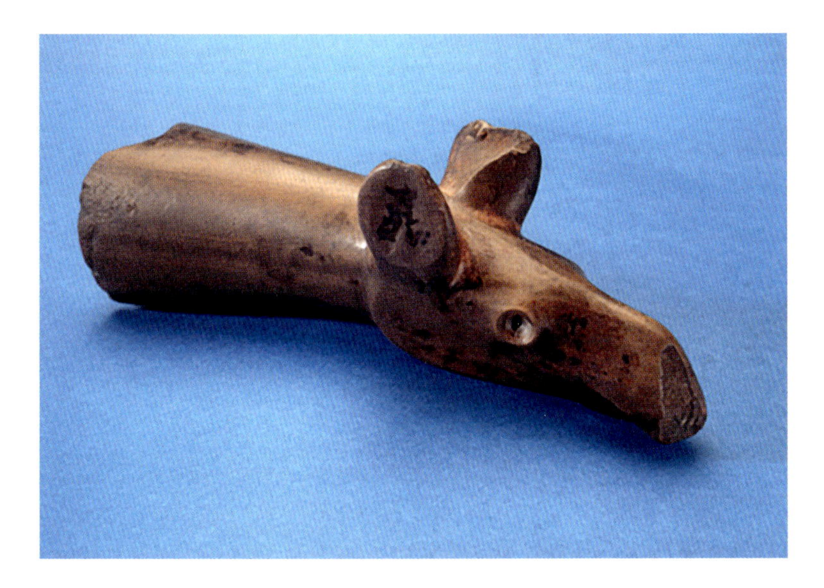

Fig. 5. A prehistoric axe with an elk head from Säkkijärvi (Russia, former Finnish city or village of Southern Karelia or Viborg Region; photo M. Haverinen, Museovirasto [Finnish National Board of Antiquities]: KM4909:1, Photographic Collection of Archaeology [Arkeologian kuvakokoelma Diakokoelma]: AKD46927:1, public domain).

Fig. 6. "Humanlike bears", photo by Susanna Lendiera, 2018. In Karelia and Finland some women refused to eat the bear meat, considering the bear "too humanlike" or a bewitched human transformed into a bruin. Eating its meat was dangerous or a "sin", like anthropophagy. Some physical features of the bear were considered especially "humanlike". Finnish and Karelian hunters noticed that after the skinning of the fur the anatomy of the corpse of a bear was "like a human body". The capacity of bears to stand on two feet was also considered a proof of their "humanlike" status in the animal realm. The bear was considered a being "in between" the animal realm, the world of the haltias *(forest guardian spirits) and humanity. The difficulties in categorising the bear required several rites to eat its flesh and the strict ritualisation of the whole bear feast.*

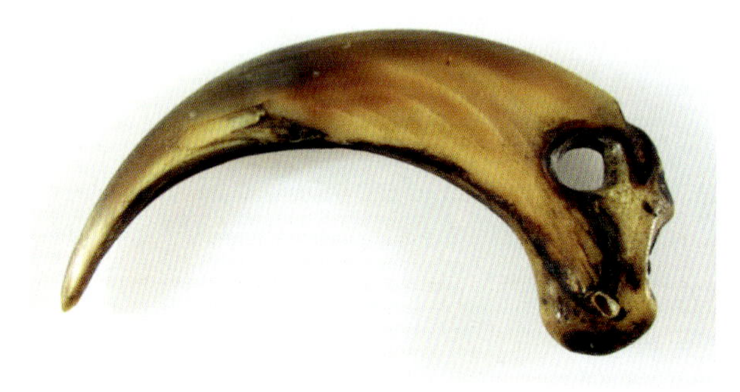

Fig. 7. Bear claws were used by hunters and tietäjäs *(seers) in many rites, from protective and healing rituals to the ones that bring sexual or wedding luck. This claw from Pomarkku (Region of Satakunta, Finland), 7 x 2 cm, was used against toothache and in rituals to take revenge against sorcerers and envious people (photo Satakunnan Museo [Museum of Satakunta] SME7852, Museovirasto [Finnish National Board of Antiquities], public domain).*

Fig. 8. Tietäjäs *(seers) and hunters used bear teeth as powerful amulets against bears or the dangerous "force of the forest"* (metsän väki). *This bear tooth from Honkajoki (Region of Satakunta, Finland), 7 x 2 cm, was used to protect horses during travels in the forest (photo Satakunnan Museo [Museum of Satakunta] SME9357, Museovirasto [Finnish National Board of Antiquities], public domain).*

The Finno-Karelian bear skull rituals: Bringing the bruin home to ensure its regeneration

By Vesa Matteo Piludu

Keywords: Finno-Karelian bear skull ritual, sacred pine, return of the bear to its birth land, "sending-off" rituals, regeneration of the bear, treatment of bones, testimony of the bear

*Abstract: In this chapter I analyse the third and last part of the Finno-Karelian bear ceremonialism. In a solemn procession the hunters carried the bear's skull and bones to a sacred pine in the forest. The skull was hung on one of its branches, and the bones were buried under its roots. During the bear skull rituals, the hunters stressed that the skull was not thrown on the ice, it was hung on a good and sacred tree, and it was brought to the mythic place where the bear was born – in the sky, nearby the constellation of the Plough (*Ursa Maior*), or in the sacred forest. The bear was brought back to the world of its guardian spirits (*haltias*) and parents, which were able to ensure the bruin's regeneration from its bones. All the honours shown to the bear skull were fundamental to achieve its regeneration and to obtain good luck in the hunt for the future, as the bruin was supposed to tell the spirits about its treatment. In this chapter, I will also compare the Finno-Karelian, Ob-Ugrian, Siberian and North American traditions about the "sending-off" rituals, the regeneration of game animals from bones, and the bear's testimony about its treatment.*

Bringing the bear skull back to the sky

In the last part of the Finnish and Karelian bear ceremonialism the hunters carried the bear's skull and bones in a solemn procession to a sacred pine (see Fig. 1). These final rites followed the bear hunt and the bear feast, in which all the meat and organs of the bear were eaten.[1] The bear skull ritual is called the "disposal of remains" in Hallowell's model of the phases of the bear ceremonialism (Hallowell 1926, 135–144).[2]

The skull (see Figs. 2–3) was hung on one of the tree's branches, and the bones were buried at its foot (Piludu 2019, 267–271). The hunters reassured the bear's soul, singing that they hung its skull in

1 For an analysis of the songs and rites of the bear hunt, the bear feast, the source materials, and the vernacular conceptions of the forest, see Piludu, this volume, on the songs and rituals of the Finno-Karelian bear hunt, and on the Finno-Karelian bear feast and wedding.

2 For an analysis of the theories of Hallowell, see Piludu, this volume, on the songs and rituals of the Finno-Karelian bear hunt.

Bear and Human: Facets of a Multi-Layered Relationship from Past to Recent Times, with Emphasis on Northern Europe, ed. by Oliver Grimm (Turnhout: Brepols, 2023), pp. 745–760 BREPOLS PUBLISHERS 10.1484.M.TANE-EB.5.134359

nected to the idea of the regeneration of the bear, rarely described in the songs. If the bear was able to watch the stars and the moon, it was "roaming" again in the world of the living.

The call to learn the stars comprising the Plough was also connected with the idea of awakening and travelling somewhere, as the constellation was important for orientation during the night. In Ladoga Karelia, the bride was advised to be very labourious and wake up before dawn "to learn the stars of the Plough" (SKVR XII2/4358, 6, 1890). Before being shot, the bear was also invited to wake up from the den and hibernation to observe the sun and the moon (Piludu 2019, 191). Leaving for the forest, the Viena Karelian hunters asked the forest spirits to teach them to travel observing the stars and the arc of sky (SKVR I4/1107, 6–7; I4/1253, 16–17).

The impression is that the travel of the bear did not end on the tree branch; therefore, the bruin should recognise the constellation to travel somewhere, to awake from his death and to watch the sun and the moon, as a human newborn does after birth.

The sacred pine, Hongotar, and the bear's return to its birth land in the forest

Other *Bear Songs* contain evidence that the bear skull ritual has the goal to bring back the bear to one of its mythical homelands: the woodland of forest spirits, called Metsola (or Mehtola), Tapiola or Pohjola (or Pohja, the Northland),[14] and the pine where it was born. In the Finnish and Karelian traditions, there was a huge quantity of *The Birth of the Bear* incantations, with many variations (Piludu 2019, 101–153): In the incantations sung by the hunters, the bear was generally born in a mythical forest.

During the bear skull ritual, the hunters sang that they did not discard the skull on the road or left it on the ice and in the snow (SKVR VII2/3396, 1–8, Ilomantsi, North Karelia, Finland, 1846). Antti Huttunen from Iisalmi sang that the hunters did not hang the skull on a rowan or a willow, but "rather on an old pine, the good tree" (*Vaan honkahan hyvähän puuhun*; SKVR VI2/4919, Iisalmi, North Savo, Finland, 1876) or on a "pure tree" (SKVR VII2/3034, Ilomantsi, North Karelia, Finland, 1846).

Matti Sarmela mentioned that no information or songs on the ritual of the bear skull were collected in Viena Karelia (Sarmela 1991, 224; 2009, 79). This information is incorrect. There are songs and statements of singers demonstrating that the ritual of the bear skull was performed in Viena Karelia, too (SKVR I4/1245b, I4/1245c; SKVR I4/1253: 16–17, I4/1943).

Sometimes the bear skull tree could be a spruce or a birch (see Fig. 3), but generally it was a pine (see Fig. 2). The anthropologist Francesco Spagna[15] noted that the pine, being a tall evergreen tree, is deeply connected with concepts of immortality or regeneration (Spagna 1998, 185). It remained a beautiful tree even at the end of winter, when the skull rite was performed. The rowan and the willow were not allowed as bear skull trees: They are not evergreen, they can be sad-looking in the wintertime, and they were not suitable as bear skull trees. In some of *The Birth of the Trees* incantations, the willow is described as a bad tree, created from evil, by a pagan (Krohn 1917, 51), or by the devil (Anttonen 1998, 142).

14 Metsola or Mehtola means "forestland" and it is often associated to Mielikki, the forest mistress or the mistress of the forestland (*metsölän emäntä*), the most important female forest spirit; Tapiola means forestland or the "land of Tapio", the master of the forest, the most important male forest spirit. Pohjola or Pohja means "the Northland"; it was associated with the mistress of Pohjola, an ambiguous and dangerous being able to give birth not only to bears and wolves, but also to lethal illnesses. In epic songs, the Northland is a dangerous, "dark", "man-eating" and "hero-drowning" place (Tarkka 2013, 398), and the mistress of Pohjola has magic powers, beautiful daughters and a magic object forged by the smith Ilmarinen – the Sampo. In the songs for the bear ceremonials, she is a strongly associated with Hongotar, a "mother" and protector of bears. All these placenames are associated to spirits that generate or protect bears.

15 Francesco Spagna is an Italian cultural anthropologist of the University of Padua. He has worked intensively on the folk stories about the marriages between bear and humans, and he is an expert of Native American cultures in Canada.

Only certain pines, spruces, or birches were suitable for the bear skull. Kaarle Krohn observed that the chosen tree was generally a large one (see Fig. 1), and it was measured by being hugged (Krohn 2008, 150). Martti Haavio also stressed that the bear skull tree should be huge (Haavio 1967, 31). Sometimes the tree was situated near the house of the hunter who had performed the circling of the bear, but more often the tree was in the deep forest, possibly in a place where people had not come before (Krohn 2008, 150). The latter tradition indicates the necessity of a spatial separation between the skull tree and the human sphere. Sometimes the pine tree was even more isolated from the villages, being on a small island in a lake. Such was the case of Karhunpääsaari (Bear's Head Island) in Kangasniemi, in the Finnish region of South Savo (Haavio 1967, 31).

Karelian and Finnish hunters sang that they hung the bear skull on a middle height: On top of the tree, the cold wind could harm the skull, on a low branch, the black ants would eat or ruin it (SKVR I4/1245c, 5–10, Vuonninen, Viena Karelia, Russia, 1888; SKVR XII2/6572, 1834; XII2/6574). This detail indicates that the skull should be well protected.

In Pielisjärvi the hunters sang that the skull was situated nearby a lake full of fish: "I put [it] on the oldest of the pines, / by its side a shore of whitefish, / beside it, a salmon fishery; / by your side the whitefish swim, / beside you the salmons spawn" (*Panen puuhun vanhimpaan, / sivullaan siikaranta, / luonahan lohiapaja, / sivullasi siiat uipi, / luonasi lohet kutee*; SKVR VII5/3403, 40–44, Pielisjärvi, North Karelia, Finland, 1838, name of singer not mentioned in the manuscript by Elias Lönnrot). This song is very interesting because the hunters seem to deny that the bear had been killed – the bruin's life was described as continuing in a place rich with fish and food.

Furthermore, people believed that the skull and its pine continued to be powerful and dangerous many years after the skull ritual. In Kangasniemi (South Savo, Finland) it was believed that a man who dared to hit a bear skull pine with an axe would go crazy (Haavio 1967, 31). Haavio noted that it was prohibited to touch the skull or take it away from the tree (Haavio 1967, 32). People interested in ruining a bear skull tree could be other hunters, which were jealous or envious of the success of the bear killers.

In Finno-Karelian folklore, the competition between hunters was generally hard. Antti Huttunen of Iisalmi sang curses against those who would dare to remove a skull: "Twist his head, / wind his nose aside!" (*Väännä päätä väärällehen, / nokka syrjähän syseä!*; SKVR VI2/4920, 5–8, Iisalmi, North Savo, Finland, 1876, Antti Huttunen). Another singer from Iisalmi sang a similar curse: "Wrench his hand into a hook" (*Kättä koukkuun kokkoo*; SKVR VI2/7407 a, 29, Iisalmi, North Savo, Finland, 19th century, year and singer unknown, manuscript by Kaarle Krohn). We can suppose that a dreadful spiritual force remained present and vigilant on the spot, protecting the tree and the skull against other hunters who dared to fell or ruin them.

Several songs suggest that this ritual return of the bear skull and soul to the forest and the sacred pine was a prelude of its regeneration. The songs performed at the beginning of the procession of the bear skull ritual included a description of the boundary between the profane village and the sacred forest. The Savonian singer Antti Huttunen exhorted the bear to wander "along that golden alley, / a silver street, / in a golden cup, / in a copper basket, / across the hill of Pohjola" (*Tätä kultaista kujoa, / hopeaista tietä myöten, / kultaisessa kuppisessa, / vaskisessa vakkasessa, / poikki Pohjolan mäkien*; SKVR VI2/4919, 5–9 Iisalmi, North Savo, Finland, 1876). By passing over that hill, being carried in a newborn's basket (*vakka*), the bruin was returned to the mythical land of its birth (Pohjola, or Pohja, the Northland). The reference to the gold and silver alleys and streets indicate that the place is very uncommon and mythical. At the beginning of the bear hunt, the North Karelian hunter Honkanen sang that the bear was born "in the dark Pohjola, / in the rigorous Tapiola" (*Pimeessä Pohjolassa, / tarkassa Tapiolassa*; SKVR VII5/3386: 2–3, Kivilahti, Kauvonniemi, Ilomantsi, North Karelia, Finland, 1885, T. Honkanen). In Nurmes the hunters sang that the bear was born "in the honeyed Mehtola" (*Metisessä Mehtolassa*; SKVR VII5/3385, 9, Nurmes, North Karelia, Finland, 1832, name of singer not mentioned in the manuscript by Elias Lönnrot). In hunting songs, when Pohjola, Tapiola and Metsola/Mehtola were combined through parallelism, they were names for an otherworldly forest (Franssila 1900, 383; Karhu 1947, 115).

The forestland where the bear was born was not a natural place in the modern sense; it was a mythic environment inhabited by supernatural beings, the *haltias* (guardian spirits) of the forest. In the *The Birth of the Bear* incantations sung by hunters, the bruin is often born under a small pine (or a spruce, another evergreen tree), or on its roots (SKVR VI2/4886, 13; Piludu 2019, 114–120). The hunters sang that the bear was born "under a small pine" (SKVR VI2/4886) or that Mielikki, the Mistress of the forest, conceived the bear "under the spruce's flower-top" (SKVR XII2/6480, 30–36). The hunters gave back the bear's skull and bones, setting them in the exact places where the bruin was born. In different *Bear Songs* the names of the bear's mothers and protectors varied; the most common ones are Mielikki, Hongotar, the Crone or Mistress of Pohjola (Pohja), the Mistress of the forest.

One of the typical female *haltias* protecting the bear was Hongotar, the "Pine Lady" (Piludu 2019, 120–121), and perhaps the hunters gave back the bear's bones to this spirit by hanging the bear skull on a pine tree. The hunters seemed to send the bear's bones back to their birth land, so that they were in place for the regeneration of the animal. Haavio considered Hongotar (also Hongatar, Honkatar, Hongas) to be the original mother of the bear, because her name relates to the pine tree (*honka*) where the bear skull was hung (Haavio 1967, 25). She is mentioned in some of the oldest *Bear Songs* – the *Cantio Ursina* (SKVR IX4/1101, 6, Rautalampi, North Savo, Finland, 1675, anonymous singer) and the *Text of Viitasaari* (after 1750). In the latter, the hunter sang to the bear: "Your family is of the pine grove, / Hongotar is of your family" (*Hongincosta sinun sucusi / Hongotar sinun sugusi*; SKVR IX4/1096, 37–38, Viitasaari, central Finland*).* In a *The Birth of the Bear* incantation from Kuusamo, the bruin was born "from the den of Petäjätär (the Pine Lady), / the room of Hongatar (the Pine Lady)" (*Petäjättären pesästä, / hongattaren huonehesta*; SKVR XII2/6867, 3–4, Kuusamo, North Ostrobothnia, Finland). This idea could be related to an actual situation: Bears often build their den in digging a hole nearby the roots of a pine or under spruce or pine branches, and she-bears often give birth in their dens. Maybe the songs added a mythical dimension to a natural phenomenon, defining the den as the dwelling place of the forest spirit and that spirit as the mythic mother of the bear.

Martti Haavio considered Hongotar a personification of the pine tree, where the first bear skull was hung (Havio 1967, 25–26). However, this interpretation seems to be a bit too restricted. In the incantations and *Bear Songs*, Hongotar or Hongatar was a typical forest spirit, not only the protector of the pine of the bear skulls. She was also supposed to help the humans who respected her performing rituals and honouring her in the songs. The hunters invoked her to guide them towards the bear den. The hunters prayed to Hongatar to carve signs on the trees to mark the place where the den or the bear could be found (SKVR VII5/3312, 4–10; VII5/3313, 7–12). The cattle herders often invoked Hongatar in incantations to protect the cattle from bears during the grazing season (SKVR VI2/5350, 4–6; VI2/5352, 4–6; VII5/3849, 19–23; VII5, Metsäsuomalaiset 314, 1; XII 2/6740, 37–39).

The regeneration of the bear in Karelian and Siberian traditions

A *Bear Song* by the Karelian singer Iivana Malinen[16] demonstrates the presence of vernacular conceptions about the regeneration of the bear in the Karelian tradition: "The crooked-claw of Pohja, / crooked-claw, bone-hunch, / felt about ten claws, / felt ten claws, / on the north side of the river, /

16 Iivana Malinen was the grandson of Ontrei Malinen, one of the most skilled Karelian singers, and the son of Jyrki Malinen. He learned his *Bear Songs* from his grandfather.

on the sunny side of the hill, / on the root of a wrenched young spruce. / Take from there, my *otso* [bruin], / claws to replace your claws, / the best shirt from the bundle, / the most desired of the teeth" (*Pohjan on kyyttö kynsi, / kyyttö kynsi, luu hamura, / vanu kynttä kymmenisen, vanu kynttä kymmenkunnan, / pohjois-puolella jokea, / päivän-puolella mäkeä, / juuressa nyry-närehen. / Ota sieltä otsoseni, / kynsiä kynsien sijahan, / paita pakasta parahin, haluisimmat hampahista*; SKVR I4/1244 e, 1–11, Vuonninen, Viena Karelia, Russia).

These lines were sung after the skinning of the bear, not during the skull rituals. However, the text is probably the most precise Finno-Karelian description of how a dead bear would be able to regenerate himself. "Crooked-claw" and "bone-hunch" were epithets for the Crone of Pohja or the Mistress of Pohjola (SKVR I1/90, 1; I4/1112, 3–4; I4/1413, 1–2), who felted new claws in Pohjola "on the root of a young spruce". In Iivana Malinen's *The Birth of the Bear* incantation, the Crone of Pohjola was the mother of the bear.[17] She gave birth to the bear "on the root of a wrenched young spruce" (SKVR I4/1191, 12; Piludu 2019, 129–136). Thus, the regeneration of the bear happened at the same spot where the bruin had been born, and it seems that it was put in motion by the mother of the bear. The killed bruin was requested to reach back to its birthplace and take new claws, teeth and the best "shirt," probably a new fur.

The word "rebirth" was never mentioned in the song of Iivana Malinen, who, like all the Finno-Karelian hunters, sang about the death of the bear as little as possible. The procedure in which the bear was involved seemed rather a regeneration, an "exchange" of body parts or the healing of them, than a resurrection. In Iivana's song, the killed bruin is described as a living being able to travel back to the land of its birth. Malinen apparently sang about partial regenerations: new claws for old claws, new teeth for old teeth. However, the claws and the teeth of the bear in the Finno-Karelian tradition represented the force of the whole animal; they were often preserved by hunters and used as protective amulets in many rituals (Piludu 2019, 260–262). By taking new claws and teeth from Pohjola, the force of the bear was fully restored. They were "hard" body parts and the "weapons" of the bear; like swords or knives, they could be "changed". In hunting rituals, the regeneration of animals was believed to start from hard body parts, like claws, teeth, bones, or skulls, not from meat or other organs that decomposed.

The folklorist Matti Kuusi considered the bear skull ritual as directly connected to the idea of regeneration from the bones: "The bones were given back to the forest mistress or a divine being living near the copper mountain or the sky mountain, who was able to make with them the building material of the new bears" (Kuusi 1963, 49). Kuusi made a hypothetical reconstruction, but his theory seems to be based on Iivana Malinen's songs.

The ethnologist Ivar Paulson[18] stated that northern Eurasian peoples ritually preserved the bones of bears, reindeer, elk, deer, foxes, hares, sables, wolves, lynxes, wolverines, walruses, seals, swordfish, and various species of birds and fish. Paulson added that scholars collected more ethnographic data about the preservation of bear bones, "but the rituals performed with them are not fundamen-

17 The anthropologist Matti Sarmela stated that the myth of the bear's birth from the Mistress of Pohjola was a depravation of the ancient *The Birth of the Bear in the Sky*. Sarmela describes the Mistress of Pohjola as the incarnation of evil, as she is also defined as a "whore" and the mother of illnesses in healing incantations. In his theory, this *Birth of the Bear* was the product of an agrarian society that fully demonised the bear, as it could kill the cattle in the grazing season. I am not fully convinced by his interpretation. It is correct that the Mistress of Pohjola was a dangerous being, and she was dreadful in Iivana Malinen's *The Birth of the Bear*, too. However, she was not fully evil, but very ambiguous, able to damage or save people and animals. The herders could invoke her to protect the cattle, and she was able to regenerate bears. Her role in the folk beliefs was not only destructive, but also generative and protective.

18 Ivar Paulson (1922–1966) was an Estonian ethnologist and scholar of Siberian and Finnic religions. He worked in Hamburg and Stockholm and published intensively in German. He was particularly interested in the problem of rituals about regeneration of game animals from their bones.

tally different from those where other animal bones were involved" (Paulson 1968, 451). Among the Mistassini Cree (Canada) hunters, the inedible parts of game animals have "intrinsic power" (Tanner 1979, 141).

The main difference is that among the Finno-Karelians, the Sámi (Fennoscandia), the Mansi (Western Siberia), the Khanty (Western Siberia), the Nivkhi (Eastern Siberia and Sakhalin Island), the Ainu (Japan) or the Cree (Canada), the skull ritual was an integral part of the bear ceremonials and feast. In the circumboreal area, however, the hunting of big sea mammals (whales, walruses, and seals), elk or great quantities of salmon included feasts and ceremonials, too (Lantis 1938; Fienup-Riordan 1994, 88–142; Watanabe 1994). Watanabe identifies ceremonials as "sending-off" rituals, performed before the animal spirits entered the village as guests of honour of the festival and then afterwards, when they were sent back to their environments, to ensure that they would reappear the next hunting or fishing season (Watanabe 1994, 67; Wiget/Balalaeva 2011, 139–140).

Paulson stressed that the general concept of the revival of the bones can be found in many Eurasian traditions, but the hunters themselves "think rather of the animal's continued existence than a resurrection in the proper sense of the word" (Paulson 1968, 455). I will dare to suppose that the scholarly obsession with discussing the resurrection of the animals could also be influenced by the Christian background of several scholars. Resurrection is a concept influenced by centuries of Christian theology, and it implies that the animal experienced a socially recognised death. The problem here is that the bear's death was almost negated during the rituals, and it was transformed into the visit of a guest of honour to the village.

Scholars have made clear statements about the resurrection of animals from bones, but the hunters tried to avoid speaking about the issue. The ethnologist Eveline Lot-Falck[19] stated that the Siberian hunters did not offer clear explanations about the destiny of the animal: "What happened to the animals killed during the hunt? The animal came, it left, so it will come back. Where does it go? How will it come back? No clarification about that. Certainly, in our times, if the word death is pronounced, it is to continue the fictional tale of the voluntary visit" (Lot-Falck 1961, 211). Lot-Falck stressed that Siberian hunters had a different conception of death, compared to the modern one: "Death is not a break. It does not have an irrevocable nature, and it remains a transitory state. A passage between one world to another often leaves the possibility of a return, so the absence of the soul could be temporary" (Lot-Falk 1961, 211). This statement fits well with the Finno-Karelian folk beliefs, in which the death of the bear did not signify the complete absence of life and the dissolution of the bruin's self, but involved the bruin's immediate and voluntary participation in the bear feast (Piludu 2019, 216–224) and the regeneration and continuation of the bear's life after the skull ritual.

Both the Finno-Karelians and Siberians did not appreciate anyone speaking or singing about the kill or the resurrection of the game animals. Siberian hunters had similar beliefs about the necessity of avoiding mention of the bear's death or rebirth. The Ainu (Japan) did not say to the bear that it had been killed, but that it was being "sent away" to its relatives in the mountains (Zolotarev 1937, 123; Batchelor 1901, 206–207).

The Tungusic Orochs (self-designation: Nani; eastern Siberia and Sakhalin Island) people said to the killed bear: "Go fast, go to your masters, put on a new fur, and come back the next year so that I may look at you" (Shternberg 1933, 439; Zolotarev 1937, 123). Lot-Falck stated that the Siberians believed that life continued to inhabit the bones, in different parts of the skeleton, which, at a certain point, could cover itself again with flesh. Life remained and was always potentially present in ritually

19 Eveline Lot-Falck (1918–1974) was a French ethnologist, professor of religions of North Asia and an expert at the Asian Department of the Musée del'Homme of Paris.

preserved bones. As every part potentially represented the whole, a single but powerful bone – such as the skull – was able to generate an entire new body (Lot-Falk 1961, 212).

In the Finno-Karelian and Oroch traditions, the hunters refrained from speaking about the bear's death and resurrection; they instead presented the bear's destiny as a travel back to its birth land for the acquisition of new body parts (fur, teeth, claws). They emphasised the positivity of this bodily regeneration, glossing over the violent death.[20]

The bear's testimony about its treatment

Can we find in Finland or Karelia some accounts of the continuation of the bear's life after its death? In Suomussalmi (Kainuu, Finland) the travel of the bear did not end in the branches of the pine. After the skull ritual, the hunter exhorted the bear to go to Mehtola (Metsola, the Forestland), the land of the forest *haltias* (guardian spirits), to tell that it had been kindly treated during the feast and the skull ritual: "Tell, after having left from here, / after having gone to Mehtola[:] / ["]Here I was not treated badly, / they fed me with mead, / they let me drink honey-drinks.["] (*Sano täältä saatuasi, / mehtolahan mentyäsi: / ei täällä pahoin pietty, / simoa täällä syötettihin, / mesijuomat juotettihin*; SKVR XII2/6572, 11–15, Kylmäsalmi, Suomussalmi, Kainuu, Finland, 1834, name of singer not mentioned in the manuscript by Elias Lönnrot).

According to Kaarle Krohn, the bear that came back to the mythical forestland (Metsola, Mehtola) would persuade other bears to participate in similar feasts in the village (Krohn 2008, 157). The bear probably told them and the forest spirits about the nice treatment in the human village during the bear feast[21] (Piludu 2019, 231–239) and the bear skull rituals. The song is interesting because it reveals that the bear was believed to be able to speak, and the speaking ability is a typical attribute of personhood.

In Finland and Karelia, the alcoholic drinks offered to the bear during the bear feast and the ritual of the bear skull acquired special importance: The bear was seen as being treated well because the hunters offered mead, ale and barley spirits (*paloviina*, a kind of Finnish *vodka* made of barley, *brännvin* in Swedish). At the end of the bear ceremonial, the hunters drank ale from the nostrils of the bear skull. In the 20th century, some informants still remembered that in the past people drank from the skull. In 1912, the informant Mikko Laitinen told that: "When a bear was killed, it was brought to the roots of a sacred tree, where the funerals were held. For this reason, ale was made, and it was drunk from the holes of the nostrils of the skull of the bear" (SKS KRA Oskari Nousiainen 85, 1912, Kangasniemi, Ohensalo, South Savo, Finland, Mikko Laitinen).[22]

The ritual of drinking from the skull is mentioned in old sources on Finnish bear ceremonialism, the *Text of Viitasari* (SKVR IX4/1096, after 1750; Pentikäinen 2007, 74; Piludu 2019, 267–271) and *Historisk och œconomisk beskrifning öfver Calajoki sock uti Österbotn* (1754) by Christian Salmenius[23] (Haavio 1967, 16). The oldest description of the ritual of drinking from the bear skull was in-

20 In Karelia, also human death was not a complete break from life. In the graveyards, the wooden grave markers were carved in a form resembling a house, and on the Day of the Dead, the people offered food and drinks to the souls of the deceased. This tradition is still alive in rural orthodox Karelia and Russia.

21 During the bear feast, the hunters offered the bear ale and barley spirits; see Piludu, this volume, on the Finno-Karelian bear feast and wedding.

22 The sources from the SKS KRA (Archive of Folk Poetry of the Finnish Literature Society) are mentioned indicating: the name of the collector; the number of his manuscript and its year; the village and its parish; the name of the informant and, sometimes, some basic information about him (age and profession).

23 Christian Salmenius (1700–1791) was a Protestant pastor of Lapua. He was interested in local history and wrote about the topic.

cluded in the sermon given by Bishop of Finland Isak Rothovius[24] for the inauguration of Academia Åboensis (Åbo or Turku[25], Finland, July 15, 1640): "When they catch a bear, a party is held in the dark, and they drink a toast for the bear out of its skull, and groan just as the bear does. Thus, would they gain a greater good fortune!" (Pentikäinen 2007, 131; 2014, 429; Siikala 2016, 380).

In the skull rite, the ale acted as an instrument to have a direct, physical connection with the bear skull and bear force. The ale was offered to the bear, but at the same time the hunter also drank it; it was a shared drinking. By offering ale to the bear skull, its soul was pleased, and it happily transmigrated to the sky or its protecting spirits, waiting for its regeneration (Apo 2001, 74). The ale passed through the skull, a powerful container of bear force, and ended up in the mouth, stomach, and veins of the hunter.

The story of the bear's testimony about its treatment is present also in other Northern indigenous cultures. Ivar Paulson stated that among the Sámi, Khanty, Nivkh, and Ainu, after the hunters had ritually treated the bear's bones, the bruin was requested to tell other bears about the good treatment and the honours it had received among the humans (Paulson 1965, 12). Sometimes the bear was supposed to tell other bears to let themselves be caught by the hunters that had treated it so nicely. Lot-Falck stated that Siberian hunters believed that the soul of the killed bear told the other bruins all the ritual procedures used in dealing with its flesh and bones. The animal person, like the human person, was not a unit that could be separated from the group. Thus, an offense to one bear reached the community of the whole animal species, which would then avenge its mistreated member (Lot-Falk 1961, 205).

The Nivkh of Sakhalin Island and the Amur River (eastern Siberia) exhorted the bear's soul to go to the Master of the Forest or the Master of the Mountain (Paulson 1968, 453). The Eastern Cree (Canada) told the bear to go to the Bear Master (*Memekwesiw*) and tell him how nicely the hunters had treated it (Rockwell 1991, 36). Among the Tungusic Olcha (or Ulch, eastern Siberia), the bear was sent back to its relatives, the forest men, or the masters of the mounds, and there it related the details of the ceremonial. If that was arranged properly, the men of the forest would be satisfied, and they would send "happy hunting" to the humans (Zolotarev 1937, 123).

In Finland a correct performance of the bear ceremonialism assured general luck in the hunt. In 1640, as stated above, Isak Rothovius, Bishop of Finland, underscored that the Finns who performed the bear ceremonials believed that a correct ritual gave the hunters good luck in hunting (Haavio 1967, 15; Pentikäinen 2007, 131; 2014, 429; Siikala 2016, 380).

Conclusions: The importance of honouring the bear and its bones

In Finno-Karelian bear ceremonials the destiny of the bear after its death is not precisely described. There were many possibilities surrounding the "posthumous life" of the bear: a) its skull and soul were oriented towards the Plough, following the myth of *The Birth of the Bear* in the Sky; b) the skull was hung on a sacred pine and the bones were buried on its roots, following the myth of *The Birth of the Bear* in a mythical forest; c) the bear was exhorted to go back to one of his mythical birthlands (Pohjola, the Northland) to take new claws and teeth made by its mother, the Crone of Pohjola, d) the bear was requested to go to the mythical forestland (Metsola, Mehtola, Pohjola, Tapiola) to speak about the good treatment and the drinks the humans had offered during the whole ceremonial.

24 Isak Rothovius (1572–1652) was Bishop of Finland and Åbo (Turku) in 1627–1652, a Chancellor of the Academia Åboensis. He was very active for the improvement of teaching in Finland, a land that he considered "barbarian".

25 The old capital of Finland has two names: Åbo in Swedish and Turku in Finnish.

What is certain is that in most of the cases the bear should go back to one of his mythic homelands, or to the exact spot where it was supposed to be born, and that its life "regenerated" or "continued" there. What was relevant for the hunters was not the precise definition of all the phases of its regeneration, but the fact that the bruin and forest spirits understood and remembered how respectfully the "guest" (bear) had been treated during the whole ceremonial, along with its bones in the last phase of the ritual.

The hunters sang the minimum necessary about the death of the bear, giving a brief deluding justification after its kill (Piludu 2019, 200–203), but they sang several lines reminding of all the honours reserved to it. The goal was to avoid the revenge of the bear and forest spirits and to please them, ensuring their favour. The bear ceremonialism reveals a circular vision of animal life: The bear died, but its life continued, thanks to the accuracy of all the rituals.

The most relevant aspect of the whole bear ceremonial was how the bear was treated in all the ritual phases. In the bear feast it was relevant how the bear – presented as a guest of honour and a groom – was invited inside the house of the village. The hunters emphasised that the bruin had a place of honour at the table, and they stressed the abundance of alcoholic drinks offered during the feast (Piludu 2019, 231–239).[26] During the bear skull rituals, the hunters pointed out that the skull was not thrown on the ice, it was hung on a good and sacred tree, it was brought to the place where the bear was born, and it was oriented facing the Plough. The honours reserved to the bear skull were fundamental to achieve the regeneration of the animal and to obtain good luck in the hunt for the future, as the bruin was supposed to tell the forest spirits about its treatment. The proper treatment of the bear throughout the whole ceremonial was the prelude to the return of the satisfied guest and groom in the village and subsequently it ensured a successful bear hunt in the future, followed by yet another joyful feast to please the bruin and the forest spirits.

Acknowledgements

I'm grateful to the Niilo Helander Foundation (*Niilo Helanderin Säätiö*) for having generously funded my research work at the University of Helsinki. I would like to express my special thanks of gratitude to Susanna Lendiera for the permission to publish her bear photo. I would like to thank Lotte Tarkka, Professor of Folklore Studies at the University of Helsinki, for her valuable assistance for the interpretation and translation of the Viena Karelian *Bear Songs* when she was the supervisor of my PhD thesis. This chapter is dedicated to the memory of Enrico Comba (1956–2020), former Professor of Anthropology of Religions at the University of Turin, for his constant interest in my research work.

Bibliography

Anttonen 1998: V. Anttonen, Pihlaja, naisen kiima ja kasvuvoiman pyhä locus. In: J. Pöysä/A.-L. Siikala (eds.), Amor, Genus & Familia. Kirjotuksia kansanperinteestä, Tietolipas 158 (Helsinki 1998) 136–147.

Apo 2001: S. Apo, Viinan voima. Näkökulma suomalaisten kansanomaiseen alkoholiajatteluun ja -kulttuuriin (Helsinki 2001).

Batchelor 1901: J. Batchelor, The Ainu and Their Folk-Lore (London 1901).

Fienup-Riordan 1994: A. Fienup-Riordan, Boundaries and Passages: Rule and Ritual in Yup'ik Eskimo Oral Tradition (London 1994).

26 See Piludu, this volume, on the Finno-Karelian bear feast and wedding.

Franssila 1900: K. A. Franssila, Kansanrunouden tutkimuksia 1. Iso tammi liitteineen (Helsinki 1900).

Harva 1933: U. Harva, Altain suvun uskonto (Porvoo 1933).
Haavio 1967: M. Haavio, Suomalainen mytologia (Porvoo, Helsinki 1967).
Hallowell 1926: I. A. Hallowell, Bear Ceremonialism in the Northern Hemisphere. American Anthropologist 28(1), 1926, 1–175.
Honko et al. 1993: L. Honko/S. Timonen/M. Branch (eds.), The Great Bear: A Thematic Anthology of Oral Poetry in the Finno-Ugrian Languages (Helsinki 1993).

Karhu 1947: J. Karhu, Karhun synty: vertaileva tutkimus (Mikkeli 1947).
Krohn 1917: K. Krohn. Suomalaiset syntyloitsut (Helsinki 1917).
Krohn 2008: K. Krohn, Suomalaisten runojen uskonto (Helsinki 2008; originally published 1915).
Kuusi 1963: M. Kuusi, Karhunpeijaiset. In: M. Kuusi (ed.), Suomen Kirjallisuus 1. Kirjoittamaton kirjallisuus (Helsinki 1963) 41–51.

Lantis 1938: M. Lantis, The Alaskan Whale Cult and its Affinities. American Anthropologist 40, 1938, 438–464.
Lot-Falck 1961: E. Lot-Falck, Riti di caccia dei Siberiani (Milano 1961).

Paulson 1965: I. Paulson, Karhunkallon riitistä Kalevalassa ja arktisissa kansanuskoissa. Kalevalaseuran Vuosikirja 45, 1965, 115–142.
Paulson 1968: I. Paulson, The Preservation of Animal Bones in the Hunting Rites of Some North Eurasian Peoples. In: V. Diószegi (ed.), Popular Beliefs and Folklore Tradition in Siberia. Uralian and Altaic Series 57 (Bloomington 1968) 451–457.
Pentikäinen 2007: J. Pentikäinen, Golden King of the Forest: The Lore of the Northern Bear (Helsinki 2007).
Pentikäinen 2014: J. Pentikäinen, The Songs of the Bear in Hunt Rituals. In: V. M. Piludu/Frog (eds.), Kalevala: epica, magia, arte, musica / Kalevala: Epic, Magic, Art, Music (Viterbo 2014) 415–443.
Piludu 2019: V. M. Piludu, The Forestland's Guest: Mythical Landscapes, Personhood and Gender in the Finno-Karelian Bear Ceremonialism (Helsinki 2019). https://helda.helsinki.fi/handle/10138/282220.
Piludu 2022: V. M. Piludu, Comparative and Contextual Approaches to the Study of Finno-Karelian and Ob-Ugrian *The Birth of the Bear in The Sky* incantations and songs. In: H. Rydving/K. Kaikkonen (eds.), Religions Around the Arctic: Source Criticism and Comparisons (Stockholm 2022) 77–110.

Rockwell 1991: D. Rockwell, Giving Voice to Bear: North American Indian Rituals, Myths, and Images of the Bear (Niwot 1991).

Sarmela 1982: M. Sarmela, The Death of the Bear: An Old Finnish Hunting Drama. The Drama Review 26(3), Scandinavian Theatre (Fall 1982), 1982, 57–66.
Sarmela 1991: M. Sarmela, Karhu ihmisen ympäristössä. In: P. Laaksonen/S.-L. Mettömäki (eds.), Kolme on Kovaa Sanaa. Kirjoituksia kansanperinteestä. Kalevalaseuran vuosikirja 71 (Helsinki 1991) 209–250.
Sarmela 2009: M. Sarmela, Finnish Folklore Atlas. Ethnic Culture of Finland 2. Published on the author's webpage: http://www.kotikone.fi/matti.sarmela/folkloreatlas.pdf.
Siikala 2008: A.-L. Siikala, Myytit, riitit ja tietäjän toimet. In: R. Räsänen (ed.), Savo ja sen kansa (Helsinki 2008) 110–186.
Siikala 2016: A.-L. Siikala, Itämerensuomalaisten mytologia (Helsinki 2016, originally published 2012).
Shternberg 1933: L. I. Shternberg, Гиляки, Орочи, Гольды, Негидальцы, Айны: статьи и материалы [Giliaki, Orochi, Gol'dy, Negidal'tsy, Ainy: Stat'i i Materialy]. Ed. A. Koshkin (Khabarovsk 1933).
Spagna 1998: F. Spagna, L'ospite selvaggio. Esperienze visionarie e simboli dell'orso nelle tradizioni native americane e circumboreali (Torino 1998).

Tanner 1979: A. Tanner, Bringing Home Animals: Religious Ideology and Mode of Production of the Mistassini Cree Hunters (London 1979).
Tarkka 2005: L. Tarkka, Rajarahvaan laulu. Tutkimus Vuokkiniemen kalevalamittaisesta runokulttuurista 1821–1921 (Helsinki 2005).
Tarkka 2013: L. Tarkka, Songs of the Border People: Genre, Reflexivity and Performance in Karelian Oral Poetry. Folklore Fellows Communications 152 (Helsinki 2013).

Watanabe 1994: H. Watanabe, The Animal Cult of Northern Hunter Gatherers: Patterns and Their Ecological Implications. In: T. Irimoto/T. Tamada (eds.), Circumpolar Religion and Ecology: An Anthropology of the North (Tokyo 1994) 47–67.
Wiget/Balalaeva 2011: A. Wiget/O. Balalaeva, Khanty: People of the Taiga (Fairbanks 2011).
Wiget/Balalaeva 2022: A. Wiget/O. Balalaeva, Valuing Difference: Bear Ceremonialism, the Eastern Khanty, and Cultural Variation among Ob-Ugrians. Sibirica 21(1), 2022. https://doi. 10.3167/sib.2022.210103.

Zolotarev 1937: A. M. Zolotarev, The Bear Festival of the Olcha. American Anthropologist 39, 1937, 113–130.

Vesa Matteo Piludu
PhD, Postdoctoral researcher
University of Helsinki
Indigenous Studies,
Department of Finnish, Finno-Ugrian and Scandinavian Studies
Faculty of Arts
Helsinki
Finland
vesa.piludu@helsinki.fi

Fig. 1. The pine of Timinkylä (Hämeenkyrö, Finland) is a particular karhunkallohonka *– a bear skull pine tree. It is a 400-year-old pine and probably the oldest tree of the region of Pirkanmaa. The hunters hung bear skulls on its branches with coloured stripes. The same tree was also a* uhripuu *(tree for offerings): The local people offered their fields'* primitiae *(first grain or milk of the year) on its roots. Until* c. *1930 people also offered money putting coins in the holes and cracks of the pine. According to Finnish folk beliefs, the* haltia *(guardian spirit) of the tree assured good luck for the cattle and other rural activities if the people performed these offerings. But if the people forgot to perform rituals, the* haltia *took a hard revenge and people could get sick. In the 19th century almost every household in eastern Finland has his own* uhripuu*: The healthier the tree was, the richer and more prosperous was the household. Finally, this pine was also a* parantava puu *(healing tree), as its powerful* haltia *was believed to heal the toothache. Nowadays the tree is protected, it is a local natural attraction, and it has even a personal post box: Visitors can leave written messages describing the feelings they had in front of the huge tree (photo V. M. Piludu, 2019).*

Fig. 2. Bear skull and hunters near a bear trap in Viena Karelia (Russia). Photo taken by I. K. Inha, 1894 (photo Museovirasto [Finnish National Board of Antiquities], Kuopion kulttuurihistoriallinen museo [Museum of Cultural History of Kuopio]: KHMKUVV 243:44, public domain).

Fig. 3. A bear skull on a birch tree in Inari (Lapland, Sámi village, Finland). Generally, the bear skull was hung on a pine tree or a spruce in Finland (photo S. Paulaharju, 1914, Museovirasto [Finnish National Board of Antiquities], Kansatieteen kuvakokoelma [Ethnologic Photographic Archive]: KK3490:1797, public domain).

Fig. 4. "Bear", photo by Susanna Lendiera, 2018. Many rituals of the Finno-Karelian bear ceremonies focused on the head and skull, which was considered the most powerful part of the bruin. The head and the skull contained the soul of the bear. During the bear feast, all the meat and the organs (eyes, ears, tongue) of the head were ritually eaten by the hunters. Doing so, they believed to obtain the supernatural senses and skills of the bear, as the ability of singing a "kindred song", a song in a shared language, which was understood by bears, humans and haltias *(guardian spirits). The preservation of the skull on the branches of the sacred pine ensured the regeneration of the bear from its skull and bones and the possible return of the bruin as a prey. Rituals for the regeneration of the animal avoided the revenge of bears and pleased the* haltias.

The human-bear relationship among swidden cultivators and forest peasants in Savonia, Finland, and central Scandinavia

By Marja-Liisa Keinänen

Keywords: Human-bear relationship, bear ceremonies, bear in magic, swidden cultivation, animal husbandry, Finland, central Scandinavia

Abstract: Building on the social anthropologist Matti Sarmela's valuable tradition-ecological work, this paper examines the relationship between swidden cultivators and forest peasants and the bear. Geographically, it centres on southern and northern Savonia as well as on the old Savonian settlements of northern Tavastia in Finland and central Scandinavia, notably in northwest Värmland and the bordering side in Norway. The folkloristic source material largely covers the 19th century. The paper investigates what kind of "bearhood" the bear songs and charms produced and what these data tell us about the relationship between bear and man. With increasing animal husbandry, the bear came to be seen as a threat to the cattle that were pastured in woodlands and forest villages. Also, the human-bear relationship changed profoundly. Even though the bear was still honoured and treated with respect, it increasingly came to be seen as a beast of prey with a less honourable origin than was the case among earlier hunters. In spite of these changes in attitude, the songs and charms establish an intersubjective and relational stance between bears and humans. Through this communication, the bear was established as a person, who was perceived as an intentional, sentient being with the agency and capacity to make choices, with whom the cattle-minder could reason.

Introduction

Bear ceremonialism has been an integral part of many circumpolar hunting cultures, in which the bear was perceived as a sacred animal to be treated with respect. The bear, seen in Ob-Ugrian traditions as a son of a heavenly god, was ritually slain and honoured through ceremonials. After the celebration, the bear's soul was sent back to its heavenly place of origin. These practices were to guarantee future hunting luck and the continuity of the bear stock (Honko 1993, 120–125). The rich traditions of Finnish-Karelian bear songs, charms, and ritual practices have their roots in an earlier hunting context, but have undergone fundamental changes over time, when adapted to the ecological and cultural conditions of swidden cultivation and animal husbandry (see Piludu, this volume).

According to social anthropologist Matti Sarmela's tradition-ecological analysis, the swidden cultivator's worldview and attitude towards the bear differed quite radically from that of the northern hunters and was in some respects quite the opposite of it. For an eastern Finnish swidden cultivator, the bear was an ecological competitor, a beast of prey who posed a threat to the crops and the livestock that grazed in forestlands (Sarmela 2009, 39). The next stage in Sarmela's evolutionary sketch

Bear and Human: Facets of a Multi-Layered Relationship from Past to Recent Times, with Emphasis on Northern Europe, ed. by Oliver Grimm (Turnhout: Brepols, 2023), pp. 761–776 BREPOLS PUBLISHERS 10.1484.M.TANE-EB.5.134360

relationship between human and wolf – the latter was not ascribed the similar degree of personhood as was the bear, and the human-wolf relationship remained distant.

"Bearhood" – the bear as a person and a relative

Probing deeper into the question of how the humans' view of the bear changed among the aforementioned groups of swidden cultivators and forest peasants, I will build on the critical findings of the recent studies of indigenous religions, which have challenged the premises of the earlier studies on so called animism. Proceeding largely from A. Irving Hallowell's research (Hallowell 1926), these studies have particularly questioned the underlaying anthropocentrism, ontologies, and epistemologies of the previous scholarship and have further elaborated the notion of person and personhood in indigenous religions.

They argue, among other things, that ontologies and epistemologies of indigenous religions are characterised by relationality. According to this view, the universe is populated by various categories of persons, which in the Savo-Karelian case comprised human persons, animal persons, and various kinds of spirit persons. Even the forest was treated in certain contexts as a person. Personhood implies here that various categories of persons, and even certain things, were assumed to possess the "same qualities of sentience, will, rationality, and emotionality that characterize man himself" (Morrison 2000, 28). The universe populated by different person categories was perceived as intersubjective, meaning that these persons communicated with each other in various ways, and through this communication established themselves and each other as persons. Also, causality was understood in personal rather than impersonal terms. Persons shared causal agency and exercised power, i.e. could influence each other in different ways (Morrison 2000, 28).

Although the human-bear relationship radically changed during the long transition from hunting to swidden culture, the intersubjective and relational stance can still be discerned in Savo-Karelian folklore materials. Lotte Tarkka's findings, based on her White Sea Karelian data, also have bearing in Savonia. The forest was imagined in charms and songs as a sociomorphic world modelled on human community. The woodlands were populated by various kinds of spirits ruled by Tapio, the master of the forest, who reigned over the forest animals. He was pictured in songs and incantations as living in a cottage with his family. Concerning bear traditions, too, the forest was imagined as the bear's home, where it lived in a house and slept in a bed. Tapio, and some female forest spirits, also had the power to offer the animals to the hunter or deny them to him. In some contexts, Tapio even personified the forest, as did the bear (Tarkka 1994, 257–259; Siikala 2008, 136–137).

According to R. E. Nirvi's analysis of word taboos (Nirvi 1944), the noun *forest* was used as a synonym for a bear, and the phrase "the forest moves" (*metsä liikkuu*) not only indicated that the bear was on the move, but was a euphemism for the beast also having caused damage. A malicious person was assumed to be able to set the forest/bear into motion: "Rise forest against the cow, / from alder thicket the cow-eater, / the angry cat from a grove" (SKVR IX4: 1359). Circumlocutions such as "the forest visited the cattle" and "the forest cuts (*leikkaa*)" were expressions for the bear or a predator having injured the cattle (Nirvi 1944, 27–28; SMSK s.v. *metsä* 3). A spell, "May the forest eat your cattle!", was said in order to set a bear on someone else's cattle; the forest denoting here a bear (SKVR IX4: 1360). A moving forest had to be "locked" or "shackled" by a wise person (SMSK s.v. *metsä* 3:1bb). The agency ascribed to the forest also comes forth in the idea of the forest as a cover; the forest could actively hide a cow or a human person (Tarkka 1994, 82–84).

A hunter-livestock keeper communicated with the animate forest, its supernatural owners, and the animals by means of songs and incantations. In this communication, the bear was addressed as if it were a sensing and conscious person with the desire and ability to communicate and act with intent.

Nirvi maintains that the bear was assumed to hear and understand speech, as well as to respond to it (Nirvi 1944, 74). Due to the generative, performative nature of indigenous languages, speech was assumed to impact the material world (cf. Morrison 2000, 34–35; see Hautala 1960, 16, 19). The fear of mentioning certain potentially dangerous matters, such as a bear, the devil, fire, etc., was based on the idea that a mere uttering of these words, or the acts they were associated with, would imply calling upon them. The fear of conjuring up a bear comes forth in a comment by a Savonian informant, who concurred to talk about his encounter with a bear only briefly, fearing that the animal might materialise (SKVR VI2: 5440). A bear was also believed to get annoyed if its proper appellation was not used (Nirvi 1944, 55, 74). For these reasons, we find a great number of circumlocutory designations for the bear.

Since kinship in Savonia was the cornerstone of social organisation, the relationship to various person categories was often imagined in terms of kinship. The hunter-livestock keeper's communication with a bear established the animal not only as a bear-person, i.e. an intentional, sentient being with the agency and capacity to make choices, but also as a relative when engaging in and maintaining a relationship with it (cf. Bird-David 1999, 73; Morrison 2000, 28). A Savonian hunter appealed in his hunting songs to various female spirits of the forest in a highly self-assured and erotic tenor. His relationship to the quarry, sometimes even called "the daughter of the forest", was analogous to that between the suitor and the courted maiden (Tarkka 1994, 257; Siikala 2008, 136–137).

According to Sarmela, in kinship-based swidden societies of Savonia, rituals or marriage were a common means of "creating legal alliance relationships". A farmstead's shortage of manpower was usually solved by an incorporation of outsiders into the kinship group, e.g. as foster children or as members of an extended family, such as in-marrying husbands. Analogically, by means of the bear feast (*kouvon päälliset*), which in central Finland was staged as a wedding, the bear was joined by marital bonds to the family of humans (Sarmela 2009, 43; for the wedding see Piludu, this volume). The Finnish appellation *kouko* (gen. *kouvon*) used of a bear presumably originally denoted "a father", "a grandfather" or "an old man" (Nirvi 1944, 41). It is often assumed that the word was originally used to denote a forefather in a totemic clan context (Honko 1993, 126). However, the word seems to have lost this denotation in 19th-century parlance and was simply used as a circumlocution for the tabooed word "bear". But there were other endearing and honouring appellations for a bear that were generally used when addressing highly respected persons, such as parents, grandparents, and elders as well as chiefs ("King of the Forest"; Nirvi 1944, 79).

The intensified efforts by the authorities to put an end to the bear ceremonies led to their gradual decline from the 18th century onwards. The changes in the subsistence economy also contributed to changes in the human-bear relationship. The celebrations of "weddings" declined, but we still can discern the intersubjective and relational attitude in our oral bear traditions from the 19th century.

The status of the bear

According to Sarmela, the transition from hunting to swidden cultivation and animal husbandry implied that the animal of celestial origin, who was honoured by hunting cultures, was degraded to a creature of evil stock (Sarmela 2009, 95). The bear was symbolically removed from the forest farmer's environment and its world was to be kept apart from that of man. The objective of the bear ceremonies was no longer to ensure that the bear was reincarnated but instead was sent back to its place of origin, which was no longer considered to be the celestial sphere but the ill-reputed Northland (Pohjola, Pohja), governed by its matron, a harlot. In Finnish-Karelian exorcisms, Pohjola was pictured as "the cradle of evil, the place to which diseases were banished", and now also the origin of the bear (Sarmela 2009, 95, 103).

However, Sarmela's generalisation is somewhat problematic given the fact that the attitude to the bear varied contextually. As Tarkka has observed, hunting charms address the bear with endearing and enticing words, whereas the view of the bear in the context of cattle-minding turns into its opposite (Tarkka 1994, 258). Some sources show that attitudes to the bear also varied during one and the same ritual occasion. One Juho Eskelinen (b. 1839) from Sonkajärvi, northern Savonia, who was quite knowledgeable about the bear ceremonies, characterised the tenor of the feast as quite solemn, with participants being on their best behaviour. The singer was assigned an honorary place at the table, treating the bear respectfully, but also scolding and reproaching it in his songs (SKVR VI2: 4926). Thus, the bear songs and rituals were characterised by ambivalence, the tone altering from awe and respect to scolding and reproaching, a phenomenon we also find among the hunting people (Honko 1993, 120).

Moreover, Sarmela's interpretation is complicated by the fact that the incantations are often fragmentary, muddled, and sometimes even contradictory. One and the same charm may combine several themes of the bear's origin without a clear, logical connection. Also, the meaning of Pohjola/Pohja may vary contextually. Sometimes it denotes the direction (north) or, as mentioned, the place where the bear in our case was expelled to. But a comparison of bear songs and incantations in other hunting lore from the same geographical area shows that Pohjola also appears without evil connotations. In some contexts, it simply represents the otherworldly, enchanted forest (Piludu 2019, 115). Also, the valourisation of the various places of origin may vary contextually.

The function of origin themes in incantations was to gain control over a certain phenomenon through a wise person's demonstration of knowledge of its mythical origin. In charms, the themes of origin were used to heal or restore damages, generally called "wraths" (*vihat*, pl.), caused by certain natural phenomena, objects, or animals (Siikala 2002, 86–90). The majority of the bear's origin themes were incorporated in incantations used in livestock-keeping. Thus, by revealing the bear's origin, cattle-minders either sought to send the potential "cow-eater" back to its mythical home, often the otherworldly forest, or just sought to render it harmless.

The bear's celestial origin is mentioned in only five cattle-keeping charms from Savonia. The Tavastian charms of origin (six items) and the Forest Finnish ones (four items) do not refer to the bear's origin in the sky, but instead to its origin on earth. The most common motif among these groups is the bear's birth in the centre (*napa*) of the forest (Karhu 1947, 67, 97, 217). The bear was lulled under "the spruce with flowering top, / under a tiny pine tree, / on the tip of an iron bench" (SKVR VI5: 4822). The change of the place of birth from heavenly heights down to earth does not necessarily imply a degradation of the bear's status. As mentioned, the forest of the charms is not the natural, physical forest, but the mythical Pohjola, which is not only imagined as the abode of the evil, but equally often as an enchanted, otherworldly forestland.

According to J. Karhu's analysis of the origin themes, the longest and most complete charms in Savonia are those that describe the bear being fashioned out of wool (Karhu 1947, 84). Building on Karhu's interpretation, Sarmela sees the wool motif as belittling since these supposedly petty materials of which the bear of the charms was moulded were abandoned by Pirjotar (St Brigit; see Sarmela 2009, 98). However, the reasons for claiming St Brigit's primacy in this function and the negative interpretation of the wool theme (e.g. Karhu 1947, 230; Haavio 2020, 461–463) may be questioned. Wool was also coupled to other saints of the vernacular religion, for instance the Virgin Mary, Kati (St Catherine), and Yrjänä (St George). All of these were connected to livestock and their protection. The idea of fashioning a furry bear out of wool is not far-fetched in a peasant community, and the metaphor of wool is quite practical when seeking to disarm the bear. When rendering a bear harmless, the wise person asked it to hide its claws and fangs in wool, or he/she wished that these sharp weapons would turn into soft wool (for example, SKVR VI2: 4884).

Moreover, in some Savonian incantations, the bear's creation out of wool was located in the primordial sea – the stage of the world's creation in the epic poems. Even though the introductory lines of a charm recorded in 1819 in Kiuruvesi, northern Savonia, can be read as somewhat derogatory due to the chant's ritual context, warding off the bear, its birth was still described in honourable terms. The creators, who tossed the wool into the sea, where it drifted for "seven summers", are the central couple of Finnish-Karelian syncretic charms: Ukko, the god of thunder, and the Virgin Mary. The wool drifted to a "heathen cape, to a point without a name" – namelessness in charms signifying otherworldliness (Tarkka 2013, 417). The Virgin Mary, accompanied by "Kuihkamo, the king of the forest", gathered the tufts in her hems and carried them to the forest where she shaped them into a bear (SKVR VI2: 5413; see Piludu 2019, 145).

However, there are also charms in my materials that ascribe evil origins to the bear or associate it with evil. A charm from Kivijärvi in central Finland calls the bear a "*hiisi* bear" (*hiien Karhu*), and *mörkö* (bugbear), the latter word being a common circumlocution for a bear. The multifarious term *hiisi* has here a negative, devilish denotation (see Siikala 2008, 148). In this charm, the bear was created as an unruly predator, who did lots of damage but gave a pledge promising to behave himself (SKVR VI2: 1303; see below).

The ambivalent view of the bear's birth comes forth also in a Forest Finnish charm titled *Karhun emu* (The bear's mother). Although the bear was born by the honoured Kati (St Catherine),[1] a couple of lines indicate that the person who bore the bruin out of wedlock was "a bad person". At the end, the enchanter banished the bear to the otherworld "across the nine seas" (SKVR VII5: 315). Thus, there is an implication that the bear was considered an illegitimate child and a bad creature who was to be sent away, just as Sarmela argues.

Moreover, in a Forest Finnish charm for healing wolf bites, Kati also appears as the wolf's mother (SKVR VII5, 206).[2] This is logical since some of the Forest Finnish charms seem to have been used interchangeably against the wolf as well as the bear. Their interchangeability may indicate the decline of the bear's status; the bear now being seen as a beast of prey on par with the wolf.

In Finnish settlements around the Norwegian-Swedish border area, where livestock-keeping gained substantial importance much earlier than in their original homelands in Savonia and northern Tavastia, the bear may have lost its honoured position quicker and have come to be seen as a beast of prey. As Karhu mentions, the golden chains of the cradle by which the heavenly bear was once lowered down onto earth were now used to shackle it in a charm (Karhu 1947, 322; SKVR VII5: 315a).

However, if we look at the motifs of the wolf's birth in the Savonian data (four items), the difference between the valorisation of these two animals becomes more striking. The wolf's dishonourable origin is pictured as quite unequivocal. A spell from 18th-century Savonia mentions Lovetar, "the harlot, mistress of the Northland" as the mother of wolf, who carried "the dog" in her womb (SKVR VI2: 5454, see also VI2: 5455–5457). Lovetar usually appears as the mother of the nine diseases (Haavio 2020, 391), and the noun "dog" in charms often personifies an evil agent, a carrier of malevolent sorcery.

It is quite logical that, in the charms used for the protection of the cattle, the bear was pictured as bad, a potential "cow-eater", who was to be chased away. Still, it does not imply that the bear of the charms was perceived as inherently evil, even though the human-bear relationship had gradually grown more distant. The bear was no longer related to humans by a close family bond, but increasingly through an imagined common Christian ancestry or a membership in a common Christian community. A charm from Nurmes in Finnish Karelia said the bear was "of Eve's and Adam's lineage", whereas some Savo-Karelian bear songs describe a bear's baptismal, whereby it joined the Christian community (Siikala 2008, 144; Piludu 2019, 149–151). In a charm from Kiuruvesi, northern

1 In vernacular Christianity, Kati was the protector of livestock, especially sheep (Järvinen 2016, 434).
2 Other female spirits and saints appear in this role in the incantations, too.

Savonia, the Virgin Mary herself acted as godmother, bringing the bear to be baptised in the River Jordan (SKVR VI2: 5413). Thus, the bear was not excluded from the Christian world, as Sarmela argues was the case in western Finnish agrarian communities, but through baptism came to be seen as part of Creation and the moral community of humans.

Negotiating space: Forest – village

Although the purpose of the cattle charms was to ward off the bear, they still recognised the forest as its legitimate territory – the bear was the forest's cattle according to a charm from Värmland (SKVR VII5: 329) –, whereas the village was designated as the domain of humans and their livestock. However, in everyday life it was not possible to uphold a stable boundary between the forest and the village. It was constantly crossed from both sides: During the outdoor season the cattle were pastured in the forestland, and bears paid unwelcome visits to the village (Tarkka 2013, 356–357).

Since bears, humans, and their livestock partly shared the same territory, the cattle-keepers sought a way to a peaceful co-existence with the bear. The spells clearly established the forest as the rightful domain for a bear to roam, and they negotiated the boundary between the bear's and the cattle's territory. They also defined the suitable undertakings for a bear in each of the spheres (SKVR IX4: 1341):

Pysy kouko kotonasi,	Kouko, stay in your home, /
Karhu ole kankahalla,	Bear, remain in the moorlands,
Älä eti elukoita,	Don't look for the livestock,
Älä kato karjan päälle,	Don't covet the cattle,
5 Syö siellä muuramia	Eat the cloudberries,
Sekä kaiva karpaloita. […]	and gather cranberries. […]
10 Kiskahele kiviä,	pull up stones,
Sekä kaiva kantosia.	dig into tree stumps.
Tormaa toukan päälle	Charge at maggots,
Sekä kuovi kusiainen.	And dig up ants./
Ole peto pesässäsi,	Stay, you beast, in your lair,
15 Sekä kaiva kanarvia,	And dig heathers,
Makoa maan kolossa. […]	lay in a hole in the ground […].
20 Oo koira kotonasi,	Stay dog at your home,
Elä emäs kanssa, […]	Live with your mother, […]
Pysy Tapion takana,	Stay behind Tapio,
Metän tummun turpahissa.	in the forest granny's turfs.
Ei ole sinun osasi	It is not your business
35 Olla ihmisten ilmossa,	To stay in the people's world,
Eikä kaivaa karjasia.	Neither to covet the cattle.

The charms could also mark the boundary in other ways. A Forest Finnish charm asked the Virgin Mary to keep "the forest's cattle" apart from the farmstead's cattle and to raise an iron fence – a standard protective device of the charms – around the cattle. The charm also pleaded with the bear itself, addressing it as "the Golden King of the Forest", asking him to leave "the horn-heads" and "milk-givers" in peace "in the armpit of the forest" (SKVR VII5: 329). Another charm courteously asked a she-bear to leave the human domain and move to "Pohja", where there was lots of space for proper bear activities, e.g. digging ant hills, etc. It also specified the consequences should she cross the line to the human domain: The forest would hold her firmly in its grip by letting her sink into the marsh up to her waist. The variant that addressed a male bear opens respectfully, calling him the "Golden King of the Forest", but then strikes a harsher note. The bear was sent to the far end of Pohjola with wordings common in exorcisms: "There is boneless meat / headless fish, / where a crone is hung / the elks are shackled there", and where there were ant hills and sand for him to dig (SKVR VII2: 5425).

In a charm from the end of the 17th century, the bear was sent back to its home in the centre of the forest (*metsän navolle*). In this mythic landscape there were "golden ant hills, pots of honey" – foodstuff that was the proper diet for a bear, which had also nourished its parents. The tone of the formula is quite conciliatory and closes with the standard wish: "Leave the cattle in peace, live in amity (*sovinto*) with the dung-thighs" (SKVR VII2: 5434).

In some Savo-Karelian incantations, we can find a bear's pledge theme; a theme that reminds us of the myths among the arctic hunters where the heavenly god stipulated the rules for the bear's behaviour, its diet, the terms of its hunting and the rules for the observance of bear ceremonies (Honko 1993, 125). A charm recorded in Kivijärvi, central Finland, in 1884 laid out the terms for a bear's visit and the hunting of it. The incantation opens with a long description of the bear's creation, the forging of it into a fierce animal, a "*hiisi* bear", a *mörkö* (bugbear) in a smithy, who through its rage created great turmoil. But suddenly a giant slapped the raging bear on the ear, and the bear took a pledge (SKVR IX4: 1303):

[…] "Anna olla minun täällä,	[…] "Let me stay here,
90 Älä minua ahista	Don't harass me. /
Minä asun metsikössä,	I will live in the forest,
Sekä kuusikossa kumajan,	in spruce forest I will rumble, /
En tule enää sinulle	I won't come to you
Pahennusta tekemähän.	To do any harm.
95 Jos minä tulen likelles,	Should I come close to you,
Pistä mua keihäällä	hit me with your spear.
Sekä ammu pyssylläsi." […]	And shoot me with your gun." […]

The bear also promised not to ravage forestland but said he would be content to dig for ants and eat berries in the forest. It would prepare a lair where it would stay all winter, keeping away from the human quarters. If the bear broke this deal, humans had the right to slay it: "Should I come close to you, / hit me with your spear. / And shoot me with your gun" (SKVR IX4: 1303). This is the only bear's pledge motif in my materials, but we find the theme in the Karelian ones, too (see Piludu 2019, 149–151).

However, some incantations recognise the forest as a shared space where both bears and humans had the right to move, but the territorial boundaries of the bear's movement in the forest were to be negotiated. As mentioned, many charms acknowledge the bear's right to move in the forestland, sometimes even in the village, but they also delineate the terms for its visits. A charm from Kuopio, northern Savonia, established: "[…] You are allowed, / three times a summer, / to wander about the cattle's land, / but don't you do any harm. […]" (SKVR VI2: 5342).

In numerous incantations, the cowbells' ringing range marked the boundary of the bear's legitimate movements in the territory, instructing it how to avoid the cattle. A cowbell served as a signal for the bear either to leave the area or to avoid the rural areas where the cattle were grazing (SKVR VI2, 5433):

Kuin sinä kuulet karjan kellon,	When you hear a cowbell,
Kellon helevän hevosen,	pealing of a horsebell,

Elä tule sille mualle!	Stay away from those lands!
Kuin ne menee mäin alate,	When they [the cattle] walk at the foot of a hill,
5 Mene sinä mäkiä myöten,	walk the brow of the hill,
Eli kun [!] ne menee mäkiä myöten,	when they walk on the brow of the hill,
Mene sinä mäin alate!	walk at the foot of the hill!

A ninety lines long charm from Kiuruvesi, northern Savonia, recorded in 1819, appealed to the bear (SKVR VI2: 5413):

[...] Kuinsas kuulet karjan kellon,	[...] When you hear the cowbell,
Helkkävän hevosen kellon,	Ringing of a horsebell,
60 Mänes toisellen mäellen,	Hurry to another hill,
Nouse toisellen norollen,	run to another brook, /
Pane maata mättähälle,	Lay down on turf,
Nurmelle nukahtamaan,	sleep in the grass,
Lyö kaxi kämmentäs	cover your both ears,
65 Kahen puolen korvilleis,	with your two paws,
Tunge turpasi kuloon,	shove your muzzle in dry hay, /
Paina pääsi pöckelöön. [...]	push your head into a decaying tree. [...]

Had a bear crossed the boundary between forest and village, i.e. had it been traced or sighted in the vicinity of a farmhouse or pastures, it was sent back to the forest by magic means. The spell involved a physical manipulation of the bear's traces, for instance lifting the lump of soil with the imprint and turning it to face away from the village (see Rantasalo 1933, II, 801–804). There were also persons who were specialised in warding off predators. In some areas in central Sweden, Finns were renowned for their assumed skills of "turning the traces" (*att vända fjäten*), and chasing away the beasts (Tillhagen 1987, 148–149).

Despite the ambivalence and changes in the forest peasants' attitudes towards the bear, it was still recognised as a person to be respected. Even though the aim of the spells was to send the bear away, the tenor of the incantations was often quite conciliatory, particularly if compared to the treatment of the wolf. The bear was perceived as a person with desires that could subject it to temptations, and it was the wise person's duty to contain them by asking the bear to exercise self-restraint and to curb its craving for flesh and blood.

In order to avoid temptation, the bear was frequently asked in charms and songs to cover its ears so it would not hear the cowbell, and to lock its jawbones. It could also be asked to conceal its fangs and claws in its fur ("wool") to render them harmless, or render its teeth and claws soft as wool. But equally often the hunter or the livestock keeper appealed to a spirit, Hongatar, the Lady of the Pine, or various saints to lock the bear's jaw or raise a protective fence around the cattle. Another common motif was a strap of willow, iron, or copper to bind the jaw (e.g. SKVR VI: 5413). Thus, the bear was treated as a person who needed to harness its desires in order not to harm the cattle.

"Raising the bear" – the bear as a weapon

In a hunting-swidden society, farming entities competed for natural resources (Sarmela 2009, 230), and the bear was sometimes employed as a weapon in this competition. A bear's aggressive behaviour was seen as a deviation from its natural behaviour and was suspected as being magically induced. Thus, a bear who behaved violently, killing cattle or pulling up trees and stumps was suspected of being a so-called *nostokarhu*; that is, a bear that had been "raised", instigated by a human being into a rage (*vimma*, *into*; cf. SKVR VI2: 5446 and 5450; Mörtberg 2011, 93). In the SKVR collection, there are a total of 134 verbal charms from Finland and Karelia that aimed at raising a bear against somebody's cattle. Sixteen of these charms were recorded in Savonia, 22 in northern Tavastia, and two in Värmland. Only a couple of these cases provide us with a more accurate explanation for this hostile act. In Rantasalo's collection, where we find a total of 72 cases of "raising a bear" (besides an occasional wolf), the motive for the act is seldom given. One source mentioned envy as a reason (Rantasalo 1934, 2013). One of the cases in SKVR suggests that a bear was used as an instrument for causing damage or for seeking revenge in disputes concerning cattle (SKVR VI2: 5714).

Savonian court records tell us that cattle gorging themselves on growing crops was a frequent source of conflict between neighbours. The courts, however, seem to have often denied the plaintiffs a recompense for their loss, referring to inadequate fencing (Saloheimo 1990, 220–222). In such cases, the harmed party may have sought to settle the score by magic means. In a couple of records, the owner himself used a bear to punish his own cattle for their having eaten standing crops. The bear did not kill the cows, just "bruised their backs" (SKVR VI2: 6038; cf. also Gottlund 1984, 185).

Some court cases in the Forest Finnish areas in Sweden provide us with contextual information about the bear-raising magic. In a couple of cases, there seems to have been a conflict between neighbours or neighbouring villages over lands or other natural resources. In Bergslagen, in central Sweden, one Matts Larson was accused in 1685 of having summoned a bear to assault a neighbour's cattle. The accusers saw this as an act of aggression as they believed that Larsson tried to obtain economic advantages at their expense.[3]

Also in Torp, Medelpad, in northwest Sweden, a bear was supposedly used as an instrument in a protracted conflict between two Finnish settlements. One Lars Andersson, a widely consulted wise man in the province, was accused at the local court of having summoned a bear against his neighbour's cattle. The bear had slain six cows, an ox and six sheep, thereby causing substantial damage for the victim. Andersson defended himself by blaming one certain Old Majsa who had first put a spell on his livestock, making them dry. In revenge, he "fixed" it so that she lost all of her cattle (Gothe 1993, 77). We also find similar kinds of conflicts of interest in later data from the northwest of Värmland. According to Karin Henriksson (b. 1855), a local man had allegedly sent a bear against his neighbours because of a disagreement over a meadow (Mörtberg 2011, 97).

The court records do not provide us with any detailed information about the methods that these alleged sorcerers used. The later folklore collections, however, make available relatively rich data in this respect. As mentioned, frenzy was considered a sign of a bear being "raised" by a malicious person. A conjurer, when reading his/her charms, accompanied them with acts that imitated a bear's aggressive behaviour, that is, digging ground with one's claws, pulling down trees, etc. (e.g. SKVR VI2: 5418; Rantasalo 1933, 786 §2; 1934, 2013–2014); in other words, he/she was behaving like the raging bear of the charms. This 18th-century verbal charm, noted down most likely in Savonia, summoned a bear to somebody's livestock (SKVR VI2: 5414):

Nouse karhu kankahasta,	Rise bear from the moorland,
Hiedasta hevoisen syöjä,	From sand the horse eater,
Viiasta vihainen kissa,	From a thicket the fierce cat,
Korvesta kovero-koura,	From backwoods the rounded paw,
Karvahassu [!] halmehista,	Hairy paw from a field,

3 Riksarkivet, Grythytte och Hällefors bergslags häradsrätt, Domböcker vid ordinarie ting, SE/ULA/10338/A I a/3 (1681–1693), bildid: C0105090_00001.

Mullikoita murtamahan,	To pester bull calves,
Vasikoita vainomahan,	Calves to hound,
Hevosia haastamahan,	Horses to hassle,
Karjan laumaa kaatamahan!	Cattle herd to fell!

It was a wise person's duty to ward off the threats that could jeopardise the enterprises of a farming unit and to safeguard the group's resources (Sarmela 2009, 230). A summoned bear was dealt with by counter magic; it was "settled down" (*asettaa*). This demanded a special state of mind from the wise person, who induced a frenzied, motoric trance (see Siikala 2002, 242–246). In Pielavesi, northern Savonia, the agitated sorcerer bellowed: "May the evil flee away, / The beast from these pastures, / Out to the wide world!" (SKVR VII2: 5446). Another way of dealing with a summoned bear was to send it back to its sender, or rather, against the person's cattle. We find a total of eight charms in the SKVR to be used for this purpose.

Despite the interpersonal view of causality, which presupposes an intentional agent behind misfortune and accidents, an aggressive bear was exempted from responsibility for its acts. A fierce bear was not seen as a being with the intention to damage, but primarily as a medium used by malicious humans in their conflicts (cf. Tarkka 2013, 360).

Summary

The aim of this paper was to examine the relationship between swidden cultivators and forest peasants and the bear and to investigate what kind of "bearhood" the communicative bear songs and charms produced, and what these tell us about the relationship between bear and man. Was the bear seen as an enemy by the forest peasants and, if so, in what respect? With increasing animal husbandry, the bear and other predators were naturally seen as a threat to forest peasants and their cattle that were pastured in woodlands and forest villages. This development led to profound changes in the human-bear relationship. Even though the bear was still honoured and treated with respect, it increasingly came to be seen as a beast of prey with a less honourable origin than was the case among earlier hunters. The purpose of the declining ceremonies was no longer to ensure its reincarnation but to send the animal back to its place of origin, Pohjola, the otherworldly enchanted forest.

In spite of these overall changes in the attitudes towards the bear, the songs and charms, which were largely collected during the 19th century, establish an intersubjective and relational stance between bears and humans. Through this communication, the bear was established as a person who was perceived as an intentional, sentient being with the agency and capacity to make choices, and with whom the cattle-minder could reason. Although the protective charms aimed at sending the bear away from the human domain, they still tended to recognise the forest as its legitimate habitat. The charms negotiated in different ways the boundaries for a bear's movements in a space that was shared by humans and bears. Moreover, our sources indicate that the bear's aggressive behaviour was considered unnatural and was suspected of being induced by malevolent people. Thus, the animal was not held responsible for its aggressive behaviour. The claim that the bear was just seen as a beast of prey by livestock keepers therefore needs to be more nuanced.

Bibliography

Bird-David 1999: N. Bird-David, 'Animism' Revisited: Personhood, Environment, and Relational Epistemology. Current Anthropology 40, 1999, 67–91.

Bladh 1995: G. Bladh, Finnskogens landskap och människor under fyra sekler. Doctoral thesis, Högskolan i Karlstad (Karlstad 1995).

GOTHE 1993: R. GOTHE, Från trolldomstro till kristendom. Studier rörande det kulturella tillståndet bland skogsfinnarna i Sverige under 16–1700-talen (Hjältanstorp, Mora 1993; originally published 1943).

GOTTLUND 1984: C. A. GOTTLUND, Dagbok öfver dess Resor på Finnskogarne i Dalarne, Helsingland och Vermland år 1817 (Falun 1984).

HAAVIO 2020: M. HAAVIO, Suomalainen mytologia (Helsinki 2020; originally published 1967).

HALLOWELL 1926: I. A. HALLOWELL, Bear ceremonialism in the northern hemisphere. American Anthropologist 28, 1926, 1–175.

HAUTALA 1960: J. HAUTALA, Sanan mahti. Jumin Keko. In: Tutkielmia kansanrunoustieteen alalta (Helsinki 1960) 7–42.

HONKO 1993: L. HONKO, Hunting. In: L. Honko/S. Timonen/M. Branch (eds.), The Great Bear. A Thematic Anthology of Oral Poetry in the Finno-Ugrian Languages (Helsinki 1993) 117–140.

JÄRVINEN 2016: I.-R. JÄRVINEN, Transformations of Saint Catherine of Alexandria in Finnish Vernacular Poetry and Rituals. In: T. M. S. Lehtonen/L. Kaljundi (eds.), Reforming Texts, Music, and Church Art in Early Modern North (Amsterdam 2016) 421–447.

KARHU 1947: J. KARHU, Karhun synty. Vertaileva tutkimus (Mikkeli 1947).

MORRISON 2000: K. MORRISON, The Cosmos as Intersubjective: Native American other-than-human persons. In: G. Harvey (ed.), Indigenous Religions. A Companion (London, New York 2000) 23–36.

MÖRTBERG 2011: T. M. MÖRTBERG, Matti Mörtbergs värmlandsfinska uppteckningar. Edited by T. Söder. Acta Academiae Regiae Gustavi Adolphi CXVIII (Uppsala 2011).

NIRVI 1944: R. E. NIRVI, Sanakieltoja ja niihin liittyviä kielenilmiöitä itämerensuomalaisissa kielissä. Suomalaisen Kirjallisuuden Seura (Helsinki 1944).

PILUDU 2019: V. M. PILUDU, The Forestland's Guests. Mythical Landscapes, Personhood, and Gender in the Finno-Karelian Bear Ceremonialism (Helsinki 2019).

RANTASALO 1933: A. V. Rantasalo, Suomen kansan muinaisia taikoja IV: Karjataikoja I and II. Suomalaisen Kirjallisuuden Seura (Helsinki 1933).

RANTASALO 1934: A. V. Rantasalo, Suomen kansan muinaisia taikoja IV: Karjataikoja III. Suomalaisen Kirjallisuuden Seura (Helsinki 1934).

SALOHEIMO 1990: V. SALOHEIMO, Savon historia II:2. Savo suurvallan valjaissa 1619–1721. Kustannuskiila Oy (Kuopio 1990).

SARMELA 2009: M. SARMELA, Finnish Folklore Atlas (Helsinki 2009).

SIIKALA 2002: A.-L. SIIKALA, Mythic Images and Shamanism. A Perspective on Kalevala Poetry (Helsinki 2002).

SIIKALA 2008: A.-L. SIIKALA, Uskomusmaailma. In: R. Räsänen (ed.), Savo ja sen kansa. Suomalaisen Kirjallisuuden Seura (Helsinki 2008) 110–186.

SKVR VI2: SAVON RUNOT. Suomalaisen Kirjallisuuden Seura. https://skvr.fi.

SKVR VII5: SKANDIVANIAN METSÄSUOMALAISTEN RUNOT. Suomalaisen Kirjallisuuden Seura. https://skvr.fi.

SKVR IX4: POHJOIS-HÄMEEN LOITSUT. Suomalaisen Kirjallisuuden Seura. https://skvr.fi.

SMSK 2022: SUOMEN MURTEIDEN SANAKIRJA. Kotimaisten kielten keskuksen verkkojulkaisuja 30. URN:NBN:fi: kotus-201110. https://kaino.kotus.fi/sms/.

SOININEN 1974: A.M. SOININEN, Vanha maataloutemme. Maatalous ja maatalousväestö Suomessa perinnäisen maatalouden loppukaudella 1720-luvulta 1870-luvulle. Historiallisia tutkimuksia 96 (Helsinki 1974).

TARKKA 1994: L. TARKKA, Metsolan merkki – metsän olento ja kuva vienalaisrunostossa. In: P. Laaksonen/S. L. Mettomäki (eds.) Metsä ja metsänviljaa (Helsinki 1994).

TARKKA 2013: L. TARKKA, Songs of the Border People: Genre, Reflexivity, and Performance in Karelian Oral Poetry (Helsinki 2013).

TILLHAGEN 1987: C.-H. TILLHAGEN, Allmogejakt i Sverige (Stockholm 1987).

WEDIN 2007: M. WEDIN, Den skogsfinska kolonisationen i Norrland (Falun 2007).

WIRILANDER 1960: K. WIRILANDER, Savon historia III. Savo kaskisavujen kautena 1721–1870 (Kuopio 1960).

Associate Prof. Marja-Liisa Keinänen
Department of History of Religions
Stockholm University
Stockholm
Sweden
marja-liisa.keinanen@rel.su.se

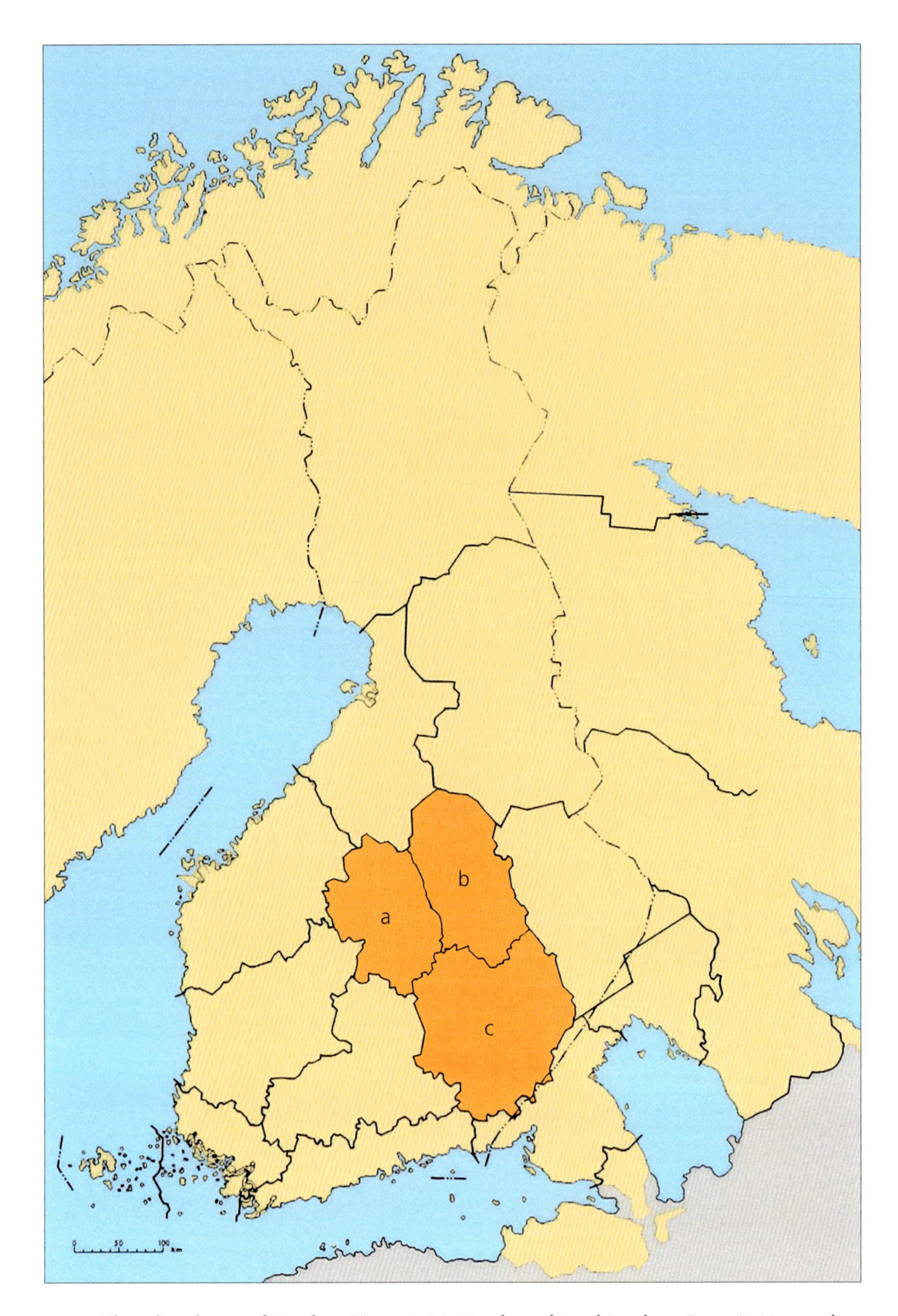

Fig. 1. The cultural areas of Northern Tavastia (a), Northern (b) and Southern Savonia (c) according to The Finnish Literature Society's mapping (map C. Lux-Kannenberg, ZBSA, after a template by M. Sarmela).

Fig. 2. Värmland in southwest Sweden. Östmark parish and the surrounding area on both sides of the border between Sweden and Norway had substantial Forest Finnish settlements (map C. Lux-Kannenberg, ZBSA, after a template).

Karhurokka – traditional bear meat soup and other bear meat recipes from Finland

By Tuija Kirkinen

Keywords: Brown bear, food culture, recipes, Finland, historical time

Abstract: In present-day Finland, bear meat is served in special restaurants as an exotic delicacy. However, the bear has been a valuable game animal in Finland for ages, and the ritual hunt and the sharing of its meat among villagers have been documented in ethnographic sources. This paper summarises the preparing of bear meat soup in the ritual feast and presents two modern recipes, which carry the memories of historical bear meals.

Introduction

The bear is a large predator, with a weight of *c.* 60–300 kg, which provides its hunter with a large amount of meat. This meat is dark, low-fat and coarse-grained, and its strong taste varies according to the animal's age and the food it has consumed during its lifetime. In general, a bear carcass is treated in the same way as a pig's, and it needs to be gutted soon after the kill. The meat has to be cooked properly because there might be trichinae (*Trichinella spiralis*) in its muscles (Kolmonen 1988; Aaltonen/Arkko 1998).

In present-day Finland, bear meat is served in special restaurants as an exotic delicacy. However, the bear has been a valuable game animal in Finland for ages. Even during the 19th century and at the beginning of the 20th century, it was hunted in its winter den in late February or March, when there was a shortage of food. The act of hunting has been described in detail in Finnish epic sources, which carry legacies of pre-Christian rituals and mythology (see Piludu, this volume). The core of this tradition was the idea of the bear's divine origin and its special relationship to humans. Therefore, the respectful killing and eating of the bear needed to be controlled by rituals. The ceremony culminated in a feast, in which bear meat soup, *karhurokka*, was shared and enjoyed (see Haavio 1967, 16–41; Honko 1993; Tarkka 2005, 272–274; Pentikäinen 2007, 65; Krohn 2008, 146–164; Sarmela 2009, 79–94; Siikala 2012, 368–370).

The earliest literary reference to the eating of bear meat is from 1320 in the Black Book of the Turku Cathedral (*Registrum ecclesiae Aboensis*), describing that some of the priests used to collect bear meat, i.e. bear shoulder or smoked bear-gammon, called *böste*, for taxation (Aaltonen/Arkko 1998, 168; Seppälä 2009, 57–58, 134). At the same time, bear skins, donated by hunters, were used by the Church as carpets in front of the altar (Korhonen 1982a; b; see also Jahnsen, and Korhonen, this volume). These acts might be connected to the aim of the Church to secularise the bear, which was valued in folk religion as a possessor of supra-normal qualities (Korhonen 1982a; b; also in Greenland: see Østergård 2009, 120–121).

Bear and Human: Facets of a Multi-Layered Relationship from Past to Recent Times, with Emphasis on Northern Europe, ed. by Oliver Grimm (Turnhout: Brepols, 2023), pp. 777–780 BREPOLS PUBLISHERS 10.1484.M.TANE-EB.5.134361

The slaying of the bear in its winter den was considered a heroic act, and bear killers were honoured and admired. Alongside the modernisation of the world at the end of the 19th and the beginning of the 20th century, big game hunting became more of a sport for the upper class in Finland, Russia, and Germany. For example, Aleksander II of Russia visited Finland for bear hunting several times during the 1870s (Peltonen 2020, 129–133). Gradually, environmentalism changed the attitudes against the killing of the bear in its den as well as the use of foothold traps, which were banned in 1964 (Peltonen 2020, 133–134). During the hunting season in 2020/2021, a total of 342 bears were killed[1] and their meat was cooked by hunters or sold to restaurants. The current bear population of Finland amounts to *c.* 2,000 individuals (2,020–2,130 animals before the hunting season in autumn 2019[2]).

In the following, the traditional way of sharing bear meat soup, *karhurokka*, as well as two modern recipes are presented. The first one of these, bear palms (Fig. 1), might have similarities with the bear feast tradition, in which it is mentioned that paws were eaten. The second one, *böste* (bear ham), refers to the medieval way of smoking bear ham, as mentioned above. However, the role of the bear and its consumption among the Sami are omitted from this paper.

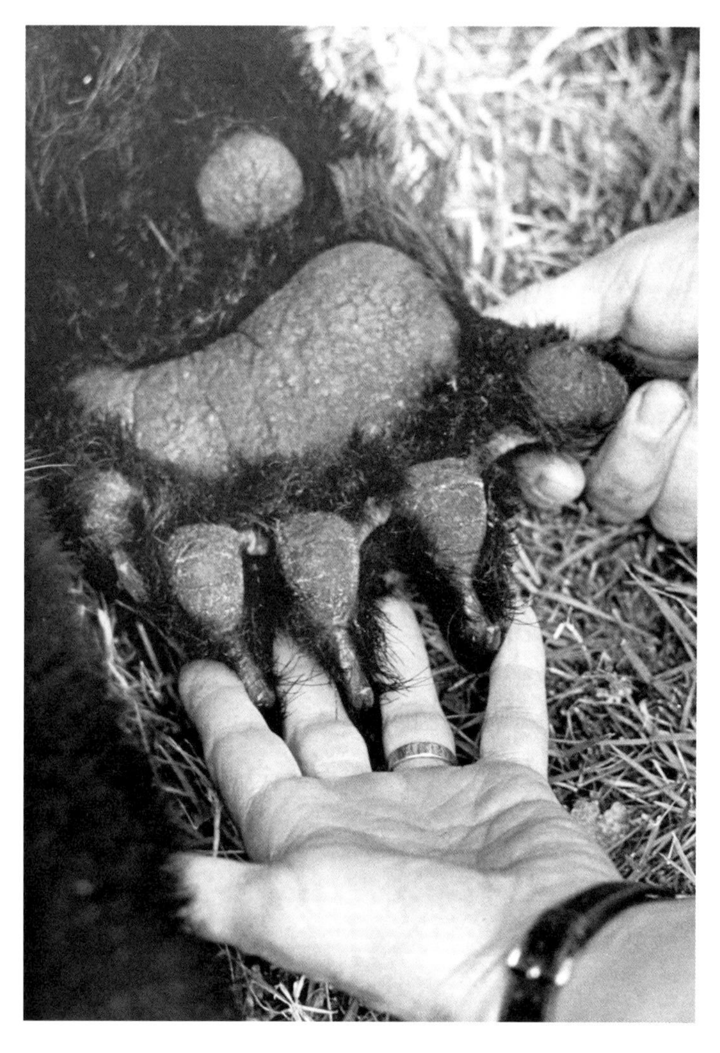

Fig. 1. Bear palm (photo E. Tuominen, The Hunting Museum of Finland, photo archive).

1 Cf. oma.riista.fi/static/bear/bear.html (accessed: 01.03.2021).

2 Cf. suurpedot.fi/lajit/karhu/esiintyminen-ja-maara.html (Finnish), or: largecarnivores.fi/species/brown-bear/number-and-distribution-of-bears.html (English) (accessed: 01.03.2021).

Traditional bear meat soup *karhurokka*

The bear hunting ritual has been described in its entirety e.g. by Sarmela (2009, 80–88) and by Piludu (this volume). Here, the focus is on the consuming of bear meat and the treatment of the carcass.

The ritual slaying of the bear took place in late February or March, in its winter den where the bear hibernated, possibly with its cubs. The den was first identified by men who then marked the location soon after the bear had retired to sleep. After the kill, the sharing of bear meat and skin followed a hierarchy in which the owner of the land on which the den was situated and the slayer of the bear got the most valuable parts. However, a bear is a big animal, and the sharing of food was a communal act in which all villagers had their part.

After the ritual killing (see further in Sarmela 2009), the bear was carried to the village and skinned. This moment was honoured by the women, who expressed mourning by singing dirges in the same way as in the case of human funerals or weddings (Tarkka 2005, 273; cf. Pentikäinen 2007, 81). The cutting of the bear's meat needed to be done carefully without breaking the bones, which were meant to be returned to the circle of life in the forest at the end of the feast (Sarmela 2009, 80–81, 85). The following description is based on M. Sarmela (2009, 79–106): In the hunting feast, the bear was the guest of honor, which was marked by placing the skin at the end of the table. A meal the sources especially mention is soup, which was made of the head and sometimes also the paws of the bear. The head was considered the most valued part, and its nose, ears, and eyes were considered to have special powers, i.e. the bear's senses of smell, sight, and hearing. Accordingly, by eating the bear's brains, its wisdom was assimilated by the consumers. By eating the paws, the bear's power was transferred to the consumer and, in the case of the claws, their sharpness.

After cooking, the head was served on a table and all the meat was picked off it until the skull was clean. We do not have any detailed information about the ingredients of the soup other than that it was forbidden to add agricultural products, such as cattle meat or vegetables, to it. This might originate from the idea that the bear was a threat to cattle and that these worlds, domestic and wild, were meant to be kept separate from each other. At the meal, the skull dish was placed at the top of the table, followed by the other bear dishes in anatomical order. The head was consumed by the men who had slain the bear, and the rest of the bear was shared between the villagers. Finally, all bones were collected and carried to a forest, where the skull was placed in a sacred pine and the bones were buried under it.

Two modern bear meat recipes

Bear palms

The edible palms can be cooked with flavourings (e.g. herbs, juniper berries). Others marinate the palms before cooking. For example, sweet and sour spice broth made with honey softens the paws before cooking.

1. Skin the bear's palms.
2. Crush the berries and mix all the ingredients.
3. Rinse the palms and place them in spice broth for two days.
4. Put the palms, water and spices in an ovenproof dish. Cook the palms in the oven at 150 degrees until their cartilage is soft, for at least four, but maybe even seven hours. It is important for the taste that the finger cartilages have softened.
5. Cut the fingers apart. You can offer them stewed in this way or fry them in a pan with butter. One can also marinate them before stewing (after Kolmonen 1988, 113–114).

Böste (bear ham)
The bear ham is removed from the carcass in the same way as pork ham; the membranes are carefully removed by scraping. The ham is marinated for 24 hours and must be turned frequently. It is then placed in a baking bag with other ingredients strained from the marinade and cooked in an oven at 245 degrees for 80 minutes per kilogram. The frying broth is retained for the sauce. Serve with potatoes, together with cranberry and rowanberry jelly. You can also smoke the ham in the sauna slowly and enjoy it in the same way as the churchmen did centuries ago (after Aaltonen/Arkko 1998, 170).

Acknowledgements

I am deeply grateful to Pekka Allonen, Taru Liukkonen, Perttu Matero, and Jukka Peltonen (all: The Hunting Museum of Finland), and Maarit Knuuttila, Teppo Korhonen, and Johanna Mäkelä (all: University of Helsinki) for their kind help with this paper.

Bibliography

Aaltonen/Arkko 1998: T. Aaltonen/M. Arkko, Eränkävijän ruokakirja (Helsinki 1998).

Haavio 1967: M. Haavio, Suomalainen mytologia (Porvoo 1967).

Honko 1993: L. Honko, Hunting: Introduction. In: L. Honko/S. Timonen/M. Branch (eds.), The Great Bear (Helsinki 1993) 117–140.

Kolmonen 1988: J. Kolmonen, Jaakon parhaat erikoiset (Helsinki 1988).

Korhonen 1982a: T. Korhonen, Saaliseläimen talja kirkkouhrina. Suomen Museo 89, 1982, 45–68.

Korhonen 1982b: T. Korhonen, Saaliseläimen taljan ja kotieläimen vuodan käyttö uhreina. Suomen antropologi 3, 1982, 96–121.

Krohn 2008: K. Krohn, Suomen suvun uskonnot. Suomalaisten runojen uskonto (Helsinki 2008).

Peltonen 2020: J. Peltonen, Karhunkierrosten myynti käynnisti metsästysmatkailun. In: T. Kirkinen (ed.), Kar jalan eräperinne. Luovutetun Karjalan metsästyksen ja kalastuksen kulttuurihistoria (Helsinki 2020) 128–134.

Pentikäinen 2007: J. Pentikäinen, Golden King of the Forest: the Lore of the Northern Bear (Helsinki 2007).

Sarmela 2009: M. Sarmela, Finnish Folklore Atlas. Ethnic Culture in Finland 2 (partially revised version in English). Suomalaisen Kirjallisuuden Seura 587 (Helsinki 2009). Published first in Finnish, 1994: https://www.sarmela.net/_files/200000116-8d4a98e455/folkloreatlas.pdf.

Seppälä 2009: S. Seppälä, Viljana, nahkoina, kapakalana. Talonpoikien maksamat kruununverot Suomessa vuosina 1539–1609. Bibliotheca Historica 125 (Helsinki 2009).

Siikala 2012: A.-L. Siikala, Itämerensuomalaisten mytologia. Suomalaisen Kirjallisuuden Seuran toimituksia 1388 (Helsinki 2012).

Tarkka 2005: L. Tarkka, Rajarahvaan laulu: Tutkimus Vuokkiniemen Kalevalamittaisesta runokulttuurista 1821–1921. Suomalaisen Kirjallisuuden Seuran toimituksia 1033 (Helsinki 2005).

Østergård 2009: E. Østergård, Woven into the Earth. Textiles from Norse Greenland (Aarhus [2]2009).

Dr. Tuija Kirkinen
Department of Cultures, Archaeology
University of Helsinki
Helsinki
Finland
tuija.kirkinen@helsinki.fi

Bear skins as a church offering

By Teppo Korhonen

Keywords: Finland, bear skin, hunting prey, church offering, saints protecting cattle

Abstract: In the north, it was long legal to hunt bears anytime and anywhere. From the mid-Middle Ages until 1891, livestock owners were obligated to participate in common hunts, which comprised the whole parish and were led by a hunt bailiff. Until 1640, the bear skins acquired in these hunts could be donated to the church, but after this they belonged to the hunt bailiff of the jurisdictional district. According to Olaus Magnus, the peasantry of the Nordic countries promised the skins to the church to avoid being savaged by the beasts. It seems that the skins were given in the hope that bears would not kill the farmers' domestic animals. The calendar of the saints included about ten names whose bearers were thought to be able to control the savaging beasts, or to increase the well-being of the livestock. St George, St Margaret, and St Michael, who had each defeated a dragon, were seen as more powerful than the others. This article follows the journey of the bear skin, acquired in a hunt, from the pre-Christian bear killing festivities to the altar of the Christian church, to the kneeling stool and the pulpit. There it served mainly as a warmer under the priest's feet in the cold church, but possibly also under the wedding couple's feet to strengthen their wedding vows and, on etymological grounds, later also as the mass chasuble of the priest. At the peak of this tradition, there could be four skins in one and the same church. The Finnish data consists of information from 32 churches. The earliest mention of the custom is from the 1400s, the latest from 1936. There are only a couple of mentions of a bear skin being replaced by a wolf or seal skin.

Introduction

In his *History of the Nordic Peoples* Olaus Magnus explains how the tradition was practiced in the early 1500s (Olaus Magnus 19, XVI:20; transl. by the author of this paper):

When the Nordic people leave their remote villages to go to church, they take only a crossbow, a sword and an axe as weapons with them. The bow is used to defend them against wild animals, most of all huge bears and greedy wolves, which during the three months of January, February and March are usually wilder than in other seasons. And before leaving they have a habit of giving God a promise, which they also honourably keep, to give the skins of the wild animals that they manage to kill as offerings to the church to be used as a cover for the altar stool under the priest's feet when he is performing the mass. The custom of priests standing on a bear skin in the freezing cold when reading the mass stems from that. If, instead, they have got a wolf, a lynx, a fox or some other wild animal with a net or a trap, the money from the skin is piously spent on wax candles.

Bear and Human: Facets of a Multi-Layered Relationship from Past to Recent Times, with Emphasis on Northern Europe, ed. by Oliver Grimm
(Turnhout: Brepols, 2023), pp. 781–794 BREPOLS PUBLISHERS 10.1484.M.TANE-EB.5.134362

punki in the 1700s). All in all, there are 14 data entries that locate a skin in a church: Six of them are from the 1600s, seven from the following century, and the most recent one is from 1893.

The information about the more exact locations of the skins reveals that they were located *somewhere around the altar* (Pertteli, 1690; Luumäki, 1734). The more exact expressions are divided into: *in front of the altar* (Piikkiö, 1707; Aura, 1730; Kisko, 1758; Dalarna; Boëtius 1932, 124) or *as an altar rug* (Loimaa, 1685; Masku, 1701; Maaria, 1723; Jääski 1893; Salenius 1870, 97; Oja 1938, 25), and they are mentioned five times as being, generally, *on the altar* (e.g. Lieto, 1730; Riska 1964, 267), in the Dalarna Mickelsfjärd chapel in 1744 *fanns en biörn hud inuti altaret* (i.e. "a bear skin on the altar"; Boëtius 1932, 144), and *on the steps of the altar* (Valkjärvi; Salenius 1870, 97). In a judicial conflict that took place in Turku in 1685, professional furriers accused a certain furrier of acting without a license, threatening to take a bear skin away from him. The accused then promised to give the skin to the poor house to act as the cover of the altar stool (Pylkkänen 1970, 107). A skin is twice mentioned as being *next to the altar* (Muurla, 1791, Uusikirkko Vpl., 1767; Oja 1938, 25; Riska 1966, 116, 175), as well as being the cover of the *kneeling stool* of the altar, but as being *in the pulpit* only once (Lieto, 1730; Angelniemi, 1757; Riska 1964, 267, 285). A few churches had several skins – up to four (Elimäki, 1745; Kokkonen 1931, 80), which was enough for the altar, the kneeling stool and the pulpit. Perhaps the rug near the altar was also situated under the bride and bridegroom's feet. This practice derived from old beliefs in which the vows sworn standing on a bear skin were considered stronger than other vows. Skins were replaced in the 1900s by woven *rya* rugs, which were donated by home societies and handicraft societies to churches (Korhonen 1999, 160–173).

Also, the priest's original mass gown could be an animal's furry skin, thrown over his shoulders. When, in 1927, Toivo Kaukoranta pursued the meaning of the Finnish word *hakuli*, of which the most commonly known seemed to be "a child's single thin hair" or "a sheep's first wool", he simultaneously, hesitatingly, introduced an etymology for the word. The root is the Gothic *hakuls* "an animal's fleece, skin", which also was the root for the ancient Swedish *hakul* and also e.g. *messhakel* (mass chasuble), and from these the modern Swedish word *mässhake* or "mass chasuble" was derived (Kaukoranta 1927, 55). To support this assumption, we can use a certain verse from White Sea Karelia (Viena Karelia). Since it is a bear poem, also discussing the fate of the bear skin, we can notice that apparently not any skin would do as a chasuble; it had to be the most powerful kind (Krohn 1914, 160):

> You won't be put in a bad place
> but in a good place,
> as the bishop's long sleeves,
> as the clothes of the authorities.

In this connection, it is also interesting to note that in Siberia the attire of an haruspex included bear claws, bones and skins until recently (Harva 1933, 339, 345). In the Late Middle Ages, bear skins were accepted as taxes and also bought by the crown but, nevertheless, there is no information on them from Finland in connection to the attire of the aristocracy; the same also applies to the following Renaissance period (Pylkkänen 1956, 87, 96; Jokipii 1967, 25–27). It is not until the Baroque period that we know that textiles made of bear skins belonged, in particular, to the aristocracy's winter sleigh accessories, such as fur rugs, covers, foot bags, muffs, mittens and even caps (NBA: HTKA Mäntyharju; Pylkkänen 1970, 107; Korhonen 2020, 62–65). Skins were also given to the priest to be laid on the guest room floor. Among the peasantry, pieces of bear skin were most commonly seen in eastern Finland on top of the shaft bow of a horse harness, a so-called *umber*, which originally had a protective, magical purpose (Vilkuna 1940, 43–54). Possibly the men from Pyhtää in the aforementioned case were cutting their bear skin for this kind of use when the sub-hunt master, Jordan, interfered in the matter.

According to Olaus Magnus, the bear skins that were not donated to the church were sold, and wax candles were bought with the resulting funds (Olaus Magnus 1916, XVI:20). However, the Raisio church bought a wolf skin to serve as the altar rug for six bucks in 1670. A little more than half a century later (1728), a wolf skin is mentioned as being the cover of the kneeling stool of the Kaarina church. The wolf skin donated by the Luonnonmaa peasants was also in similar use ten years later (1738) in the Naantali church, but was probably abandoned in 1766. At that time, the widow of a magistrate, Maria Åsten, donated a red damask spread for the stool; this was trimmed with silk and gold thread fringes and decorated with the donator's initials and the year 1766. The decline in value of wolf skins can also be seen in a piece of information according to which the wolf skin acquired in a hunt in Parainen in 1752 was sold due to the request of the governor and the money was spent on new gunpowder for beast hunting (Riska 1964, 34; Lilius et al. 1972, 72, 121). In the Kalmar province in the 1700s, the skins of wolves acquired in common hunting were placed in the church chest until they were sold at an auction. The revenue was spent primarily on rewarding the person who had first spotted the beast. It had thus become some kind of bounty (Hammarstedt 1920, 117).

Why were some wolf skins not sold but used in interior church decoration? Perhaps bear skins were not available at the time and people wanted to place some kind of skin in front of the altar. This explanation is partly supported by information from the outer Turku archipelago, where a large seal skin was purchased for the altar floor of the local church in 1733 (Nikula 1973, 226). This use of local, available resources is emphasised among the Sami people, where the most important church sacrificial skin is a deer or reindeer hide (Itkonen 1913; Hammarstedt 1920, 115–116).

The skins of the smallest fur animals also seem to have been given to the church, at least in northern Ostrobothnia and in Perä-Pohjola. In Oulunsalo, even non-locals put sacrificial furs in the branches of the candelabrum on the designated date in 1736. This kind of practice might have been based on the regulation made by Queen Christina, according to which it was permissible and even advisable to remember the House of the Lord and the poor with, for example, candles and furs, when God had saved the person in question from misfortune (Suolahti 1912, 20; 1919, 139).

The decrease in skin entries in the late 1700s and the data on the selling of them must probably be seen partly as an accomplishment of purism. The skins were seen as futility items and emblems of paganism or Catholicism. A certain Bishop Getzelius' letter refers to this, and Swedish visitors commented on it more than once (Suolahti 1912, 20; Hammarstedt 1920, 114–116). Along with new, beautiful textiles, noblemen and women were also able to emphasise their piousness through their donations in this domain as well. This is probably partly also due to the decrease in the beast population, when plantations expanded and the apparently overlarge steady bear population at the time of the Great Northern War was reduced to normal numbers.

The nature of the offering

In the sources I have used, it is not once mentioned that any skin was given in return for the donor's life being saved from a beast attack. Thus there is no evidence of them being given as a thanks offering in the same way we know votive ships have been donated to the church.

Therefore, the offerings were probably given in advance, as a form of preventive magic, at least for the most part. The need for prevention was not directed towards the donor himself but his property, in this case livestock. A former study (Oja 1938) ended in a cautious presumption about bear skins being donated to the church in the name of St Bartholomew. This was, on the one hand, based on the statement that some skins were donated between August 20th and 30th, between which dates St Bartholomew's Day (August 24th, Finnish: *Pärttyli's Day*) happens to be, and, on the other hand, on the fact that the symbol of the apostle in question is his own skin or a skinning knife (Oja

1938, 29–30). But to make such a generalised presumption, the precisely-dated (exact day of donation) material is too scarce. When, however, we look more closely at those four mentions that the presumption was based on, we can observe that in one case (Lohja, 20.8.1646) it is the donation day in question, whereas in the other three cases (Lohja, 27.–29.8.1706; Vahto, 30.8.1727; Paattinen, 23.8.1731) it is the day of the common hunt in question (Oja 1938, 25; Ylikangas 1973, 246). According to the church archives of the Skultuna church in Västmanland: "A bear skin was placed on the altar on the 7th of November in 1673. It originated from a bear that had been caught with great effort and with three men being hurt in the Eckieby forest in the hunt arranged on the 6th of July" (Hammarstedt 1920, 117; Erixon 1921, 90). When we further remember that the skin in Turku Cathedral in the 1400s was on the altar of St George, not in the apsis of St Bartholomew, which was located in the same church, it becomes evident that we need to look at both the dates, the donations and on which days the beasts were hunted, and at the connection of the saints to beasts, especially the bear.

The old laws in Sweden, some of which have also been applied in Finland in some places, ordered everyone to hunt for the bear and the wolf whenever they had caused damage. Of the regional laws, only the more recent one in Western Götaland County defined four fixed bear hunting days in the year: 1) on the fourth day after Easter, 2) on the fourth day after Pentecost, 3) five days after St Michael's mass, and 4) on the fifth day after Christmas.

The country laws did not give any kind of instructions on the date of hunting but apparently the hunt was held at least around Pentecost, because the inspection of the nets used in the hunt was prescribed in the country law of Magnus IV of Sweden to be held at exactly that time, simultaneously with one hunt prescribed in Western Götaland regional law (SLL 1946, 357; Konung Magnus Eriksons Landslag, 156). In the patent granted by Charles IX in 1608, one compulsory hunt was prescribed to be held annually, but without defining the date any more precisely (Pentecost?), and on other days as needed. In the so-called Rosengren's Suggestion related to the patent, it was recommended that hunts be held in autumn and in spring and according to need (Kongl. Stadgar 1706, 126; Rosengren 1864, 381). At some places in the Satakunta region, three hunts were carried out in the autumn of 1628 (Pirkkala ja Karkku, NA nn 2: 20 v. 45).

The hunting regulations of 1647 and 1664 did not contain any kind of statutes about the dates or the number of hunts, and the common realm law of 1734 also acquiesced to just recommend particular activity during the littering time of the beasts (SRL 1934, 87–88). The dates and number of compulsory hunts were the "hot topics" of parliamentary discussions in the late 1600s and the early 1700s, because the sensibleness of them greatly depended on local circumstances. They were legislated in the directives of the governors related to the realm law in 1734. Hunts were allowed to be organised during the busiest working time or in good winter weather only in an extreme emergency (when the beasts had caused great damage); also holidays were prohibited (Samling af Instructioner 1852, 384; 42 §; Korhonen 1977, 57).

Even though it is only randomly possible to verify the attaching of the communal bear hunt to a certain time in the yearly cycle on a legislative level, on the other hand, we know from the old calendar tradition that rites were carried out on certain days to protect the livestock. This is where the saints come in again.

Saints as the protectors of livestock and the enemies of beasts

In the yearly cycle, January was the time when people's traditions began to focus on the bear. When hibernating, the bear was believed to turn over on his side on *Henry's Day* (January 19th/20th), stating "Night halfway gone, hunger in my stomach, a ball of pitch in my rectum". After this, it resumed sucking on its paw and went back to sleep. On Henry's Day, children went around the houses in the village and hit the walls with a beater or mallet to mark the middle of the winter, after which the bells and jingle bells of the livestock were sounded, and metal dishes were rattled and drummed to drive the beasts away from the vicinity. Henry's commemoration day happened to be close to the middle of winter, and the bear turning and the dividing of the winter in two have remained the traditions of that day (VILKUNA 1969, 36–37).

In the eastern Finnish cultural area, the matron hung a cowbell on the coal hook on *Maundy Thursday* and tied a knife, a sickle, shears and coal- and shingle pliers to her waist. Then she went around the yard with an axe and a burning torch in her hand. At the corners of the buildings she hit the log with the end of the axe, saying: "Evil spirit, go to the woods!". With the help of this measure, borrowed from the use of incense during the Catholic church service, people believed that evil forces could not get into the house and yard in the summer to do harm. The lighting of bonfires had the same aim. The Ostrobothnian Easter fires had elements that showed they had another purpose, too, apart from the scaring away of witches; straw – the growth of the field – and leaf-branches gnawed by sheep, i.e. the food of the livestock, were burnt on the bonfires built on the fields. Bonfires were lit in some places in Finland at Shrovetide, at Easter, on Summer Nights, on Ascension Day, at Pentecost, and at Midsummer, the last of which is the oldest and most widely spread of the bonfire lighting days. The most important purposes of the lighting of the annual bonfires were the securing of a good harvest and a plentiful livestock yield (VILKUNA 1969, 88–95; SARMELA 1996, 94–98).

Legends and folk beliefs mention several saints, both men and women, who had the power to prevent beasts from harming the livestock. The nights before the commemoration days of Julian, Justinian, and Tiburtius (April 12th–14th) were known as *Summer Nights*. Shepherds climbed up the hills early in the morning in southeastern Finland to play bugles at the four compass points. In the Savo region, they had *Hollering Nights* when they called out loud: "Dear George, St George, tie up your hounds from Summer Nights to Winter Nights". Simultaneously, they lit bonfires on the hills. The controller of the beasts mentioned in the refrain was St George. He was known by many other names like *Jyrki, Jegor, Yrjö, Yrjänä,* or *Ilja* in Finland. The commemoration day of St George was celebrated on April 23rd in memoriam of his decapitation in 303 by ruling of the emperor, Diocletianus. According to the legend that was popular all over Christendom in the late 1300s, George had defeated a dragon that demanded people and livestock as offerings. Additionally, he recreated a poor plowing man's ox and got a poor widow back her sheep (MANSIKKA 1943, 190; VILKUNA 1947; BRAUN 1964, 283–287). Naturally, such a benefactor was asked to protect the livestock from more familiar beasts, too. There was a bear skin on his altar in Turku Cathedral already in the 1400s.

Verbal pleas, in which the saints were asked to put the wolves and bears in an iron shackle, are familiar from western Finland to Russian Karelia. The custom even spread along with Forest Finns to mid-Scandinavia, where Swedes assumed the custom in some places but instead addressed the plea to Peregrinus (May 16th), whose commemoration day was the same day the livestock was let out into the fields. St George's Day is mentioned as the day the livestock was let out into the fields in central and eastern Finland, Karelia and Ingria. In White Sea Karelia, St George was thought to especially protect horses from bears (MANSIKKA 1943, 185, 191–192, 194; VILKUNA 1969, 122–123).

In Georgia, the cult of St George is as old as Christianity. It spread from Asia Minor to southeastern and eastern Europe. Although St George was, above all else, seen as the lord of the wolves, his might extended to bears as well. The Ukrainian Slavs asked St George on the day the livestock was let out (MANSIKKA 1943, 185):

> You, our brave George
> protect our livestock in the fields
> and outside the fields,
> in the forest and outside of it

from the greedy wolf, from the angry bear,
the treacherous wild beast.

In eastern Finland, the commemoration day of St George was so established as the day the livestock was let out that people asserted: "George will even carry the cows on his back". Although there was still snow on the ground and the water was frozen, the custom was at least ceremonially adhered to by letting the cows out of the cowshed to stretch their legs for a while. Depending on the circumstances, the day they were actually let out could be delayed by a week or even by three weeks, though it would still be on the same weekday as the actual St George's Day. On the same day people in the eastern Finnish cultural area and in East Karelia hung cowbells around their necks and ran around the houses sounding the bells and making other kinds of noise and banging the trees in the nearby woods to shoo away the beasts (Salmelainen 1852, 127–128; Mansikka 1943, 169, 177–178, 190; Vilkuna 1947).

The bear had two commemoration days, one of which was related to birth and the other to killing. As to the saints, the killing day was the same as the birthday, which meant being reborn to a new life that happened at the moment of death. Another commemoration day of the saints was the day of moving their bones. The bones of the bear – the skull especially – were moved to a bear cemetery during a funeral feast and were lifted onto a tree close to good hunting grounds. The memories of the commemoration day of the bear (July 13th) have become nearly completely buried under Christian traditions. An example of this is thought to be the location of the church named after St Margaret on Bear Island in the Ii parish in northern Ostrobothnia, where a pine with a bear skull once stood. The bear killing of the old times, with its festivities and bear weddings, was concentrated on the midwinter and the midsummer festivities. Every year, on these days, people sang certain poems to protect the herdsmen and the livestock from bear damage and to ensure the bear killing would be successful. The successful bear killing was followed by feast poems (Haavio 1967, 457–460; Vilkuna 1968, 176–178; Sarmela 1996, 38–43).

To obfuscate the pagan celebration of the old commemoration day of the bear in the summer (July 13th), the Catholic priesthood relocated *St Margaret's Day* (Finnish: *Marketta*) to a week earlier in the countries of the northern Dominican missionary district in the Baltic region, Finland and Sweden. According to legend, Margaret beat the accosting devil, which was disguised as a dragon, with the help of her cross and her faith. In medieval church murals, the saint was represented fighting the dragon, holding a triple-forked spear or a cross-headed staff that she was cramming in the beast's maw. The people understood the weapon to be a rake because the saint's commemoration day happened to be at the beginning of haymaking time. She was asked, as was St George, to close the mouths of both wolves and bears with her triple-forked spear (Vilkuna 1969, 206–210).

In Estonia, the whole of July was sometimes called "bear month". "Bear day" was the bear's birthday, which was celebrated by singing poems aimed at protecting herdsmen and livestock and by making offerings. The Catholic church was again able to undertake measures to replace bear worship by remembering St Margaret. The church of the Karuse (meaning "bear") jurisdictional district in Estonia (dating to 1271) was dedicated to St Margaret, and July 13th was celebrated here with great festivities. Even after the Reformation in 1683, it was marked with regret in the minutes of one parish that St Margaret's day was continually held sacred in fear of bears otherwise killing the livestock. One description from 1732 states that people avoided working on Margaret's Day/the commemoration day of the bear so the bear would not cause damage. According to the vicar, Christfrid Ganander, the same forbiddance of work applied to Olaf's Day (July 29th) in Ostrobothnia in the late 1700s. Later, the inhabitants of the Virumaa coast believed that they could protect themselves from bear damage by pushing a copper coin as an offering into a slit in the church floor on Margaret's Day (Vilkuna 1968, 174–179, 187–190). Mall Hiiemäe has pointed out that some notes relating to the "Day of the

Bear" (*karusepäev*) are specially related to farming, not to hunting culture. In this connection, she sees, as does K. VILKUNA (1969), the link between St Margaret's Day and the hairy caterpillars that damage cultivated fields (HIIEMÄE 1998, 158–159). Additionally, both in Finland and in Estonia in ancient times the surface of a field and the weed damaged by caterpillars were tended with a multi-spiked wooden harrow. The appearance of the object resembled a bear skin, which gave rise to the Finish name *risukarhi* and, in Estonian, *karuäke* (RÄNK 1955, 39; ANTTILA 1968, 14–23).

St Bartholomew (August 24th) died as a martyr when he was skinned alive. The procedure resulted in the pious man's symbols in church paintings being a large knife and his own skin (HARVA 1935, 14; BRAUN 1964, 118–119). Only two of the medieval Finnish churches (Pertteli, Laitila) were dedicated to St Bartholomew. Murals depicting him can additionally be found in the cathedrals of Hattula, Lohja (with his skin as his attribute; 1646, 1708) and Turku, where his attribute is a skinning knife, and in Rymättylä, Sauvo (1757) and Kumlinge, where his attribute is again the skin. In the painting in the Pyhtää church, the saint does not have a symbol at all. As mentioned, in Turku cathedral, a bear skin had been placed on the altar of St George in the 1400s, although there was also the altar of St Bartholomew. Perhaps this reflected the reciprocal precedence of the altars. Bird hunting started at the earliest on Perttu's (= Bartholomew's) Day, but the hunting of furred animals did not usually start until Simo's Day (October 14th). Overall, the beginning of the hunting season varied within a four-week-period, according to the climatic and geographical circumstances of Finland (VILKUNA 1969, 237–241).

St Anna (September 9th and December 15th), *Annikki* in Finnish, the mother of the Virgin Mary, was – according to folk beliefs – also the matron of the forest and Tapio's (the patron of the woods) maid, and the beasts were believed to be her dogs. Thus she was the female counterpart to St George. The Greek Orthodox people of the Setumaa region in Estonia asked Anna to protect the livestock grazing on the forest fields, because in the eastern church Anna's Day is on September 9th. In western countries, Anna's Day is not until December 15th, when the livestock is already inside, out of the reach of the beasts (KROHN 1914, 203; VILKUNA 1968, 277–283).

St Michael (September 29th), *Mikko* in Finnish, became the lord of the beasts as the third saint along with St George and St Margaret, who had defeated their dragons. He was the counterpart to St George, since on his commemoration day the summer herding of the livestock that had begun on St George's Day ended. He was believed to open the mouths of the wolves that St George (*Jyrki*) had closed in the spring (KROHN 1914, 19l). The ecclesiastical St Michael's celebration happened to be close to the old autumnal concluding celebration of harvest time, and that was why it quickly became rooted in the annual celebration traditions. The herdsmen returning from the pastures with the livestock on St Michael's Eve hooted their horns when arriving in the village, guiding the lead cow, which was equipped with a garland and a cowbell. Also, horses were taken to the courtyard from the outer pastures (VILKUNA 1969, 255–260).

St Birgitta's Day (October 7th) started the inside feeding season for the livestock in some places. According to a medieval legend, the bear was believed to have been born as a result of Birgitta's incompetence – from wool she had thrown into water. The saint thus occupied the place of Hongotar (see PILUDU, this volume) who, based on folk belief, was the bear's new ruler and controller. She got her power from the fact that people used to put the skull of the bear on an old pine, Finnish *honka*. Hongotar is a feminine form of the "owner" of this tree (VILKUNA 1969, 267–168; HAAVIO 1967, 461–462, 464).

St Martin (November 11th) was known among the Swedish-speaking population of Finland as the protector of livestock, along with St Michael. In other parts of Finland, this saint, who lived in the 300s, appeared merely as the protector of horses. He had belonged to the cavalry when he was young, and later converted Franks as bishop of Tours (KROHN 1914, 191; VILKUNA 1969, 293–294).

St Catherine (November 25th) lived in Alexandria in the early 300s. People first tried to torture her to death on a dismembering wheel, which, however, broke due to the power of her prayers. When the maiden was then beheaded with a sword, milk instead of blood burst from her veins. A breaking wheel and an open book were the martyr's attributes, the first of which the people understood as a spinning wheel and the other as the beaters used in carding wool. Catherine thus became the protector of livestock, especially sheep, both in the cowshed and in forest fields (Vilkuna 1969, 298–302).

In continental Europe, *St Nicholas* (December 6th) was believed to make the bear docile and to give it full latitude only on the first three days of September (Fuchs 1924, 250–252).

The most important days of letting out livestock among the old commemoration days were St George's Day (April 23rd, or May 5th, according to the Julian calendar), in eastern Finland and Ingria, and Vappu Day (May 1st), i.e. Sofia's Day in the orthodox calendar, and the Day of the Cross (May 3rd) in the Häme region and partly in the Savo region in the sphere of influence of the old church of the Holy Cross in Hattula. The aforementioned dates were too early in most of Finland and, often, everywhere else as well. The ground was still frozen in some places and the snow had not melted in the shadows of the forest, so the cows could not live on the scarce food outside. When choosing the release day, people also took into account the phase of the moon, the winds and the weekday. So, often corresponding weekdays a week or two later were chosen, provided that they were suitable with respect to other signs.

Eric's Day (May 18th) was already the favoured release day of the livestock in parts of Scandinavia and in the Swedish regions in Finland in the 1700s. Later special days of release were the third and fourth Pentecost days, possibly because they were put into use only after the ecclesiastical nature of these holidays had been removed in 1772. Other dates were the days named after Caroline (May 20th), Urban (May 25th), Wilhelmina (May 28th), Nicodemus (June 1st) and Gustav (June 6th), all of which tell us about either how far to the north these places were located, or, vice versa, intensified inside feeding of the livestock in more recent times, when feed stocks were larger in number and the cows already milked nearly throughout the winter.

The overview of the legends of the saints presented above and the folk beliefs deriving from them show that special days related to beasts and the protection of cattle or the beginning of the hunting season can be observed a few months apart throughout the year. A great many of these days associate the bear with livestock, whereas the connection between the commemoration days and bear hunting remains anecdotal. The explanation lies in the mostly agricultural economy that already prevailed in the Middle Ages and, above all else, in the saints with their legends, who – apart from St Birgitta and St Olaf – were borrowed from southern agricultural countries. Nevertheless, it is useless to speculate about the connection between the donation day of a bear skin and the possible commemoration day of a saint close to it, unless there is a more definite reference to it in the written heritage, such as, for example, the skin being placed on the altar of St George in Turku Cathedral.

The whole village negotiated when the livestock should be let out into the fields. It was a celebrated moment, visible also in the foods and the merrymaking. In Karelia, the matrons prayed when letting the livestock out in the field again for the first time (Salminen 1914, 2):

> Old bearded man of the forest,
> the golden king of the hill,
> please make up with us,
> let's make summer peace,
> this summer of Jesus
> the great summer of God.

My bruin, my bird,
leave the robust bulls
and mucky cows alone.

When you hear the cowbell,
put your head in the turf,
stick a nail in your fur,
so your teeth won't break,
nor your jaws crack.

Virgin Mary
dear merciful mother,
come here if we need you
to protect my gleam,
in the summery lanes.

Like you did in the courtyard,
so do in the livestock pastures.

It is a prayer of the person who lets out the livestock (Salminen 1914, 7–8; Haavio 1967, 17), as is the following poem, written down among the Forest Finns of Värmland (SKVR VII, 5, 332):

George tie up your dogs
with iron chains
with golden chains.
Close your mouth,
gag your mouth.
Hide your nails in the fur,
your teeth in your gums.

The title says the poem was read to protect the livestock from the bear, not the wolf, as is usually the case with St George (Kirchberger 1970). When the transition from hunter-gatherer economy to agriculture began, animal husbandry also became more common, for which, among others, glades created by slash-and-burn clearance were excellent pastures. This procedure was observed in the Battle Axe culture and the Kiukainen culture in Finland (Huurre 1979, 74–77, 87, 99; Siiriäinen 1982) – despite the fact that no domestic animal bones predating the Bronze Age have been found –, which is exactly when a strong increase in the making of bear-headed axes can be seen (Carpelan 1974; 1977). The provision of manure was necessary to settled agriculture; manure was an important product of animal husbandry along with meat, hides and milk. When the livestock population increased, people had to pay even more attention to securing its safety. The significance of the bear as a quarry was emphasised. The securing of the livestock now became an essential task. Bear-headed weapons emerged beside elk-headed ones, among other ancient weapons, but they did not end up in rock paintings, due to the individual nature of the hunting (Taavitsainen 1978, 180, 184).

Instead, old rite conventions – i.e. bear killing celebrations – retained features that derived from the hunter-gatherer period when the brown one was a taboo, the creator, or the son of God in heaven etc., but new ideas intertwined in now, and the bear became the enemy of the livestock. As I see it, the development did not end here, since after the bear population had been reduced in certain areas to occasional individuals, or they had retained their taboos and so were no good as quarries, sacrificial practices shifted to domestic animals. An anticipatory protection magic was created in which the skins of possible domestic quarries of the bear were sacrificed in the sacred groves, or to the church, and feasts were held where their meat was eaten. Such animals were the ox, sheep and reindeer (Korhonen 1999, 125–135; cf. Sarmela 1996, 48–51).

Conclusion

The essential question is how people managed to pass the bear skin off as a church offering instead of a feast artefact. When we mark on the map (Korhonen 1982, 46, 65) the area where the hunt bailiff institution had become established before 1640 – that is, before the skin was assigned to be the compensation of the hunt master – we can notice it covers exactly the regions where we know people donated skins to the church. The most remarkable difference seems to be in the Satakunta region. But this is just an ostensible contradiction, because, although we do not have many facts about skins in churches in this area, the aforementioned complaint of the peasantry of the upper Satakunta region, addressed to the King, proves the custom was commonly also known there. The statutory beast hunting, which had its roots in medieval times and had been revived again in the early 1600s, led by hunt bailiffs, brought bear skins to Finnish churches. Since it was a legal activity, its results, i.e. the fate of the bear skin, could be decided by the authorities. Although legislation does not have instructions on this, it would feel natural to donate the skin to the church on the recommendation of the authorities. Thus, this pagan ritual object could also be assimilated into the ceremonies of the new religion. It is revealing, however, that the skin was simultaneously moved into a subordinate position both symbolically and specificly, relocating it from the pagan "altar", the feast table, to under the feet of the representative of the more powerful religion, the priest.

In the end, the church was indifferent as to which saint's name the peasantry wanted to associate with the skins to protect their livestock. Although such a skin undoubtedly had great significance as a thermal insulation under the priest's feet (Hammarstedt 1920, 115; Oja 1938, 29) in the cold churches of the past – especially since the ecclesiastical proceedings included more kneeling prayers than nowadays – it seems indisputable that this aspect was originally not the only reason for acquiring the skins. Apparently, though, it was this reason that was emphasised later, when the church did not want to stress the sacrificial nature of the skins.

Bibliography

Archival sources

HTKA: Kirkonarkistojen kopiokokoelma

KA: Kihlakunnanoikeuksien renovoidut tuomiokirjat

NA: The National Archives (Kansallisarkisto)

NBA: National Board of Antiquities (Museovirasto)

RR: Registratuurajäljennökset

Secondary sources

Antell 1956: K. Antell, Pernå sockens historia I (Helsingfors 1956).

Anttila 1968: V. Anttila, Äes Suomessa. Suomi 114,1 (Helsinki 1968).

Boëthius 1932: G. Boëthius, Sveriges kyrkor 37, Dalarne 1.3 (Stockholm 1932).

Braun 1964: J. Braun, Tracht und Attribute der Heiligen in der deutschen Kunst (Stuttgart 1964).

Carpelan 1974: C. Carpelan, Hirven- ja karhunpääesineitä Skandinaviasta Uralille. Suomen Museo 1974, 29–88.

Carpelan 1977: C. Carpelan, Älg- och björnhuvudforemål frå Europas nordliga delar. Finskt Museum 1975, 1977, 5–67.

Erixon 1921: S. Erixon, Skultuna bruks historia (Stockholm 1921).

Fuchs 1924: A. D. Fuchs, Beiträge zur Kenntnis des Volksglaubens der Syrjänen. Finnisch-ugrische Forschungen XVI(2/3), 924, 1-78.

Haavio 1967: M. Haavio, Suomalainen Mytologia (Porvoo 1967).

Hamberg/Berlin-Hamberg 1966: G. Hamberg/U. Berlin-Hamberg, Sveriges kyrkor 107. Härjedalen 1.3 (Stockholm 1966).

Hammarstedt 1920: N. E. Hammarstedt, Hudar och skinn som offer. Fataburen 1919, 114–122.

Harva 1933: U. Harva, Altain suvun uskonto (Porvoo 1933).

Harva 1935: U. Harva, Varsinais-Suomen henkistä kansankulttuuria. Varsinais-Suomen historia III,1 (Porvoo 1935).

Hausen 1881–1883: R. Hausen, Bidrag till Finlands historia. Första delen (Helsingfors 1881–1883).

Hiiemäe 1998: M. Hiiemäe, Der estnische Volkskalender. Folklore Fellows Communications CXXII (Helsinki 1988).

Huurre 1979: M. Huurre, 9000 vuotta Suomen esihistoriaa (Keuruu 1979).

Itkonen 1913: I. Itkonen, "Kirkkouhreista" Inarissa. Kotiseutu 1913, 96–97.

Jokipii 1967: M. Jokipii, Metsästyksestä Satakunnassa 1500- ja 1600-luvulla. Satakunta XVIII, 1967, 24–67.

Kaukoranta 1927: T. Kaukoranta, Tunnetaanko sana hakuli? Sanastaja 4, 1927, 55.

Kirchberger 1970: M. Kirchberger, Pyhän Yrjänän eläimet. Proseminaariesitelmä. Folkloristiikan laitos. Unpubl. paper, Helsinki University (Helsinki 1970).

Kokkonen 1931: H. Kokkonen, Elimäen pitäjän historia II (Helsinki 1931).

Kongl. Stadgar 1706: Kongl. Stadgar / Förordningar / Bref och Resolutioner Ifrån Åhr 1528 intill 1701 Angående Justitiae- och Executions-Ährenden, edited by J. Schmedeman (Stockholm 1706).

Konung Magnus Erikssons Landslag: Corpus iuris sue-gotorum antique X, utg. af D. C. J. Schlyter (Lund 1862).

Korhonen 1977: T. Korhonen, Jahtivoutilaitos Suomessa 1600-luvulla. Hallintohistoriallinen tutkimus. Laudaturtyö. Historian laitos. Unpubl. paper, Helsinki University (Helsinki 1977).

Korhonen 1982: T: Korhonen, Saaliseläimen talja kirkkouhrina. Suomen Museo 1982, 45–68.

Korhonen 1999: T. Korhonen, Saaliseläimen talja ja kotieläimen vuota uhreina; Hääryijy – käsityötaidetta ja uskomuksia. Tietolipas 162, 1999, 106–180.

Korhonen 2020: T. Korhonen, Karhuverkosta susipantaan. Karhun ja suden metsästys Suomessa keskiajalta 2000-luvulle. Kansatieteellinen arkisto 62 (Tampere 2020).

Koskimies 1908: J. Koskimies, Lapun pitäjän historia (Oulu 1908).

Krohn 1914: K. Krohn, Suomalaisten runojen uskonto (Porvoo 1914).

Lilius et al. 1972: H. Lilius/S. Nikula/T. Riska, Naantalin rovastikunta. Suomen kirkot. Turun arkkihiippakunta VI (Helsinki 1972).

Luukko 1945: A. Luukko, Nuijasodasta isoonvihaan. Etelä-Pohjanmaan historia III (Vaasa 1945).

Mansikka 1943: Y. J. Mansikka, Keväänalkajaiset ja Yrjön päivä. Virittäjä 1943, 166–198.

Nikula 1973: S. Nikula, Borgå stift, del I. Suomen kirkot. Åbolands prosteri I (Helsingfors 1973).

Oja 1938: A. Oja, Karhuntalja entisaikain kirkossa. Kotiseutu 1938, 24–34.

Olaus Magnus 1916: Olaus Magnus, Historia om de nordiska folken. Tredje delen. Tolfte-sextonde boken (Stockholm 1916; originally published 1555).

Perälä 1959: V. Perälä, Paattinen 1359–1959 (Turku 1959).

Pylkkänen 1956: R. Pylkkänen, Säätyläispuku vanhemmalla Vaasa-ajalla 1550–1620. Suomen Muinaismuistoyhdistyksen Aikakauskirja 55 (Helsinki 1956).

Pylkkänen 1970: R. Pylkkänen, Barokin pukumuoti Suomessa 1620–1720. Suomen Muinaismuistoyhdistyksen Aikakauskirja 71 (Helsinki 1970).

REA: Registrum Ecclesiasiae Aboensisis eller Åbo Domkyrkans Svartbok. Utg. av R. Hausen (Helsingfors 1890).

Riska 1959: T. Riska, Vehmaan rovastikunta. Suomen kirkot. Turun arkkihiippakunta I (Helsinki 1959).

Riska 1961: T. Riska, Mynämäen rovastikunta. Suomen kirkot. Turun arkkihiippakunta II (Helsinki 1961).

Riska 1964: T. Riska, Turun tuomiorovastikunta. Suomen kirkot. Turun arkkihiippakunta III (Helsinki 1964).

Riska 1966: T. Riska, Perniön rovastikunta II. Suomen kirkot. Turun arkkihiippakunta IV (Helsinki 1966).

Riska 1979: T. Riska, Loimaan kirkot. Suomen kirkot. Turun arkkihiippakunta 7. Loimaan rovastikunta I (Helsinki 1979).

Riska 1980: T. Riska, Alastaron, Mellilän ja Metsämaan kirkot. Suomen kirkot 10. Turun arkkihiippakunta 8. Loimaan rovastikunta II (Helsinki 1980).

Riska/Sinisalo 1968: T. Riska/A. Sinisalo, Perniön rovastikunta. Suomen kirkot. Turun arkkihiippakunta V (Helsinki 1968).

Rosengren 1864: C. Rosengren, Asiakirjat Valtiopäivillä Porvoossa vuonna (Helsinki 1864).

Ränk 1955: G. Ränk, Vanha Viro – kansa ja kulttuuri (Helsinki 1955).

Salenius 1870: J. M. Salenius, Valkjärven pitäjän kertomus. Suomalaisen Kirjallisuuden Seuran Toimituksia 47. Pitäjänkertomuksia II (Helsinki 1870).

Salmelainen 1852: E. Salmelainen, Vähäinen kertoelma Muinais-Suomalaisten pyhistä menoista (Helsinki 1852).

Salminen 1914: V. Salminen, Eräs karhurunoja sisältävä vanha muistiinpano. Suomalais-ugrilaisen Seuran Toimituksia XXXV,12 (Helsinki 1914).

Samling af Instructioner 1852: Samling af Instructioner för högre och lägre tjenstemän I Sverige och Finland. Föranstaltad af Kongl. Samfundet för utgifvande af handskrifter rörande Skandinaviens historia, C.G. Styffe (utg.) (Stockholm 1852).

Samzelius 1915: H. Samzelius, Jägeristaten. Anteckningar om svenska väldets Skogs- och Jaktväsen (Stockholm 1915).

SKVR: Suomen Kansan Vanhat Runot VII, 5. Raja- ja Pohjois-Karjalan runot 5. Somalaisen Kirjallisuuden Seuran Toimituksia 143. osa (Helsinki 1933).

SLL 1946: Svenska landskapslagar 5. Tolkade och förklarade för nutidens svenskar av Åke Hålmbäck och Elias Wessén (Uppsala 1946).

SRL 1934: Sveriges Rikses Lag gillad och antagen på riksdagen åhr 1734. Minneskrift ägnad 1734 års lag av jurister i Sverige och Finland III (Stockholm 1934).

Sarmela 1996: M. Sarmela, Suomen perinneatlas. Suomen kansankulttuurin kartasto 2. Toinen painos. Suomalaisen Kirjallisuuden Seuran Toimituksia 587 (Helsinki 1996).

Siiriäinen 1982: A. Siiriäinen, Recent Studies on the Stone Age Economy in Finland. Fennoscandia Antiqua I, 1982, 17–26.

Suolahti 1912: G. Suolahti, Suomen pappilat 1700-luvulla (Porvoo 1912).

Taavitsainen 1978: J.-P. Taavitsainen, Hällmålningarna – en ny synvinkel på Finlands förhistoria. Suomen Antropologi 3(4) 1978, 179–195.

Vilkuna 1940: K. Vilkuna, Karhunnahkainen harjus. Suomen Museo XLVI, 1939, 43–53.

Vilkuna 1947: K. Vilkuna, 'Bind dina hundar'. Rig 30(4), 1947, 157–167.

Vilkuna 1968: K. Vilkuna, Vuotuinen ajantieto. Vanhoista merkkipäivistä sekä kansanomaisesta talous- ja sääkalenterista enteineen. Toinen korjattu ja lisätty painos (Helsinki 1968).

Vilkuna 1969: K. Vilkuna, Finnisches Brauchtum im Jahreslauf. Folklore Fellows Communications 206 (Helsinki 1969).

Ylikangas 1973: H. Ylikangas, Lohjalaisten historia I (Helsinki 1973).

Teppo Korhonen, PhD Docent
Ruralia-Institute
University of Helsinki
Helsinki
Finland
korhoset.mikkeli@gmail.com

Bears in churches: Skins, paws, and claws from Norway

By Jahn Børe Jahnsen

Keywords: Norway, church history, bear, amulet, local tradition

Abstract: In medieval Norwegian churches, bear skins were often placed in front of the altar, whereas bear paws and claws have been found beneath church floors. Remarkably, in a number of cases, such bear remains still exist, and modern scientific analyses of these materials have led to sometimes surprising insights into their age and origin. Skins, paws, and claws were used in ecclesiastical contexts, the former as foot-warmers for priests, who stood in front of the altar, and the latter as amulets, which were hidden beneath church floors in the belief that they would recharge their power, but were never recovered. In neither case are there any ascertained pieces of evidence that the use of bear remains went back to older, prehistoric traditions.

Introduction

The present author has lived in the inner east-Norwegian district of Valdres (Innlandet) for decades; in this district, a bear skin is still preserved in Hedalen stave church (dated 1163), a bear paw is exhibited in the medieval Ulnes stone church (dated 1265), and a claw was found during excavations in the ruins of the medieval Mo church (*c.* 1215). These finds were the point of departure for an analysis of the number and meaning of bear skins, paws, and claws in Norwegian churches (Fig. 1; cf. Jahnsen 2012). The gathering of data has taken many years, and modern scientific analyses of preserved organic materials will provide additional, and sometimes surprising, information.

All in all, bear skins have been recorded for 23 Norwegian churches, of which five skins, quite remarkably, still exist (Fig. 2; cf. Jahnsen 2012; see also Nodermann 2009, 156–157). In turn, bear paws (one still preserved) and claws are less numerous, being recorded for eleven churches. Most of the available information concerns the bear skins, which is why they are the focus of this article, followed by the paws and claws. In contrast, three figures from the outside of eastern Norwegian churches, the interpretation of which as bears is not compelling, will be omitted. Finally, when it comes to the church bears of Finland, Sweden, and Iceland, these will be only briefly mentioned (see further below; cf. Korhonen, this volume, on Finland).

On the age and origin of bear skins in churches

All in all, there is knowledge of bear skins from 23 Norwegian churches. These finds are mainly distributed in eastern and southwestern/western Norway, whereas they are less numerous in the middle and northern part of the country and, surprisingly, entirely absent in the south. In five cases,

Bear and Human: Facets of a Multi-Layered Relationship from Past to Recent Times, with Emphasis on Northern Europe, ed. by Oliver Grimm (Turnhout: Brepols, 2023), pp. 795–810 BREPOLS PUBLISHERS 10.1484.M.TANE-EB.5.134363

the skins still exist, with no less than four from eastern Norway, and a fifth that is from the very northern edge of middle Norway.

Recently, and remarkably, scientific analysis of carbon isotopes in the preserved organic materials has been carried out for all surviving bear skins (plus the surviving paw from Ulnes; see below). ^{14}C-dating will allow a chronological allocation of the given object (Jahnsen 2012, 86), whereas ^{13}C-analysis may point to a particular area of origin, based on the composition of the diet of the animal in question.

The bear skin from the Hedalen stave church in inner eastern Norway (Innlandet, erected in 1163), has been radiocarbon-dated to the time span of 1290–1370; this is worth mentioning since the local legend (see below) relates to a bear shot after the Black Death in the church (Figs. 3–4). The skin from Viker church in Ringerike (Viken), likewise in inner eastern Norway, is dated to the period of 1310–1405, so it is more or less the same age as the one from Hedalen. There is a local tradition that this skin used to be a part of the one from Hedalen, but ^{13}C-analysis could neither verify nor negate this (personal communication with F. A. Grøndahl, Lands Museum, Dokka, November 2011). The remaining ^{14}C-datings for the skins from Kvikne (Innlandet), Nissedal (Vestfold og Telemark), and Gløshaug (Trøndelag) relate to rather modern finds, post-1500, and in cases like these the chronological allocation by radiocarbon dating is not that reliable (Jahnsen 2012, 86).

Apart from the aforementioned five bear skins, which still exist in Norwegian churches, there is a considerable number of such skins, a little less than 20, that are only known from written sources (in the following after Jahnsen 2012). Dating back to the period from the late 13th to the mid-16th centuries, there are several records. These start with the will, around 1280, of a certain Arnbjørn from Hebnes, a wealthy man from southwestern Norway, who gave skins to three churches in the area in which he lived: St. Olav's church in Stavanger, the Utstein monastery church, and the Jelsa church (cf. Jahnsen 2012, 26). The records continue with inventory lists from the early 14th century, which mention bear skins in Hålandsdalen (Holdhus) church and Ylmheim (Ølmheim) church, both in Vestland. Also, inventory lists from the early 18th century testify to bear skins in churches in Årdal (Rogaland), Røldal (Vestland), and Hen (Møre and Romsdal). Furthermore, there is a traveller's account left by an Italian, Pietro Querini, one of the Venetians who was shipwrecked in northern Norway in 1432; he saw a bear skin in St. Olav's church in Trondheim, middle Norway (Storm 1890/1891; Helland 1900).

One may expect a high degree of credibility for the above-mentioned cases, owing to the very nature of the sources. The situation is quite different for the following legends. There are two different kinds of "wandering legends" with a link to bear skins in churches. In the first case, the narration pretends to provide a long look back at bears and churches in the period after the Black Death, whereas the other one refers to rather recent dispatchings of "killer bears". Neither of the two will be analysed in detail; it will suffice to concentrate on the information provided about the origin of the bear skins.

The legend about Hedalen stave church in Valdres, eastern Norway, written down in 1743, reads like this: After the big plague, the Black Death, in 1349–1350, the Hedalen valley was deserted and was like this for a long time. The forest grew over the many deserted farms and the church. Long after this, a hunter came walking through the forest. He shot an arrow at a bird on a tree top, but missed. The arrow continued and hit something that gave a fine sound. The hunter ran for the sound and found a stave church in the middle of the forest. The arrow had hit the church bell. The hunter believed this was the church of the underground people, and threw his fire iron over the church to protect himself. (To throw steel over the underground people was well-known advice against witchcraft.) The hunter entered the church, where he found a roaring bear with its lair in front of the altar. This time the hunter did not miss, but shot the bear. He skinned the bear, and the bear skin was put up on the wall where it is still hanging (Landt 1951, 335; Jahnsen 2002, 190–194).

There are different variants of this legend about Hedalen, and in one of these, remarkably, the bear skin first came to lie in front of the altar, before it was later moved to a storage building near the church (Jahnsen 2012, 13). This legend is also associated with a few other churches in Norway (Hemsedal, Tuft, Heddal, all in eastern Norway), possibly in the form of a "wandering legend", written down (much) later, as was the case with Hedalen. Repeatedly, lively bears are described with their lair at the altar, whereas, after their killing, their skins were hung on inner church walls rather than placed in front of the altar (Jahnsen 2012).

In fact, there is a second and more numerous type of legend attached to bear skins in Norwegian churches; that of "killer bears" – ruthless killers of humans and animals –, which were finally dispatched by hunters in rather recent times. The narration may be considered a "wandering legend", too. The lengthy Nissedal legend was printed for the first time by M. B. Landstad in the 1880s (see reprint: Landstad 2002, 210; cf. also Åsen 1986, 437–439; Jahnsen 2012, 23–25). One variation has to do with the belief in a human in a bear's cloth/skin. Folk did not dare to name that bear in the correct manner. In contrast, the bear was given the name of a man. Another variation sees in the largest bears bewitched sons of kings, and yet another one has the bear killed by means of a silver bullet shot by an elder, weak man. Finally, it is worth mentioning that, in the case of the Sollia church in eastern Norwegian Hedmark, the legend about a killer bear was written down by none other than the Nobel laureate in literature, Sigrid Undset (*Lykkelige Dager*, Undset 1947, 179–180; *Happy Times in Norway*, Undset 1942), but elsewise there is no trace of that skin. Remarkably, in a number of churches linked with "killer bears", the skins of the animals were placed in front of the altar (for example Høle, southwestern Norway, and Saltdal, northern Norway).

On the interpretation of bear skins from churches

As has already been mentioned to some extent, there is an old connection between churches and bear skins in Norway. For around 1280, it is recorded in a testament that one bear skin was supposed to be given to St. Olav's church in Stavanger, or, more precisely, St. Mary's altar. Much later, three inventory lists from the early 18th century testify to bear skins placed in front of altars. This position is also indicated, to some extent, in the "wandering legends" about bears and churches in the time after the Black Death and the rather recent dispatchings of "killer bears".

Of particular interest in this regard are traveller's accounts. In the early 15th century, the aforementioned Italian Pietro Querini saw a white bear skin, 14½ feet long (thus originating from more than just one bear), on the floor in front of the archbishop's chair in St. Olav's church in Trondheim (Storm 1890/1891; Helland 1900). According to Pierini, the reason why skins served as floor rugs was the intense cold in Norway. The second record goes back to the well-known Swedish archbishop, Olaus Magnus, who visited Nidaros cathedral in Trondheim in 1518/1519 (Kolsrud 1914, 60). In 1555, in his *Historia de gentibus septentrionalibus* (The History of the Nordic People), he wrote that hunters used to give, after a promise they had made, white bear skins to the main altars in cathedrals or other churches, so that the priest should not have freezing feet in the terrible cold when officiating the service (Olaus Magnus 1555, Book 4, chapter 15; Book 16, chapter 20; Book 18, chapter 24).

It is recorded in only one instance (Ullensvang church in western Norway; cf. Jahnsen 2012, 28) that the act of skin-giving to the church as performed by a hunter was preceded by the prayer of a priest who asked for good luck in hunting. This kind of donation after the dispatching of the animal, with the skin placed in front of the altar or, in Trondheim, placed before the Archbishop's seat in the cathedral, may be assumed as the background for the early mentions of bear skins and for both types of the aforementioned considerably younger legends. In these, sometimes other skin use is described, but this may reflect variations, when the original use of skins had already been forgotten.

On skins in Norwegian churches: Polar bears vs. brown bears

The repeated mentions of polar bears in the early records of Norwegian church bears is striking (see above). They are mentioned in connection with the church at Hålandsdalen (Holdhus) in western Norway (inventory list, early 14th century) and St. Olav's church in middle Norwegian Trondheim (traveller's account from the Italian Pietro Querini from 1432). Olaus Magnus mentions white bear skins, too, in the middle of the 16th century. Finally, there is scientific evidence for the paw from eastern Norwegian Ulnes, which is from a polar bear that lived either in the 9th or the early 10th century AD (!) (see below).

As a matter of fact, polar bears were native in Greenland, whereas they came to Iceland only rarely on pack ice, and in Norway they were entirely foreign. Thus, such skins had a particular value, since they were foreign goods and the outcome of a "heroic deed" – the killing of a bear – which demanded particular courage (Oehrl 2013). As one may assume, the skins came to Norway as trading goods rather than as skins of animals hunted by Norwegians in Greenland or, less likely, Iceland.

The important point is that without doubt the overall number of bear skins in Norwegian churches came from brown bear. As regards the provenance of bear skins, paws, and claws from the churches, there is a match between their area of origin and information in oral tradition and written sources as to the whereabouts of persons being wounded or killed by predators like bears and wolves during the last 400 years (Fig. 5; cf. Furseth 2005). Thus, human and bear had a "shared history" in the respective areas of Norway, which also points to the dispatching of native bears and the use of their remains in a local context. An exception from the rule are several islands along the western Norwegian coast, where bears were foreign, but bear parts are represented in churches (cf. Fig. 1; e.g. Kinn in western and Alstahaug in northern Norway).

On the age and origin of bear paws and claws in churches

About half of the eleven Norwegian churches with bear paws and claws are situated in the eastern part of the country, and the other ones in the area between the west and the north. The number of these objects is too small to allow any reliable comparison with the distribution pattern of the 23 bear skins from churches (see above), but nevertheless it is worth mentioning that two out of three paws come from the east (and possibly yet another one that, however, has not been properly recorded).

The paw from Ulnes church in inner eastern Norway holds a special position inasmuch as it has survived until today (Figs. 6–7). In ten other churches, bear paws and claws came to light beneath the floors: three paws, seven claws, and three animal claws for which no species allocation is given (Jahnsen 2012, 63). The latter may be from bears, as is the case in a number of churches, but only archaeozoological analysis could tell us so. One find spot has to be omitted from further consideration; five burnt claws found beneath Mære church in middle Norway most likely belong to a Migration period cremation burial (Jørgensen 2008, 31; cf. Mansrud, this volume, on the prevalence of bear claws in Migration period cremation graves in Norway).

In the case of the existing paw from Ulnes church (erected around 1265), the results of scientific analyses were remarkable. As it turns out, the paw belongs to the time span of AD 820–920, according to a ^{14}C-dating, which makes it 300–400 years older than the actual church from which it originates (Jahnsen 2012, 86). In addition, ^{13}C-analysis has shown that the Ulnes bear had a "marine diet", which consisted of salt water fish or animals, typical for a polar bear, but not a brown one (personal communication F. A. Grøndahl, Lands Museum, Dokka, November 2011; see also Jahnsen 2012, 64–65). Thus, in the given case, it is both an exotic animal and a particularly early one, with an unknown history before one of its paws was hidden beneath the choir of Ulnes church.

For any further analysis of bear paws and claws, it is essential to have at hand as much information as is possible from excavation documents about the find circumstances of the respective bear claws and paws; where exactly were the objects found, within the church (horizontally), but also beneath its floor (vertically)?

Any such attempt, however, faces the problem that beneath the floors there is often only mixed-up earth which originates, amongst other things, from medieval or later burials placed below the nave and to some extent even the choir (Magnussen 2007, 24). Thus, no reliable find context can be given for objects that came to light beneath church floors; they were not found in original, undisturbed layers and may thus be rather young or (much) older. In addition, it is only in recent church excavations, from the 1980s onwards, that all finds were kept – not just the coins, which had received the main attention so far (Magnussen 2007, 11).

The excavation in the east Norwegian Uvdal church in 1978 is worth a mention inasmuch as it sheds further light on bear claws (Fig. 8). Approximately 3,000 objects were uncovered beneath the church floor, among them *c.* 550 coins, but sadly a proper analysis of the finds from the animal realm seems to be missing (briefly: Christie 1992, 17; cf. also find no. C34866 of KHM/Kulturhistoriske museer, Oslo, Norway). All in all, eleven claw/skin remains were found, among which there are seven from the choir and four from the nave. Only four out of seven claws definitely originate from bears; there are three more without classification, and one that is questionable. Among the unspecified ones, there is one with undetermined skin remains (nave), one that has been drilled (nave), and possibly yet another one (choir). Since three skin remains, which remain unanalysed, carry related find numbers, they may all originate from the same area in the choir. As stated in a legend, a wolverine was shot in the church in the period after the Plague, not a bear, as is usually reported (Jahnsen 2012, 21–22). Today, a sheep skin lies on the church floor close to the altar (Fig. 9). Thus, Uvdal church yields very interesting materials, in both archaeological and written form, but there are quite a number of open issues. This also relates to the question of whether a paw was found beneath the floor in Uvdal church (affirmative: Christie 1992, 17, but with no such entry in the find catalog).

The western Norwegian church at Kaupanger and its excavation in the 1960s is an interesting example as regards bear paws (in the following according to Magnussen 2007, 55–56, 79). Among the *c.* 1,500 objects that came to light during excavations beneath the church floor, there were more than 1,400 coins, followed by 36 pieces of jewellery or clothing, plus other types of finds, which are represented only rarely, found mainly in the nave. Among the salvaged objects, there are three bear claws, one still attached to a paw, and one bird claw (grouse or domestic chicken). One bear claw was found beneath the church floor in the area of the eastern choir; the other bear remains and the bird claw are recorded for yet another area in the choir, to the east, but with no further details given. In the present case, with the choir, the holiest part of the church that is closed to ordinary churchgoers, as the area of origin and some of the bear remains and the bird claw probably recovered from the same spot, one may suspect an intentional deposition, in particular for the paw which cannot simply be considered as accidentally lost.

To sum up, the record of find circumstances in the churches and the identification of animal remains found therein is to some extent insufficient, but the latter could be improved by future analyses. Here, the unspecified claws from Uvdal (eastern), Moster (western) and Alstadhaug (northern Norway) would matter (Jahnsen 2012, 63). Bear claws were recovered in a total number of seven Norwegian churches in both choir and nave, whereas in the case of the paws, two originate from the choir (Ulnes, Kaupanger; possibly also Uvdal) and one from the nave (Lom). The paws and claws from choirs demand special attention; the objects were not simply lost, they were presumably placed ("offered") deliberately beneath floors in the part of the building restricted to clerics. For the claws in the nave, one cannot be as certain about the circumstances; there may have been cases of accidental loss. However, at any rate, these claws must have had some importance for their owners.

The bear claws are in need of closer analysis. During the aforementioned Uvdal church excavation, two claws (bear?) came to light that had been drilled, which suggests that they were worn on a string/chain. This, however, cannot be generalised, since Uvdal is the only church, among seven with identified claws, where such drilled claws occur. Interestingly, the use of undrilled claws is revealed by the 10th-century female grave IV in Rösta, Ås parish, Jämtland, in northern Sweden; here, two claws were found in a bag, placed by the waist of the buried person (see Jordahl et al., this volume). Since bags of organic material are likely to decompose, they will usually defy identification in archaeological materials. However, and quite remarkably, Norwegian churches have yielded organic bags in some numbers. These bags, sometimes called *pjoterposer*, allegedly contained small objects that were believed to possess magic powers (Bø 1978, 191; Christie 1992, 21). When opened, however, such bags did not produce any objects such as, for example, bear claws, but only specks of dust (Magnussen 2007, 57).

As has been suggested, claws and paws hidden beneath the church floor, on holy ground, were believed to "recharge" their power as amulets, in the belief that the remains (claws) of wild animals (be they bears or birds) brought with them particular strength (Christie 1992, 17). If one follows this idea, the objects found during excavations – in particular those from choirs – might have been placed beneath the church floor with the idea of future recovery which did not happen.

Owing to the special physical quality ascribed to the bear, its remains, such as teeth, claws, paws, and skin, had a special role in folk medicine for healing, regeneration, fertilising and strengthening (see Böldl and Mansrud, this volume). For example, a tooth could help against toothache, or would provide protection for the hunter. Regarding the use of bear paws, there is a remarkable record for western Norway. Two bear paws, kept in Sogn Folkemuseum, were once sent from farm to farm and are said to have had a curative effect against pain and suffering for persons and animals (Fig. 10; cf. Jahnsen 2012, 66–68). In the case of difficult births, such a paw was carried beneath the clothing or moved over the stomach and genitalia of a pregnant woman. In Veitastrond, this still happened until after 1900 (note from Aud Ross Solberg, Sogn Folkemuseum, March 30th 2006).

Final remarks

There is a considerable number of bear skins, paws, and claws associated with medieval Norwegian churches. As demonstrated, it is only the bear skins for which there is sufficient archaeological and written records, whereas for the paws and claws the find records are often imprecise or a species identification of the skins and claws is missing. In turn, the question of bear figures has been omitted from the paper, since their interpretation is not compelling (see Fig. 11 for the cases in question).

Broader analyses would be needed. What do we know about the representation of other animals in Norwegian churches? In this respect, the aforementioned Uvdal church provides an interesting case; once, it allegedly housed a wolverine hide, but, in more recent times, this has been replaced by a sheep skin. And what do we know about church bears in the north of Europe but outside Norway? To mention this only briefly, Norway is special owing to the paws and claws found in churches, since elsewise only skins have been recorded (in *c.* 50 Swedish, *c.* 20 Finnish, and likewise *c.* 20 Icelandic churches; see Jahnsen 2012, 59–61; cf. Nodermann 2009, 152–157). Furthermore, it is mostly Norway that yields comparatively early mentions of bear skins with details about their origin, placement in the church, and use (late 13th to mid-16th centuries).

What role did bears have in (late) pre-Christian times, and up to what extent did this influence the use of bear skins, paws, and claws in churches? According to written accounts, both scaldic poetry and sagas, bears had no prominent position among the animals in Old Norse religion, except for the so-called berserks, members of warrior bands who wrapped themselves in bear or wolf skins (see Lombardi, Ney and Sundquist, this volume). Somewhat contradictorily, Frösö in Jämtland, northern Sweden, has yielded a late pre-Christian offering site beneath the choir of the local stone church with a considerable amount of bear remains close to a tree stump (see Magnell, this volume). This may reflect that bears were hunted quite often in the area in question and some parts were offered, possibly following some sort of "bear ceremonial", which was meant to show respect towards the hunted species and in particular the mighty and powerful bear. In a way, the Frösö offerings of bear remains and the bear skins donated to Norwegian churches may be considered in a similar vein; as the thanksgiving of hunters for the luck they had in dispatching such an animal.

To return to bear paws and claws; according to folk religion (medicine), they would bring about help against pain or suffering by animals and humans or difficult childbirth. Particular emphasis should be placed on paws and claws from the choir. Quite possibly, there was the belief that, while they were in place beneath the church floor, paws and claws would "recharge" their power as amulets, but they were then not recovered by their owners. This kind of amulet use may have been rooted in a Christian setting. It does not necessarily lead back to older times, though this cannot be ruled out.

When it comes to bear skins, their use as floor rugs for priests is recorded for the first time around the middle of the 15th century, and from then onwards quite regularly. Here, one may pose the question of whether this was the original or only a secondary use, and, if it was original, was there any symbolic meaning attached to it, that went further back in time? In this context, it is worth mentioning that, according to the regulations of the nunnery of Vadstena in Sweden, the nuns were not supposed to use feathers for (stuffing of) bedclothes but bear skins (Nodermann 2009, 145). Also, Holy Birgitta is said to have rested upon a bear skin (ibid.) In contrast, there is no proof that the nuns of Gudhems kloster in Swedish Västergotland, as has been suggested, were laid in their graves on bear skins (personal communication Maria Vretemark, Västergötlands museum). With this knowledge, based on Swedish sources, the bear skins that served as floor rugs for warming the feet of the priest during the church service would be yet another facet of the practical use of the skins, here in ecclesiastical contexts, with no ascertained connection to any special importance ascribed to bear skins in pre-Christian times.

Finally, Olaus Magnus has pointed out a special use, not for bear but wolverine hides. In "How to honour guests with coverlets of wolverine hides", he tells us that such hides were not trading goods, but were kept as bedding for important guests (Olaus Magnus 1555, chapter 8; cf. Sigvallius 1994, 76). This immediately brings to mind the aforementioned church from east Norwegian Uvdal, which has not only yielded bear skin remains and bear claws, but also a legend, according to which a wolverine was shot in the church in the period after the Plague; a variant of the "wandering legend" about the church and the bear (Jahnsen 2012, 21–22). Today, a sheep skin lies on the church floor close to the altar, possibly in the same position as the wolverine hide before. If one is to follow Olaus Magnus, wolverine hides saw a secular use, whereas bear skins, paws, and claws are found embedded in an eccleasiastical context.

Acknowledgement

The author would like to thank Oliver Grimm for his help in editing the text.

Bugge 1954: A. Bugge, Heddal stavkirke (Oslo 1954).

Bø 1978: O. Bø, Kyrkjefunn og folketru. Foreningen til norske fortidsminnesmerkers bevaring, Årbok 1978, 191–198.

Christie 1992: H. Christie, Kirkebygningens historie: Undersøkelsene i 1978 gav ny kunnskap. In: N. Friis (ed.), Uvdal stavkirke forteller: funn og resultater etter undersøkelsene av stavkirken i 1978 (Uvdal 1992) 5–25.

Furseth 2005: A. Furseth, Drept av ulv og bjørn (Oslo 2005).

Gihle 1978: P. Gihle, Hoff kirke. In: P. Gihle (ed.), Frå gammalt. Festskrift i anledning Pål Gihles 70 års dag 21. februar 1978 (Gjøvik 1978) 49–94.

Helland 1900: A. Helland, Venetianeren Querini paa Røst i 1432. In: Norges Land og Folk, Nordlands Amt (Kristiania 1900) 865–891.

Jahnsen 2002: J. B. Jahnsen, Sagn i Valdres – Helgener og haugafolk, troll og tjuver (Oslo 2002).

Jahnsen 2012: J. B. Jahnsen, Bjørnen i kirken. Om bjørneskinn, bjørnelabber, bjørneklør og bjørnefigurer i norske kirker (Vallset 2012).

Jørgensen 2008: G. Jørgensen, Blant gull og bjørnklør – glemte funn fra Mære kirke. Spor 2, 2008, 28–31.

Kolsrud 1914: O. Kolsrud, Olavskyrkja i Trondheim. Norske Folkeskrifter 63 (Oslo 1914).

Landstad 2002: M. B. Landstad, Mystiske og andre sagn (Oslo 2002).

Landt 1951: P. J. Landt, Ourdahls kald 1743. Tidskrift for Valdres Historielag 1951, 321–338.

Magnussen 2007: A. R. S. Magnussen, Funn under kirkegulv. En arkeologisk analyse av gjenstandsfunn fra Borgund, Kinsarvik og Kaupanger kirker. Unpubl. Hovedfagoppgave i arkeologi, Universitetet i Bergen (Bergen 2007).

Nodermann 2009: M. Nodermann, Björnens fäll vid altaret och i graven. Saga og sed 2009, 133–159.

Oehrl 2013: S. Oehrl, Svá beitum vér björnuna á mörkinni norðr – Bear hunting and its ideological context (as a background for the interpretation of bear claws and other remains of bears in Germanic graves of the 1st millennium AD). In: O. Grimm/U. Schmölcke (eds.), Hunting in northern Europe until 1500 AD. Old traditions and regional developments, continental sources and continental influences. Papers presented at a workshop organized by the Centre for Baltic and Scandinavian Archaeology (ZBSA) Schleswig, June 16th and 17th, 2011. Schriften des Archäologischen Landesmuseums, Ergänzungsreihe 7 (Neumünster 2013) 297–332.

Olaus Magnus 1555: Olaus Magnus, Historia de gentibus septentrionalibus (Rome 1555); Transl.: Historia om de Nordiska Folken [...], Commentary by J. Granlund (Östervåla 1976).

Pettersen 1914: F. Pettersen, Bamble stenkirke – Bratsberg amt. In: Foreningen til Norske Fortidsminnesmerkers Bevaring (Kristiania 1914) 156–170.

Sigvallius 1994: B. Sigvallius, Funeral pyres. Iron Age Cremations in North Spånga. Theses and Papers in Osteology 1 (Stockholm 1994).

Storm 1890/1891: G. Storm, Venetianerne paa Røst i 1432. Det Norske Geografiske Selskabs Aarbog 1890/1891, 1–22.

Undset 1942: S. Undset, Happy Times in Norway (New York 1942).

Undset 1947: S. Undset, Lykkelige dager (Oslo 1947).

Åsen 1986: K. Åsen, Nissedal bygdesoge – kultursoga frå dei eldste tider til år 1900 (Treungen 1986).

Cand. phil. Jahn Børe Jahnsen
Retired senior curator and archivist at Valdres Museums
Fagernes
Norway
jbjahnsen@gmail.com

Fig. 1. The author's book on bear remains in Norwegian churches (cover drawing G. Helgen).

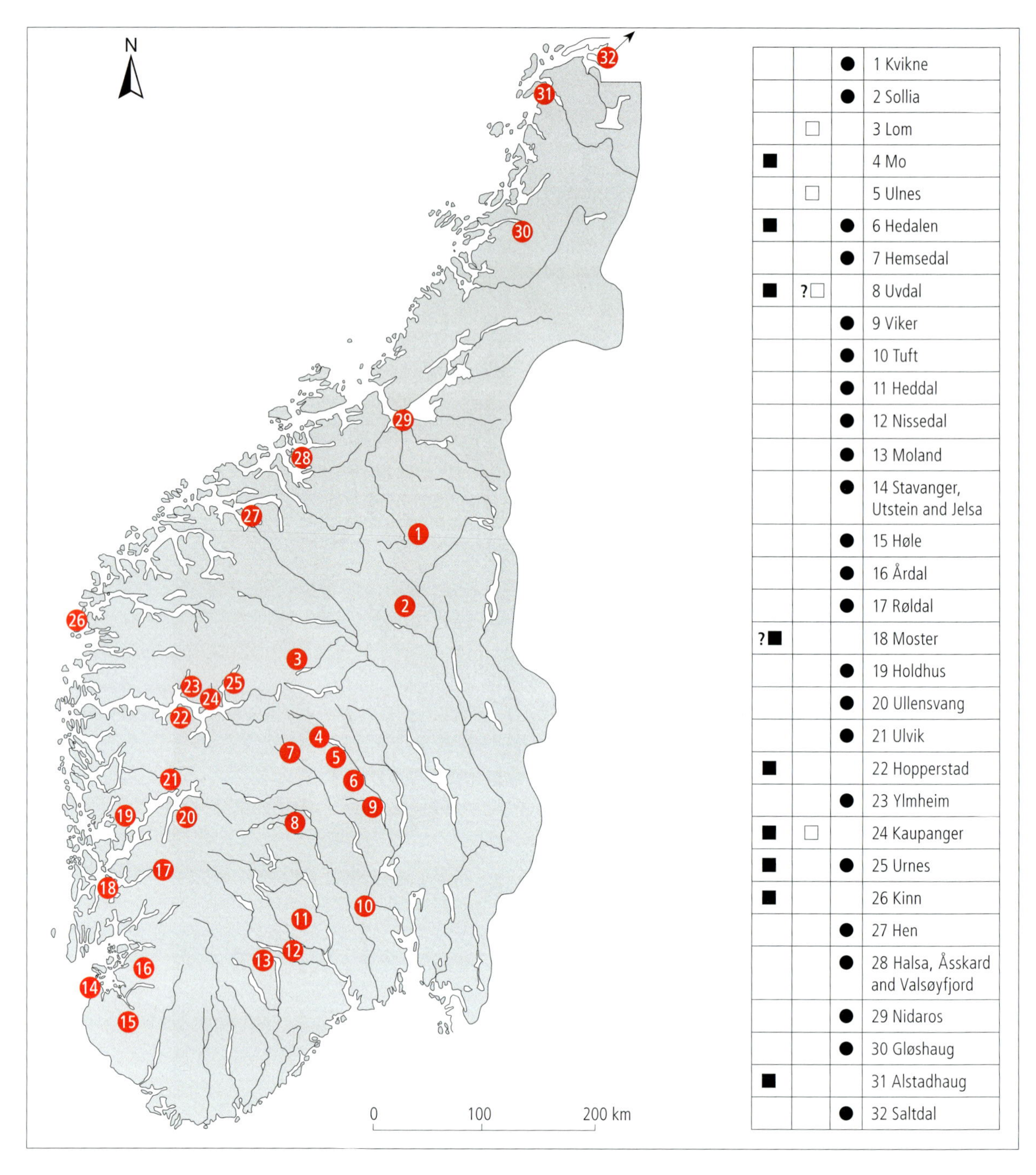

Fig. 2. Distribution of bear skins, paws and claws in medieval Norwegian churches. Circle: bear skin, open square: bear paw, full square: bear claw (map M. Bolte, ZBSA, after a draft by J. B. Jahnsen).

Fig. 3. Stave church in Hedalen, Valdres (Innlandet), in eastern Norway. Built in 1163 and rebuilt in 1699 (photo J. B. Jahnsen).

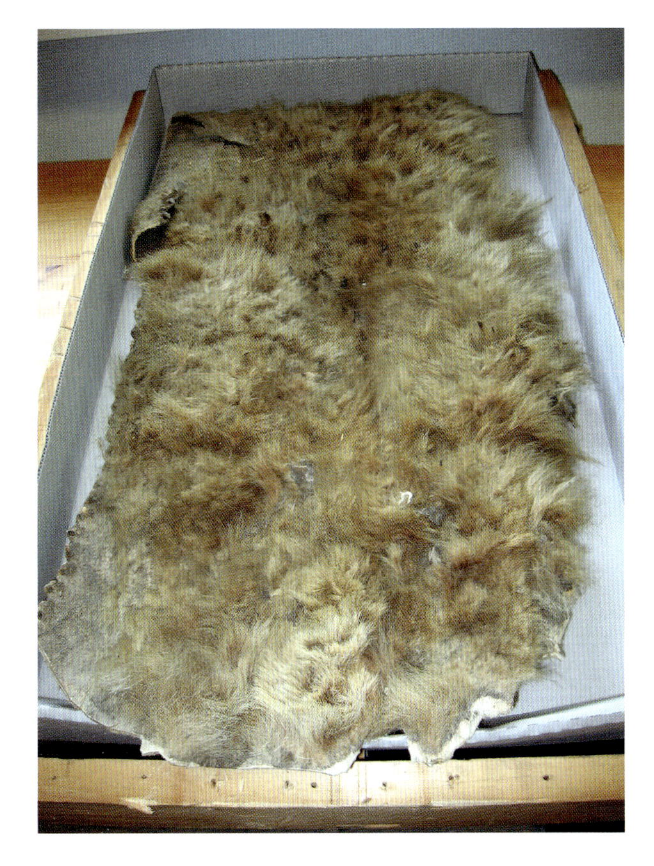

Fig. 4. Premodern bear skin from the medieval stave church in Hedalen, Valdres (Innlandet), in eastern Norway (photo J. B. Jahnsen).

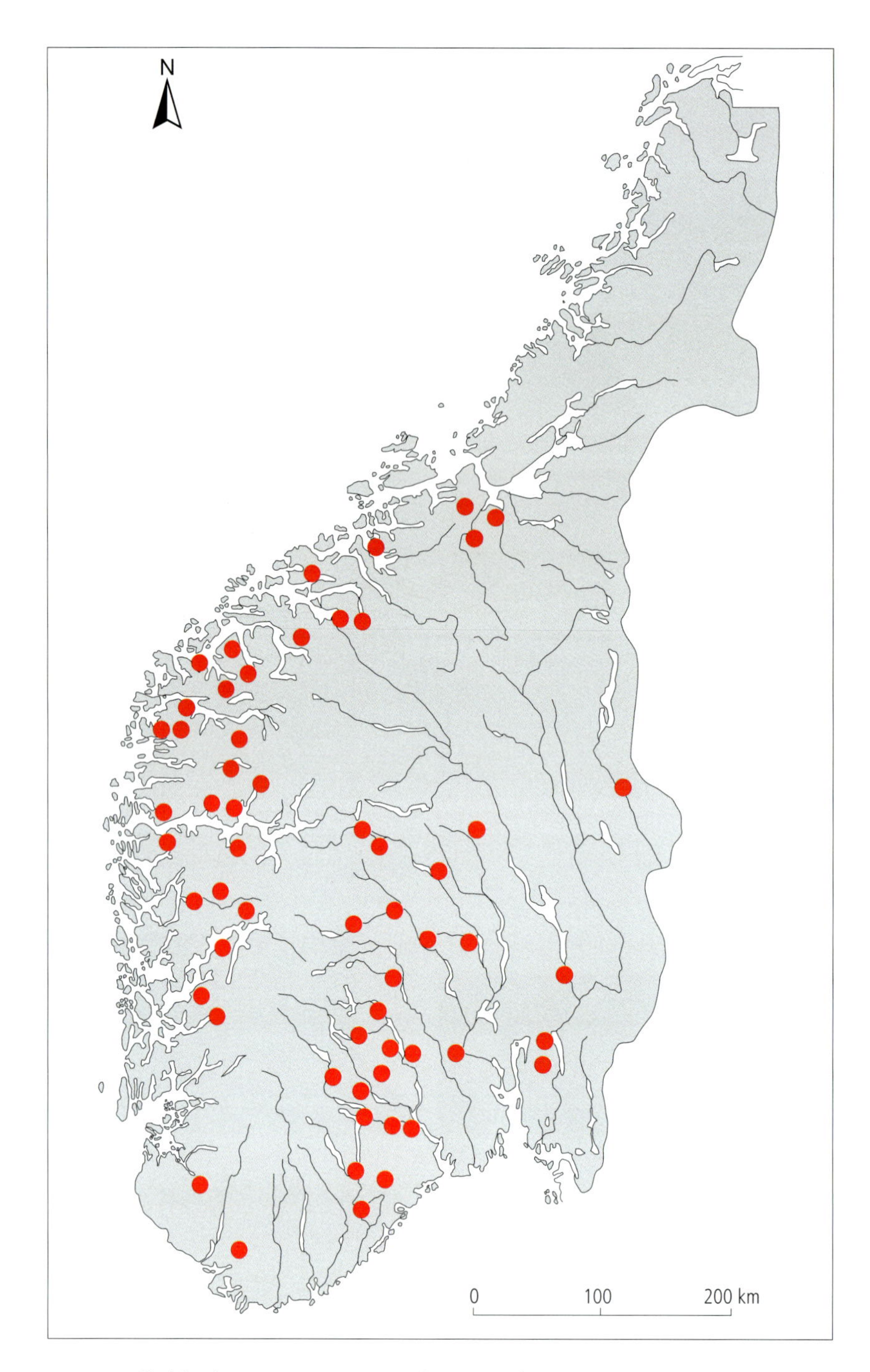

Fig. 5. Killed by bear. Norwegian incidences in the past 400 years, with northern Norway excluded (map M. Bolte, ZBSA, after Furseth *2005, 26).*

Fig. 6. Stone church in Ulnes, Valdres (Innlandet), in eastern Norway. Built c. 1265 (photo J. B. Jahnsen).

Fig. 7. Premodern bear paw from the medieval stone church in Ulnes, Valdres (Innlandet), in eastern Norway (photo J. B. Jahnsen).

Fig. 8. Stave church in Uvdal, Numedal (Viken), in eastern Norway (photo J. B. Jahnsen).

Fig. 9. Sheep skin in front of the altar in the medieval stave church in Uvdal, Numedal (Viken), in eastern Norway (photo J. B. Jahnsen).

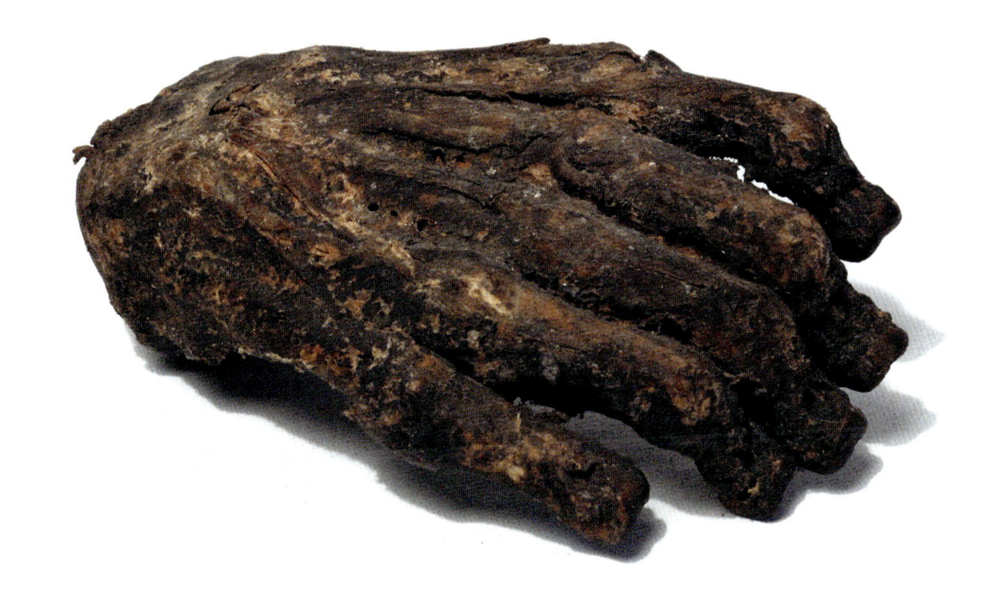

Fig. 10. Bear paws from western Norway, once used in the case of disease or birth (photos Sogn Folkemuseum).

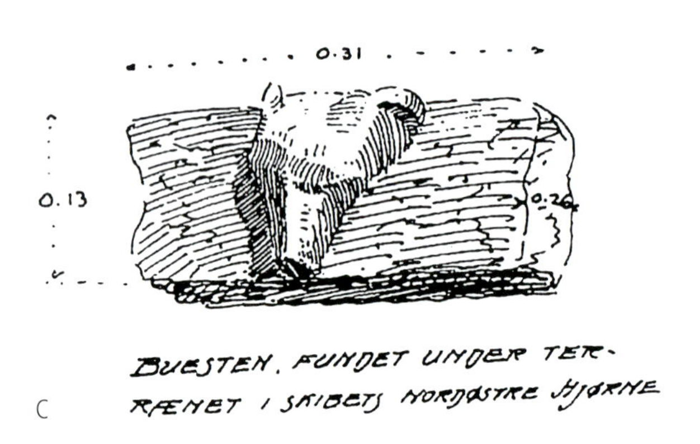

Fig. 11. Disputable bear figures from eastern Norwegian medieval churches. a: Heddal stave church (Vestfold and Telemark, after Bugge *1954, 37; photo B. Mogan Lindheim); b: Hoff stone church, Toten (Innlandet; after* Gihle *1978, 55); c: Bamble (Vestfold and Telemark, after* Pettersen *1914, 169).*